FRONTIERS OF ASTRONOMY
SECOND EDITION

FRONTIERS OF ASTRONOMY

SECOND EDITION

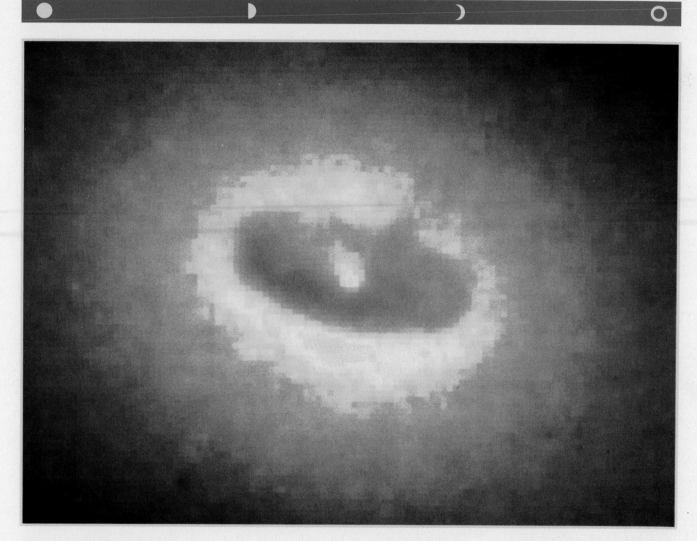

DAVID MORRISON
Chief, Space Science Division
NASA Ames Research Center

SIDNEY C. WOLFF
Director
National Optical Astronomy Observatories

Saunders Golden Sunburst Series
SAUNDERS COLLEGE PUBLISHING
Harcourt Brace College Publishers

Fort Worth   Philadelphia   San Diego   New York   Orlando   Austin   San Antonio   Toronto   Montreal   London   Sydney   Tokyo

Requests for permission to make copies of any part of the work should be mailed to Permissions Department, Harcourt Brace & Company, 8th Floor, Orlando, Florida 32887.

Text Typeface: Palatino
Compositor: General Graphic Services
Publisher: John Vondeling
Associate Editor: Lloyd W. Black
Managing Editor: Carol Field
Project Editor: Anne Gibby
Copy Editor: Kim LoDico
Manager of Art and Design: Carol Bleistine
Art Director: Caroline McGowan
Text Designer: Nanci Kappel
Cover Designer: Lawrence Didona
Text Artwork: George V. Kelvin, Science Graphics
Director of EDP: Tim Frelick
Production Manager: Charlene Squibb
Marketing Manager: Sue Westmoreland

Cover: The Cocoon Nebula is an emission cloud powered by the hot stars near the center. These stars radiate powerfully in the ultraviolet region, and the nebula gas absorbs the starlight and converts it to visible light. The gas radiates most strongly at wavelengths typical for hydrogen, the dominant gas in the nebula. To create this image, infrared light was combined with the visible so as to emphasize the colors and brightnesses of the faint stars that power the nebula. But because the dominant emission from hydrogen gas now falls in the band colored green instead of red, the nebular emission is now biased toward blue-green. These techniques of color enhancement are called spectrally augmented color. (Photo courtesy of Rudolf Schild/Harvard-Smithsonian Center for Astrophysics)

Title page: Hubble Space Telescope image of NGC 4261, a giant elliptical galaxy in the Virgo cluster, located 45 million light years away. This close-up view of the galaxy's center shows a giant disk of cold gas and dust that fuels a possible black hole at the core of the galaxy. Estimated to be 300 light years across, the disk is tipped enough to provide astronomers with a clear view of the bright hub, which presumably harbors the black hole. (STScI)

Printed in the United States of America

Frontiers of Astronomy, Second Edition

ISBN: 0-03-093933-X

Library of Congress Catalog Card Number: 93-086196

4567 032 987654321

PREFACE

Astronomy is one of the most exciting intellectual pursuits of the 20th century. Each period of history is identified with a few great achievements that define for future generations what society at that time valued and what it contributed to the continuing human legacy. The achievements of early societies have come to us largely in the form of buildings, works of art, and literary masterpieces, but today the discoveries of science and technology must be added to our accomplishments. It is likely that one of the most important legacies of the late 20th century will be what we have learned about the universe and its origin. For the first time in human history we can visit the other planets, probe the structure of the cosmos, and provide some answers to the most fundamental questions of the origin and ultimate fate of the universe and of life itself.

The challenge, the excitement, the fascination of this voyage of discovery need not be limited to professional researchers. The public can and should share in this great adventure. The kinds of questions that astronomers are now attempting to answer have been asked throughout human history. How far does space extend? Where did the universe come from? How will it end? Why do stars shine? What is the ultimate fate of the Sun and solar system? How old is the Earth? Are we unique? Or are there other planets with life or even other civilizations? These questions are at the frontiers of current research, but they are also the subject of introductory college courses in astronomy.

An introductory astronomy course should do more, however, than expose students to current discoveries in astronomy. Even more important is to communicate a sense of the scientific process. One of the most fundamental problems in our complex, technological society is the pervasive public lack of critical thinking, coupled with innumeracy—the inability of so many intelligent people to use numbers or understand quantitative reasoning. Comprehending the sizes of numbers and learning to read a simple graph are skills that can be learned in an astronomy course, and stimulating critical thinking is perhaps the most important objective of science education for the non-science student.

FEATURES

While we hope *Frontiers of Astronomy* will be read with interest and pleasure, we recognize that it is primarily a text, aimed at a one-semester or one-quarter college survey course. In writing this particular introductory text, we have chosen to keep the book short and focus on the currently most exciting areas of astronomical research. We want the readers of this book to be able to understand the reports of new astronomical discoveries that constantly reach us through the media. We also want those who use this book to share the wonder and awe that we feel about the universe.

We have tried to make *Frontiers* accessible, with a strong narrative thread and a "user friendly" writing style. While the coverage of astronomy is broad, there are a few major themes that stand out—highlighted in a handful of key questions that accompany each chapter. We have tried to tell a coherent story about each theme by exploring it in enough depth to show the chain of thought and experiment that has led to our present understanding. Details not relevant to the chosen themes have been omitted, and technical jargon excised wherever possible. The technical terms we have included are introduced because they are needed to understand the material, not to be memorized as an end in themselves. All of these technical terms are identified by boldface when they first appear, carefully defined in the text, and later repeated in chapter summaries and a glossary.

The subject matter will be best mastered if students go beyond reading to work with the material presented. Accordingly, we have developed questions at the end of each chapter in three categories: *review* questions that recapitulate the main themes, *thought* questions that expand upon these ideas, and *problems* that call for simple numerical applications using a pocket calculator. We do not expect students to know algebra or trigonometry, but we do believe they should learn to work their way through numerical applications of the ideas they learn, such as Kepler's laws. Extensive lists for further reading are provided, including recent magazine articles that represent a variety of perspectives on current discoveries.

CHANGES TO THE SECOND EDITION

For this second edition, we have profited from comments and suggestions by many readers and instructors. While keeping a broad perspective on modern astronomy, we have strengthened our topical focus by adding key questions in the margins, to remind the readers of the most important themes, and including additional thought questions and problems. To help bridge the gap between the classroom and the "real" world, new essays called "Observing Your World" have been added to encourage students to make their own simple observations (without special equipment and even in polluted urban environments), in addition to more extensive and updated reading lists to stimulate use of library resources.

The list of changes to this second edition of *Frontiers of Astronomy* includes:

- A revised order of presentation in the first three chapters that allows students to move quickly to "real" astronomy. Coverage of telescopes has been moved to Chapter 8, logically positioned next to the discussion of light and other forms of electromagnetic radiation.

- New material about Mercury's recently discovered ice caps, Magellan's results from its survey of Venus; perspectives on the role of impacts in the history of the Earth and other planets; global warming and the Antarctic ozone hole; Comet Shoemaker-Levy's impending impact on Jupiter; asteroids, including Galileo's flyby of Gaspra and radar imaging of Toutatis; planets orbiting pulsars; the latest discussion on dark matter; revised theories of galaxy formation; and analysis of COBE observations of the cosmic background radiation.

- More tables that summarize important facts in an easy-to-see and digest format.

- New "Observing Your World" essays that encourage students to observe astronomical phenomena for themselves by giving them small activity projects.

- Revised short essays on topics at the frontier that provide added detail about the latest research and discoveries.

- Revised Suggestions for Additional Reading.

ANCILLARY MATERIALS

Available to all adopters of the *Frontiers of Astronomy* second edition are the following ancillaries:

Astronomy Now!—A Special Issue of *Discover Magazine* *New for 1994*—A collection of 14 astronomy articles from past issues of *Discover* are presented in this special full-color issue. Interesting and at times provocative articles help broaden students' perspectives of astronomy and offer instructors fresh opportunities for discussion topics. A chart in the issue maps the articles to places in the textbook where associated material is present. This unique supplement is *free*, shrink-wrapped with each copy of the text.

A Reader's Guide to Astronomy Now! *New for 1994*—Andrew Fraknoi, past Executive Director of the Astronomical Society of the Pacific and now Professor of Astronomy at Foothill College, has prepared a set of critical thinking questions to help students get the full benefit from reading the articles in the special issue of *Discover*. This question booklet is also free, shrink-wrapped with the textbook.

Instructor's Manual This completely revised manual by Andrew Fraknoi contains course outlines, answers to all end-of-chapter questions and problems, additional readings, and other sources of information about astronomy-related activities.

Saunders Astronomy Transparency Collection   This comprehensive set contains 205 color overhead transparency acetates of conceptually based artwork. Over half are enlarged reproductions of figures from the textbook. The others are supplemental illustrations that complement the text figures. A detailed guide arranged by topic accompanies the collection.

Saunders Astronomy Transparency Collection Supplement   *New for 1994*—This supplemental collection of 25 new images from the Hubble Space Telescope and major observatories is available free to adopters.

Saunders Astronomy Slide Collection   The set of 205 transparency acetates is also available in 35-mm slide format.

Saunders Astronomy Videodisc Library: Volume I: The Solar System
New for 1994—This exclusive videodisc offers a comprehensive overview of the solar system via a mix of motion video, animations, still photographs, and artwork. Many of the photographs and drawings are taken directly from the textbook. All images, whether motion or still, are arranged by topic in the traditional earth-outward order, reflecting the way most instructors teach the course. Free to qualified adopters.

LectureActive™ Interactive Videodisc & Lecture Presentation Software   *New for 1994*—This distinctive software (available in IBM/Windows® and Macintosh formats) helps instructors customize lectures by giving them quick, efficient access to the video clip and still frame data on *Volume I: The Solar System* of the Saunders Astronomy Videodisc Library. All video clip and still-frame data from the videodisc are entered and listed on the software, so it is easy to customize lectures swiftly and simply by calling up the desired footage or images. Free to qualified adopters.

Saunders Astronomy Video Tape   Free to all adopters, this 93-minute VHS tape presents five NASA/JPL "movies," including a segment on images obtained from the Magellan Venus Orbiter.

NOVA Videotapes   Qualified adopters can choose from a selection of NOVA videotapes. See your sales representative for details.

Printed Test Bank   This bank contains 1,000 questions in multiple choice and other formats, including essay questions.

ExaMaster™ Computerized Test Bank for IBM and Macintosh   All the questions from the printed test bank are presented in a computerized format that allows instructors to edit, add questions, and print assorted versions of the same test. Also included in the testing package are these advanced features: graph generation via a built-in graphing program; graphical image input via a built-in paint program; an easy-to-use algorithm-based question generator; an on-line testing (OLT) module; a full-function gradebook with statistical capability.

RequesTest™   Instructors without access to a personal computer may contact Saunders Software Support Department at (800) 447-9457 to request tests prepared from the computerized test bank. The test will be mailed or faxed to the instructor within 48 hours.

For more information about these ancillaries contact your local Saunders sales representative or call Harcourt Brace College Publishers at (800) 237-2665 or (708) 647-8822. In Canada call (416) 255-4491.

ACKNOWLEDGMENTS

We would like to thank the following professors who read and critiqued the first edition prior to writing the second edition manuscript:

Jorge Cossio, Miami-Dade Community College
Donald Foster, Wichita State University
Padmanabh Harihar, University of Massachusetts at Lowell
William M. Hussong, College of DuPage
James C. LoPresto, Edinboro University of Pennsylvania
David McDavid, University of Texas at San Antonio
Thomas M. Stadelmann, Massasoit Community College

Valuable, cogent reviews of the second edition manuscript were provided by:

Carlson R. Chambliss, Kutztown University of Pennsylvania
Jorge Cossio, Miami-Dade Community College
Tom Harrison, University of North Texas
Thomas Hockey, University of Northern Iowa
William M. Hussong, College of DuPage
James C. LoPresto, Edinboro University of Pennsylvania

Rudolf Schild of the Harvard-Smithsonian Center for Astrophysics graciously provided the cover image of the Cocoon Nebula in Cygnus. Andrew Fraknoi's contributions to the project have been numerous, from revising the Suggestions for Additional Reading to his work with the ancillaries. His suggestions and improvements are most appreciated. We also wish to acknowledge our continuing debt to George Abell. While this text is very different in style and approach from the books originally written by Abell and subsequently revised by us, we have used some historical and explanatory material written by him. We hope that this text would meet the very high standards he set for accuracy and clarity.

Thanks also to Janet L. Morrison, who reviewed the manuscript and made many helpful suggestions, and to our publisher, John Vondeling; sponsoring editor, Lloyd Black; project editor, Anne Gibby; marketing manager, Sue Westmoreland; and production manager, Charlene Squibb, at Saunders for their encouragement and support.

David Morrison
Sidney C. Wolff
January 1994

▌CONTENTS OVERVIEW

▶ CONTENTS

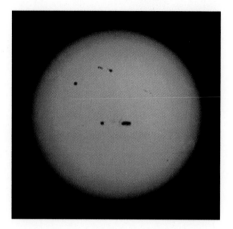

FINDING OUR PLACE IN THE UNIVERSE

Star trails over Kitt Peak National Observatory. *(National Optical Astronomy Observatories)*

Astronomy is an observational *science*. Astronomers are unable to carry out experiments or to measure directly the properties of celestial objects. All we have to work with is the faint light collected by our telescopes, light that has traversed vast distances to reach us. Despite this limitation, astronomers have been able to build up an impressive picture of the universe, its content, and its evolution. That picture is the subject of this book.

We begin with a summary of the tradition of astronomy in European civilization, tracing the story of how improved observations and theory have changed our view of what astronomy is, and how they changed our view of the Earth and its human population in relation to the universe.

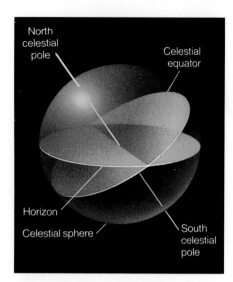

Figure 1.1   The celestial sphere; the celestial poles, equator, and ecliptic are shown. The observer views the upper half of the celestial sphere from inside at its center; the Earth itself blocks our view of the lower hemisphere.

What is the apparent daily motion of the Sun and stars?

1.1

THE HEAVENS ABOVE

Our concept of the cosmos—its basic structure and origin—is called **cosmology.** Before the invention of telescopes, humans had to depend on the simple evidence of their senses for a picture of the universe. Ancient peoples developed cosmologies that combined their direct view of the heavens with a rich variety of philosophical, mythological, and religious symbolism.

The Celestial Sphere

Our study of astronomy begins with the view of the heavens above us, a view identical to that available to peoples all over the world before the invention of the telescope. If we look up on a clear night, we get the impression that the sky is a hollow spherical shell with the Earth at the center. The early Greeks regarded the sky as just such a **celestial sphere** (Figure 1.1). Some apparently thought of it as an actual sphere of crystalline material, with the stars embedded in it like tiny jewels.

The Sun, Moon, and stars rise and set as the Earth turns within this imaginary sphere. In ancient times, of course, people did not realize that the Earth was a planet, spinning about its axis. Their concept of the Earth was restricted to the apparently flat world that they could see with their own eyes. It is easy to recreate this viewpoint just by finding a dark, quiet spot from which to look at the stars above. If we watch the sky for several hours, we see that the celestial sphere appears gradually to turn around us. Stars rise in the east and set in the west, moving completely across the vault of heaven in the course of the night. The ancients, unaware of the Earth's rotation, imagined that the celestial sphere rotated about an axis that passed through the Earth. As it turned, the celestial sphere carried the stars up in the east, across the sky, and down in the west. The best way to learn the apparent motions of the stars is to spend a few hours under a dark sky; alternatively, we can see these motions recreated by projecting an image of the sky onto the dome of a planetarium.

As the celestial sphere rotates, all of the objects in the sky maintain their positions with respect to each other. A grouping of stars like the Big Dipper has the same shape wherever we see it in the sky, although its apparent orientation with respect to the terrestrial foreground shifts during the night. Even objects that we know have their own motions, such as planets or the Moon, seem fixed relative to the stars over the period of a single night. Only the meteors—brief "shooting stars" that flash into view for just a few seconds—move appreciably with respect to the celestial sphere, and they are located within the atmosphere of the Earth.

The pole or point about which the celestial sphere appears to pivot lies along an extension of the line through the Earth's North and South Poles. As the Earth rotates about its polar axis, the sky appears to turn in the opposite direction about the **north celestial pole** and **south celestial**

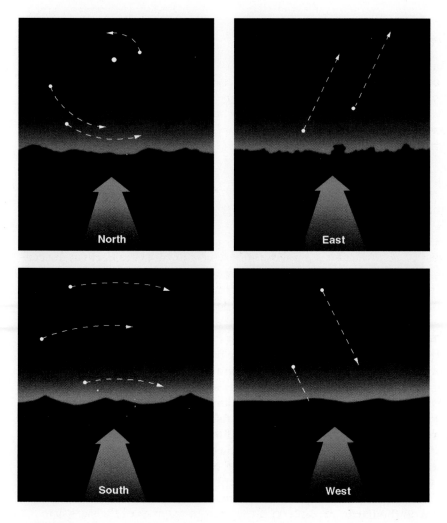

Figure 1.2   The apparent motion of the sky as seen facing each of the four cardinal directions for an observer in the temperate latitudes of the northern hemisphere.

pole (Figure 1.2). Halfway between the poles, and separating the sky into its northern and southern halves, is the **celestial equator**—just like the Earth's equator, which separates the Northern and Southern Hemispheres of our planet.

Terrestrial and Celestial Coordinates

We specify positions of places on the Earth by giving their latitude and longitude, both of which are in angular measure (e.g., degrees of angle). The latitude is the angle measured along the surface of the sphere north or south of the equator (which is at latitude of 0 degrees). At the North Pole the latitude is 90 degrees N; at the South Pole it is 90 degrees S. Longitude is the distance east or west of the north-south line (called the prime meridian) that runs through the old Royal Observatory at Greenwich, England. In the United States, for example, Chicago has a latitude of 42 degrees N and a longitude of 88 degrees west of Greenwich.

Figure 1.3   The elevation in the sky of the north celestial pole is equal to the observer's latitude.

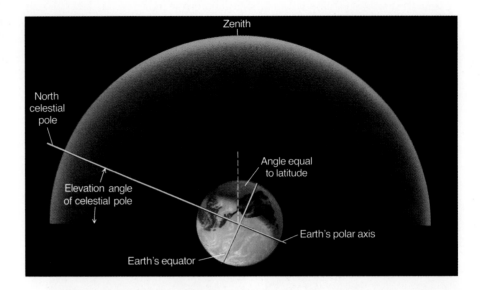

The apparent motion of the celestial sphere depends on the latitude of the observer. Someone at the North Pole of the Earth sees the north celestial pole directly overhead, at the position called the **zenith.** The stars would all circle about the sky parallel to the horizon, with none rising or setting. Conversely, an observer at the Earth's equator sees the celestial poles at the north and south points on the horizon. As the sky turns about these points, all the stars rise straight up along the eastern horizon and set straight down toward the west. For an observer in Europe or the United States, the north celestial pole appears above the northern horizon at an angular height, or altitude, equal to that observer's latitude (Figure 1.3). The sky projector in a planetarium can be adjusted to show the apparent motions of the stars as seen from any place on the Earth.

Astronomers have worked out a system to indicate the positions of stars on the celestial sphere that is very similar to the latitude-longitude system for positions on the Earth. The angular distance of a star north or south of the celestial equator (the equivalent of latitude) is called *declination*. The distance east or west (equivalent of longitude) is called *right ascension;* it is measured with respect to the position on the celestial sphere called the vernal equinox.

Constellations and Star Names

Right ascension and declination provide a precise way to specify the position of any star or other object on the celestial sphere. Often, however, we simply wish to know the general region of the sky in which a star is found, just as we might note the state or country where a city is located on Earth. These regions of the sky are called **constellations.**

The ancient Greeks, Chinese, and Egyptians all divided the sky into constellations, associated with apparent patterns of stars (Figure 1.4).

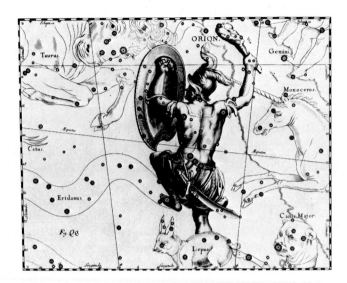

Figure 1.4 The winter constellation of Orion, the hunter, as illustrated in the 17th century atlas by Hevelius. *(J.M. Pasachoff and the Chapin Library)*

Well-known examples, drawn from Greco-Roman mythology, include Orion, Hercules, and Andromeda. Sometimes the brightest stars of a constellation form a prominent, easily recognized figure like the Big Dipper. More often, however, a constellation bears no more resemblance to the animal or person for whom it is named than does the outline of Washington State resemble George Washington. The modern boundaries between the constellations are imaginary lines in the sky running north-south and east-west, so that each point in the sky falls within the borders of one of the 88 constellations.

A few of the brighter stars have individual names, drawn primarily from the Arabic language. Much of ancient astronomical tradition reached Europe through the Islamic culture of North Africa and Moorish Spain, which flourished during the Dark Ages of Europe. Other bright stars are designated by letters of the Greek alphabet, while the myriad of fainter stars bear only numbers or have no designation at all. An example of a named star is Sirius, the brightest star in the night sky located in the constellation of Canis Major. The nearest star, in contrast, is known simply as Alpha Centauri, the brightest star in the southern constellation of Centaurus. (Alpha, the first letter of the Greek alphabet, is usually assigned to the brightest star in each constellation.)

The ancient Greeks not only named many stars and constellations, but they also bequeathed to us a system for designating the apparent brightness of the stars. In this system, the brightest dozen or so stars visible from Greece were called stars of the **first magnitude.** Sirius and Alpha Centauri are examples of first magnitude stars. Fainter stars were called second magnitude, third magnitude, and so on down to the sixth magnitude stars, which are at the limit of visibility on a clear moonless night. This basic magnitude system is still used in astronomy; we will define it more precisely in Chapter 10, when we discuss the properties of stars.

1.2
SUN, MOON, AND PLANETS

The stars of the celestial sphere provide a backdrop for the motions of the Sun, Moon, and planets. Understanding these more complex motions was the central problem of astronomy in the ancient world.

Rising and Setting of the Sun

We have described the appearance of the night sky. The situation during the day is similar except that the brilliance of the Sun renders the stars and planets invisible. (The Moon is still easily seen in the daylight, however.) We can think of the Sun as being located at a position on the hypothetical celestial sphere. When the Sun rises—that is, when the rotation of the Earth carries the Sun above the horizon—sunlight scattered about by the molecules of the atmosphere produces the blue sky that hides the stars that are also above the horizon.

For thousands of years, astronomers have been aware that the Sun gradually changes its position, moving each day about 1 degree to the east relative to the stars. The Sun's apparent path around the celestial sphere, which reflects the revolution of the Earth about the Sun, is called the **ecliptic.** Each day the Sun rises about 4 minutes later with respect to the stars; the Earth must make just a bit more than one complete rotation (with respect to the stars) to bring the Sun up again.

As we look at the Sun from different places in our orbit, we see it projected against different stars in the background, or we would, at least, if we could see the stars in the daytime. In practice, we must deduce which stars lie behind and beyond the Sun by observing the stars visible in the opposite direction at night. After a year, when the Earth has completed one trip around the Sun, the Sun will appear to have completed one circuit of the sky along the ecliptic. We have a similar experience if we walk around a campfire at night; we see the flames appear successively in front of each person seated about the fire.

Fixed and Wandering Stars

The Sun is not the only object that moves among the fixed stars. The Moon and each of the five planets visible to the unaided eye—Mercury, Venus, Mars, Jupiter, and Saturn—also slowly change their positions from day to day. The Moon, being the Earth's nearest celestial neighbor, has the fastest apparent motion; it completes a trip around the sky in about one month. During a single day, of course, the Moon and planets all rise and set, as do the Sun and stars. But like the Sun, they have independent motions among the stars, superimposed on the daily rotation of the celestial sphere.

The Greeks of two thousand years ago distinguished between what they called the fixed stars, the stars that maintain fixed patterns among themselves throughout many generations, and the wandering stars or planets. The word planet means "wanderer" in Greek. Today, we do not regard the Sun and Moon as planets, but the ancients applied the term to

How did the ancient Greeks interpret the complex apparent motions of the Sun and planets?

all seven of the moving objects in the sky. Much of ancient astronomy was devoted to observing and predicting their motions. In the Romance languages, the planets give the names for the seven days of our week, although modern English retains only Sunday (Sun), Monday (Moon), and Saturday (Saturn).

The Calendar

One of the most important functions of astronomy in the ancient world was the determination of the calendar—that is, the days of the week, month, and year. There are two traditional functions of any calendar. First, it must keep track of the passage of time, allowing people to anticipate the cycle of the seasons and to honor special religious anniversaries. Second, in order to be useful to a large number of people, it must address the problem of the relationships among the basic natural time intervals defined by the motions of the Earth and Moon.

The natural units of the calendar are the *day*, based on the period of rotation of the Earth; the *month*, based on the period of revolution of the Moon about the Earth; and the *year*, based on the period of revolution of the Earth about the Sun. Difficulties have resulted from the fact that these three periods are not commensurable—that is, one does not divide evenly into any of the others.

The rotation period of the Earth is, by definition, 1.0000 days. The period required by the Moon to complete its cycle of phases, called the lunar month, is 29.5306 days. The basic period of revolution of the Earth, called the tropical year, is 365.2422 days. Clearly, the ratios of these numbers (1.0000/29.5306/365.2422) are not convenient for calculations. Our natural clocks run on different time. This is the historic challenge of the calendar, dealt with in various ways by different human civilizations.

Even the earliest cultures were concerned with the keeping of time and the calendar. Particularly interesting are monuments left by Bronze Age people in northwestern Europe, especially in the British Isles. The best preserved of the monuments is Stonehenge (Figure 1.5), located

(text continued on p. 11)

What is a calendar, and how are calendars related to astronomy?

Figure 1.5   Stonehenge, an ancient monument that served in part as an observatory and calendar-keeping device, located in the Salisbury Plain of England.   *(David Morrison)*

ESSAY   Eclipses of the Sun and Moon

The Sun and Moon have nearly the same apparent size in the sky. Although the Sun is about 400 times larger in diameter than the Moon, it is also about 400 times farther away, so both appear to be the same size—about 0.5 degree across. (You can test this statement by covering the Sun and the Moon with your thumb held at arm's length, because for most people the width of the thumbnail at this distance is about 1 degree). As a result of this coincidence, we are fortunate to be able to see an eclipse of the Sun when the Moon is between us and the Sun. In addition, we get an eclipse of the Moon when the Moon moves into the shadow of the Earth.

Any solid object in space casts a shadow by blocking the light of the Sun from a region behind it. An eclipse occurs whenever any part of either the Earth or the Moon enters the shadow of the other. When the Moon's shadow strikes the Earth, people on Earth within that shadow see the Sun covered up by the Moon; that is, they witness a *solar eclipse.* When the Moon passes into the shadow of the Earth, people on the night side of the Earth see the Moon darken—a *lunar eclipse.*

The geometry of a total solar eclipse is illustrated in Figure 1A. If the Sun and Moon are properly aligned, the Moon's shadow will intersect the ground at a small point on the Earth's surface. Anyone on the Earth within this small area covered by the tip of the Moon's shadow witnesses a total solar eclipse. Conversely, within a larger area of the Earth's surface, one sees part but not all of the Sun eclipsed by the Moon—a partial solar eclipse. It does not take

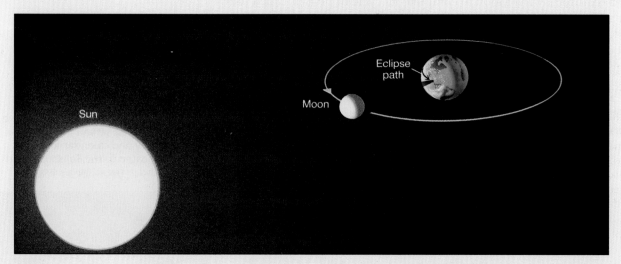

Figure 1A   Geometry of a solar eclipse.

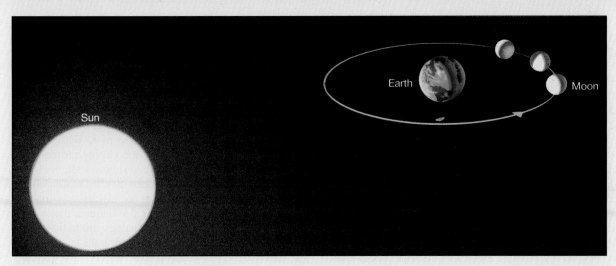

Figure 1B Geometry of a lunar eclipse.

long for the Moon's shadow to sweep past a given point on Earth. The duration of totality may be only a brief instant; it can never exceed about 7½ minutes.

A lunar eclipse occurs when the Moon enters the shadow of the Earth. The geometry of a lunar eclipse is shown in Figure 1B. Unlike a solar eclipse, which is visible only from certain local areas on the Earth, a lunar eclipse is visible to everyone who can see the Moon at that time. Because a lunar eclipse can be seen (weather permitting) from the entire night side of the Earth, lunar eclipses are observed far more fre-quently from a given place on Earth than are solar eclipses.

Total eclipses of the Moon occur, on the average, about once every two or three years. Total solar eclipses take place every year, but unless you are willing to travel, you are not likely to witness one during your lifetime. If you ever do have the chance to see a total eclipse of the Sun, by all means do so; it is one of the most spectacular sights of nature and a once-in-a-lifetime experience!

ESSAY   Astrology: An Ancient Religion

The science of astronomy and the religion of astrology are unrelated except in the sense that both had common origins in the Greco-Roman world (and in other societies too); yet many otherwise-informed people still confuse the two.

Modern research has shown that all matter in the universe is composed of atoms—and the same kinds of atoms. Our probes that landed on Mars and Venus and our telescopic studies of the light from the most remote galaxies indicate that they are made of the same elements that make up our own bodies. Most ancient cultures, however, thought of the heavens as quite separate and distinct from the terrestrial world. They assumed that the luminous orbs in the sky, the stars and planets, are made of "heavenly" substances; perhaps this is why they attributed special powers to the wandering planets. These planets were associated with the gods of ancient mythologies; in some cases, they were themselves thought of as gods.

In ancient China and the Middle East, it was thought that the planets and their motions influenced the fortunes of kings and nations. Eventually, the idea arose that the planets bore their influence on every individual. In particular, the doctrine called natal (birth) astrology taught that the configuration of the planets at the moment of a person's birth (called a horoscope) affected his or her personality and fortune. It is this ancient religion, older than Christianity or Islam, that is practiced by today's astrologers.

A modern variant of natal astrology is sun-sign astrology, which uses only one aspect of one "planet": the sign occupied by the Sun at the time of a person's birth. Although even professional astrologers do not place much trust in such a limited scheme, which tries to fit everyone into just 12 groups, sun-sign astrology is the mainstay of newspaper astrology columns and party games, and apparently many people take it seriously. A recent poll showed that more than half of U.S. teenagers said they "believed in astrology." There is evidence that a surprising number of Americans make judgments about people—who they will hire, associate with, even marry—on the basis of astrological information.

Astrological prejudice is a significant source of bias in modern American society—a judgment of people based not on who they are or what they have accomplished, but instead on an accident of their birth.

Today, with our knowledge of the nature of the planets as physical bodies as well as our understanding of human genetics, it is hard to imagine that the directions of these planets in the sky at the moment of one's birth could have anything to do with one's personality or future. There are no known forces, not gravity or anything else, that could cause such effects. Nor does the idea that one-twelfth of all people in the world will have similar life histories, as implied by newspaper astrology predictions, make much sense. If the tenets of astrology are to be taken at all seriously, we would need compelling evidence of its validity. In the tens of centuries of astrology, no such evidence has been presented.

During the past few years, a number of careful statistical tests have been carried out to test astrology's predictive power. The simplest of these examine sun-sign astrology to determine whether some signs are more likely than others to be associated with such objective measures of success as winning Olympic medals, earning high corporate salaries, or achieving elective office or high military rank. You can make such a test yourself with, for example, the birthdates of members of Congress, or of members of the U.S. Olympic Team. One example is a study of the birth signs of all U.S. Marines who re-enlisted over a period of several years; presumably such a group might have some common traits, but an analysis of their "signs" showed a completely random distribution. More sophisticated studies have also been done, involving horoscopes calculated for thousands of individuals. (With modern computers, the once laborious process of calculating a horoscope is practically instantaneous.)

The results of all of these studies are the same: There is no evidence that natal astrology has any predictive power, even in a statistical sense. Astrology is not a science, it is not scientific in its methods, and it has failed every test so far applied to it.

about 13 km from Salisbury in southwest England. It is generally believed that at least one function of the monument was connected with the keeping of a calendar.

The calendar we use today derives from Greek calendars dating from at least the 8th century B.C., subsequently modified by the Romans at the time of Julius Caesar and then again by the Catholic Church in the sixteenth century. We begin with a year that normally has 365 days. By adding an extra day periodically (in leap years), our calendar approximates the year by 365.2425 days, which is correct to about 1 day in 3300 years. We have abandoned the task of reconciling this year with the period of the Moon, but other lunar calendars persist, such as the Islamic calendar used to set religious festivals. The phase of the Moon also figures in the calculation of the dates of the Christian Easter, the Jewish Passover, and the Chinese New Year.

1.3
EARTH IN THE UNIVERSE: CHANGING PERCEPTIONS

The greatest contribution of astronomy to the European Renaissance was the overthrow of the traditional view that the Earth is stationary and located at the center of the universe. The discovery that the Earth is one of the planets represented a true revolution in our thinking about the universe and ourselves.

The Traditional View

Our senses tell us that the Earth is the center of the universe—the hub around which the heavens turn. This **geocentric** view was held almost universally until the European Renaissance. It is simple, logical, and seemingly self-evident. Further, the geocentric perspective reinforces philosophical and religious systems that teach the unique role of man as the central focus of the cosmos. However, the geocentric view happens to be wrong. One of the great themes of intellectual history is the overthrow of the geocentric perspective and the successive steps by which we have re-evaluated the place of our world in the cosmic order.

Although they firmly believed that the Earth was the hub of the universe, many scientists and philosophers of the ancient past recognized that the Earth is not flat. There are a variety of lines of evidence to support the conclusion that the Earth is spherical. For example, the lower parts of a ship disappear first as it sails away over the horizon. But circular or flat, the Earth was considered immobile. Solid and substantial, it was thought to dominate the cosmos.

In this view, which reached its highest development in the Egyptian city of Alexandria during the second century A.D., all of the observed phenomena of the universe could be accounted for by combinations of circular motions with the stationary Earth at the center. Adopted by the early Christian Church, this Alexandrian cosmology remained unchallenged until the sixteenth century.

The Copernican Revolution

How did Copernicus revolutionize our concept of the place of the Earth in the universe?

Figure 1.6 Nicholas Copernicus (1473–1543), Polish cleric and scientist. Copernicus developed the heliocentric cosmology that revolutionized astronomy. However, he was so cautious about these new ideas that he only permitted their publication when he was dying.

One of the most important events of the Renaissance was the displacement of the Earth from the center of the universe. This intellectual revolution was initiated by a Polish lay monk, Copernicus (Figure 1.6). Nicholas Copernicus (1473–1543) was born in Torun on the Vistula. His training was in law and medicine, but Copernicus' main interests were astronomy and mathematics.

Copernicus' great contribution to science was a critical reappraisal of the existing theories of planetary motion and the development of a new Sun-centered, or **heliocentric,** model of the solar system. Copernicus concluded that the Earth is a planet, and that the planets Mercury, Venus, Earth, Mars, Jupiter, and Saturn all circle the Sun. Only the Moon was left in orbit about the Earth. Although Copernicus could not prove that the Earth is in motion, he evidently found something orderly and pleasing in the heliocentric system, and his defense of it was elegant and persuasive. He also found that calculations of planetary positions based on the heliocentric model were simpler and more accurate than those derived from the previous geocentric perspective.

One of the objections raised about the heliocentric theory was that if the Earth were moving, we would all sense or feel this motion: Solid objects would be ripped from the surface, a ball dropped from a great height would not strike the ground directly below, and the like. But a moving person is not necessarily aware of that motion. We have all experienced seeing an adjacent train, car, or ship appear to move, only to discover that it was we who were moving.

Copernicus argued that the apparent annual motion of the Sun about the Earth could be represented equally well by a motion of the Earth about the Sun. He also reasoned that the apparent rotation of the celestial sphere could be accounted for by assuming that the Earth rotates while the celestial sphere is stationary. To the objection that if the Earth rotated about an axis it would fly into pieces, Copernicus answered that if such motion would tear the Earth apart, the still faster motion (because of its great size) of the celestial sphere required by the alternative hypothesis would be even more devastating.

In Copernicus' time, few people thought that ways existed to *prove* whether the heliocentric or the older geocentric system was correct. A long philosophical tradition, going back to the Greeks and defended by the Catholic Church, held that pure human thought combined with divine revelation represented the path to truth. Nature, as revealed by our senses, was suspect. In this environment, there was little motivation to carry out observations or experiments to try to distinguish between competing cosmological theories (or anything else). It should not surprise us, therefore, that the heliocentric idea was debated for more than half a century without any tests being applied to determine its validity. (In fact, the older geocentric system was still taught at Harvard University in the first years after it was founded in 1636). Contrast this with the situation today, when scientists rush to test each new theory. In

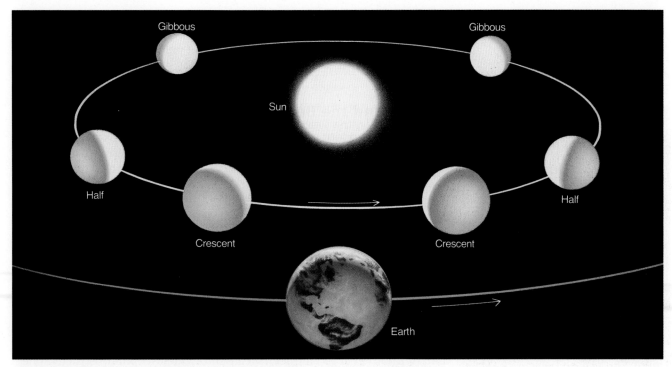

Figure 1.7   The phases of Venus as seen from the Earth in the heliocentric system. Galileo's observation that Venus went through this exact cycle of phases provided confirmation that the Sun, not the Earth, was the center of planetary motion.

the five centuries since Copernicus, we have learned that nature is orderly and consistent, and that experiments and observations can therefore be used to check theories. We have developed sophisticated technologies to aid in our probing of the natural world.

When a new hypothesis or theory is proposed in science, it must first be checked for consistency with what is already known. Copernicus' heliocentric idea passed this test, for it allowed planetary positions to be calculated at least as well as the geocentric theory. The next step is to see what predictions the new theory makes that differ from those of competing ideas. In the case of Copernicus, one example is the prediction that if Venus circles the Sun, it should go through the full range of phases just as the Moon does, whereas if it circles the Earth, it should not (Figure 1.7). But in those days, before the telescope, no one could have imagined testing this prediction.

Today, the tests of competing theories are rarely so simple as this example of the phases of Venus. Because most predictions in science are *quantitative,* involving numerical values that can be calculated from mathematical equations, the checks usually require accurate measurements of the phenomena being studied. Often it is only through increasing the precision of the observations that a distinction between two

What is the scientific method?

competing theories can be made. For example, ordinary gravitational theory and Einstein's theory of general relativity make identical predictions about everyday experience, diverging only under extreme conditions, such as speeds near the speed of light.

Galileo and the Beginning of Modern Science

Many of the modern scientific concepts of observation, experimentation, and the testing of hypotheses through careful quantitative measurements were pioneered by a man who lived nearly a century after Copernicus. Galileo Galilei (1564–1642), a contemporary of Shakespeare, was born in Pisa, Italy (Figure 1.8). Like Copernicus, he began training for a medical career, but he had little interest in the subject and later switched to mathematics. He held faculty positions at the Universities of Pisa and Padua, and eventually he became scientific advisor to the Grand Duke of Tuscany in Florence.

It is not certain when the principle was first conceived of combining two or more pieces of glass to produce an instrument that enlarged images of distant objects, making them appear nearer. The first telescopes that attracted significant attention were made by the Dutch spectacle maker Hans Lippershey in 1608. Galileo heard of the discovery in 1609 and, without ever having seen an assembled telescope, constructed one of his own with a three-power magnification, which made distant objects appear three times nearer and larger (Figure 1.9).

On August 25, 1609, Galileo demonstrated a telescope with a magnification of 9× to officials of the Venetian government. By a magnification of 9×, we mean that the linear dimensions of the object being viewed appeared nine times larger, or alternatively, that the objects appeared nine times closer than they really were. There were obvious military advantages associated with a device for seeing distant objects. For his invention, Galileo's salary was nearly doubled, and he was granted lifetime tenure as a professor. His university colleagues were outraged, particularly because the invention was not even original.

Figure 1.8 Galileo Galilei (1564–1642), considered to be the father of both modern physics and modern astronomy. His telescopic observations verified the ideas of Copernicus, but his outspoken advocacy of the heliocentric system offended the Roman Catholic Church; in 1616 he was brought before the Inquisition, threatened with torture, and forced to deny that the Earth moved about the Sun.

Figure 1.9 Two telescopes used by Galileo.

The First Astronomical Telescope

Before using his telescope for astronomical observations, Galileo had to devise a stable mount, and he improved the optics to provide a magnification of 30×. Galileo also needed to acquire confidence in the telescope. At that time, the eyes were believed to be the final arbiter of truth about sizes, shapes, and colors. Lenses, mirrors, and prisms were known to distort distant images by enlarging them, reducing them, or even inverting them. Galileo undertook repeated experiments to convince himself that what he saw through the telescope was identical to what he saw up close. Only then could he begin to believe that the miraculous phenomena that were revealed in the heavens were real. Although Galileo was convinced of the validity of what he saw, others were not. One unbelieving colleague said that he "...tested this instrument of Galileo's in a thousand ways, both on things here below and on those above. Below, it works wonderfully; in the sky it deceives one, as some fixed stars are seen double." Another scholar refused even to look through the telescope because doing so gave him a headache.

Beginning his astronomical work late in 1609, Galileo found that many stars too faint to be seen with the naked eye became visible with his telescope. In particular, he found that some nebulous blurs resolved into many stars and that the Milky Way was made up of multitudes of individual stars. He found four satellites or moons revolving about Jupiter with periods ranging from just under two days to about seventeen days. This discovery was particularly important because it showed that there could be centers of motion that in turn are in motion themselves. Defenders of the geocentric view had argued that if the Earth were in motion, the Moon would be left behind because it could hardly keep up with a rapidly moving planet. Yet here were Jupiter's satellites doing exactly that!

With his telescope, Galileo was also able to carry out the test of the Copernican theory mentioned earlier. Within a few months he had found that Venus goes through phases like the Moon, showing that it must revolve about the Sun; thus, we see different parts of its daylight side at different times (see Figure 1.7). These observations could not be reconciled with any model in which Venus circled about the Earth.

Galileo observed the Moon and saw craters, mountain ranges, valleys, and flat, dark areas that he thought might be water (the dark maria, or "seas," on the Moon were thought to be water until long after Galileo's time). These discoveries show that the Moon might not be so dissimilar to the Earth, which suggested that the Earth, too, could belong to the realm of celestial bodies.

The discoveries of Copernicus and Galileo revolutionized our concept of the cosmos. Contrary to previous belief, they showed that space is vast and that the Earth is relatively small. They demonstrated that the Earth is a planet and raised the possibility that the other planets might be worlds themselves, perhaps even supporting life. They demoted the Earth (and humanity) from the center of the universe. We take these things for granted, but four centuries ago such concepts were frighten-

What discoveries did Galileo make with his telescope?

OBSERVING YOUR WORLD   THE SKY AT NIGHT

In this book, we describe a number of activities that will help you to understand what you see when you look at the night sky. These activities require no equipment at all, and so you can discover for yourself many of the phenomena that were first described by ancient astronomers. You have the advantage that we now understand the explanations of these phenomena, but astronomers in ancient times had the advantage that the skies, especially in and near cities, were much darker.

Some simple techniques will help you to enjoy the night sky. The first is *take your time*. Some events, such as the changes in the phases of the Moon or the motion of the Sun relative to the background stars, simply take a long time to occur. There is also much to see, and you cannot expect to discover all the beauty of the heavens in a few minutes. Weather will frustrate you, as it does the professional astronomer. Thick clouds will hide all of the stars. Thin clouds may be barely visible but will make faint stars difficult to see, so it is necessary to be aware of sky conditions. Bright moonlight will swamp the light of all but the brightest stars, and so serious stargazing is best done when the Moon is not present.

City lights also mask the light from stars. From large cities in the U.S., only the Moon and the brightest stars and planets can be seen. Observing even those few objects and noting the changes in their positions from day to day can be fun. If you live in a big city, however, try to take advantage of vacations and other trips to see more of what is in the sky.

Activities

1. Experiment to see how long it takes your eyes to adapt so that you can see better in the dark. The size of the pupil of your eye determines how much light is admitted. In bright light, the pupil contracts so that only a small amount of light can enter it. When you go outside from a brightly lit room, it takes some time for your pupil to expand and to become *dark adapted*. Go outside and note how many stars you can see. After about 5 minutes you should begin to see many more stars, and full dark adaptation should be achieved after about 15 minutes. In order to identify constellations, you will probably need to use a flashlight and star charts. Keep the flashlight as dim as possible or, better yet, use a red light (or red cellophane over a white light) because red light does not affect your dark adaptation as much as white light will.

2. The sensitivity of the eye depends on just where light falls on the retina. You can actually see slightly fainter stars if you look not directly at them but just slightly to one side. Try using *averted vision* to see

ing and heretical for some, while immensely stimulating for others. The pioneers of the Renaissance started the European world along the path toward science and technology that we still tread today. For them, Nature was rational and ultimately knowable, and science provided the means to reveal its secrets.

1.4
MODERN ASTRONOMY

Astronomy led the sciences in its efforts to formulate universal laws. Ever since the time of Newton, we have looked to science for clear, quantitative predictions, and nowhere have these expectations been bet-

fainter objects. Pick a star that you can just barely see when you look directly at it. Then try looking at various spots nearby and see how its apparent brightness changes.

3. For some of the activities described in later chapters of this book, it is useful to be able to estimate distances in the sky. The units normally used are degrees, and the hand provides a way of estimating the distance between two objects in degrees (Figure 1A). For most people, a thumb as seen with the arm fully extended is about 1 degree wide; the first two knuckles of the fist are separated by about 2 degrees; the fist itself is about 10 degrees across.

Verify the estimate for the width of the fist by measuring the distance from the horizon to the zenith in terms of the width of your fist extended at arm's length. Estimate the diameter of the full Moon. How far apart are the stars at the end of the bowl of the Big Dipper? A line drawn between the two stars at the end of the bowl of the Big Dipper and extended beyond the Dipper points to Polaris, the north star. How far is Polaris in degrees from the bowl of the Big Dipper?

4. The north celestial pole appears at an altitude above the horizon that is equal to the observer's latitude. Polaris, the north star, lies very close to the north celestial pole. Measure its altitude by using

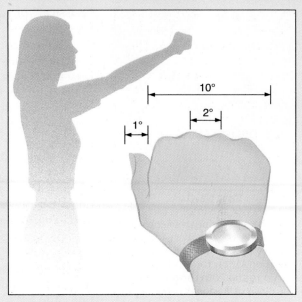

the width of your fist. Compare this estimate with your latitude. The next time you travel several hundred miles north or south determine the altitude of Polaris again. Can you detect a difference? (This experiment cannot easily be performed in the Southern Hemisphere because Polaris itself is, of course, not visible. There is no bright star near the south celestial pole.

ter served than in the calculation and prediction of planetary orbits. Today we go further, however, in aspiring to understand the detailed physical and chemical processes that describe the workings of planets, stars, and galaxies.

Newton and the Mechanistic Universe

It was the genius of Isaac Newton (1643–1727) that unified the natural laws discovered by Galileo and his contemporaries. Newton (Figure 1.10) was born in Lincolnshire, England, one year after Galileo's death. He entered Trinity College at Cambridge in 1661, and eight years later he was appointed Professor of Mathematics at Cambridge.

Figure 1.10   Isaac Newton
(1643–1727) was the intellectual giant
of 18th century European science. His
work on the laws of motion,
gravitation, optics, and mathematics
laid the foundation for much of
science up to the 20th century, and
the influence of his ideas gave birth to
the modern concept of human
progress through advancing
technology.

What were the great contributions
of Newton?

Newton's greatest achievements are described in his book, *The Mathematical Principles of Natural Philosophy,* published in 1687. At the beginning of this book, Newton states the laws that describe mathematically the forces that determine the motions of all objects, on the Earth and in space. These laws, together with the new mathematical techniques also developed by Newton, still form the basis of what is called "classical" physics. But Newton's grandest concept is that of universal gravitation.

Newton could not say why there is a gravitational force between all material bodies in the universe (nor can we), but he brilliantly described how that force operates. He established that a general force of attraction holds the planets in their orbits. He not only suggested this general idea, but also used observations and experiments to derive the mathematical formula to describe this universal gravitation and permit exact calculations of celestial motions. In Chapter 2, we discuss the Law of Gravitation and its applications to planetary motions.

Newton was the intellectual giant of his age. His laws of motion and of gravitation brought together the celestial and terrestrial realms, demonstrating that both are subject to the same scientific principles. He also inaugurated a century of developments in mathematics that allowed scientists to use these laws to predict with remarkable accuracy the motions of the planets. Astronomical science—at least the science of celestial motions—emerged as increasingly mechanistic and deterministic. From the positions of the planets today, one could predict their motions into the future. Similarly, the configuration of the planets now was the product of their mutual gravitational interactions in the past, all developing according to strict mathematical laws. Like a clock, according to this viewpoint, the universe once created would continue forever, never running down or deviating from its course.

During the two centuries following Newton, mathematical astronomy dominated the field. But the increasing power of astronomical telescopes also contributed. In addition to their studies of planetary motions, astronomers began to look outward to the stars. Gradually, evidence accumulated that these myriad points of light in the sky were themselves suns, as intrinsically luminous as our own Sun but unimaginably distant. But if the stars are suns, and the Sun but one star among millions, it does not make much sense to treat the Sun as the center of the universe. Thus, heliocentrism gradually gave way to a much more magnificent view of a universe of such vastness as to dwarf the Sun, the Earth, and humankind in a galaxy of other worlds.

The Astrophysical Perspective

About a century ago, astronomy entered a new stage in which theories of physics, based on extensive laboratory experiments, were applied to the celestial realm. Until that time, astronomers concerned themselves primarily with the discovery of new objects (planets, moons, comets,

Figure 1.11 The Milky Way Galaxy; the location of the Sun is shown, far from the center of the system.

stars) and the analysis of their motions. However, in the late nineteenth century it became apparent that we could also address questions of the *physical* and *chemical* nature of celestial objects. What materials are stars and planets composed of? How big are they and how luminous? What are their temperatures? Are other stars different from the Sun, and if so in what ways? Might the surfaces and atmospheres of Mars or Venus be similar to those of the Earth, and can they support life? How distant are the stars, and can we use their distribution in space to map the large-scale structure of the universe?

Larger, more powerful telescopes, together with the new analytical tools derived from physics, brought answers within reach. This combination of physics with astronomy generated the new science of **astrophysics,** a word that has become almost synonymous with astronomy.

Early in this century, astronomers made several discoveries that further demoted the Earth from its once central place in the cosmos. First, they identified the basic structure of our Milky Way Galaxy, the assemblage of more than 100 billion stars of which we are a part. Within the Milky Way, the Sun is just an ordinary star, not particularly luminous or remarkable in any way. It is not even located near the center of the Milky Way Galaxy, but out near one edge of the disk, in a sort of "celestial backwater" (Figure 1.11). The Milky Way Galaxy was then itself identified as just one of many millions of such stellar systems.

As our vision of space expanded, so did our concept of time. Observations hinted that the universe was very old, having formed 10 to 15 billion years ago. Over time spans of billions of years, stars form, live their lives, and die. The Sun and Earth are only 4.5 billion years old, relative latecomers in the Galaxy. We find ourselves embedded in a continuum of time and space stretching away to the limits of observation.

Why is modern astronomy so closely associated with astrophysics?

▶ SUMMARY

1.1 The **cosmologies** of ancient people were based on the idea of a stationary Earth beneath a rotating **celestial sphere.** The apparent rotation of the celestial sphere can be described in terms of **celestial poles, equator,** and **zenith.** Positions of celestial objects are given by their right ascension and declination;

the sky is also divided into 88 regions called **constellations.** The brightest stars are said to be of the **first magnitude.**

1.2 The Sun's annual motion around the **ecliptic,** and the still more complex apparent motions of the Moon and planets, are more difficult to understand from a **geocentric** perspective, and much of ancient astronomy was devoted to such problems. Astronomers also established calendars, which kept track of the passage of time and attempted to reconcile the conflicting lengths of the day, month, and year.

1.3 The geocentric perspective of the ancient world was replaced in the European Renaissance by the **heliocentric** cosmology of Copernicus, which displaced the Earth (and humans) from their central position in the universe. The telescopic discoveries of Galileo and a new appreciation of the scientific method led to the rapid progress of astronomy.

1.4 In the late seventeenth century, Newton introduced the modern methods of mathematical astronomy, dealing primarily with celestial motions. About a century ago, another transition occurred to modern **astrophysics,** which seeks a more comprehensive understanding of the nature of celestial objects and the processes that influence them.

▶ REVIEW QUESTIONS

1. From where on Earth could you observe all of the stars during the course of a year? What fraction of the sky can be observed from the North Pole? Describe a practical way to determine the constellation in which the Sun is located at any time of the year.

2. Explain how stars are named, grouped into constellations, and designated in position on the celestial sphere.

3. Summarize some of the things that Galileo discovered by looking at the sky with his telescope.

4. Draw a picture that explains why Mercury should go through phases like the Moon, according to the heliocentric cosmology. Does Jupiter also go through phases as seen from the Earth?

▶ THOUGHT QUESTIONS

5. Show with a simple diagram how the lower parts of a ship disappear first as it sails away from you on a spherical Earth. Use the same diagram to show why lookouts on old sailing ships could see farther from the masthead than from the deck. Would there be any advantage to posting lookouts on the mast if the Earth were flat? (Note that these nautical arguments for a spherical Earth were quite familiar to Columbus and other mariners of his time.)

6. (a) Where on the Earth are all stars at some time visible above the horizon? (b) Where on the Earth is only half the sky ever above the horizon?

7. What is the latitude of (a) the North Pole? (b) the South Pole? (c) a point two-thirds of the way from the equator to the South Pole?

8. Tell where you are on the Earth from the following descriptions:
(a) The stars rise and set perpendicular to the horizon.
(b) The stars circle the sky parallel to the horizon.
(c) The celestial equator passes through the zenith.
(d) In the course of a year, all stars are visible.

9. Suppose you were asked to develop a calendar in which every month had 30 days. What could you do with the extra days to make the months come out even? Suppose that we wanted to make life even simpler by making each month exactly 4 weeks, or 28 days long. Outline the advantages and disadvantages of such a scheme.

10. Consider three cosmological perspectives: (1) the geocentric perspective; (2) the heliocentric perspective; and (3) the modern perspective in which the Sun is a minor star on the outskirts of one galaxy among billions. Discuss some of the cultural and philosophical implications of each point of view.

▶ PROBLEMS

*11. According to the *geocentric* theory of the orbits of the planets, Venus moves on a circle that always lies between the Earth and the Sun (see diagram). What phases would Venus go through? How does this prediction differ from that of the heliocentric theory?

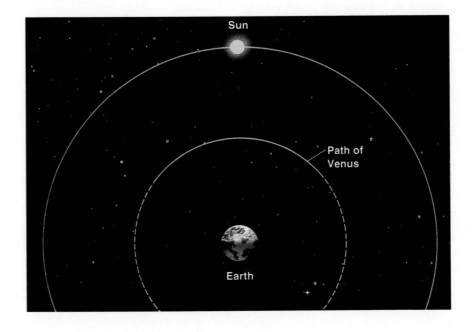

▶ SUGGESTIONS FOR ADDITIONAL READING

Books

Boorstin, D. *The Discoverers.* Random House, 1983. Essays on the development of astronomy, timekeeping, and geography, and their effects on culture.

Ferris, T. *Coming of Age in the Milky Way.* Morrow, 1988. History of our efforts to understand the scale of the cosmos and our place within it.

Krupp, E. *Echoes of the Ancient Skies.* Harper & Row, 1983. Excellent discussion of archaeoastronomy.

Brown, H. *Man and the Stars.* Oxford U. Press, 1978. Includes good sections on the history of time-keeping and the calendar.

Christianson, G. *This Wild Abyss.* Free Press, 1978. A good history of Renaissance astronomy.

Culver, R. & Ianna, P. *Astrology: True or False.* Prometheus Books, 1988. A skeptical analysis of astrology, by two astronomers.

Articles

Fraknoi, A. "Your Astrology Defense Kit" in *Sky & Telescope,* Aug. 1989, p. 146.

Kunitzsch, P. "How We Got Our Arabic Star Names" in *Sky & Telescope,* Jan. 1983, p. 20.

Gingerich, O. "How Galileo Changed the Rules of Science" in *Sky & Telescope,* Mar. 1993, p. 32.

Rosen, E. "Copernicus' Place in the History of Astronomy" in *Sky & Telescope,* Feb. 1973, p. 72.

OTHER WORLDS: PLANETS AND THEIR ORBITS

CHAPTER 2

The Saturn system.   *(Voyager photo)*

Since the time of Copernicus we have known that it is the Sun, not the Earth, that dominates the solar system. The Sun, a rather ordinary star, contains more than 99 percent of all of the mass of the solar system, and its luminous energy far exceeds the reflected light of any planet. Yet the planets, for all of their smallness, are our nearest neighbors in space. In this chapter, we take a general look at the planetary system, with special attention to the motions of the planets. Then, in the next five chapters, we explore several of the better-known members of that system in more detail.

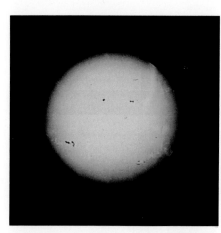

Figure 2.1   The Sun, which is the nearest star. The most massive object and the source of energy for the other members of the system, the Sun dominates the planetary system.

2.1
OVERVIEW OF THE PLANETARY SYSTEM

The solar system consists of the Sun and many smaller objects: the planets; their satellites and rings; and such "debris" as asteroids, comets, and dust. It is believed that most of these objects formed together with the Sun about 4.5 billion years ago. They represent aggregations of material that condensed from a hot cloud of gas and dust; the central part of this cloud became the Sun, and a small fraction of the material in the outer parts eventually formed the other objects.

Inventory

The Sun (Figure 2.1) is by far the most massive member of the solar system, as shown in Table 2.1. It dominates the system through both its gravity and the energy that it generates by nuclear reactions deep in its interior. As a star, the Sun is discussed later in this book. For the present, we address the planets and the smaller objects that make up the planetary system.

Most of the material in the planetary system is concentrated in the largest planet, Jupiter. The nine planets include the Earth, five other planets known to the ancients (Mercury, Venus, Mars, Jupiter, and Saturn), and three discovered since the invention of the telescope (Uranus, Neptune, and Pluto).

The planets all revolve about the Sun in approximately the same plane. Each of the planets also rotates about an axis running through it, and in most cases, the direction of rotation is the same as that of revolution about the Sun. The exception is Venus, which rotates backward (that is, in a **retrograde** direction). Uranus and Pluto also have strange rotations, each spinning about an axis tipped nearly on its side.

The four planets closest to the Sun (Mercury, Venus, Earth, and Mars) are called the inner or **terrestrial planets;** often the Moon is also discussed as a part of this group, bringing the total of terrestrial bodies

TABLE 2.1 Mass of Members of the Solar System	
Object	**Percentage of Mass**
Sun	99.80
Jupiter	0.10
Comets	0.05
All other planets	0.04
Satellites and rings	0.00005
Asteroids	0.000002
Meteoroids and dust	0.0000001

TABLE 2.2   The Planets

Name	Distance from Sun (AU)	Revolution Period (yrs)	Diameter (km)	Mass (10^{23} kg)	Density (g/cm^3)
Mercury	0.39	0.24	4878	3.3	5.4
Venus	0.72	0.62	12102	48.7	5.3
Earth	1.00	1.00	12756	59.8	5.5
Mars	1.52	1.88	6787	6.4	3.9
Jupiter	5.20	11.86	142984	18991	1.3
Saturn	9.54	29.46	120536	5686	0.7
Uranus	19.18	84.07	51118	866	1.2
Neptune	30.06	164.82	49660	1030	1.6
Pluto	39.44	248.60	2300	0.01	2.1

to five. The terrestrial planets are relatively small worlds, composed primarily of rock and metal. The remaining five planets (Jupiter, Saturn, Uranus, Neptune, and Pluto) are called the outer planets. These five outer planets are not physically similar. Jupiter, Saturn, Uranus, and Neptune are very large and are composed primarily of lighter ices, liquids, and gases. We call these four the jovian or **giant planets.** Little Pluto is neither a terrestrial nor a jovian planet; it is similar in size and composition to the satellites of the outer planets. Table 2.2 summarizes some of the main facts about these nine planets.

What are the fundamental differences between the terrestrial and giant planets?

Most of the planets are accompanied by one or more satellites; only Mercury and Venus are alone. There are 62 known satellites (see Appendix 10), and undoubtedly many other very small satellites remain undiscovered. Each of the jovian planets also has a system of rings, which are remnants of the original building blocks for the planets.

A Scale Model of the Solar System

Astronomy often deals with dimensions and distances that far exceed our ordinary experience. What does a billion kilometers really mean to anyone? Sometimes it helps to visualize the system in terms of a scale model.

In our imagination, let us build a scale model of the solar system, adopting a scale factor of one billion (10^9)—that is, we reduce the actual solar system by dividing every dimension by a factor of 10^9. The Earth then has a diameter of 1.3 cm, or about the size of a grape. The Moon is a pea orbiting this grape at a distance of 40 cm, or a bit over 1 ft. The Earth-Moon system fits into an attache case.

In this model the Sun is nearly 1.5 meters in diameter, about the average height of an adult, at a distance of 150 m—about one city block. Jupiter is 5 blocks away, and its diameter is 15 cm, about the size of a very large grapefruit. Saturn is 10 blocks away, Uranus is 20 blocks away, and Neptune and Pluto are each 30 blocks away (Pluto is now closer to the Sun than is Neptune). Most of the satellites of the outer solar system are the size of seeds of various kinds orbiting the grapefruit, oranges, and lemons we use to represent the outer planets.

In our scale model, a human is reduced to the dimensions of a single atom, and autos (and spacecraft) are reduced to the size of molecules. Sending the Voyager spacecraft to Neptune is like navigating a single molecule from the Earth-grape toward a lemon that is 5 km away using an accuracy that is equivalent to the width of a thread in a spider's web. In this fruity model, the nearest stars are tens of thousands of kilometers from Earth.

2.2
PLANETARY DISTANCES AND ORBITS

Copernicus, and other astronomers of the Renaissance, recognized that the Earth was a planet, and that the distances separating the planets were thus much larger than had been thought previously. They determined that the planets orbited the Sun, but initially they were not sure of the exact shapes of these orbits, or of the rules that governed the spacing of these orbits. In the seventeenth century, the most important problems in astronomy concerned the distances to the planets and the calculation of their orbits.

How Far Away Are the Planets?

Copernicus determined *relative* distances from the Earth (and Sun) to the planets known at that time, but he still had only a very rough idea of their true or absolute distances. Knowing the relative distances to the planets represented a big step forward, however. To find the distances to all of these objects, it was necessary to measure the actual distance to only one of them. The situation was analogous to the road maps that we use today. A road map is a scale drawing of the locations of highways and cities. To determine the actual distance between any two locations, it is necessary to know the scale of the map—the number of miles to the inch or the number of kilometers per centimeter. Copernicus had drawn a scale map of the solar system, and it was seventeenth-century astronomers who made the measurements necessary to convert distances on that scale map to kilometers.

The scale of distances within the solar system is given by the average distance of the Earth from the Sun—called by astronomers the **astronomical unit (AU)**. Modern determinations of the astronomical unit are based primarily on radar observations of Venus (Figure 2.2). Radar signals are sent from the Earth to Venus, where they are reflected back to Earth. Radar signals travel at the speed of light, which has been accurately measured to be 299,792,459 meters per second. By measuring

How big is the solar system? How do we measure distances?

Figure 2.2 Radar telescope of the NASA Deep Space Net in the Mojave Desert of California. *(NASA/JPL)*

the time it takes radar to make the round trip, it is possible to calculate the distance to Venus accurately and, from the scale map of the solar system, to determine the length of the astronomical unit, which is now known to a precision of one part in a billion.

The astronomical unit is related to the metric unit of length, the meter. The metric system was officially established in Napoleonic France in 1799. It replaced earlier measures of distance based on human dimensions: the inch as the distance between knuckles on the finger, or the yard as the span from the extended index finger to the nose of the king. The meter was originally defined as 1/10,000,000 of the distance along the Earth's surface from the equator to the pole, but was later redefined in terms of the speed of light. Other units of length are derived from the meter. Thus, 1 km equals 1000 m, 1 cm equals 1/100 m. Even nonmetric units, such as the inch and the mile, are defined in terms of the meter.

The length of 1 AU can be measured from radar data as corresponding to a light (or radar) travel time of 499.00485 seconds, or about 8.3 light minutes. In other words, light requires about 500 seconds or 8.3 minutes to travel from the Sun to the Earth. Using the definition of the meter in terms of the speed of light yields a value of 1 AU = 149,597,892,000 m, or approximately 150 million km.

The Problem of Orbits

We all know that the planets "circle" the Sun, but what does this mean? Do planets (and other members of the solar system) literally follow circular paths around the Sun? No, nature is not that simple. If planetary orbits were really circular, understanding celestial motions would not have presented so formidable a problem to early astronomers, and the history of science might have been quite different.

When Copernicus proposed his heliocentric theory of the solar system, he remained true to the traditional concept that celestial motions must be uniform and circular, or at least that they were made up of combinations of perfectly circular motions. Yet even four centuries ago it was clear to astronomers that combining circular motions was at best an awkward way to construct an accurate theory of planetary motion. The situation became worse in the century following Copernicus, when more accurate observations of planetary positions revealed further discrepancies between theory and observation.

What is a planetary orbit?

If we could look down on the solar system from somewhere out in space, far above the plane of the planets' orbits, the problem of interpreting planetary motions would be much simpler. The solar system always looks orderly when illustrated in books (Figure 2.3, on the next page)! But the fact is that we must observe the positions of all the other planets from our own moving planet, and scientists of the Renaissance did not know the nature of the Earth's orbit any better than that of the other planets. Their problem, therefore, was to deduce the nature of the Earth's motion as well as that of each of the other planets, using only observations of the apparent positions of the other planets as seen from our own shifting vantage point.

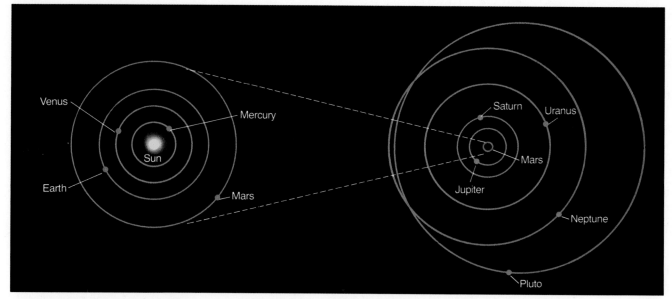

Figure 2.3   Diagram showing the relative sizes of the orbits of the planets, as they would be seen from far above the plane of the solar system. Different scales are used to show the inner and outer planets.

Figure 2.4   Johannes Kepler (1571–1630) discovered the laws of planetary motion that still bear his name. He lived in turbulent times: His mother was once arrested for witchcraft, and he helped support himself by casting horoscopes for wealthy people.

Orbits According to Kepler

The first scientist to provide an accurate interpretation of these observations was the German contemporary of Galileo, mathematician Johannes Kepler (Figure 2.4). Kepler (1571–1630) lived in Germany during the turmoil of the Thirty Years War. While studying for a theological career, he learned the principles of the Copernican system and became converted to the heliocentric hypothesis. Supported under royal patronage in Prague between 1600 and 1620, Kepler correctly deduced the nature of planetary orbits, and he summarized his conclusions in three simple statements known as *Kepler's Laws.*

Kepler began his research with the idea that the orbits of planets were circles, but the observations contradicted this assumption. Working with data for Mars, he eventually discovered that the orbit of that planet had the shape of a flattened circle, or **ellipse.** Next to the circle, the ellipse is the simplest kind of closed curve, belonging to a family of curves known as conic sections (Figure 2.5).

An ellipse is a curve for which the sum of the distances from any point on the ellipse to two points inside the ellipse is the same. These two points inside the ellipse are called its **foci** (singular: **focus**). This property suggests a simple way to draw an ellipse. The ends of a length of string are tied to two tacks pushed through a sheet of paper into a drawing board, so that the string is slack. If a pencil is pushed against the string, so that the string is held taut, and then slid against the string around the tacks (Figure 2.6), the curve that results is an ellipse. At any point where the pencil may be, the sum of the distances from the pencil to the two tacks is a constant length—the length of the string. The tacks are at the two foci of the ellipse.

The maximum diameter of the ellipse is called the major axis. Half this distance, that is, the distance from the center of the ellipse to one end, is the **semimajor axis,** which is the quantity usually given to specify the size of the ellipse. For example, the semimajor axis of the orbit of Mars, which is also the planet's average distance from the Sun, is 228 million km. The shape (roundness) of an ellipse depends on how close together the two foci are, compared to the major axis. The ratio of the distance between the foci to the major axis is called the **eccentricity** of the ellipse. If an ellipse is drawn as just described, the length of the major axis is the length of the string, and the eccentricity is the distance between the tacks divided by the length of the string.

If the foci (or tacks) coincide, the eccentricity is zero, and the ellipse is a circle; thus a circle is an ellipse of eccentricity zero. Ellipses of various shapes are obtained by varying the spacing of the tacks (as long as they are not farther apart than the length of the string). The greater the eccentricity, the more elongated is the ellipse, up to a maximum eccentricity of 1.0.

The size and shape of an ellipse are completely specified by its semimajor axis and its eccentricity. Kepler found that Mars has an orbit that is an ellipse and that the Sun is at one focus (the other focus is empty). The eccentricity of the orbit of Mars is only about 0.1; the orbit, drawn to scale, would be practically indistinguishable from a circle, yet the difference is critical for understanding planetary motions.

Kepler's first law describes the elliptical shape of planetary orbits. The second law deals with the speed with which each planet moves along the ellipse. Working with observations of Mars, Kepler discovered that the planet speeds up as it comes closer to the Sun and slows down as it pulls away from the Sun. He expressed this relation by imagining that the Sun and Mars are connected by a straight, elastic line. As Mars travels in its elliptical orbit around the Sun, the areas swept out in space by this imaginary line are always equal in equal intervals of time (Figure 2.7). This is also a general property of the orbits of planets. A planet in a circular orbit always moves at the same speed, but in an eccentric orbit, the planet's speed varies considerably.

Laws of Planetary Motion

Kepler's first two laws of planetary motion describe the shape of a planet's orbit and allow us to calculate the speed of its motion at any point in the orbit. Kepler was pleased to have discovered such fundamental rules, but they did not satisfy his quest to understand planetary motions. He wanted to know why the orbits of the planets were spaced as they are, and to find a pattern in their orbital periods—a "harmony of the spheres." For many years he worked to discover mathematical relationships governing planetary spacings and periods of revolution.

In 1619, Kepler succeeded in finding the simple algebraic relation that links the semimajor axes of the planets' orbits and their periods of revolution. The relation is now known as Kepler's third law. It applies to all of the planets, including Earth, and it provides a means for calculat-

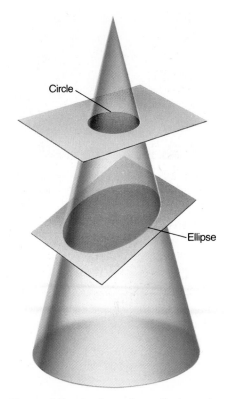

Figure 2.5   Conic sections: Circle and two ellipses of different eccentricity. All orbits involving the mutual motion of two bodies under the influence of gravitation follow curves that are conic sections.

What are Kepler's Laws of planetary motion?

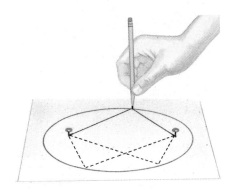

Figure 2.6   Drawing of an ellipse, using a string, two tacks (the foci of the ellipse), and a pencil held taut against the string.

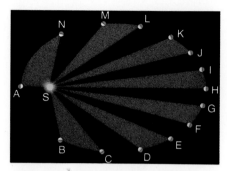

Figure 2.7 Kepler's law of equal areas. A planet moves most rapidly on its elliptical orbit when it is at position A, nearest the Sun (S), at one focus of the ellipse. The orbital speed of the planet varies in such a way that, in equal intervals of time, it moves distances AB, BC, CD, and so on, so that the regions swept out by the line connecting the planet to the Sun (alternating shaded and clear zones) are always the same in area.

ing the relative distances of the planets from the Sun, given their periods of revolution.

Kepler's third law takes its simplest form when the period is expressed in years (the revolution period of the Earth) and the semimajor axis of the orbit is expressed in astronomical units (AU). Kepler's third law is then:

$$(\text{distance})^3 = (\text{period})^2$$

For example, this relationship tells us how to calculate the average distance from the Sun (the semimajor axis of its orbit) to Mars from its period of 1.88 years. The period squared is $1.88 \times 1.88 = 3.53$, and the cube root of 3.53 is 1.52, which is equal to the semimajor axis in AU.

Kepler's three laws of planetary motion can be summarized as follows:

- *Kepler's first law: Each planet moves about the Sun in an orbit that is an ellipse, with the Sun at one focus of the ellipse.*
- *Kepler's second law: The straight line joining a planet and the Sun sweeps out equal areas in space in equal intervals of time.*
- *Kepler's third law: The squares of the periods of revolution of the planets are in direct proportion to the cubes of the semimajor axes of their orbits.*

These three laws provided a precise geometric description of planetary motion within the framework of the Copernican system. With these tools, it is possible to calculate planetary positions with undreamed-of precision. But Kepler's laws are purely descriptive; they do not help us understand what forces of nature constrain the planets to follow this particular set of rules. That step was left to Newton.

2.3
UNIVERSAL GRAVITATION

Isaac Newton provided the conceptual framework for understanding the natural laws discovered by Galileo, Kepler, and others. His development of the concept of universal gravitation gave a physical basis to Kepler's laws and ultimately provided the means to calculate interplanetary trajectories for spacecraft as well as to study the orbits of planets with remarkable accuracy.

The Law of Gravitation

Newton first determined the fundamental laws of motion that govern all objects, whether in the laboratory or in the heavens above. He recognized that, left to themselves, objects at rest stay at rest, and those in motion continue in uniform motion in a straight line; thus, it is the *straight line*, not the circle, that defines the most natural state of motion. In the case of the planets, some *force* must act to bend their paths from straight lines into ellipses. That force is **gravitation.**

It is obvious that the Earth exerts a gravitational force upon all objects at its surface. Newton reasoned that the Earth's gravity might extend as far as the Moon; he further speculated that there is a universal attraction between all bodies everywhere in space. The precise strength of that gravitational force must compel the planets to follow orbits described by Kepler's three laws. Moreover, the law of gravitation must predict correctly the behavior of falling bodies on the Earth. How then must the gravitational force depend on distance for these conditions to be met?

Answering this question required the use of a type of mathematics not yet developed. But this did not deter Isaac Newton, who invented what today we call calculus. Through the use of calculus, he demonstrated that the gravitational force drops off with increasing distance between the Sun and a planet in proportion to the *inverse square* of their separation. That is, if the distance is doubled, the force drops off by 2×2, or 4 times; if the distance increases by a factor of 3, the gravitational force declines by 3×3, or a factor of 9.

Newton also concluded that the gravitational attraction between two bodies must be proportional to their masses. *Mass* is an important concept. It is a quantity that characterizes the total amount of material in a body. In the metric system, mass is expressed in units such as grams, kilograms, or tons (see Appendix 3).

Expressed as a formula, the gravitational attraction between any two objects is given by:

$$\text{Force} = \frac{GM_1M_2}{R^2}$$

where M_1 and M_2 are the masses of the two objects and R is their separation. The number represented by G is called the Constant of Gravitation. With such a force and the laws of motion, Newton showed mathematically that the only orbits permitted were exactly those described by Kepler's laws.

Mass, Volume, and Density

It is important not to confuse mass, volume, and density. Mass is a measure of the amount of material in an object. Volume, in contrast, is a measure of the physical space occupied by a body, say in cubic centimeters or liters. In short, the volume is the "size" of an object—it has nothing to do with its mass. A penny and an inflated balloon may both have the same mass, but they have very different volumes (Figure 2.8).

The penny and the balloon are also very different in density, which is a measure of how much mass is contained within a given volume. Specifically, density is the ratio of mass to volume:

$$\text{density} = \text{mass}/\text{volume}$$

Note that often in everyday language we use "heavy" and "light" as indications of density (rather than weight), as for instance when we say that iron is heavy or that a puff pastry is light.

What is Newton's Law of Gravitation?

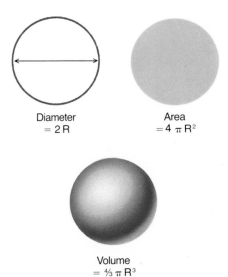

Diameter
= 2 R

Area
= 4 π R²

Volume
= ⅓ π R³

Figure 2.8 Four measures of the "size" of a spherical object like the Earth: its diameter, surface area, volume, and mass.

TABLE 2.3	Densities of Materials
Material	Density (g/cm³)
Gold	19.3
Lead	11.4
Iron	7.9
Earth (bulk)	5.5
Rock (typical)	2.5
Water	1.0
Wood (typical)	0.8
Insulating foam	0.1

The units of density that will be used in this book are grams per cubic centimeter (g/cm^3) or, alternatively, metric tons per cubic meter. Familiar materials span a considerable range in density, from gold ($19 \ g/cm^3$) to artificial materials such as plastic insulating foam (less than $0.1 \ g/cm^3$) (Table 2.3). In the astronomical universe, much more remarkable densities can be found, all the way from a comet's tail ($10^{-16} \ g/cm^3$) to a neutron star ($10^{15} \ g/cm^3$).

To sum up then, mass is "how much," volume is "how big," and density is "how tightly packed."

Gravitation and Orbital Motion

How do astronomers measure the masses of the planets?

Now that we have seen how Newton developed the theory of universal gravitation, let us look again at the orbits of the planets. Kepler's laws are descriptions of the orbits of objects moving according to Newton's laws of motion and the law of gravitation. Knowing that gravity is the force that attracts planets toward the Sun, however, allowed Newton to introduce an important modification to Kepler's third law. As before, we express distances in AU and periods in years, but we also introduce the masses of the two objects, both expressed in units of the Sun's mass. If D is the distance, M is the mass, and P is the period, Newton's law of gravitation can be used to show mathematically that:

$$D^3 = (M_1 + M_2) \, P^2$$

In the solar system, where most of the mass is in the Sun itself, the sum of the masses ($M_1 + M_2$) is very nearly equal to M_1, the Sun's mass. This is the reason that Kepler did not realize that both masses had to be included in the calculation. There are many cases in which we do need to include the two mass terms, however, and for this very reason observations of motions of objects acting under their mutual gravitation permit the astronomer to deduce their masses.

2.4
APPLICATIONS OF KEPLER'S AND NEWTON'S LAWS

Kepler and Newton provided scientists with powerful tools to analyze the motions of planets, satellites, and modern spacecraft. All of these objects follow the same basic rules that govern the motion of any object subject to the force of gravity. We now look at some applications of these rules.

Artificial Satellites

Gravitation and Kepler's laws describe the motions of Earth satellites and interplanetary spacecraft as well as the planets. Sputnik, the first artificial Earth satellite, was launched by the U.S.S.R. on October 4, 1957. Since that time, more than 1000 satellites have been placed into orbit around the Earth, and spacecraft have also orbited the Moon, Venus, and Mars.

How is a satellite launched and what kinds of orbits can it follow?

Once an artificial satellite is in orbit, its behavior is no different from that of a natural satellite, such as our Moon. If the satellite is high enough to be free of atmospheric friction, it will remain in orbit forever, following Kepler's laws in a perfectly respectable way. However, although there is no difficulty in maintaining a satellite once it is in orbit, a great deal of energy is required to lift the spacecraft off the Earth and accelerate it to orbital speed.

To illustrate how a satellite is launched, imagine a gun on top of a high mountain, firing a bullet horizontally (Figure 2.9a as adapted from a similar diagram by Newton, Figure 2.9b). Imagine, further, that the

Figure 2.9 Firing a bullet into a satellite orbit (a) as described in the text, and (b) as depicted in a similar diagram by Newton published in 1731. *(Crawford Collection, Royal Observatory, Edinburgh)*

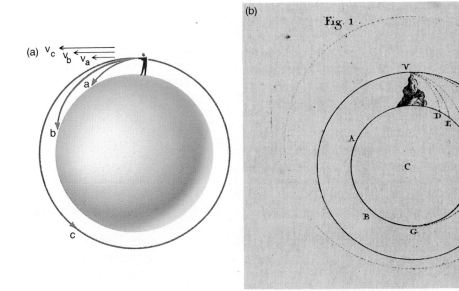

friction of the air could be removed, and that all hindering objects, such as other mountains, buildings, and so on, are absent. Then the only force that acts on the bullet after it leaves the muzzle is the gravitational force between the bullet and Earth.

If the bullet is fired with velocity v_a, it continues to have that forward speed, but meanwhile the gravitational force acting upon it pulls it downward so that it strikes the ground at a. However, if it is given a higher muzzle velocity v_b, its higher forward speed carries it farther before it hits the ground, for, regardless of its forward speed, the downward gravitational force is the same. Thus, this faster-moving bullet strikes the ground at b. If the bullet is given a high enough muzzle velocity v_c, as it falls toward the ground, the curved surface of the Earth causes the ground to dip out from under it so that it remains the same distance above the ground and "falls around" the Earth in a complete circle. This speed of a satellite orbiting the Earth is about 8 km/s.

Suppose that a rocket is fired to an altitude of a few hundred miles, is then turned so that it is moving horizontally, and finally is given a forward horizontal thrust. It proceeds in an orbit the size and shape of which depend critically on the exact direction and speed of the rocket at the instant of its "burnout," that is, the instant when the thrust supplied by its fuel is shut off. For simplicity, suppose that it is moving exactly horizontally, or parallel to the ground, at burnout. The possible kinds of orbits it can enter are shown in Figure 2.10.

If the rocket's burnout speed is less than 8 km/s, its orbit is an ellipse, with the center of the Earth at one focus of the ellipse. The apogee point of the orbit, that point farthest from the center of the Earth, is the point of burnout; the perigee, the point of closest approach to the center of the Earth, is halfway around the orbit from burnout. For the more

Figure 2.10 Various satellite orbits that result from different initial speeds parallel to the Earth's surface. A is an orbit that is intercepted by the solid Earth (like that of a ballistic missile); C is a circular orbit; D through G are orbits of increasing energy, but all with the perigee at the point of injection.

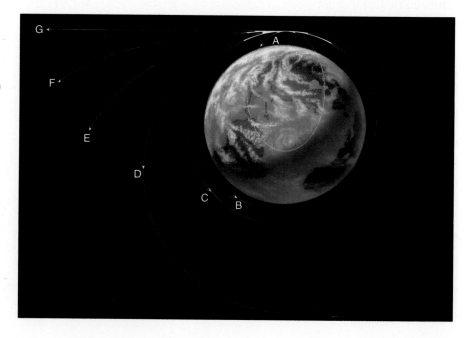

common case of an orbit around the Sun, the closest point is called **perihelion** and the farthest is **aphelion.**

If the burnout speed is substantially below the circular satellite velocity, most of the rocket's elliptical orbit lies beneath the surface of the Earth (orbit A), where, of course, the satellite cannot travel. Consequently, it will traverse only a small section of its orbit before colliding with the surface of the Earth. This is the kind of trajectory followed by a ballistic missile. If the burnout speed is just slightly below the circular satellite velocity, the rocket may clear the surface of the Earth (orbit B), although its orbit will probably lie too low in the atmosphere for the satellite to be long-lived.

If the burnout speed were exactly the circular satellite velocity, a circular orbit centered on the center of the Earth would result (orbit C). A slightly greater burnout speed produces an elliptical orbit with perigee at the burnout point and apogee halfway around the orbit (orbit D).

Interplanetary Flight

Most satellites are launched into low Earth orbit, since this requires the minimum launch energy. At the orbital speed of 8 km/s, they circle the planet in about 90 minutes. These orbits are not indefinitely stable, since the drag generated by friction with the thin upper atmosphere eventually leads to a loss of energy and "decay" of the orbit. Upon reentering the denser parts of the atmosphere, most satellites are burned up by atmospheric friction, although some solid parts may reach the surface.

To escape the Earth entirely, a spacecraft must achieve **escape velocity,** which is about 11 km/s. After reaching escape velocity, it can coast to distant targets such as the Moon or Mars. For any planet, the escape velocity is equal to the circular or orbital velocity multiplied by $\sqrt{2}$ (about 1.4). Most of what we have learned about the planets has been achieved by small robotic spacecraft sent to fly past or orbit distant targets. The most successful such spacecraft was Voyager 2 (Figure 2.11),

Figure 2.11 The Voyager planetary spacecraft, which has a mass of approximately 1 ton. Voyager 2 has flown to Jupiter, Saturn, Uranus, and Neptune. *(NASA/JPL)*

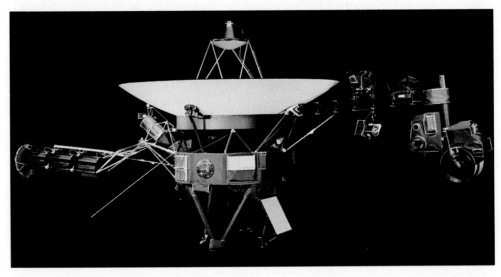

which visited Jupiter, Saturn, Uranus, and Neptune—a grand tour of the solar system.

Discovery of the Outer Planets

Most ancient civilizations knew the five planets that are easily seen with the unaided eye—Mercury, Venus, Mars, Jupiter, and Saturn. The initial recognition of these objects is lost in the haze of prehistory. However, the three outer planets—Uranus, Neptune, and Pluto—are all *discovered* objects. In their times, each discovery of a new planet created widespread public interest and brought fame to the discoverer. Even today, we frequently read newspaper articles speculating on the possible existence of a tenth planet or "Planet X."

Uranus was discovered on March 13, 1781, by the German-English musician and amateur astronomer William Herschel, who was making a systematic telescopic survey of the sky in the constellation of Gemini. Herschel noted that through his telescope one object did not appear as a stellar point but seemed to present a small disk. He believed it to be a comet and followed its motion for some weeks. Later, its orbit was computed and found to be nearly circular, lying beyond that of Saturn. Herschel proposed to name the newly discovered planet Georgium Sidus, in honor of George III, England's reigning king. Others suggested the name Herschel, after the discoverer. The name finally adopted, in keeping with the tradition of naming planets for gods of Greek and Roman mythology, was Uranus, father of the Titans and grandfather of Jupiter.

By 1790, an orbit had been calculated for Uranus on the basis of observations of its motion in the decade following its discovery. Even after allowance was made for the gravitational effects of Jupiter and Saturn, however, it was found that Uranus did not move on an orbit that fitted exactly the earlier observations of it made since 1690. In 1843, John Couch Adams, a young Englishman who had just completed his work at Cambridge, began an analysis of the irregularities in the motion of Uranus to see whether they could be produced by an unknown planet. His calculations indicated the existence of a planet more distant from the Sun than Uranus. In October 1845, Adams delivered his results to George Airy, the Astronomer Royal, informing him where in the sky to find the new planet. Adams's predicted position for the unknown body was correct to within 2 degrees.

Meanwhile, Urbain Jean Joseph Leverrier, a French mathematician, unaware of Adams or his work, attacked the same problem and published its solution in June 1846. Airy, noting that Leverrier's predicted position for the unknown planet agreed to within 1 degree with that of Adams's, suggested to James Challis, director of the Cambridge Observatory, that he begin a search for the new object. The Cambridge astronomer, having no up-to-date star charts of the region of the sky in Aquarius, where the planet was predicted to be, proceeded by recording the positions of all the faint stars he could observe with his telescope in that location. It was Challis's plan to repeat such plots at intervals of several days, in the hope that the planet would reveal its presence and dis-

---

How were gravitational perturbations used to discover the planet Neptune?

tinguish itself from a star by its motion. Unfortunately, he was negligent in examining his observations; although he had actually seen the planet, he did not recognize it.

About one month later, Leverrier suggested to Johann Galle, an astronomer at the Berlin Observatory, that he look for the planet. Galle received Leverrier's letter on September 23, 1846, and, possessing new charts of the Aquarius region, he found and identified the planet that very night. It was less than a degree from the position Leverrier had predicted. The discovery of the eighth planet, now known as Neptune (Latin for the Greek Poseidon, god of the sea), was a major triumph for gravitational theory because it dramatically confirmed the generality of Newton's laws. The honor for the discovery is properly shared by the two mathematicians, Adams and Leverrier (although not astronomer Galle, who actually found it at the telescope).

Pluto, the final planetary discovery, was found as the result of a careful, systematic search, in which tens of millions of star images were examined on wide-field photographs of the sky. In February 1930, Clyde Tombaugh, of Lowell Observatory, found a faint moving image on photographs exposed a few weeks earlier. The object was confirmed as a planet and announced to the world on March 13, 1930.

Are there other planets beyond Pluto? If so, they must be very small—smaller even than Pluto. Perhaps if such objects are found they will be called asteroids or comets rather than planets.

▶ SUMMARY

2.1 In orbit around the Sun we find inner **terrestrial planets**, larger jovian or **giant planets,** and little Pluto. Smaller objects include numerous asteroids and comets. The Sun dominates this system, but among the planets Jupiter is larger than all others combined. Most of the planets rotate in the normal direction, but Venus has **retrograde** spin, and Uranus and Pluto are tipped on their side.

2.2 Distances within the solar system are measured by radar and are expressed in terms of the **astronomical unit.** The nature of planetary orbits and the laws of planetary motion were determined by Kepler early in the seventeenth century. Orbits are **ellipses,** described in terms of **semimajor axis** and **eccentricity,** with the Sun at one **focus.** Kepler's third law provides a mathematical relationship between distance from the Sun and period of revolution ($D^3 = P^2$).

2.3 Newton derived the fundamental laws of motion and the law of **gravitation,** which allow detailed understanding of celestial motions. Gravitational force depends on mass and the inverse square of the distance (GM_1M_2/R^2). Recognizing that orbits depend on mass, Newton modified Kepler's third law and provided astronomers with a fundamental tool for measuring the masses of astronomical objects ($D^3 = (M_1 + M_2)P^2$).

2.4 The laws of Kepler and Newton allow us to understand the launching and the various possible orbits of Earth satellites. Orbits have a **perihelion** (closest to the Sun) and an **aphelion.** Massive objects have high **escape velocity** (about 11 km/s for the Earth). The planet Neptune was discovered as a result of its gravitational influence on the orbit of Uranus. However, the other two discovered planets, Uranus and Pluto, were the result of careful searches rather than gravitational predictions.

▶ REVIEW QUESTIONS

1. Describe the orbits of the planets and their relative distances from the Sun.

2. Draw ellipses with eccentricities of 0.0, 0.2, and 0.5, all with the same semimajor axes, and label one focus.

3. Use the data in Table 2.1 to verify the correctness of Kepler's third law for the planets of our solar system.

4. Earth satellites in low orbits (like the U.S. Shuttle) require 90 minutes for each orbit. Calculate their speed, and note that this circular satellite velocity is equal to the escape velocity divided by $\sqrt{2} = 1.4$.

▶ THOUGHT QUESTIONS

5. Suppose that a planet was composed of equal combinations of rock and iron. From the data in Table 2.3, what would be its density? What would be the density of a satellite composed of equal parts of ice and rock?

6. What is the major axis of a circle with a radius of 13 cm? What is its semimajor axis?

7. Use a history book or an encyclopedia to find out what was happening in England during Newton's lifetime, and discuss what trends of the time might have contributed to his accomplishments and the rapid acceptance of his work.

8. Newton showed that the periods and distances in Kepler's third law depended on the masses of the objects. What would be the period of revolution of the Earth (at 1 AU from the Sun) if the Sun had twice its present mass? What would be the period of revolution of the Earth if the Sun kept its present mass but expanded to three times its current diameter?

▶ PROBLEMS

9. What is the period of revolution (in years) of an asteroid with a semimajor axis of 10 AU?

10. What is the distance of an asteroid from the Sun (in AU) if it has a period of revolution of eight years?

11. What is the period of revolution for a hypothetical planet that orbits halfway between Mercury and the Sun? What is it for a planet with twice the semimajor axis of Pluto?

12. The four Galilean satellites of Jupiter form a miniature solar system. Look up the periods of revolution and the distances from Jupiter of these four satellites (Appendix 10).Verify that the square of the periods is proportional to the cube of the distances, as predicted by Kepler's third law.

13. (a) Calculate the total mass of 10,000 asteroids each with an average diameter of 10 km and a density of 2.5 g/cm³. (b) Calculate the total mass of the largest asteroid, Ceres, with a diameter of 1000 km and a density of 2.5 g/cm³. (*Hint:* The volume of a sphere is given by $4/3\pi R^3$). Which mass is greater?

*14. Comet Halley has a period of 76 years. What is its semimajor axis? If its closest point to the Sun (perihelion) is 0.6 AU, what is its farthest distance from the Sun (aphelion)? (*Hint:* Remember the definition of semimajor axis.)

*15. Although the Galilean satellites display the correct proportionality of period and distance (see Question 12), their periods are not the same as those that would be found for planets orbiting the Sun. The reason is that Jupiter, the center of their motion, does not have the same mass as the Sun. Use Newton's reformulation of Kepler's third law and the data for the Galilean satellites to calculate the mass of Jupiter relative to that of the Sun.

*16. The simplest spacecraft trajectory from one planet to another is not a straight line. Instead, it is best to follow an elliptical orbit that lies between the orbits of the two planets, touching the inner planetary orbit at the perihelion of the spacecraft orbit and the outer planetary orbit at the aphelion. For such a trajectory from Earth to Jupiter, calculate the perihelion, the aphelion, the semimajor axis, and the period of revolution of the spacecraft orbit.

▌ SUGGESTIONS FOR ADDITIONAL READING

Books

Henbest, N. *The Planets: A Guided Tour*. Viking Penguin, 1993. Illustrated layperson's overview.

Beatty, J., *et al*, eds. *The New Solar System*, 3rd ed. Sky Publishing and Cambridge U. Press, 1990. A collection of articles on the worlds in the solar system, written by leading planetary scientists. Occasionally technical.

Koestler, A. *The Sleepwalkers*. Macmillan, 1959. A readable history of the beginnings of modern science, with extensive discussions of the work of Copernicus, Galileo, and Kepler.

Sheehan, W. *Worlds in the Sky*. U. of Arizona Press, 1992. A good history of how we learned about the solar system.

Preiss, B., ed. *The Planets*. Bantam, 1985. Science articles and related science fiction stories about each member of the solar system.

Articles

Downs, H. "Modeling the Universe in Your Mind" in *Sky & Telescope*, Oct. 1993, p. 24. On useful scale models you can make (mentally) to help give you perspective.

Wilson, C. "How Did Kepler Discover His First Two Laws?" in *Scientific American*, Mar. 1972.

Cohen, I. "Newton's Discovery of Gravity" in *Scientific American*, Mar. 1981.

Gore, R. "Between Fire and Ice" in *National Geographic*, Jan. 1985, p. 4. A beautifully illustrated overview of the solar system.

Bennett, J. "The Discovery of Uranus" in *Sky & Telescope*, Mar. 1981, p. 188.

Moore, P. "The Discovery of Neptune" in *Mercury*, July/Aug. 1989, p. 98.

Tombaugh, C. "The Discovery of Pluto" in *Mercury*, May/June 1986, p. 66; July/Aug. 1986, p. 98.

3

THE EARTH: OUR HOME PLANET

The Earth in space. *(NASA)*

In this and the following two chapters we will look in detail at five rocky worlds: Earth, Moon, Mercury, Mars, and Venus. One purpose of studying the other terrestrial planets is to gain a better understanding of our own planet Earth. In a similar way, we can interpret the new data about other worlds only if we begin with a sound knowledge of the basic processes at work on our own planet. Thus we introduce this comparative discussion with the Earth, the planet we know best.

3.1
PLANETARY COMPOSITION

The Earth and other terrestrial planets are composed of different materials from those that make up the objects in the outer part of the solar system. Understanding these compositional differences is important if we wish to determine the processes that made each planet the way it is.

Chemistry of the Planets

The planetary system is composed primarily of matter in solid and liquid form, generally at temperatures far lower than those encountered in the Sun and stars. Matter at high temperatures is relatively easy to understand, since it consists of individual atoms or fragments of atoms in the gaseous state. But, under planetary conditions, atoms interact with each other to produce molecules and minerals, and we must deal with the complexities of chemistry.

What are planets made of?

The solid matter in the planetary system is made up of compounds involving most of the 92 naturally occurring elements. However, three simple kinds of matter dominate: (1) metal (such as iron and nickel), often found in planetary cores; (2) rock, which is the name we give to a wide variety of compounds of silicon, oxygen, magnesium, aluminum, iron, sulfur, and other elements; and (3) ices or frozen gases, the most common of which are water (H_2O), carbon dioxide (CO_2), ammonia (NH_3), and methane (CH_4). As we look at individual planets and satellites, we see that they are composed of various proportions of these together with gaseous atmospheres for the larger bodies. Table 3.1 lists the most common elements in the universe; these are the primary building blocks for the planets.

Because oxygen and hydrogen are abundant and chemically reactive elements, they tend to dominate the chemistry of the solar system. Much of the chemical evolution of the planets and their satellites therefore depends on the relative proportions of these two elements.

On Earth, hydrogen is relatively rare, since this light gas escapes easily from the upper atmosphere. Oxygen therefore dominates, and we live in an **oxidized** environment. All of the terrestrial planets are chemically oxidized. They are composed primarily of iron, silicon, oxygen,

| TABLE 3.1   The Cosmically Abundant Elements |||
Element	Symbol	Number of Atoms Per Million Hydrogen Atoms
Hydrogen	H	1,000,000
Helium	He	68,000
Carbon	C	420
Nitrogen	N	87
Oxygen	O	690
Neon	Ne	98
Magnesium	Mg	40
Silicon	Si	38
Sulfur	S	19
Iron	Fe	34

and sulfur in various proportions. The most common rocks are chemical combinations that include silicon and oxygen; these are called **silicate** rocks.

What are the differences between oxidizing and reducing chemistry?

When there are two or more hydrogen atoms present for each atom of oxygen, the hydrogen dominates and the chemical environment is said to be **reduced.** Any available oxygen combines with hydrogen to produce water (H_2O), and the left-over hydrogen combines with other elements to produce an entirely different set of compounds, such as ammonia (NH_3) and the hydrocarbons (compounds of hydrogen and carbon). As we will see in Chapter 6, the giant planets all have chemically reduced atmospheres, with plentiful free hydrogen.

Temperatures in the Solar System

Within the solar system, the dominant source of energy is the Sun. In general, the temperature of each object depends on its distance from the Sun, with Mercury and Venus being the hottest planets and Neptune and Pluto being the coldest.

In astronomy, as in most of science, we find it convenient to express temperature in the Kelvin, or absolute, scale. Kelvin temperatures are equivalent to the more common Celsius scale except that they are set equal to zero for the absolute zero of temperature, rather than the freezing point of water. **Absolute zero** is the lowest possible temperature—the temperature at which all atomic and molecular motion ceases. On the Celsius scale, absolute zero is at $-273°$ C. On the Kelvin scale, which we normally use in this book, absolute zero is equal to 0 K, and no negative temperatures are possible.

Common "room temperature" on Earth is 20° C (68° F) or 293 K. In the outer solar system, temperatures are much lower: typically 100 K in the Saturn system and down to 50 K at Pluto. It is because the temperatures are so low that water ice and other frozen gases are stable in the outer solar system. Thus, the distinction between oxidizing conditions among the terrestrial planets and reducing conditions in the outer solar system is partly a consequence of temperature differences.

3.2
EARTH AS A PLANET

Before the space age, few people thought of the Earth as a planet. Not until the first Apollo trip around the Moon, in 1968, did the general public begin to realize the implications of the fact that we live on a fragile globe, a small shining blue ball in the blackness of space.

The Shape and Size of the Earth

How can we determine the shape and size of the Earth?

The idea that the Earth is round can be traced back to the ancient Greeks and Romans. Many philosophers of that time believed in a spherical Earth, based in part on their liking for the *idea* of the sphere—the most "perfect" shape possible. The first fairly accurate determination of the Earth's diameter was made about 200 B.C. by Eratosthenes, a Greek living in Alexandria in Egypt. His method was a geometric one, based on observations of the Sun. The Sun is so distant from the Earth compared to its size that the Sun's light rays intercepted by all parts of the Earth approach it along essentially parallel lines. If people all over the Earth who could see the Sun were to point at it, their fingers would all be essentially parallel to each other.

Eratosthenes noticed that at Syene, Egypt (near modern Aswan), on the first day of summer, sunlight struck the bottom of a vertical well at noon, which indicated that Syene was on a direct line from the center of the Earth to the Sun. At the corresponding time and date in Alexandria, he observed that the Sun was not directly overhead but slightly south of the zenith, so that its rays made an angle with the vertical equal to 1/50 of a circle (about 7 degrees). Yet the Sun's rays striking the two cities are parallel to each other. Therefore, Alexandria must be 1/50 of the Earth's circumference north of Syene (Figure 3.1). Alexandria had been measured to be 5000 stadia north of Syene (the stadium was a Greek unit of length); thus the Earth's circumference must be 50 × 5000, or 250,000 stadia.

It is not possible to evaluate precisely the accuracy of Eratosthenes' solution because there is uncertainty as to length of the stadium he used, but his result was certainly within 20 percent of the correct value of 40,000 km for the Earth's circumference. The diameter is equal to the circumference divided by π (about 3.14), or approximately 13,000 km. The size and some other basic properties of the Earth are summarized in Table 3.2.

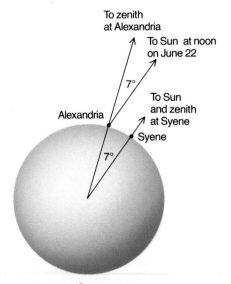

Figure 3.1   Eratosthenes' method of measuring the size of the Earth, based on the angles observed for the position of the Sun at noon as seen from both Alexandria and Syene, in Egypt.

Earth's Interior

The interior of a planet is difficult to study—even for our own Earth—and its composition and structure must be determined indirectly. Our only direct experience is with the outermost skin of the Earth's crust, a

TABLE 3.2 Some Properties of the Earth	
Semimajor axis	1.00 AU
Period	1.00 year
Mass	5.98×10^{24} kg
Diameter	12,756 km
Escape velocity	11.2 km/s
Rotation period	23h 56m 4s
Surface area	5.1×10^8 km^2
Atmospheric pressure	1.00 bar

What is the internal composition
and structure of the Earth?

layer no more than a few kilometers in depth. It is important to remember that in many ways we know less about our own planet a few kilometers beneath our feet than we do about the surfaces of Venus or Mars.

The Earth is composed largely of silicate rock and metal, partly in a solid state and partly molten. The structure of the material in the interior has been probed in considerable detail from measurements of **seismic waves,** produced by natural earthquakes or by artificial impacts or explosions. The seismic waves travel through a planet rather like sound waves through a bell, and the response of the planet to various frequencies is characteristic of interior structure just as the sound of a bell reveals its size and construction. Some of these vibrations travel along the surface; others pass directly through the interior.

Seismic studies have shown that most of the interior of the Earth is solid and that it consists of several distinct layers of different density and composition. The top layer is the crust, the part of the Earth we know best. The oceanic crust, which covers 55 percent of the surface, is typically about 6 km thick and is composed of volcanic rocks called **basalts.** The continental crust, which covers 45 percent of the surface, is from 20 to 70 km thick and is predominantly made of a different volcanic class of silicates called granites. The crust makes up only about 0.3 percent of the mass of the Earth.

The major part of the Earth, which is called the mantle, stretches from the base of the crust down to a depth of 2900 km. The mantle is more-or-less solid, but at the temperatures and pressures found there, the mantle rock can deform and flow slowly. Beginning at a depth of 2900 km we encounter the dense metallic core of the Earth. The core, which is composed primarily of iron, has a diameter of 7000 km (substantially larger than the planet Mercury). The Earth's magnetic field is generated by moving material in the Earth's metallic core, obtaining its energy from the slow escape of interior heat and the rotation of the planet.

Planetary Differentiation

How did the Earth acquire its layered structure? The answer goes back to the early history of the solar system, when the larger bodies in the planetary system differentiated. **Differentiation** is the name given to the process by which a planet organizes its interior into layers of different density and composition. Differentiation takes place when the planet is heated, either during its formation or subsequently as a result of natural radioactivity in its rocks. Once the planet melts, the heavier metal tends to sink to form a core, whereas the lightest minerals float to the surface to form a crust. When the planet cools, this layered structure is preserved (Figure 3.2).

In order for a rocky planet to differentiate, it must be heated to the melting point of rocks, typically above 1200 K. Each of the inner planets was heated to this degree at some time during its early history.

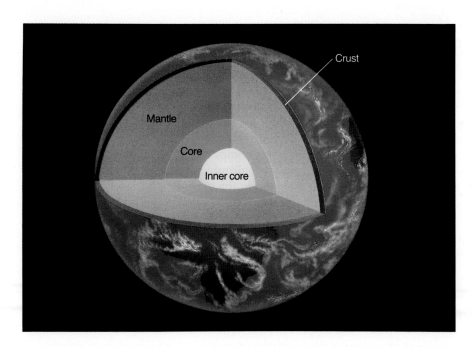

Figure 3.2   Interior structure of the Earth. All of the differentiated inner planets show a similar division into core, mantle, and crust.

3.3
THE CRUST OF THE EARTH

The Earth's crust is a dynamic place. Volcanic eruptions, erosion, and large-scale movements of the continents constantly rework the surface of our planet. Geologically, ours is the most active of the terrestrial planets. Many of the geological processes described in this section take place on other planets as well, but usually with less intensity or confined to periods in the planet's distant past.

Composition of the Crust

The crust of the Earth is made up of the oceanic basalts and the continental granites. These are both examples of **igneous** rock, which is the term used for any rock that has cooled from a molten state. There are two other kinds of rock with which we are familiar on the Earth, although it turns out that neither is common on other planets. Sedimentary rocks are made of fragments of **igneous** rocks or of living organisms. On Earth, these include the common sandstones, shales, and limestones. Metamorphic rocks are produced by the chemical and physical alteration of igneous or sedimentary rocks at high temperature and pressure. Metamorphic rocks are produced on Earth because geological activity carries surface rocks to considerable depths and then brings them back up to the surface.

There is a fourth, very important category of rock not represented on the Earth that can tell us much about the early history of the planetary system. These are **primitive** rocks, which have largely escaped chemical modification by heating. Primitive rock represents the original material out of which the planetary system was made. There is no primitive material left on the Earth because it was strongly heated early in its history. To find primitive rocks, we must look to smaller objects: comets, asteroids, and small planetary satellites.

A block of marble on Earth is composed of materials that have gone through all four of these stages: beginning as primitive material before the Earth was born, it was heated in the early Earth to form igneous rock, subsequently eroded and redeposited (perhaps many times) to form sedimentary rock, and finally transformed several kilometers below the Earth's surface into the hard, white metamorphic stone we see today.

Plate Tectonics

Geology is the study of the crust of the Earth, particularly of the processes that have shaped the surface throughout history. Heat escaping from the interior provides the energy for the formation of mountains, valleys, volcanoes, and even the continents and ocean basins themselves.

Plate tectonics is a theory that explains how slow motions within the mantle of the Earth move large segments of the crust, resulting in a gradual drifting of the continents as well as the formation of mountains and other large-scale geologic features. It is a concept as basic to geology as evolution by natural selection is to biology or gravitation is to understanding the orbits of planets.

The Earth's crust and upper mantle (to a depth of about 60 km) are divided into about a dozen major plates that fit together like the pieces

What is primitive material and why is there none on Earth?

How does plate tectonics explain the large-scale geology of the Earth?

Figure 3.3 Tectonic plates on the Earth. Arrows indicate the direction of motion of the major plates.

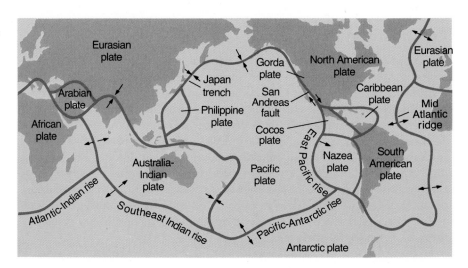

of a jigsaw puzzle (Figure 3.3). In some places, such as the Atlantic Ocean, these plates are moving apart, while in others they are forced together. The motive power for plate tectonics is provided by slow **convection** of the mantle, a process by which heat escapes from the interior through the upward flow of warmer material and the slow sinking of cooler material.

Four basic kinds of interactions between crustal plates are possible at their boundaries: (1) they can pull apart; (2) they can push together with one plate burrowing under another; (3) they can collide; or (4) one plate can slide along parallel with another. Each of these activities is important in determining the geology of the Earth.

Rift and Subduction Zones

Plates pull apart from each other along **rift zones** such as the Mid-Atlantic Ridge, driven by upwelling currents in the mantle (Figure 3.4a). A few rift zones are found on land, the best known being the central African rift, an area in which the African continent is slowly breaking apart. Most rift zones, however, are in the oceans. The new material that rises to fill the space between the receding plates is basaltic lava, the kind of igneous rock that forms most of the ocean basins.

Figure 3.4 The boundaries between plates in the Earth's crust: (a) A rift zone where plates are separating and new ocean crust is being formed; (b) A subduction zone where crust is being destroyed at an ocean trench.

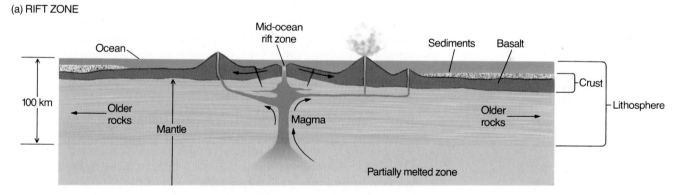

(a) RIFT ZONE

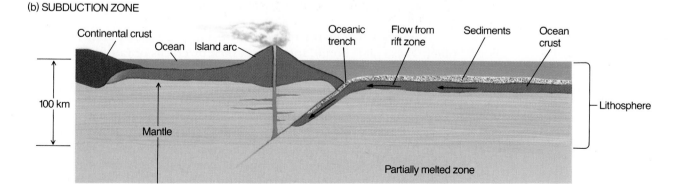

(b) SUBDUCTION ZONE

From a knowledge of sea-floor spreading, we can calculate the average age of the oceanic crust. About 60,000 km of active rifts have been identified, with average separation rates of about 4 cm per year. The new area added to the Earth each year is about 2 sq km, enough to renew the entire oceanic crust in a little more than 100 million years. This is a very short interval in geological time, less than 3 percent of the age of the Earth. The present oceans are among the youngest features on the planet!

When two plates press together, one plate often dives down beneath another in what is called a **subduction zone** (Figure 3.4b). Often a subduction zone is marked by an oceanic trench, a fine example being the deep Japan Trench along the coast of Asia. The subducted plate is forced down into regions of high pressure and temperature, eventually melting several hundred kilometers below the surface. Its material is recycled back into a downward-flowing convection current, ultimately balancing the material that rises along rift zones.

A part of the subducted material reaches the surface more directly through volcanic eruptions. All along the subduction zone, earthquakes and volcanoes mark the death throes of the plate. Some of the most destructive earthquakes in history have taken place along subduction zones, including a 1976 earthquake that leveled the Chinese city of Tangshan and resulted in more than half a million deaths.

Faults and Mountains

When two continental masses are pushed together by the motion of the crustal plates, the Earth buckles and folds, forcing some rock deep below the surface and raising other folds to heights of many kilometers. This is the way most of the mountain ranges on Earth were formed; as we will see, however, quite different processes produced the mountains on other planets.

At the same time that a mountain range is being formed by upthrusting of the crust, its rocks are subject to the erosional force of water and ice. The sharp peaks and serrated edges that are characteristic of our most beautiful mountains (Figure 3.5) have little to do with the forces that make mountains but are instead the result of the processes that tear them down. Ice is an especially effective sculptor of rock. In a planet without moving ice or running water, mountains remain smooth and dull.

Along much of their lengths, the crustal plates slide parallel to each other. These plate boundaries are marked by cracks or **faults.** Along active fault zones, the motion of one plate with respect to the other is several centimeters per year, about the same as the spreading rates along rifts. One of the most famous faults is the San Andreas Fault (Figure 3.6), lying on the boundary between the Pacific Plate and the North American Plate. This fault runs from the Gulf of California in the south to the Pacific Ocean just west of San Francisco in the north. The Pacific Plate, to the west, is moving northward, carrying Los Angeles, San Diego, and parts of the southern California coast with it. In several million years, Los Angeles will be an island off the coast of San Francisco.

Figure 3.5 The Alps, a young region of the Earth's crust where sharp mountain peaks are being sculpted by glaciers. *(David Morrison)*

Unfortunately for us, the motion along most fault zones does not take place smoothly. The creeping motion of the plates against each other builds up stresses in the crust that are released in sudden, violent slippages, generating earthquakes. The longer the interval between earthquakes, the greater the stress that accumulates and the larger the amount of the energy released when the surface finally moves. For example, the part of the San Andreas Fault near the central California town of Parkfield has slipped about every 22 years during the past century, moving an average of about 1 m each time. In contrast, the average interval between major earthquakes in the Los Angeles region is about 140 years, and the average motion is about 7 m. The last time the San Andreas slipped in this area was in 1857; tension has been building ever since, and sometime soon it is bound to be released.

Volcanoes

Volcanoes mark locations where molten rock rises to the surface. One example is the mid-ocean ridges, which are volcanic features formed by mantle convection currents at plate boundaries. A second, major kind of volcanic activity is associated with subduction zones, and volcanoes sometimes also appear in regions where continental plates are colliding.

Another location for volcanic activity on our planet is found above so-called mantle hot spots, areas far from plate boundaries where heat is nevertheless rising from the interior of the Earth. One such hot spot is

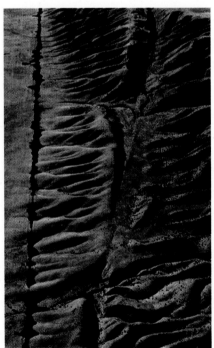

Figure 3.6 The San Andreas Fault, located north of Los Angeles, is a very active region where one crustal plate is sliding sideways with respect to the other. *(U.S. Geological Survey)*

Figure 3.7   Mauna Loa, a large volcano in Hawaii, in eruption. The stars of the Southern Cross are visible to the left of the volcanic plume. *(Dale P. Cruikshank)*

under the island of Hawaii, where it supplies the energy to maintain three currently active volcanoes, two on land and one under the ocean (Figure 3.7). The tallest Hawaiian volcanoes are up to 100 km in diameter and rise 9 km above the ocean floor; however, as we shall see, other planets have volcanoes that are considerably larger.

Not all volcanic eruptions produce mountains. If the lava is very fluid and flows rapidly from long cracks, it can spread out to form lava plains. The largest known terrestrial eruptions, such as those that produced the Snake River Basalts in the northwestern U.S. or the Deccan Plains in India, are of this type.

3.4
THE EARTH'S ATMOSPHERE

We live at the bottom of the ocean of air that envelops our planet. The atmosphere, weighing down upon the surface of the Earth under the force of gravitation, exerts a pressure at sea level of one **bar,** which equals the weight of 1030 kg (1.03 tons) over each square meter. The total mass of the atmosphere is about 5×10^{18} kg, or about a millionth of the total mass of the Earth. Thus the atmosphere represents a smaller fraction of the Earth than the share of your mass represented by the hair on your head. Yet we are interested in the atmosphere, because it constitutes our immediate environment.

What are the composition and structure of the Earth's atmosphere?

Structure of the Atmosphere

The structure of the Earth's atmosphere is illustrated in Figure 3.8. Most of the atmosphere is concentrated near the surface, within about the bottom 10 km. That is where clouds form and airplanes fly. Within this region, called the **troposphere,** warm air, heated by the surface, rises and is replaced by descending currents of cooler air—another example of *convection*. This atmospheric circulation generates clouds and other manifestations of weather. Within the troposphere, temperature drops rapidly with increasing elevation to values near 50° C below freezing (223 K) at its upper boundary, where the stratosphere begins. Most of the **stratosphere,** which extends to about 80 km, is cold and free of clouds.

Near the top of the stratosphere is a layer of **ozone** (O_3), a form of oxygen having three atoms per molecule instead of the usual two. Ozone is formed from ordinary oxygen by absorption of ultraviolet light. Because ozone itself is a good absorber of this light, it protects the surface from some of the Sun's dangerous ultraviolet radiation, making it possible for life to exist on land. Because ozone is essential to our sur-

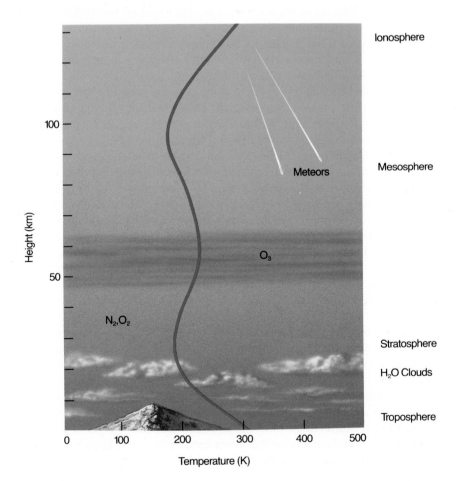

Figure 3.8   The structure of the Earth's atmosphere. The curving line shows the temperature (scale at bottom of figure).

vival, we react with justifiable concern to news that the industrial chemicals called CFCs (chlorofluorocarbons) are depleting atmospheric ozone to a significant degree.

At heights above 100 km, the atmosphere is so thin that orbiting satellites can pass through it without being dragged down. At these elevations, individual atoms can occasionally escape completely from the gravitational field of the Earth. Thus there is a continuous slow leaking of atmosphere from the planet, especially of lightweight atoms, which move faster than heavy atoms. The Earth's atmosphere cannot, for example, hold on to hydrogen or helium, which escape into space.

Atmospheric Composition and Origin

At the Earth's surface, the atmosphere consists of 78 percent nitrogen (N_2), 21 percent oxygen (O_2), and 1 percent argon (A), with traces of water vapor (H_2O), carbon dioxide (CO_2), and other gases. Variable amounts of dust particles and water droplets are also found suspended in the air.

Chemically, the gases nitrogen and argon are relatively inert. It is oxygen that sustains animal life on Earth, by allowing animals to oxidize their food to produce energy. Oxygen is also required for all forms of combustion (rapid oxidation) and thus is necessary for most heat and power production. In the process of photosynthesis, green plants absorb carbon dioxide and release oxygen, which helps to replenish the oxygen consumed by humans and other animals.

A complete census of the Earth's atmosphere should look at more than the gas now present. Suppose, for example, that our planet were heated to above the boiling point of water (100° C, or 373 K); the oceans would boil and this water vapor would become a part of the atmosphere. To estimate how much water vapor would be released, we note that there is enough water to cover the entire Earth to a depth of about 3000 meters. Since the pressure exerted by 10 meters of water is about equal to one bar, the average pressure at the ocean floor is about 300 bars. Since water weighs the same whether it is in liquid or vapor form, the atmospheric pressure of water if the oceans boiled away would also be 300 bars. Therefore, water would dominate the Earth's atmosphere, with nitrogen and oxygen reduced to the status of trace constituents.

On a warmer Earth a source of additional atmosphere would be found in the sedimentary rocks of the crust. These minerals contain abundant carbon dioxide, that, if released by heating, would generate about 70 bars of CO_2, far more than the current CO_2 pressure of only 0.0005 bar. Thus, the atmosphere of a warm Earth would be dominated by water vapor and carbon dioxide, with a surface pressure close to 400 bars.

We are not sure of the origin of the Earth's ancient atmosphere. Today we see that CO_2, H_2O, SO_2, and other gases are released from volcanoes. Much of this apparently new gas, however, is recycled material that has been subducted through plate tectonics. The original, primordial atmosphere could have had a different source. The atmosphere could have been formed with the rest of the Earth, as it accumulated from the debris left over from the formation of the Sun; or it could have been derived from impacts by comets or other icy materials from the outer parts of the solar system. Current opinion favors the cometary hypothesis, but both mechanisms probably contributed.

Weather and Seasons

All planets with atmospheres have "weather," which is a name we give to the circulation of the atmosphere. The energy that powers the weather is derived primarily from the sunlight that heats the surface. The atmosphere and oceans redistribute the heat from warmer to cooler areas: from day to night and from warm latitudes toward the poles. Weather on any planet represents the response of its atmosphere to changing inputs of energy from the Sun.

The Earth's orbit is not quite circular, its distance from the Sun varying by about 3 percent. However, the changing distance of the Earth from the Sun is not the cause of the seasons, which exist because the plane in which the Earth revolves (the ecliptic plane) is not the same as the plane of the Earth's equator. The equator and the ecliptic are inclined to each other by about 23 degrees, the tilt of the rotational axis of the Earth. The Northern Hemisphere is inclined toward the Sun in June and away from it in December.

Figure 3.9 shows the Earth's path around the Sun. The line EE' is in the plane of the celestial equator. We see that on about June 22 (the date of the summer solstice in the Northern Hemisphere), the Sun shines down most directly upon the Northern Hemisphere of the Earth. It

What causes the Seasons?

Figure 3.9   The seasons are caused by the inclination of the plane of the Earth's orbit to the plane of the equator.

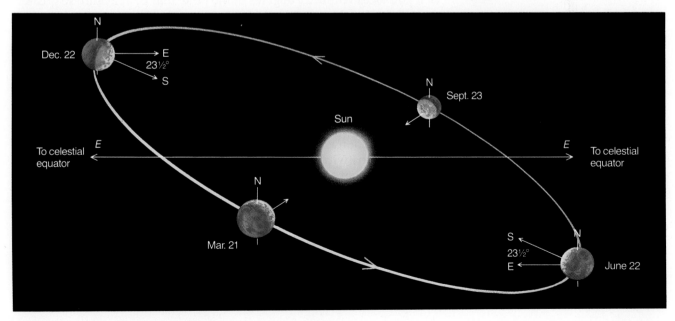

Figure 3.10 The Earth on June 22, the summer solstice in the northern hemisphere.

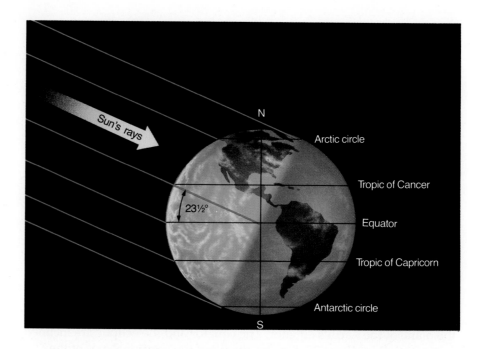

appears 23 degrees north of the equator and thus, on that date, passes through the zenith of places on the Earth that are at latitude 23 degrees N.

The situation is shown in detail in Figure 3.10. To a person at a latitude 23 degrees N (in Hawaii, for example), the Sun is overhead at noon. This latitude, at which the Sun can appear at the zenith at noon on the first day of summer, is called the Tropic of Cancer. We see also in Figure 3.9 that the Sun's rays shine down past the North Pole on this date. In fact, all places within 23 degrees of the pole, that is, at a latitude greater than 67 degrees N, have sunshine for 24 hours on the first day of summer. The Sun is then as far north as it can get; thus, 67 degrees is the southernmost latitude in which the Sun can be seen for a full 24-hour period (the midnight Sun). That circle of latitude is called the Arctic Circle. During this time, the Sun's rays shine obliquely on the Southern Hemisphere. All places within 23 degrees of the South Pole—that is, south of latitude 67 degrees S (the Antarctic Circle)—do not see the Sun at all.

The situation is reversed six months later, on about December 22 (the date of the winter solstice), as shown in Figure 3.11. Now it is the Arctic Circle that has a 24-hour night and the Antarctic Circle that has the midnight Sun. At latitude 23 degrees S, the Tropic of Capricorn, the Sun passes through the zenith at noon. It is winter in the Northern Hemisphere and summer in the Southern Hemisphere.

Finally, we see in Figure 3.9 that on about March 21 and September 23 the Sun appears to be on the celestial equator; on these dates, the equator itself represents the daily path of the Sun. Every place on the Earth then receives 12 hours of sunshine and 12 hours of darkness. The points where the Sun crosses the celestial equator are called the vernal

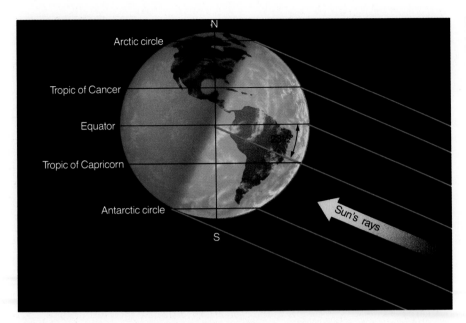

Figure 3.11 The Earth on December 22, the winter solstice in the northern hemisphere.

(spring) and autumnal (fall) equinox, from the fact that day and night are equal.

Except near the equator, most solar energy is deposited during the spring and summer months. A typical spot in the United States, on the first day of summer (about June 22), receives 14 to 15 hours of sunshine. At this time, the Sun appears high in the sky, so the sunlight is more direct and thus is more effective at heating than it is in the fall and winter, when the Sun appears lower in the sky. We receive only 9 to 10 hours of sunshine in the fall and winter. Also, the Sun is low in the sky; a bundle of its rays is thus spread out over a larger area than in summer, and it is less effective in heating the ground.

The seasonal effects are different at different latitudes. At the equator, for instance, all seasons are much the same. Every day of the year, the Sun is up half the time so there are always 12 hours of sunshine and 12 hours of darkness. As one travels north or south, the seasons become more pronounced, until we reach extreme cases in the Arctic and Antarctic. At the North Pole, celestial objects that are north of the celestial equator are always above the horizon and, as the Earth turns, circle around parallel to it. The Sun is north of the celestial equator from about March 21 to September 23; thus at the North Pole, the Sun rises when it reaches the vernal equinox and sets when it reaches the autumnal equinox. Each year there are six months of sunshine at each pole, followed by six months of twilight or darkness.

Climate Changes

Climate is a term used to refer to the effects of the atmosphere on a time scale of decades and centuries. The best documented changes in the Earth's climate are the great ice ages, which have periodically lowered

OBSERVING YOUR WORLD   THE MOTION OF THE SUN WITH RESPECT TO THE BACKGROUND STARS

Because the Earth's rotation axis is tilted with respect to its orbit, first one hemisphere of the Earth is tilted toward the Sun, and then 6 months later the other hemisphere is tilted toward the Sun. The days are longer in the hemisphere tilted toward the Sun, and the Sun is higher in the sky. Both effects lead to warmer temperatures. The apparent motions of the Sun in the sky as one hemisphere tilts first toward and then away from the Sun are easily measured.

Activities

1. The motions of the Sun north or south are best measured at sunset or sunrise. Select a place from which you can see the western horizon and sketch the foreground objects—trees, hills, buildings, power poles, etc. Look up the time of sunset in your local newspaper, and then observe the location where the Sun disappears below the horizon. Mark the location on your sketch and note the date. Repeat this observation from exactly the same spot every other week or so. From your knowledge of the seasons, predict in advance whether you expect the Sun to be moving north or south. After a month or so, determine whether or not your prediction is correct. Are the days getting longer or shorter?

2. Based on the results obtained above, would you expect the midday Sun to be higher or lower at the end of the experiment? Verify your prediction by measuring the length of the shadow of a ruler at noon. *(Do not look directly at the Sun because doing so may damage your eyes.)*

3. Determine whether the Sun moves west or east relative to the background stars. Select a bright star visible in the western sky at sunset and a second visible in the eastern sky. Estimate their altitudes using your fist as a measuring device. Note the approximate altitudes and the time of measurement. Repeat this measurement approximately every other week for several weeks at the same time of night. How does the altitude of each star change? Because you are making the measurement at the same time of night, the Sun is in approximately the same position below the horizon at the time of each measurement.

the temperature of the Northern Hemisphere over the past million years or so. Today we are in a relatively warm period, interpreted by many scientists as a fairly short-lived interglacial interval between major ice ages.

It is generally believed that the ice ages are primarily the result of changes in the tilt of the Earth's rotational axis, produced by the gravitational effects of the other planets. This idea of an astronomical cause of climate changes was first proposed in 1920 by Serbian scientist Milutin Milankovich, and these periodic changes are sometimes called Milankovich cycles. As we see in Chapter 5, there is also evidence of periodic climate changes on Mars, and modern calculations suggest that these also have an origin in slow changes of the orbit and rotational axis of that planet.

3.5
LIFE AND CHEMICAL EVOLUTION

Earth is the only inhabited planet in the solar system, as we know from the investigation of other planets by spacecraft. Terrestrial life has played an important part in the story of our planet. Life arose early in

Earth's history, and living organisms have been interacting with their environment for billions of years. We all recognize that life forms have evolved to adapt themselves to the environment on Earth, and now we are beginning to realize that the Earth itself has been changed in important ways by the presence of living matter.

Origin of Life

The record of the birth of life on Earth has been lost in the restless motions of the crust. By the time the oldest surviving rocks were formed, life already existed (Figure 3.12), but abundant fossils have been produced only during the past 600 million years—less than 15 percent of the planet's history. Thus, any theory of the origin of life must be partly speculative, since there is little direct evidence. All we really know is that the atmosphere of the early Earth, unlike that today, contained abundant carbon dioxide but no oxygen gas. In the absence of oxygen, many complex chemical reactions are possible that lead to the production of amino acids, proteins, and other chemical building blocks of life. Organic chemicals were also transported to our planet as part of the influx of comets early in our history. Sometime during the first half-billion

(a)

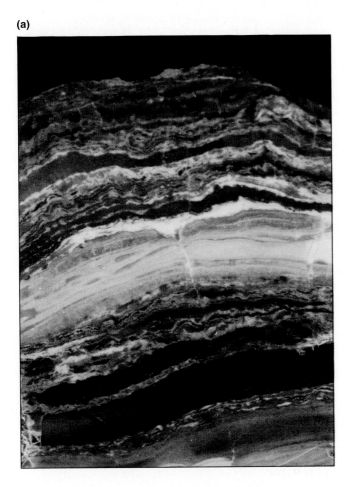

(b)

Figure 3.12   Cross-section of stromatolites, both fossil and contemporary. These colonies of micro-organisms are among the oldest evidence of life on Earth, dating back more than 3 billion years. *(NASA/Ames)*

years after the formation of the planet, life arose from these chemical building blocks.

For tens of millions of years after its formation, life (perhaps little more than large molecules like the viruses of today) probably existed in warm, nutrient-rich seas, living off accumulated organic chemicals. Eventually, however, as this easily accessible food became depleted, life began the long evolutionary road that led to the proliferation of different organisms on Earth today. As it did so, life influenced the chemical evolution of the atmosphere.

Evolution of the Atmosphere

How has life altered the composition of our atmosphere?

Studies of the chemistry of ancient rocks show that the atmosphere of the Earth lacked oxygen until about two billion years ago, despite the presence of plants releasing oxygen by photosynthesis. Chemical reactions with the rocks of the crust apparently removed the oxygen gas as quickly as it formed. Slowly, however, the increasing evolutionary sophistication of life led to a growth in the plant population, finally reaching the point where oxygen was produced faster than it could be removed, and the atmosphere became more and more oxidizing.

The appearance of free oxygen between one and two billion years ago eventually led to the formation of the Earth's ozone layer, which protects the surface from lethal solar ultraviolet light. Before that, it was unthinkable for life to venture outside the protective oceans, and the land masses of Earth were barren. The presence of oxygen and, therefore, ozone thus allowed the colonization of the land. It also made possible a proliferation of animals who lived by oxidizing the organic materials produced by plants. When the animals moved to the land, they developed techniques for breathing oxygen directly from the atmosphere. Thus life produced oxygen, and the oxygen made the land safe for life.

On a planetary scale, one of the most important consequences of life has been a decrease in atmospheric carbon dioxide. In the absence of life, Earth would probably have a much more massive atmosphere dominated by CO_2. But today, most of the carbon dioxide is locked up in sediments composed of the shells of marine creatures, which extract CO_2 from the water. The CO_2 remains trapped on the ocean floor until it is subducted, when much of it returns to the atmosphere in volcanic eruptions. Life also removes and traps CO_2 by producing deposits of the fossil fuels coal and oil. These substances are primarily carbon, for the most part extracted from atmospheric CO_2 hundreds of millions of years ago, when the first great forests populated the land.

The Greenhouse Effect and Global Warming

We have a special interest in the carbon dioxide content of the atmosphere because of the special role this gas plays in retaining heat from the Sun, through a process called the **greenhouse effect.** To understand how the greenhouse effect works, imagine the fate of the sunlight that strikes

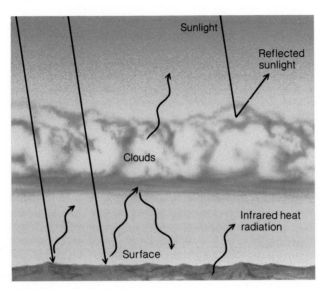

Figure 3.13   The operation of the greenhouse effect. Sunlight that penetrates to the lower atmosphere and surface is reradiated in the infrared where the atmosphere restrains this heat from escaping. The result is an elevated surface temperature.

the surface of the Earth. The sunlight is absorbed, heats the surface, and is reradiated in the infrared part of the spectrum (Figure 3.13). However, CO_2, which is a colorless transparent gas in the visible, is opaque in the infrared. As a result, it acts as a blanket, impeding the flow of infrared radiation back to space. Consequently, the surface heats up and establishes an equilibrium temperature higher than it would have had without the CO_2. The more CO_2, the higher this temperature will be.

The greenhouse effect in a planetary atmosphere is similar to the heating of a gardener's greenhouse or the inside of a car left out in sunlight with the windows rolled up. In these examples, the window glass plays the role of carbon dioxide, letting sunlight in but impeding the outward flow of infrared radiation. You can test the greenhouse heating yourself if you have a car with an enclosed trunk: leave the car in the sunlight for an hour and then compare the temperature in the main compartment, where the greenhouse effect has been at work, with the temperature in the trunk, where there is no greenhouse effect. On Earth, the current greenhouse effect elevates the surface temperature by about 20° C on the average. Without this greenhouse effect, the average surface temperature on Earth would be below the freezing point, and we would be locked in a global ice age.

Modern industrial society depends on energy extracted from burning fossil fuels. As these ancient coal and oil deposits are oxidized, additional CO_2 is released into the atmosphere. The situation is exacerbated by widespread destruction of tropical forests, reducing the planet's capacity to extract CO_2 from the atmosphere and replenish the supply of O_2. So far in this century, the amount of CO_2 in the atmosphere has increased by about 30 percent, and it is continuing to rise at 0.5 percent per year. By early in the twenty-first century, the CO_2 level is predicted to

What is the relationship between the greenhouse effect and global warming?

Figure 3.14   Increase with time of atmospheric CO_2 content, which is expected to double by the middle of the 21st century.

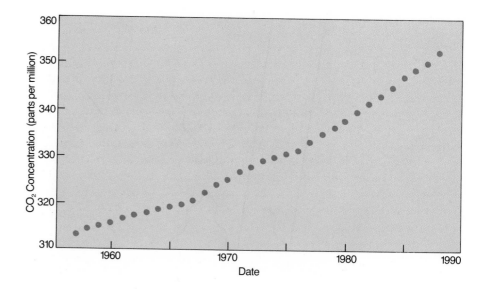

reach twice its pre-industrial value (Figure 3.14). The consequences of such an increase are being studied with elaborate computer models, but the fact is that we are not sure of the effects, and even if we were, there is no way to bring the CO_2 content of the atmosphere back down to normal levels.

Already some global warming is apparent. In the U.S. and Europe, summer temperatures throughout the 1980s reached record highs. The sea level has also risen measurably, owing to expansion of the water as its average temperature climbs. These effects, of course, are superimposed on the usual year-to-year fluctuations in weather, but most scientists are convinced that some global warming caused by an enhanced greenhouse effect is a reality. It is very difficult, however, to calculate how much additional warming is expected, and important issues in global economics depend on the answer. Sometimes in these debates people question the reality of the greenhouse effect itself, but this is foolish. The greenhouse effect is real; the issue is how the greenhouse effect may change to produce global warming.

3.6
COSMIC INFLUENCES ON THE EVOLUTION OF EARTH

Until very recently, most scientists thought that the history of our planet, including the history of life, had transpired in relative isolation from the rest of the solar system. Now, however, we are beginning to recognize external influences that have profoundly influenced the course of evolution. This in one more way is which scientists are recognizing that the Earth is indeed a planet, subject to many of the same processes that act on the other terrestrial planets.

Where Are the Craters on Earth?

In discussing the geology of the Earth in Section 3.3, we dealt only with the effects of internal forces, expressed through the processes of plate tectonics and volcanism. In contrast, when we look at the Moon we see primarily craters, produced from the impacts of interplanetary debris.

Why do we not see more evidence here of the kinds of craters that are so prominent on the Moon and other planets? Our atmosphere provides no shield against the large impacts that form craters many kilometers in diameter, nor is there any other known way to avoid these cosmic events. The difference is that on the Earth these craters are destroyed by our active geology before they can accumulate. As erosion and plate tectonics constantly rework our crust, evidence of past cratering events is destroyed.

Recent Impacts

The collision of interplanetary debris with the Earth is not a hypothetical idea. In the desolate Tunguska region of Siberia, a remarkable atmospheric explosion took place in 1908. The shock wave flattened more than 1000 sq km of forest; herds of reindeer and other animals were killed; and a man at a trading post 80 km from the blast was thrown from his chair and knocked unconscious. The blast wave spread around the world, recorded by instruments designed to measure changes in atmospheric pressure. This event was caused by the atmospheric disintegration of an impacting stony object weighing approximately 100,000 tons. The force of the blast was equivalent to a 10–20 megaton nuclear bomb. We estimate that a meteoric "airburst" of this magnitude takes place somewhere on Earth every few hundred years.

Because the incoming projectile fragmented in the atmosphere, the Tunguska event produced no crater. The most recent impact to produce a substantial crater took place 50,000 years ago in Arizona. The projectile in this case was an iron meteorite with about the same mass as the Tunguska object. Because of its great strength, this projectile did not disintegrate in the atmosphere but survived to explode at the surface. The crater, called Meteor Crater and now a major tourist attraction, is about a mile across (Figure 3.15). A total of about 150 craters have been identified on Earth, but most older craters are so eroded that only a trained eye can distinguish them.

Extinction of the Dinosaurs

The impact that produced Meteor Crater would have been dramatic indeed to any humans who witnessed it, since the energy release was equivalent to a 15-megaton nuclear bomb. But such explosions have no global consequences. Much larger (and rarer) impacts, however, can disturb the ecology of the whole planet and have a major influence on the course of evolution.

What has been the role of cosmic impacts in the history of life?

Figure 3.15 Meteor Crater in Arizona, a 50,000-year-old impact scar. Although impact craters are common on less active bodies like the Moon, this is the only well-preserved such crater on the Earth. *(Meteor Crater, Northern Arizona)*

The best-documented such impact took place 65 million years ago, at the boundary between the Cretaceous and Tertiary periods of geological history. This break in the Earth's history is marked by a **mass extinction,** when more than half of the species on our planet became extinct. Although there are a dozen or more mass extinctions in the geological record, this particular event has always intrigued paleontologists because it marks the end of the age of the dinosaurs. For more than 100 million years the dinosaurs had ruled the world. Then, suddenly, they disappeared, and thereafter the mammals began their development and diversification.

The body that collided with the Earth at the end of the Cretaceous period had a mass of more than a trillion tons and a diameter of at least 10 km (Figure 3.16). We know this because of a worldwide layer of sediment deposited from the dust cloud that enveloped the planet after the impact. First identified in 1980, this sediment layer is enriched in the rare metal iridium and other elements that are relatively abundant in asteroids and comets but very rare in the crust of the Earth. Even diluted by the terrestrial material excavated from the crater, this cosmic component is easily identified. The boundary-layer sediment also contains many minerals that are characteristic of the temperatures and pressures of a gigantic explosion.

The Cretaceous impact released energy equivalent to 5 billion Hiroshima-size nuclear bombs, excavating a crater 200 km across and deep enough to penetrate through the Earth's crust. This crater, called Chicxulub, has been located in the Yucatan region of Mexico, buried under hundreds of meters of more recent limestone sediment. The explosion lifted about a 100 trillion tons of dust into the atmosphere, as can be determined by measuring the thickness of the sediment layer formed when this dust settled to the surface. Such a quantity of material would have blocked sunlight completely from reaching the surface, plunging the Earth into a months-long period of cold and darkness. Other effects

Figure 3.16 Artist's impression of the impact of a 25-km asteroid with the Earth. *(Painting by Don Davis)*

include global acid rain and large-scale fires that probably destroyed most of the planet's forests and grasslands. Presumably these environmental effects, rather than the explosion itself, were responsible for the mass extinction, including the death of the dinosaurs.

Impacts and the Evolution of Life

Several other mass extinctions in the geological record have been tentatively identified with large impacts, although none is so dramatic as the event that destroyed the dinosaurs. But even without such specific documentation, it is clear that impacts of this size do occur and that their effects can be catastrophic for life. What is a catastrophe for one group of living things, however, may create opportunities for another group. Following each mass extinction, there is a sudden evolutionary burst as new species develop to fill the ecological niches opened by the event.

Impacts by comets and asteroids represent the only mechanisms we know that could cause global catastrophes and seriously influence the evolution of life all over the planet. According to some estimates, the *majority* of all extinctions of species may be due to such impacts—a perspective that fundamentally changes our view of biological evolution. A central issue for the survival of a species is not just its success in competing with other species and adapting to slowly changing environments, as envisioned by Darwinian natural selection. Of at least equal importance is its ability to survive random global ecological catastrophes caused by impacts.

The Earth is a target in a cosmic shooting gallery, subject to random violent events that were unsuspected a few decades ago. We owe our existence to such events. Had the impact of 65 million years ago not redirected the course of evolution, mammals might never have become the dominant large animals they are today. But even more fundamentally, if impacts with comets and asteroids had not occurred throughout our planet's history, could biological evolution in more stable conditions have produced the wondrous diversity of life that populates the Earth today?

▶ SUMMARY

3.1 The materials that make up the solid planets can be characterized as metal (mostly iron), rock (mostly **silicate),** and ice (water and other volatiles). Generally, the planetary chemistry is either **oxidized** (the inner planets) or **reduced** (the giant planets). Temperatures depend on distance from the Sun and are measured in degrees Kelvin (degrees above **absolute zero).**

3.2 Ancient Greek scientists recognized that the Earth is spherical and first measured its size. The Earth's interior composition and structure are probed using **seismic waves.** Such studies reveal the metal core and silicate mantle. The crust consists primarily of oceanic **basalt** and continental granite. Because it was once molten, the Earth is now **differentiated.**

3.3 Terrestrial rocks can be classified as **igneous,** sedimentary, or metamorphic. A fourth type, **primitive** rock, is not found on the Earth. Our planet's geology is dominated by **plate tectonics,** in which crustal plates move slowly in response to mantle **convection.** The surface expression of plate tectonics includes continental drift, recycling of the ocean floor, mountain building, **rift zones, subduction, faults,** earthquakes, and volcanic eruptions.

3.4 The atmosphere is composed primarily of N_2 and O_2 and includes such important trace gases as H_2O, CO_2, and **ozone** (O_3). Its structure consists of **troposphere, stratosphere,** and tenuous higher regions. Atmospheric oxygen is the product of life. Atmospheric circulation (weather) is driven by seasonally changing deposition of sunlight, a result of the tilt of the Earth's axis of rotation. Longer-term climatic variations, such as the ice ages, are probably caused by changes in the planet's orbit and axial tilt.

3.5 Life originated on Earth at a time when the atmosphere consisted mostly of CO_2 and there was no O_2. Later photosynthesis gave rise to free oxygen and ozone, and most of the CO_2 was trapped in oceanic sediment. Carbon dioxide in the atmosphere heats the surface through the **greenhouse effect,** and increasing atmospheric CO_2 is leading to global warming.

3.6 There are few visible impact craters on the Earth, as a result of erosion and an active geology. But impacts do take place, such as the Tunguska event and the impact that produced Meteor Crater. A much larger impact at the end of the Cretaceous period (65 million years ago) destroyed most life on Earth and redirected the course of biological evolution. Impact-generated **mass extinctions** are now thought to have played a crucial role in the history of life on our planet.

▶ REVIEW QUESTIONS

1. Why do we include a chapter on the Earth in an astronomy text? Explain whether you think more, or less, text should be devoted to our own planet.

2. Describe the main differences in composition between the inner and outer planets.

3. Consider several familiar landforms of the Earth, such as mountains, volcanoes, or canyons, and briefly describe how they were formed. Try to relate your answers to the theory of plate tectonics, which provides the basis for understanding much of terrestrial geology.

4. The energy to drive atmospheric motions on the Earth comes from solar heating. How do the changing seasons relate to our weather?

5. What is the origin of each of the main gases in the Earth's atmosphere? Explain the relationship between the composition of the atmosphere and the development of life.

6. What is the role of impacts by comets and asteroids in influencing the Earth's geology, its atmosphere, and the evolution of life?

▶ THOUGHT QUESTIONS

7. Explain why the Earth has iron in its core and rocks at its surface.

8. Suppose that Eratosthenes had found that at Alexandria at noon on the first day of summer the line to the Sun makes an angle of 45 degrees with the vertical. What then would he have found for the Earth's circumference?

9. Identify examples of igneous, sedimentary, and metamorphic rock. Explain how each is formed, and discuss why there are no primitive rocks on the Earth (except meteorites).

10. Why will a decrease in the Earth's ozone be harmful? Why will an increase in our CO_2 also be harmful?

11. Why does the Earth have so few impact craters, relative to the Moon and many other planets and satellites?

12. If all life were destroyed on Earth, would new life eventually form to take its place? Explain how conditions would have to change for life to start again on our planet.

▶ PROBLEMS

13. Two friends, one living in Chicago and one in New Orleans, decide to measure the circumference of the Earth. On Christmas day each measures the highest altitude of the Sun at noon. In Chicago, the measured altitude of the Sun above the horizon is 24 degrees, while in New Orleans the maximum Christmas altitude of the Sun is 37 degrees. From a road map the two friends determine that Chicago is 800 miles north of New Orleans. With this information, calculate the circumference of the Earth.

14. Measurements using Earth satellites have shown that Europe and North America are moving apart by 4 meters per century because of plate tectonics. As the continents separate, new ocean floor is created along the Mid-Atlantic Rift. If the rift is 5000 km long, what is the total area of new ocean floor created in the Atlantic each century? How much new area is created per year?

15. Over the entire Earth there are 60,000 km of active rift zones, with average separation rates of 4 meters per century. How much area of new ocean crust is created each year over the entire planet? (This is also approximately equal to the amount of ocean crust that is subducted, since the total area of the oceans remains about the same.)

*16. What fractions of the volume of the Earth are occupied by the core, the mantle, and the crust? (*Hint:* The formula for the volume of a sphere is $4/3\pi R^3$, where R is the radius of the sphere.)

*17. With the information from Problem 15, you can calculate the average age of the ocean floor. First, find the total area of ocean floor (equal to about 60 percent of the surface area of the Earth). (*Hint:* The formula for the area of a sphere is $4 \pi R^2$). Then compare this with the area created (or destroyed) each year. The average lifetime is just the ratio of these numbers: the total area of ocean crust compared with the amount created (or destroyed) each year.

*18. The sea-level pressure of the atmosphere (1 bar) corresponds to 1.03 tons of mass above each square meter of the Earth's surface. Calculate the total mass of the atmosphere in kilograms and in tons (one ton equals 1000 kg). Then compute the total mass of ocean, given that the oceans would be 3000 meters deep if water covered the globe uniformly. (*Note:* 1 cubic meter of water has a mass of 1 ton). Compare the mass of the atmosphere and the mass of the oceans with the total mass of the Earth to determine what percentage of our planet is represented by the atmosphere and oceans.

*19. Suppose that a major impact that produces a mass extinction takes place on the Earth once every 10 million years. Further suppose that if such an event occurred today, you and most other humans would be killed (this is true even if the human species as a whole survived). Such impact events are random, and one could take place any time. Calculate the probability that such an impact will occur within the next 50 years (within your lifetime). This is approximately equal to the probability that you will be killed by this means, rather than dying from an auto accident or from heart disease or some other "natural" cause. How do the risks of dying from an asteroidal or cometary impact compare with other risks we are concerned about?

▶ SUGGESTIONS FOR ADDITIONAL READING

Books

About Earth in General

Schneider, S. & Londer, R. *The Coevolution of Climate and Life.* Sierra Club Books, 1984. A modern discussion of weather, climate, human society, and planetary evolution.

Hartmann, W. & Miller, R. *The History of the Earth.* Workman, 1993. An overview by an astronomer and a science illustrator.

Cattermole, P. & Moore, P. *The Story of the Earth.* Cambridge U. Press, 1985. A clear introduction to our planet.

Sullivan, W. *Landprints.* Times Book, 1985. A readable account of modern geology, written by a leading science journalist.

About Impacts and Extinctions

Chapman, C. & Morrison, D. *Cosmic Catastrophes.* Plenum Press, 1989. An account of the changing ideas about the role of catastrophes in our planet's evolution.

Goldsmith, D. *Nemesis: The Death Star and Other Theories of Mass Extinction.* Walker & Co, 1985. Examines the current debate about astronomical causes of mass extinctions.

Articles

About Earth in General

Hartmann, W. "Piecing Together the Earth's Early History" in *Astronomy,* June 1989, p. 24.

Chyba, C. "The Cosmic Origins of Life on Earth" in *Astronomy,* Nov. 1992, p. 28.

Heppenheimer, T. "Journey to the Center of the Earth" in *Discover,* Nov. 1987, p. 86.

Lanzerotti, L. & Uberoi, C. "Earth's Magnetic Environment" in *Sky & Telescope,* Oct. 1988, p. 360.

Jones, P. & Wigley, T. "Global Warming Trends" in *Scientific American,* Aug. 1990, p. 84.

White, R. "The Great Climate Debate" in *Scientific American,* July 1990, p. 36. On the greenhouse effect.

Kastings, J., *et al.* "How Climate Evolved on the Terrestrial Planets" in *Scientific American,* Feb. 1988.

About Impacts and Extinctions

Morrison, D. & Chapman, C. "Target Earth: It Will Happen" in *Sky & Telescope,* Mar. 1990, p. 261.

Weissman, P. "Are Periodic Bombardments Real?" in *Sky & Telescope,* Mar. 1990, p. 266.

Alvarez, W., *et al.* "What Caused the Mass Extinction: A Debate" in *Scientific American,* Oct. 1990, p. 76.

OUR NEIGHBOR THE MOON

Astronaut on the surface of the Moon.
(NASA)

The Moon is our nearest neighbor in space as well as the best-explored body in the solar system beyond the Earth itself. Despite its proximity, however, the Moon is a very different sort of planetary object from the Earth. Because of its small mass, it lacks both an atmosphere and current geological activity. With its ancient, unchanging surface, the Moon opens windows to the early history of the solar system, periods difficult to probe in studies of our own more active planet. Mercury, which is the planet most like the Moon, is also discussed in this chapter.

4.1
APPEARANCE OF THE MOON

After the Sun, the Moon is the brightest and most obvious object in the sky, easily seen when it is above the horizon, even in the daytime. Equally obvious is the Moon's monthly cycle of phases, which results from the changing angle of its illumination by the Sun.

Lunar Phases

The Moon is said to be *new* when it is in the same direction from Earth as the Sun (position A in Figure 4.1), with its daylight side turned away from us and its night side turned toward us. In this phase it is invisible. A day or two after the new phase, the thin *crescent* first appears (position B) as we see a small part of the Moon's daylight hemisphere. The illuminated crescent increases in size on successive days as the Moon moves farther and farther around the sky away from the direction of the Sun.

Figure 4.1   Phases of the Moon. In the upper part, the Moon's orbit is viewed obliquely. The lower part shows the appearance from the Earth.

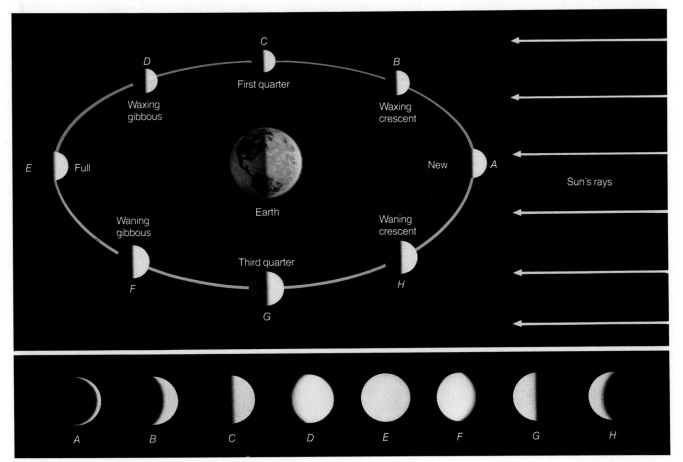

After about one week, the Moon is one quarter of the way around its orbit (position C) and is at the *first quarter* phase. Here half of the Moon's daylight side is visible. During the week after the first quarter phase, we see more and more of the Moon's illuminated hemisphere (position D), until the Moon and the Sun are opposite each other in the sky (position E). The side of the Moon turned toward the Sun is then also turned toward the Earth, and we have *full phase*. During the next two weeks the Moon goes through the same phases again in reverse order, returning to new phase after about 29 1/2 days.

If you have difficulty picturing the phases of the Moon from this verbal account, try this simple experiment: Stand about 6 feet in front of a bright electric light outdoors at night and hold in your hand a small round object such as a tennis ball or an orange. If the object is then viewed from various positions, the portions of its illuminated hemisphere that are visible represent the analogous phases of the Moon. But the best way to become fully acquainted with these lunar phases is to watch the real Moon in the sky. Observe its shape, its direction from the Sun, and its time of setting and rising.

The Moon's Revolution and Rotation

The Moon's period of revolution about the Earth, measured with respect to the stars, is a little over 27 days—27.3217 days, to be exact. The time interval in which the phases repeat—say from full to full—is 29.5306 days. The Moon changes its position on the celestial sphere rather rapidly, moving about 13 degrees toward the east each day. Even during a single evening, the Moon creeps visibly eastward among the stars, traveling its own width in a little less than 1 hour. The delay in moonrise from one day to the next, caused by this eastward motion, averages about 50 minutes.

The Moon *rotates* on its axis with exactly the same period in which it *revolves* about the Earth. As a consequence, the Moon keeps the same face turned toward the Earth; the differences in its appearance from one night to the next are due to changing illumination by the Sun, not to its own rotation.

We sometimes hear the back side of the Moon (the side we never see) called the "dark side." Of course, the back side is dark no more frequently than the front side. Because the Moon rotates, the Sun rises and sets on all sides of the Moon.

The Moon Through a Telescope

When you look at the Moon through binoculars or a small telescope (Figure 4.2), the detail you see depends on the resolution of the image. **Resolution** refers to the fineness of detail that is visible. Generally speaking, the larger the telescope aperture, the sharper the image will be. However, there are practical limits on resolution imposed by turbulence in the Earth's atmosphere, which always distorts images formed with ground-based telescopes. For most telescopes at most observato-

What causes the changing phases of the Moon?

Figure 4.2   The Moon as photographed from a spacecraft near the Earth. The hemisphere shown consists primarily of heavily cratered highlands. The resolution of such an image is several kilometers, similar to that of high-powered binoculars or a small telescope.   *(NASA)*

ries, the resolution limit is set by this atmospheric "seeing" rather than by the intrinsic resolution of the optics.

Under very good seeing conditions, the limiting resolution is equivalent to the width of a dime at a distance of three miles, or the width of one kilometer on the Moon. Because topographic features such as mountains and valleys are typically several kilometers in size, they are readily seen and photographed with even modest-sized telescopes at good sites. Useful geological studies of the Moon can therefore be made using telescopic photographs. In contrast, the nearest planets (Mars and Venus) are more than 150 times farther away, with correspondingly lower telescopic resolution: typically about 150 km. At a resolution of 150 km, no topographic information can be obtained. Indeed, the best views of Mars or Venus possible before the space age were equivalent to the Moon as seen without any optical aid. No wonder astronomers had such a difficult time trying to understand the nature of our planetary neighbors in the many years before we sent the first spacecraft to view the planets up close!

4.2
GENERAL PROPERTIES OF THE MOON

Galileo began the telescopic study of the Moon, identifying mountains and valleys and suggesting that the Moon was a real planetary world like our own. Generations of telescopic observers followed, but today most of what we know about the Moon derives from the U.S. Apollo program, which sent nine manned spacecraft to our satellite between 1968 and 1972 and landed 12 astronauts on its surface.

Exploration of the Moon

Space exploration of the Moon began in 1959, when the U.S.S.R. space-craft Luna 3 returned the first photos of the lunar farside. In 1966, Luna 9 landed on the surface and transmitted pictures and other data to Earth. Luna 9 was followed by several U.S. Surveyor landers, a Soviet robotic rover, and three automatic Soviet sample-return missions. However, these efforts were overshadowed once the first human set foot on the Moon on July 20, 1969 (Figure 4.3).

Table 4.1 summarizes the nine Apollo flights—six landings and three other missions that circled the Moon but did not land. The initial landings were on flat plains selected out of considerations of safety, but with increasing experience and confidence, NASA targeted the last three missions to more geologically interesting locales. The level of scientific exploration also increased with each mission, as the astronauts spent a longer time on the Moon and carried more elaborate equipment. Finally, on the last Apollo landing, NASA sent one scientist-astronaut, geologist Harrison Schmitt.

Figure 4.3 A typical lunar mare landscape, photographed by the Apollo astronauts. *(NASA)*

TABLE 4.1   Apollo Flights to the Moon

Flight	Date	Landing Site	Accomplishments
8	Dec 68	Orbiter	First human circumlunar flight
10	May 69	Orbiter	First lunar orbit rendezvous
11	Jul 69	Mare Tranquillitatis	First human landing; 22 kg samples returned
12	Nov 69	Oceanus Procellarum	First ALSEP; visit to Surveyor 3
13	Apr 70	Flyby	Landing aborted due to explosion in Command Module
14	Jan 71	Mare Nubium	First "rickshaw"
15	Jul 71	Imbrium/Hadley	First "rover." Visit to Hadley Rille; 24 km traverse
16	Apr 72	Descartes Highlands	Highland landing site; 95 kg samples returned
17	Dec 72	Taurus Mountains	Geologist present; 111 kg samples returned

In addition to landing on the lunar surface and studying it at close range, the Apollo missions accomplished three objectives of major importance for lunar science. First and most important, the astronauts collected nearly 400 kg of samples to return to Earth for detailed laboratory analysis (Figure 4.4). These samples, which are still being extensively studied, have probably revealed more about the Moon and its history than all other lunar studies combined. Second, each Apollo landing after

Figure 4.4   Lunar samples collected in the Apollo project are analyzed and stored in NASA facilities at the Johnson Space Center in Houston, Texas.   *(NASA)*

Figure 4.5 One of the unused Saturn 5 rockets built to go to the Moon, but now a tourist attraction at NASA's Johnson Space Center in Houston. *(NASA)*

the first deployed an ALSEP, or Apollo Lunar Surface Experiment Package, which continued to operate for years after the astronauts departed. (The ALSEPs were turned off by NASA in 1978 as a cost-cutting measure.) Third, the orbiting Apollo Command Modules carried a wide range of instrumentation to photograph and analyze the lunar surface from above.

The last human left the Moon in December 1972, just a little more than three years after Neil Armstrong took his "giant step for mankind." The program of lunar exploration was cut off in midstride as the result of political and economic pressures. It had cost just about $100 per American, spread over ten years—the equivalent of one pizza and a six-pack per year. The giant Apollo rockets built to travel to the Moon have been left to rust on the lawns of NASA centers at Cape Canaveral and Houston (Figure 4.5). Today no nation on Earth has the technology and the rockets to return to the Moon. Having reached our nearest neighbor in space, we humans retreated back to our own planet. How long will it be before we will venture out again into the solar system?

The scientific legacy of Apollo remains, however, as we shall see in the following sections of this chapter.

Some Basic Facts

The Moon has only one eightieth the mass of the Earth and a gravity (one sixth that of Earth) too low to retain an atmosphere. If, early in its history, the Moon outgassed an atmosphere from its interior or collected a temporary envelope of gases from impacting comets, such an atmosphere was lost before it could leave any recognizable evidence of its short existence. All signs of water are similarly absent. Indeed, the Moon is dramatically deficient in a wide range of **volatiles,** those elements and compounds that evaporate at moderately low temperatures. Examples of volatiles include the common ices, such as H_2O and CO_2 (dry ice).

The composition of the Moon is not the same as that of the Earth. With its density of only 3.3 g/cm^3, the Moon must be made almost entirely of silicate rock. Relative to the Earth, it is *depleted* in iron and other metals. We also know from the study of lunar samples that water and other volatiles are absent. It is as if the Moon were composed of the same basic silicates as the Earth's mantle and crust, but with the core metals and the volatiles selectively removed. The differences in composition between the Earth and the Moon provide important clues concerning the origin of the Moon, a topic we will return to later in this chapter.

Probes of the interior carried out with seismometers taken to the Moon as part of the Apollo program confirm the absence of a large lunar core. The Moon also lacks a global magnetic field like that of the Earth. This is a consistent result because, as far as we know, the only way to generate such a field is by motions within a liquid metal core. Because it is small, the Moon has cooled relatively rapidly, leading to solidification of its interior. Its absence of recent geological activity is a direct consequence of its small size relative to the Earth.

The crust of the Moon is thicker than that of the Earth, ranging from about 60 km on the side facing the Earth up to at least 150 km on the far side. The level of seismic activity on the Moon is much less than that on the Earth. No moonquake measured by any of the Apollo seismometers could have been *felt* by a person standing on the Moon, and the total energy released by moonquakes is a hundred billion times less than that of earthquakes on our planet.

> How does the composition of the Moon differ from that of the Earth?

4.3
THE LUNAR SURFACE

Galileo and other early lunar observers regarded the Moon as having continents and oceans and as being a possible abode of life. We know today, however, that the resemblance of lunar features to terrestrial ones is superficial. Even when they look similar, the origins of lunar features such as craters and mountains are likely to be different from those of their terrestrial counterparts. The Moon's relative lack of internal activity, together with the absence of air and water, makes most of its geological history unlike anything we know on Earth. Much of the lunar surface is also older than the rocks of the Earth's crust, as scientists learned when the first lunar samples were analyzed in the laboratory, providing us a window on the early history of the solar system.

Ages of Lunar Rocks

In order to trace the history of the Moon or of any planet, we must have a way to determine the ages of individual rocks. Once lunar samples were brought back by the Apollo astronauts, the techniques that had

Figure 4.6 Radioactive decay and half-life. See text for explanation.

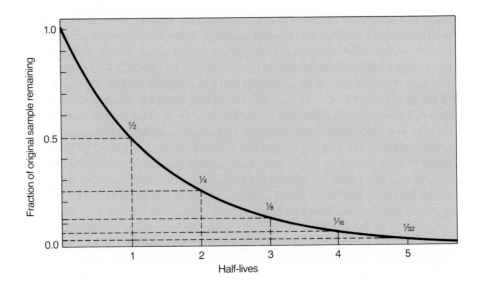

What is meant by the age of a rock, and how are ages measured?

been developed to date rocks on Earth could be applied to establish a chronology for our satellite.

The ages of rocks are measured using the properties of natural radioactivity. Radioactive nuclei of atoms are unstable and spontaneously convert to other nuclei with the emission of gamma rays or subatomic particles such as electrons. For any given nucleus the decay process is random, and it might happen at any time, but for a very large collection of identical radioactive atoms there is a specific time period, called radioactive **half-life**, during which there is a 50 percent chance that decay will occur. A particular atomic nucleus may last a shorter or longer time than its half-life, but in a large sample, almost exactly half will have decayed after a time equal to one half-life, and half of those remaining (three-quarters in all) will have decayed in two half-lives. After three half-lives, only one-eighth of the original sample remains, and so on (Figure 4.6). Thus, radioactive elements provide accurate nuclear clocks. By comparing the relative abundance of a radioactive element with that of the element it decays into, we can learn how long the process has been going on and hence the age of the sample.

Most natural rocks contain several radioactive elements with half-lives that provide suitable clocks. Among these are potassium-40[*], which decays to argon-40 with a half-life of 1.3 billion years; rubidium-87, which decays to strontium-87 with a half-life of 50 billion years; and uranium-238, which decays to lead-206 with a half-life of 4.5 billion years. In practice, as many as five separate elements are used to date a rock, and the process is carried out for several different mineral grains in the

---

[*] The number given following each element name is the atomic weight, equal to the number of protons and neutrons in the nucleus. Because isotopes of a given element differ from each other in the number of neutrons in the nucleus, we need to specify the atomic weight to identify the particular isotope being discussed.

same rock so as to further eliminate uncertainties. The final age, checked by the agreement of the different methods, is usually accurate to within a few percent, that is, to a few tens of millions of years in a rock several billion years old.

In order to interpret these radioactive measurements, it is important to recognize what the *age* represents. A radioactive age is the time during which the rock has remained in the solid form, so that both the radioactive element and its decay product are still present in each mineral grain. In most cases, a rock's age can be thought of as the time since it cooled from the molten state. It is such solidification ages that are quoted below for the Moon, as we discuss its geological features. As we will see, these ages for the Moon range from about 3.3 to 4.4 billion years—substantially older than most of the rocks on the Earth.

Radioactive dating techniques can be used to determine the date of formation of a planet as well as the solidification age of individual rocks. To obtain such dates, the total quantities of the radioactive isotopes (such as uranium-238) and their decay products (lead-206 in this example) are used. The results of this technique agree for both the Earth and the Moon; each was formed 4.5 billion years ago, with a margin of uncertainty of less than 50 million years.

Geological Features

Most of the crust of the Moon (83 percent) is heavily cratered and consists of relatively light-colored igneous silicate rocks. These regions are known as the lunar **highlands.** With ages of more than 4.0 billion years, the highlands are the oldest surviving part of the lunar crust. They represent material that solidified on the crust of the cooling Moon like slag floating on the top of a smelter. Because they formed so early in lunar history, the highlands are extremely heavily cratered, bearing the scars of billions of years of impacts by interplanetary debris (Figure 4.7).

What are the distinctions between the lunar highlands and maria?

Figure 4.7 The old, heavily cratered highlands make up 83 percent of the Moon's surface. *(NASA)*

Figure 4.8   About 17 percent of the Moon's surface consists of the maria—flat plains of basaltic lava. This view of Mare Imbrium includes numerous secondary craters and other ejecta from the large crater Copernicus, on the upper horizon.   *(NASA)*

Viewed with the unaided eye, the most conspicuous of the Moon's surface features—those familiarly called "the Man in the Moon"—are splotches of darker material of volcanic origin. These features, once thought to be liquid water, are still called *maria* (Latin for "seas"). These maria (Figure 4.8), which are much less cratered than the highlands, cover just 17 percent of the lunar surface, mostly on the side facing the Earth. They are volcanic plains, laid down in eruptions billions of years ago and partly filling huge basins or depressions caused by ancient collisions of asteroids with the Moon.

The lunar maria are all composed of basalt, very similar in composition to the oceanic crust of the Earth or to the lavas erupted by many terrestrial volcanoes (Figure 4.9). A series of such eruptions between 3.3 and 3.8 billion years ago (dated from laboratory measurements of returned samples) laid down smooth flows typically a few meters thick but extending over distances of hundreds of kilometers. Because these lavas originated at different depths, they differ slightly in chemical compositions (Figure 4.10).

Volcanic activity may have begun very early in the Moon's history, although most evidence of the first half-billion years is lost. What we do know is that the major mare volcanism, which involved the release of highly fluid basalts from hundreds of kilometers below the surface, ended about 3.3 billion years ago. After that the Moon's interior cooled, and by 3.0 billion years ago all volcanic activity ceased. Since then, our satellite has been a geologically dead world, changing only slowly as the result of random impacts.

Figure 4.9   Sample of basalt from the mare surface. The gas bubbles are characteristic of a lava rock.   *(NASA)*

Figure 4.10   Part of the far side of the Moon imaged by the Galileo spacecraft in 1992. Although highly exaggerated, the colors represent subtle differences in composition and the age of the lava flows that cover the lunar surface.   *(NASA/JPL)*

There are several distinct mountain ranges on the Moon, in addition to the widespread cratered plains of the highlands. These lunar mountains generally are found along the edges of the maria. Most of them bear the names of terrestrial ranges—the Alps, Apennines, Carpathians, and so on—but their origin is entirely different from that of their terrestrial namesakes.

The major lunar mountains are all the result of our satellite's history of impacts. The long, arc-shaped ranges that border the maria are composed of debris ejected from the impacts that formed these giant basins. These mountains have low, rounded profiles that resemble old, eroded mountains on Earth (Figure 4.11). But appearances can be deceiving; the mountains of the Moon have not been eroded except for the effects of

Figure 4.11   Mt. Hadley on the edge of Mare Imbrium, photographed by the Apollo 15 astronauts. Note the smooth contours of the lunar mountains, which have not been sculpted by water or ice.   *(NASA)*

Figure 4.12 Apollo photo of bootprint on the Moon. *(NASA)*

meteoritic impacts. They are rounded because that is the way they formed, and there has been no water or ice to carve them into cliffs and sharp peaks.

On the Lunar Surface

The surface of the Moon is covered with a fine-grained soil of tiny, shattered rock fragments. The astronauts kicked up this dark basaltic dust with every footstep, and the grit eventually worked its way into all of their equipment. The upper layers of the surface are porous, consisting of loosely packed dust into which their boots sank several centimeters (Figure 4.12).

This lunar dust, like so much else on our satellite, is the product of impacts. Each cratering event, large or small, breaks up the rock of the lunar surface and scatters the fragments. The impacts also melt tiny droplets of rock, producing spheres of impact glass that are found throughout the soil. Especially important are the multitudes of very small impacts by micrometeorites, grains of interplanetary dust that never strike the Earth's surface because they are filtered out by our atmosphere.

In the absence of any air, the lunar surface experiences much greater temperature extremes than we do on Earth, even though we are at the same distance from the Sun. Near local noon, the temperature of the dark lunar soil rises above the boiling point of water, while during the long lunar night (two Earth weeks) it drops to about 100 K ($-173°$ C). The extreme cooling is a result not only of the absence of air, but of the porous nature of the dusty soil, which cools more rapidly than would solid rock.

4.4
IMPACT CRATERS

The Moon provides an important benchmark for understanding the history of the planetary system. Most solid bodies show the effects of impacts, often extending back to the era when extensive debris from the planetary formation process was still present. On the Earth, this long history has been erased by our active geology. On the Moon, in contrast, most of the impact history is preserved. If we can understand what has happened on the Moon, we may be able to extrapolate this knowledge to the other cratered planets and satellites.

Volcanic Versus Impact Origin of Craters

Until the middle of the twentieth century, the impact origin of the craters of the Moon was not widely recognized. Because impact craters are extremely rare on Earth, it was easy for geologists to dismiss them as the major feature of lunar geology. They reasoned (perhaps unconsciously) that the craters we have on Earth are volcanic, and therefore the lunar craters must have a similar origin.

Figure 4.13 Grove K. Gilbert, the U.S. Geological Survey scientist who, in the 1890s, first presented cogent scientific arguments in favor of the impact hypothesis for the formation of lunar craters. *(U.S. Geological Survey)*

One of the first geologists to argue for an impact origin of lunar craters was G.K. Gilbert, a scientist with the U.S. Geological Survey in the 1890s (Figure 4.13). Gilbert pointed out that the large lunar craters, which are mountain-rimmed, circular features with floors generally below the level of the surrounding plains, are quite different in form from terrestrial volcanic craters, which tend to be smaller, deeper, and almost always occur at the tops of volcanic mountains (Figure 4.14). Gilbert thus concluded that the lunar craters were not volcanic.

Gilbert believed that the lunar craters were of impact origin, but he still had difficulty explaining why all of them are circular. Imagine throwing a stone into a sandbox: the pits you make are circular only when the stone hits the surface directly from above; otherwise the outlines of the sand box craters are more or less elliptical. The solution to this problem is readily seen, however, when we note the speed with which projectiles approach the Earth or the Moon. Attracted by gravity, the projectile strikes with at least escape velocity, which is 11 km/s for the Earth and 2.4 km/s for the Moon. The corresponding energy of impact leads to an explosion, and we know from experience with bomb and shell craters on Earth that explosion craters are circular. Thus, recognition of the similarity between *impact craters* and *explosion craters* removed the last important objection to the impact theory for the origin of lunar craters.

(a)   Terrestrial volcano

(b)   Lunar impact crater

Figure 4.14   Profiles of lunar impact craters (a) and terrestrial volcanic craters (b) are quite different from each other.

What caused the craters on the Moon?

The Cratering Process

Let us see how an impact at these high speeds produces a crater. When an impacting projectile strikes a planet at a speed of many kilometers per second, it penetrates two or three times its own diameter before stopping. During these few seconds, its energy of motion is transferred into a shock wave, which spreads through the target body, and into heat, which vaporizes most of the projectile and some of the surrounding target. The shock wave fractures the rock of the target, while the hot silicate vapor generates an explosion not too different from that of a nuclear bomb detonated at ground level (Figure 4.15).

Figure 4.15   Stages in the formation of an impact crater: (a) The impact; (b) The projectile vaporizes and a shock wave spreads through the lunar rock; (c) Ejecta are thrown out of the crater; (d) Most of the ejected material falls back to fill the crater and forms an ejecta blanket.

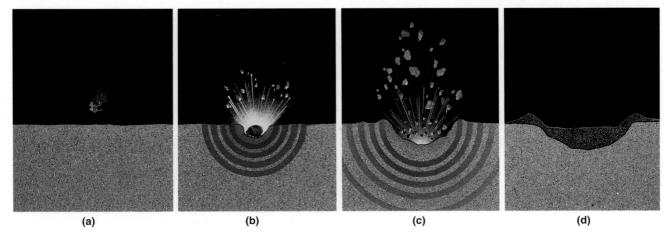

(a)   **(b)**   **(c)**   **(d)**

Figure 4.16 King Crater on the far side of the Moon, a fairly recent lunar crater 75 km in diameter, clearly showing most of the features associated with large lunar impact craters. *(NASA)*

The size of the crater that is excavated depends primarily on the energy of impact. On the Earth or the Moon, the crater is typically 10 times the diameter of the projectile. It makes little difference what the impacting body was made of; an icy comet is just as effective at producing craters as is a stony asteroid.

A high-speed impact of the sort described above leads to the characteristic kind of crater illustrated in Figure 4.16. The central cavity is initially bowl shaped ("crater" comes from the Greek word for "bowl"), but the gravitational rebound of the crust partially fills it in, producing a flat floor and sometimes creating a central peak. Around the rim, landslides create a series of terraces. The rim of the crater is turned up by the force of the explosion, so it rises above both the floor and the adjacent terrain. Surrounding the rim is a blanket of ejecta, consisting of material thrown out by the explosion. Additional, higher-speed ejecta fall at greater distances from the crater, often digging small secondary craters where they strike the surface. Some of these streams of ejecta can extend for hundreds or even thousands of kilometers from the crater, creating bright rays around some of the younger lunar craters.

Using Crater Counts to Date Planetary Surfaces

If a planet has little erosion or internal activity, like the Moon during the past 3 billion years, it is possible to use the numbers of impact craters counted on its surface to estimate the age of that surface. By *age* we mean the time since there was a major disturbance such as the volcanic eruptions that produced the lunar maria. This technique works because the rate at which impacts have occurred has been roughly constant for several billion years. Thus, in the absence of forces to eliminate craters, the number of craters is proportional to the length of time the surface has been exposed.

How are craters used to date planetary surfaces?

Estimating ages from crater counts is a little like the experience you might have walking along the sidewalk in a snowstorm when the snow has been falling steadily for a day or more. You may notice that in front of some houses the snow is deep, while next door the sidewalk may be almost clear. Do you conclude that less snow has fallen in front of Mr. Jones' house than Ms. Smith's? No, of course not. Instead, you conclude that Jones has recently swept the walk clean while Smith has not. Similarly, the numbers of craters indicate how long it has been since a planetary surface was last "swept clean."

Except for the Moon and the Earth, we have no planetary samples to permit exact ages to be calculated from radioactive decay. Without samples, the planetary geologist cannot tell if a surface is a million years old or a billion years old. The number of craters provides an important clue, however. On a given planet, the more heavily cratered terrain will always be the older (as we have defined age here). If we can calibrate the relationship between crater numbers and ages using what we know about the Moon, we may be able to estimate the ages of surfaces on other cratered planets, such as Mars or Mercury.

Cratering Rates

The rate at which craters are being formed on the Earth or the Moon cannot be measured directly, since the interval between large crater-forming impacts is longer than the span of human history. Remember that Meteor Crater in Arizona is 50,000 years old. However, the cratering rate can be estimated from the number of craters on the lunar maria. For the entire land area of the Earth, these calculations indicate that one or two 10-km craters should be produced every million years, and about one 100-km (or larger) crater every 50 million years. For the Moon the numbers are about 1/15 as great, as a consequence of the Moon's smaller total area.

ESSAY   Naming Features on the Planets

When the first maps of the Moon were published about 300 years ago, their authors undertook to provide names for lunar features. The dark maria were named for states of mind (Mare Tranquillitatis, or Sea of Tranquility), while the craters were given the names of prominent philosophers and scientists (Plato, Aristotle, Copernicus, Kepler). Curiously, only a very small crater was named for Galileo, whose telescopic discoveries initiated lunar geology. Russian names are prominent on the far side (Sea of Moscow, Tsiolkovsky Crater), since the first photographs of these regions were made by Soviet spacecraft. Today, by international agreement, small lunar craters are named for scientists from many nations and cultures.

As spacecraft were launched to other planets, it became apparent that some organized system was required to establish a distinctive nomenclature for each world. This responsibility was assumed by the International Astronomical Union, which represents astronomers from more than 100 countries.

The three most extensively mapped planets are Mars, Venus, and Mercury. For Mars, the craters are named for deceased scientists, especially astronomers who have contributed to the study of the planets (Hale, Herschel, Lowell, Milankovic). Other kinds of features have other schemes for assigning names; for example, the dry river channels are named for the name "Mars" in different languages (Ares, Huo Hsing, Nirgal, Simud). Because Venus is the only planet with a feminine name, its craters are named for famous women of history (Cleopatra, Joliot-Curie, Mead, Tubman, Tutibu). The continents of Venus take their names from the goddess of love and fertility in ancient cultures (Ishtar, Aphrodite). For Mercury, craters are named for great figures in world art and literature (Goethe, Hiroshiga, Imhotep, Rodin, Sinan, Tolstoj).

The giant planets of the outer solar system have no solid surfaces and therefore require no names. However, the Voyager photographs of the satellites of the outer planets have stimulated a number of nomenclature plans, primarily derived from the mythological context of the satellite name itself. For example, the features on Jupiter's satellite Callisto are named for gods of ice and cold in different cultures, while the volcanoes of Io are named for gods of fire.

Providing appropriate names for features on the planets is a pretty esoteric business, and most planetary maps will be used seriously by only a few hundred people. Someday, however, humans may establish permanent bases on the Moon or Mars; when that time comes, the names assigned by astronomers today may indeed become household terms known to almost everyone.

The individual maria have accumulated their current numbers of craters over timespans of 3.3 to 3.8 billion years. Earlier than 3.8 billion years ago, however, the impact rates must have been a great deal higher; this conclusion is evident when we compare the number of craters on the lunar maria with those on the highlands. Typically, there are 10 times as many craters on the highlands as there are on a similar area of maria, yet, as we have seen from radioactive dating of lunar samples, the highlands are only a little older than the maria, typically 4.2 billion years rather than 3.8 billion years. If the rate of impacts had been constant, the highlands would have had to be at least 10 times older. Therefore, they would have formed 38 billion years ago—long before our universe began.

Thus, the Moon experienced a period of much heavier bombardment previous to 3.8 billion years ago, as illustrated in Figure 4.17. This

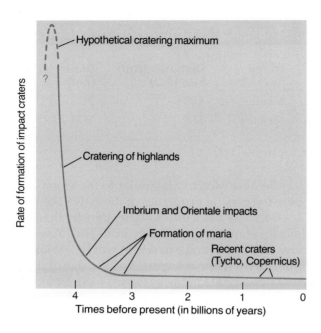

Figure 4.17 Schematic diagram of cratering flux history for the Moon, covering the past 4.3 billion years.

heavy bombardment produced most of the craters we see today in the highlands. Was this a local event, or did similar high impact rates apply throughout the solar system? A partial answer is provided by Voyager pictures of the satellites of Jupiter and Saturn, which reveal some surfaces to be as heavily cratered as the lunar highlands. Because it is impossible to accumulate this many craters *at the current rate* within the lifetime of the solar system, there must have been a period of high bombardment in the outer solar system as well.

4.5
MERCURY: THE MOON'S DECEPTIVE "TWIN"

The Moon interests us not only because it is the nearest and best studied other world, but also because it is an example of the class of small planetary bodies without atmospheres. Another small, airless world is Mercury.

General Properties of Mercury

Recall that Mercury is the innermost planet and can be viewed only near the Sun in the sky. It is never above the horizon in the dark hours of the night, and so it has proved to be a difficult object for astronomers to study. Even such a basic property as its rotation period eluded generations of telescopic observers, and its rotation was determined only when radar beams were directed toward Mercury in the 1960s. These radar measurements showed that Mercury rotates exactly 1.5 times for each orbit around the Sun; that is, its day is two-thirds as long as its year.

How does the interior of Mercury compare with that of the Moon?

Name	Distance from Sun (AU)	Diameter (km)	Mass (Earth=1)	Density (g/cm³)
TABLE 4.2	Comparison of Moon and Mercury			
Moon	1.00	3476	.012	3.3
Mercury	0.39	4878	.055	5.4

Because Mercury is similar to the Moon in so many ways, we compare the general properties of these two objects in Table 4.2.

Mercury has a much higher density than the Moon; it is composed mostly of metal, presumably iron. The planet's metallic core has a diameter of 3500 km and extends to within 700 km of the surface. We could think of Mercury as a metal ball the size of the Moon surrounded by a rocky crust 700 km thick (Figure 4.18). Unlike the Moon, Mercury has a weak magnetic field. The existence of this field suggests that at least a part of the metal core must be liquid in order to generate the observed magnetism. Thus, despite superficial similarities, the interior of Mercury is strikingly different from that of the Moon—with Mercury composed primarily of metal, and the Moon severely depleted of metal.

Geology of Mercury

The first closeup look at Mercury came in 1974, when the U.S. Mariner 10 spacecraft passed 9500 km from the surface of the planet. The photos transmitted to Earth revealed a heavily cratered world with a geological

Figure 4.18   The interior of Mercury is dominated by a metallic core about the same size as our Moon.   *(University of Arizona, courtesy of Robert Strom)*

(a)

(b)

Figure 4.19 The planet Mercury as photographed by Mariner 10 in 1974. (a) General view of the approach hemisphere. (b) Details of the heavily cratered surface. *(NASA/JPL)*

history similar to that of the Moon (Figure 4.19). Much of the surface has been saturated with impacts, including indications of a few large basins formed in the distant past. There is also some evidence to support extensive volcanism several billion years ago, although there are no clearly defined lava plains like the lunar maria.

One of the features that distinguishes Mercury from the Moon is a number of cliffs or scarps that stretch across the surface for more than 100 km and that appear to have formed after many of the impact craters

Figure 4.20 Discovery scarp on Mercury. This wrinkle, nearly 1 km high and more than 100 km long, proves that the crust was compressed after the formation of many of the craters on Mercury. *(NASA/JPL)*

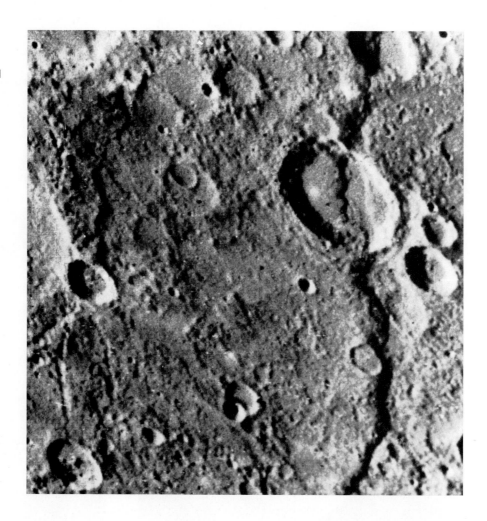

How does the surface of Mercury compare with that of the Moon?

(Figure 4.20). These cliffs, often more than 1 km in height, seem to be the result of crustal compression. Apparently this planet shrank, wrinkling the crust, and it did so after most of the craters on its surface were formed. If the standard cratering chronology applies to Mercury, this shrinkage must have taken place during the past 4 billion years. If we understood better the interior structure and composition of this planet, we could probably calculate how internal changes in temperature might have led to this global compression, which has no counterpart on the Moon or the other terrestrial planets.

Although it has been two decades since any spacecraft was launched toward Mercury, astronomers continue to study this planet from the Earth. An intriguing recent discovery is the presence of regions near both poles that are highly reflective of radar signals. Since water ice has similar reflective properties, it appears that we have discovered polar deposits of ice on Mercury, despite the overall high temperatures on the planet, which can climb to above 600 K near the equator. Curiously, there are no similar polar caps on our own Moon.

4.6
ORIGIN OF THE MOON AND MERCURY

Understanding the origin of the Moon has proved to be difficult for astronomers. Part of the difficulty is that we know so much about the Moon; the great wealth of data, particularly on the details of its composition, present a challenge to any simplified theory of origins. In addition, it is only recently that scientists have begun to appreciate the violent conditions that applied in the solar system at the birth of the planets—conditions that link the origin of the Moon with that of Mercury.

Three Simple Theories

Most of the theories that have been suggested for the Moon's origin follow one of three general ideas: (1) that the Moon was once part of the Earth but separated from it early in their history (the *fission theory*); (2) that the Moon formed together with (but independent of) the Earth, as we believe many satellites of the outer planets formed (the *sister theory*); and (3) that the Moon formed elsewhere in the solar system and was captured by the Earth (the *capture theory*).

Unfortunately, there are fundamental problems with each of these ideas. The first idea, that the Moon split off from the Earth, fails when subjected to detailed analysis using modern computers. Further, it is difficult to understand how a satellite made out of terrestrial-type material could have lost both its metals and its volatiles, as has the Moon.

Perhaps the easiest theory to reject is the capture theory. The difficulty is primarily that no one knows of any way that the early Earth could have captured a large satellite from elsewhere. One object approaching another cannot go into orbit around it without a substantial loss of energy—this is the reason that spacecraft destined to orbit other planets are equipped with retrorockets. Besides, there are too many detailed compositional similarities between the Earth and the Moon to justify seeking a completely independent origin. Scientists are therefore left with the sister theory—that the Earth and Moon formed together.

The primary problems with the sister theory arise when we look at the differences in composition between the Moon and the Earth. How could the Moon, forming out of the same raw material as the Earth, avoid accumulating its share of metals and volatiles? One way out of this difficulty is to make the Moon out of the same material as the Earth's *mantle,* after the Earth had differentiated and the metals settled into its core. Out of these concepts has come a fourth hypothesis called the **giant impact theory.**

What is the giant impact theory for the origin of the Moon?

The Giant Impact Theory

The giant impact theory envisions the Earth being struck obliquely by one of the Mars-size objects that we think were still loose in the inner solar system after most of the debris had already been gathered into the planets. The Earth by then would have differentiated, with most of the

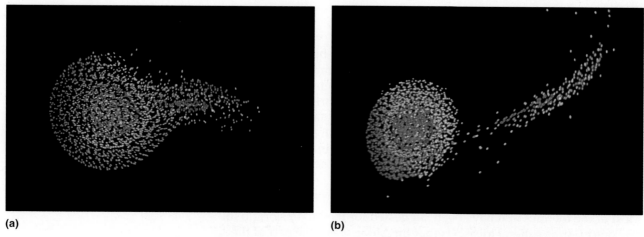

(a)

(b)

Figure 4.21 Two stages in a computer-generated simulation of an oblique giant impact on the Earth. The metal cores of the Earth and the Mars-sized projectile are shown in blue, while yellow and orange represent the silicate material of the mantle and crust. (a) Shortly after the impact, a plume of material from both the projectile and the Earth's mantle are projected into space. (b) Later, part of the ejected material begins to orbit the Earth. Note that the iron is moving more slowly and will sink again into the core of the Earth, leaving only silicate material in orbit to form the Moon. *(Courtesy of A.G.W. Cameron, Harvard Center for Astrophysics)*

metal in its core. Such an impact would nearly have shattered our planet (Figure 4.21). Calculations show that the energy of impact would eject into space a vast amount of melted or vaporized rock from the mantles of both the Earth and the impacting body. Part of this hot vapor could condense into a ring of material in orbit around the Earth, eventually forming the Moon.

The giant impact hypothesis offers potential solutions to most of the major problems raised by the chemistry of the Moon. First, because the raw material for the Moon is derived from the mantles of the Earth and the projectile, the absence of metals is easily understood. Second, most of the volatile elements could have been lost during the high-temperature phase following the impact. But by making the Moon primarily of terrestrial mantle material, it is also possible to understand the similarities between Earth and Moon.

This explanation for the origin of the Moon is representative of renewed interest in the role of large impacts early in solar system history. Current theories suggest that there were many planet-sized projectiles in the solar system during its first few million years of existence. More than one planet could have been catastrophically disrupted and then reaccumulated out of the resulting debris. These ideas are speculative, and it is difficult to prove the existence of such catastrophic events so long in the past. But the very existence of our Moon with its strange chemistry provides strong evidence that the formation of the planets was not accomplished without a measure of violence and chaos.

Formation of Mercury

The surface similarities of Mercury and the Moon are easy to understand. Both have rocky mantles and crusts, and both have been subject to billions of years of cosmic impacts. Because of their small sizes, Mercury and the Moon have each cooled internally, so there has been little geological activity to erase the long accumulation of craters. With no atmospheres, neither has experienced erosion of its surface.

In contrast, Mercury and the Moon both have anomalous bulk compositions, relative to the Earth: The Moon is depleted in metals, and Mercury is depleted in silicates. How did Mercury lose so much of its rocky material?

The most probable explanation for Mercury's loss of silicates is similar to the explanation for the Moon's lack of a metal core. Mercury is likely to have experienced several giant impacts very early in its youth, and one or more of these may have torn away a fraction of its mantle and crust, leaving a body dominated by its iron core. Paradoxically, the same process can lead to one object being depleted in metal (the Moon), and another object being depleted in silicate (Mercury), depending on the energy of the collision. This planet, like the Moon, bears testimony to the violence that must have characterized the solar system during its first few million years.

How can impacts explain the peculiar compositions of both the Moon and Mercury?

▶ **SUMMARY**

4.1 As the Moon orbits the Earth each month, its changing phases result from different angles of illumination by the Sun. The Moon rotates and revolves with the same period, always keeping the same face toward the Earth. Through a telescope, astronomers can achieve a **resolution** of about 1 km on the Moon, which is sufficient to reveal a great deal of topographic detail.

4.2 Most of what we know about the Moon derives from the Apollo program, including 400 kg of lunar samples still being intensively studied. The Moon has 1/80 the mass of the Earth and is depleted in metals and **volatiles.** It is made almost entirely of silicates like those found in the Earth's mantle and crust.

4.3 Lunar rocks are dated by their radioactivity, using atoms with different **half-lives.** Like the Earth, the Moon was formed 4.5 billion years ago. The heavily cratered **highlands** are made of rocks up to 4.4 billion years old. The darker volcanic plains of the **maria** were erupted between 3.3 and 3.8 billion years ago. Generally, the surface is dominated by impacts, including continuing small impacts that produce its fine-grained soil.

4.4 A century ago Gilbert suggested that the lunar craters were of impact origin, although the cratering process was not well understood until more recently. High-speed impacts produce explosions and excavate craters with raised rims, ejecta blankets, and (often) central peaks. Cratering rates have been roughly constant for the past 3 billion years but were much greater prior to then. Crater counts can be used to derive approximate ages for geologic features on the Moon and other planets.

4.5 Mercury is similar to the Moon in having no atmosphere and a heavily cratered surface; it differs in having a very large metal core. Its day (rotation period) is two-thirds the length of its year. The most surprising geological features are long scarps or cliffs that indicate the planet has shrunk.

4.6 There are three standard theories for the origin of the Moon: the fission theory, the sister theory, and the capture theory. All have problems, and recently they have been supplanted by the **giant impact theory,** which ascribes the origin of the Moon to the impact of a Mars-sized projectile with the Earth 4.5 billion years ago. Similarly, Mercury is now thought to have lost much of its original silicate mantle through a giant impact early in its history.

▶ REVIEW QUESTIONS

1. Explain what causes the phases of the Moon. If the Moon's rotation period is the same as its period of revolution, explain why we can never see one hemisphere.

2. What is the composition of the Moon, and how does it relate to the composition of the Earth and the planet Mercury?

3. Outline the chronology of the Moon's geological history.

4. Explain how high-speed impacts form craters, and indicate the characteristic features of impact craters.

5. Summarize the four main theories for the origin of the Moon.

6. Compare and contrast Mercury with the Moon.

▶ THOUGHT QUESTIONS

7. Describe the appearance of the Earth as seen from the Moon over the course of a month. How does the Earth move in the sky, and what (if any) phases does it pass through?

8. The lunar highlands have about ten times more craters on a given area than do the maria. Does this mean that the highlands are ten times older? Explain your reasoning.

9. Why are the lunar mountains smoothly rounded rather than having sharp, pointed peaks (the way they had almost always been depicted before the first lunar landings)?

10. Why did it take so long for geologists to recognize that the lunar craters had an impact origin rather than being volcanic?

11. Explain the evidence for a period of heavy bombardment on the Moon about 4 billion years ago. What might have been the source of this high flux of impacting debris?

12. Summarize the main weakness of each of the traditional theories for the origin of the Moon: the fission theory, the capture theory, and the sister theory.

13. The Moon has too little iron, Mercury too much. How can both of these anomalies be the result of giant impacts? Explain how the same process can yield such apparently contradictory results.

▌ PROBLEMS

14. The Moon was once closer to the Earth than it is now. When it was at half its present distance, how long was its period of revolution?

*15. The Moon requires about one month (0.08 year) to orbit the Earth. Its distance is about 400,000 km (0.0027 AU). Use Kepler's third law, as modified by Newton, to calculate the mass of the Earth, relative to that of the Sun.

*16. In any one mare there are a variety of rock ages, spanning typically about 100 million years. The individual lava flows seen by the Apollo 15 astronauts were about 4 m thick. Estimate the average interval between lava flows if the total depth of the lava in the mare is 2 km.

▌ SUGGESTIONS FOR ADDITIONAL READING

Books

On the Moon

Chapman, C. *Planets of Rock and Ice.* Scribners, 1982. Introduction to the comparative study of the Moon and other planets, written by a leading planetary scientist.

Cooper, H. *Moon Rocks.* Dial Press, 1970. A journalist's account of the Apollo program.

Hockey, T. *The Book of the Moon.* Prentice Hall, 1986. Introduction by an astronomer/educator.

French, B. *The Moon Book.* Penguin Press, 1977. Authoritative and readable introduction.

Mailer, N. *Of a Fire on the Moon.* Little Brown, 1970. Interesting account of the Apollo program by a major contemporary writer.

Murray, C. & Cox, C. *Apollo: The Race to the Moon.* Simon & Schuster, 1989. Journalists' account, assembled from many interviews.

On Mercury

Strom, R. *Mercury: The Elusive Planet.* Smithsonian Inst. Press, 1987. Nontechnical introduction, by a planetary astronomer.

Murray, B. & Burgess, E. *Flight to Mercury.* Columbia U. Press, 1977. Summary of the Mariner mission.

Articles

On the Moon

Morrison, D. & Owen, T. "Our Ancient Neighbor: The Moon" in *Mercury,* May/June 1988, p. 66, and July/Aug. 1988, p. 98.

Brownlee, S. "A 'Whacky' New Theory of the Moon's Birth" in *Discover,* Mar. 1985, p. 65.

Benningfield, D. "Mysteries of the Moon" in *Astronomy,* Dec. 1991, p. 50.

Hurt, H. "I'm at the Foot of the Ladder" in *Astronomy,* July 1989, p. 22. Twentieth anniversary retrospective about the first moon landing.

Weaver, K. "First Explorers of the Moon: Apollo 11" in *National Geographic,* Dec. 1969.

Chaikin, A. "A Guided Tour of the Moon" in *Sky & Telescope,* Sept. 1984, p. 211. Observing instruction for beginners.

On Mercury

Chapman, C. "Mercury's Heart of Iron" in *Astronomy,* Nov. 1988, p. 22.

Cordell, B. "Mercury: The World Closest to the Sun" in *Mercury,* Sept./Oct. 1984, p. 136.

Strom, R. "Mercury: The Forgotten Planet" in *Sky & Telescope,* Sept. 1990, p. 256.

EARTHLIKE PLANETS: MARS AND VENUS

Mars as photographed from the approaching Viking spacecraft in 1976. *(NASA/JPL)*

There are two planets that closely resemble our own: Mars and Venus. These are also the two nearest planets and the two most accessible to spacecraft. Not surprisingly, these fascinating worlds have been the target of most planetary exploration missions.

In this chapter we discuss some of the results of three decades of scientific exploration of Mars and Venus. What has been revealed to us is more than just a description of two new worlds. A scientific revolution has taken place, as astronomers and geologists have been able to compare these two planets in detail with each other and with the Earth. These comparisons yield some understanding of the common processes that are at work on all three planets, as well as highlighting other aspects that are unique to each.

5.1
THE NEAREST PLANETS

Mars and Venus are among the brightest objects in the night sky, and their orbits bring both of them much closer to the Earth than any of the other planets. The average distance from Mars to the Sun is 227 million km, and the average distance from Venus to the Sun is 108 million km; both planets come within about 40 million km of Earth at their closest. Typical spacecraft flight times to each are less than one year.

Appearance of Mars and Venus

Seen through a telescope, Mars is both tantalizing and disappointing. The planet is distinctly red in color because of the presence of iron oxides in its soil. At its nearest, the best telescopic resolution obtainable is about 100 km, or about the same as the Moon seen with the unaided eye. At this resolution, however, no hint of topographic structure can be detected: no mountains, no valleys, not even impact craters (Figure 5.1). On the other hand, the bright polar caps of the red planet can easily be seen, together with dusky surface markings that gradually change in outline and intensity from season to season.

Figure 5.1   These are among the best Earth-based photos of Mars, taken in 1988 when the planet was exceptionally close to the Earth. The polar caps and the detailed dark surface markings are evident, but no topographic features are visible. *(Steve Larson, University of Arizona)*

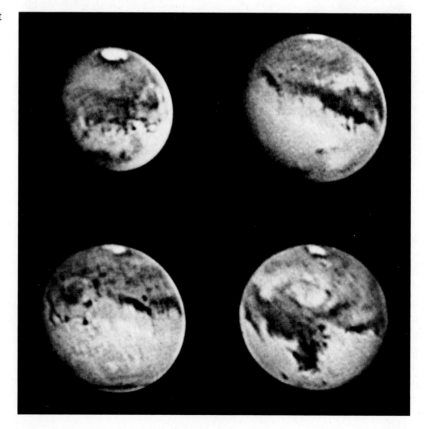

For a few decades around the turn of the twentieth century, some astronomers believed that they saw evidence of a civilization of intelligent beings on Mars. The controversy began in 1877, when the Italian astronomer Giovanni Schiaparelli announced the discovery of long, faint, straight lines on Mars that he called *canale,* or channels. In English-speaking countries the term was translated as canals, implying an artificial origin.

Until his death in 1916, the most effective proponent of intelligent life on Mars was Percival Lowell, self-made American astronomer and member of the wealthy Lowell family of Boston (Figure 5.2). In order to pursue his study of the martian canals, he established the Lowell Observatory in Flagstaff, in the Territory of Arizona. An effective author and public speaker, Lowell made a convincing case for intelligent Martians, who he believed had constructed huge canals to carry water from the polar caps in an effort to preserve their existence in the face of a deteriorating climate.

Lowell's argument for intelligent life hinged on the reality of the canals, a matter that remained in dispute among astronomers. The canals were always difficult to study, glimpsed only occasionally as

Figure 5.2   Percival Lowell in about 1910 observing with his 24-inch telescope at Flagstaff.   *(Lowell Observatory)*

Figure 5.3 The clouds of Venus, as photographed in ultraviolet light by the Pioneer Venus orbiter. In visible light, virtually no structure of any kind is visible. *(NASA)*

atmospheric conditions caused the tiny image of Mars to shimmer in the telescope. Lowell saw them everywhere, but many other observers were skeptical. When larger telescopes failed to confirm the existence of canals, the skeptics seemed to be vindicated. Astronomers had given up the idea by the 1930s, although it persisted in the public consciousness until the first spacecraft photographs clearly showed that there were no martian canals. Now it is generally accepted that the canals were an optical illusion, the result of the human mind's tendency to see order in random features glimpsed dimly at the limits of the eye's resolution.

Venus, in contrast to Mars, exhibits little of interest through even the largest telescope. As first discovered by Galileo, it displays a full range of phases as it revolves about the Sun, but the planet's actual surface is not visible because it is shrouded by dense clouds. These clouds reflect about 70 percent of the incident sunlight, a circumstance that contributes greatly to the planet's brightness but frustrates efforts to study the underlying surface (Figure 5.3).

Rotation

The presence of permanent surface markings enables us to determine the rotation period of Mars with great accuracy; its day is 24^h, 37^m, 23^s long, just a little greater than the rotation period of the Earth. This high precision is not obtained by watching Mars for a single rotation, but by noting how many turns it makes during a long period of time. Good observations of Mars date back for more than 200 years, a period during which tens of thousands of martian days have passed. The rotational pole of Mars has a tilt of about 25 degrees, similar to that of the Earth's pole. Thus Mars experiences seasons very much like those on Earth.

Because of the longer martian year, however, each season there lasts about six of our months.

Because no surface detail can be seen on Venus, the rotation period can be found only by using radar, a technique first applied to planetary studies in the 1960s. The approach is to beam a powerful radio signal to Venus; the radio waves pass through the atmosphere and are reflected from the surface. Individual surface features can thus be identified and the rotation period derived. Surprisingly, the radar observations show Venus to rotate from east to west—in the reverse direction from the rotation of most other planets—in a period of 243 days. The rotation period of Venus is about 19 days longer than its period of revolution about the Sun. The length of a day on Venus—the time between successive noons—is 117 Earth days.

Exploration by Spacecraft

The first successful spacecraft to Mars, called Mariner 4, flew by the planet in 1965 and radioed 22 photos to Earth. These pictures showed an apparently bleak landscape with abundant impact craters. Later Mariner spacecraft mapped the entire surface of Mars at a resolution of about 1 km and discovered a great variety of geological features, including volcanoes, huge canyons, intricate layers on the polar caps, and channels that appeared to have been cut by running water.

The Mariner voyages set the stage for Viking, one of the most ambitious and successful of all U.S. planetary missions, consisting of two orbiters and two landers. The Viking 1 lander touched down on the surface of Chryse Planitia (the Plains of Gold) on July 20, 1976, exactly seven years after Neil Armstrong's historic first step on the Moon. Two months later, Viking 2 landed with equal success on another plain farther north, called Utopia. Most of the information about Mars in this chapter is derived from the four Viking spacecraft.

While the U.S. was concentrating its efforts on Mars, the U.S.S.R. took the lead in the exploration of Venus in a series of missions called Venera, the Russian word for Venus. The early Venera entry probes were crushed by the high pressure of the atmosphere before they could reach the surface, but in 1970, Venera 7 successfully landed and broadcast data from the surface for 23 minutes before succumbing to the high surface temperature. Additional successful probes and landers followed, photographing the surface and analyzing the atmosphere and soil. In the 1980s, the Russian program broadened its scope: Venera 15 and Venera 16 were orbiters that mapped the surface by radar, and in 1985 the VEGA spacecraft (VEGA = Venera + the Russian equivalent of Halley) successfully deployed two instrumented balloons into the atmosphere in addition to landing on the surface, while the main VEGA spacecraft continued on to Comet Halley.

The most productive of all missions to Venus was the U.S. radar orbiter called Magellan, which arrived at Venus in 1990. By late 1992, Magellan had mapped 99 percent of the surface at a resolution of 100 m, returning a greater volume of data than all other planetary missions

Figure 5.4   The Magellan spacecraft was able to use imaging radar to create realistic views of the surface of Venus. This computer-generated image shows two large volcanoes: Sapas Mons in the center and Maat Mons on the horizon. Vertical relief is exaggerated by ten times and a false color has been added. *(NASA/JPL)*

combined. The resulting maps of the surface allowed the first detailed comparisons between the geology of Venus and that of our own planet (Figure 5.4).

Basic Properties of Mars and Venus

How do Venus and Mars compare with the Earth and Moon?

Before discussing Mars and Venus individually, it is appropriate to compare some of their basic properties. As a planet, Mars is rather small, with only 11 percent the mass of the Earth. It is larger than either the Moon or Mercury, however, and unlike them it retains a thin atmosphere. Venus is much larger, with a mass 0.82 times the mass of the Earth and an almost identical diameter to our planet. However, Venus has a high surface temperature and a much more massive atmosphere

TABLE 5.1 Properies of Earth, Venus, and Mars

Properties	Earth	Venus	Mars
Semi-major axis (AU)	1.00	0.72	1.52
Period (year)	1.00	0.61	1.88
Mass (Earth = 1)	1.00	0.82	0.11
Diameter (km)	12756	12102	6790
Density (g/cm^3)	5.5	5.3	3.9
Surface gravity (Earth = 1)	1.00	0.91	0.38
Escape velocity (km/s)	11.2	10.4	5.0
Rotation period	23.9 hours	−243 days	24.6 hours
Surface area (Earth = 1)	1.00	0.94	0.28
Atmospheric pressure (bar)	1.00	90	0.007

than the Earth's, with a surface pressure nearly a hundred times greater than ours and about 10,000 times greater than that of Mars.

Table 5.1 summarizes some of the basic data for Mars and Venus, in comparison with similar information for the Earth.

In the following sections we will look at the two planets individually, beginning with their geology. At the end of this chapter we compare Mars and Venus with each other and with the Earth, the Moon, and Mercury.

5.2
THE CRUST OF VENUS

The clouds of Venus are opaque, blocking our view of the surface beneath. Only through the use of radar can we map the topography of the surface and obtain the information required to study the geology of the planet. Imaging radar systems use highly sophisticated data processing with a powerful radar beam to reconstruct a picture of the surface, very similar in appearance to a satellite photograph of the Earth or Mars.

Global Perspective

Venus is often described as the Earth's twin because both planets have about the same size, mass, and density. The similarities in size and composition led us to expect similar geological histories for the two planets. The Magellan data do indeed show many common surface features, but there are distinctions as well. Perhaps Venus should be thought of as a fraternal twin, rather than an identical twin, of Earth.

As we saw in Chapter 3, the Earth's surface is divided into continents and ocean basins. Venus, in contrast, has no basins; most of its sur-

Figure 5.5 Global view of Venus.

Figure 5.6 Large impact craters on Venus. The rough crater rims and ejecta are excellent radar reflectors and thus appear brighter than the surrounding smooth volcanic plains in a radar image. Like most craters on Venus, these show little evidence of erosion. *(NASA/JPL)*

face consists of rolling volcanic plains (Figure 5.5). Only about 20 percent of the surface lies at higher elevations, primarily in two large continents: Aphrodite, an Africa-sized mass that stretches around the equator for about one-third the circumference of the planet and Ishtar, a smaller Australia-size continent at high northern latitude. Ishtar includes the Maxwell Mountains, which rise about 11 km above the lowland plains—similar to the height of Mt. Everest on Earth relative to sea level. Adjacent to these mountains is a broad plateau that not only resembles the Tibetan Plateau but may also have a similar origin as an area uplifted by pressures in the crust of the planet.

The number of impact craters provides a measure of the age of the surface and thus an indication of the level of geologic activity. We saw that the moderately heavily cratered lunar maria have ages of more than 3 billion years, while impact craters are rare on the younger and more dynamic Earth. One of the first results from the Magellan radar data was a determination of the number of craters on Venus. There are no small craters, since small projectiles are blocked from the surface by the planet's thick atmosphere. However, for craters larger than 30 km in diameter, the atmosphere does not shield the surface, and crater counts can be used to calculate an accurate age.

A total of almost 1000 impact craters exist on Venus; the largest is Mead, with a diameter of 280 km. These craters are generally sharp and fresh looking, because of the much lower levels of erosion than exist on the Earth (Figure 5.6). There are more craters on a given area of Venus than on an equivalent area of the Earth, but considerably fewer than on the Moon, indicating an intermediate level of activity between the Earth and Moon. Calculations show that the average surface age on the volcanic plains of Venus is about 600 million years, as opposed to less than 100 million years for the Earth's ocean basins.

The topography and generally low levels of geologic activity indicate that Venus is not experiencing large-scale plate tectonics like the Earth. There are no young ocean basins and no deep trenches that might suggest subduction zones. Despite of its large size, Venus is not as dynamic as our own planet.

Geology of Venus

The 100-m resolution of the Magellan images permits us to map a wide variety of geological features on Venus. Most of these fall into two groups: volcanic and tectonic. Virtually absent are features caused by erosion or deposition of sediment, and there is no evidence for water, either past or present.

Most of the crust of Venus consists of volcanic plains. In addition, there are many individual volcanoes ranging from small cones the size of a shopping-mall parking lot up to large mountains 300 km across. The most distinctive of these volcanoes are the pancake domes (Figure 5.7), which are low domes about 2 km high and 20–40 km in diameter. They are remarkably circular in shape, probably the product of short-lived eruptions of viscous lava. At the opposite extreme, Venus has ancient lava channels up to 7000 km long, apparently produced by eruptions of very fluid, fast-flowing lava.

Tectonic features are those resulting from compression or tension in the crust of a planet. Folded tectonic mountains form where the crust is compressed, while tectonic valleys (fundamentally large cracks) can form where parts of the crust are separating. Most of the Earth's geology is tectonic, resulting from the large-scale action of plate tectonics.

What are volcanic and tectonic processes in geology?

Figure 5.7 Pancake-shaped volcanoes on Venus. These circular domes, each about 25 km in diameter, are the result of eruptions of highly viscous lava. *(NASA/JPL)*

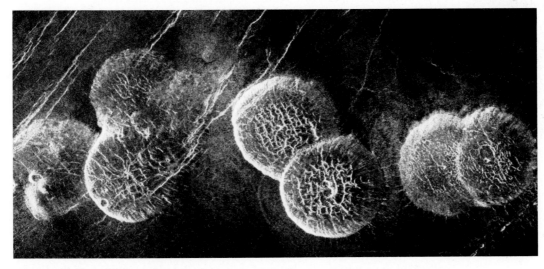

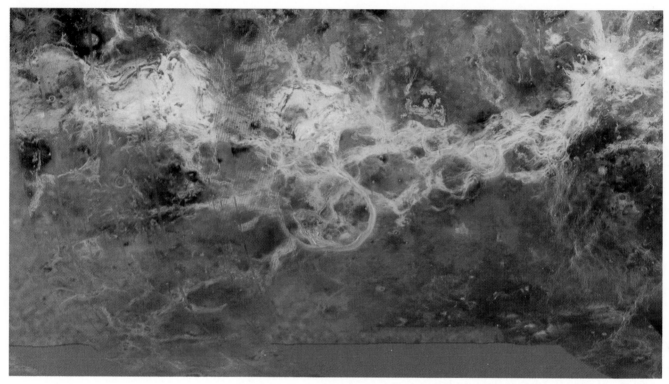

Figure 5.8 The Aphrodite continent of Venus stretches nearly the full width of this Magellan map, which shows about one-third of the circumference of the planet. Continental highlands are bright because of their fractured, folded surfaces. The orange hue is false color, added to accentuate brightness contrast. *(NASA/JPL)*

How does the geology of Venus compare with that of the Earth?

Although there is little or no plate motion on Venus, the crust is stressed by internal forces, and the results are seen in the radar images.

The mountainous, continental regions of Venus are tectonic in origin (Figure 5.8). Compression has lifted these areas and wrinkled the surface into complex mountain ridges. Because there is no water or ice erosion and no overlying vegetation, these ridges are much easier to see on Venus than on Earth. The lowland plains have also been modified by tectonics, forming patterns of small cracks and ridges. Figure 5.9 shows a striking grid pattern where the separation between ridges and cracks is about 1 km.

Among the most distinctive geological features on Venus are the large circular forms called coronae (Figure 5.10). The coronae lie above plumes of lava that rose through the mantle but never reached the surface. These coronae are both volcanic and tectonic in origin, consisting of tectonic cracks initiated by the motion of hot rock beneath the surface. Directly over the rising plume the surface is domed upward, while downflowing material (like small subduction zones) produces shallow trenches encircling the central dome. This activity has sometimes been called "blob tectonics," to distinguish it from the plate tectonics of the Earth.

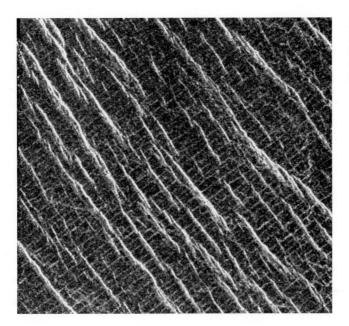

Figure 5.9 This region of the Lakshmi Plains has been fractured to produce a regular grid of tectonic cracks and ridges. The width of the radar image is 40 km.   *(NASA/JPL)*

On the Surface

The successful Venera landers found themselves on an extraordinarily inhospitable planet, with a surface pressure of 90 bars and a temperature of 730 K. (In American units, this is 860° F. For comparison, the highest temperature in a kitchen oven is about 500° F.) The surface of Venus is hot enough to melt lead and zinc.

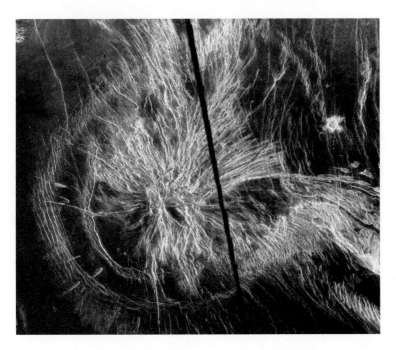

Figure 5.10 Pandora Corona, a circular feature 350 km in diameter produced when volcanic pressure built up below the surface of Venus. The black streak is due to missing radar data.   *(NASA/JPL)*

What is it like on the surface of Venus?

Despite these unpleasant conditions, the Soviet spacecraft have photographed their surroundings and collected surface samples for chemical analysis. The rock in the landing areas is igneous, primarily basalts, as we might expect from the generally volcanic nature of the planet's geology. Examples of the Venera photographs are shown in Figure 5.11. Each picture shows a desolate, flat landscape with a variety of rocks, some of them probably exposed lava flows.

The Sun cannot shine directly through the heavy, opaque clouds, but the surface is fairly well lit by diffused light. The illumination is about the same as that on Earth under a very heavy overcast, but with a strong red tint, since the massive atmosphere blocks blue light. The weather is unchanging, with calm winds and a temperature of 730 K. Because of the heavy blanket of clouds and atmosphere, one spot on the surface of Venus is similar to any other as far as weather is concerned. The explanation for the uniform conditions and blistering temperatures is to be found in the atmosphere, which we discuss in Section 5.5.

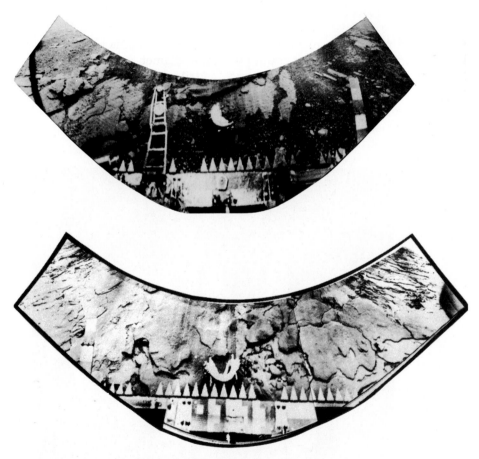

Figure 5.11   Views of the surface of Venus from the Soviet Venera landers. (*NASA/Brown University*)

5.3
THE CRUST OF MARS

As we did with Venus, we introduce the geology of Mars with a look at its large-scale surface topography. Does Mars have continents and deep basins like the Earth, indicating a possible history of plate tectonics? Does it have large impact basins like the lunar maria? Or is its surface dominated by volcanic plains like those of Venus? The answers, as revealed from spacecraft mapping, are a partial "yes" to all three questions. Despite some similarities with other planets, Mars is a unique world.

Global Properties

The surface of Mars divides into two major terrains. Approximately one-half of the planet, lying primarily in the southern hemisphere, consists of ancient cratered uplands, analogous to the lunar highlands. The other half, which is primarily in the north, contains younger, lightly cratered volcanic plains at an average elevation about 4 km lower than the uplands. These lowlands resemble the maria of the Moon or the widespread lava plains of Venus.

Unlike Earth or Venus, the surface of Mars is old enough to retain a few large impact basins, all to be found in old southern uplands. The largest basin, called Hellas, is about 1800 km in diameter and 6 km deep, larger than the largest basin on the Moon. Smaller but better preserved is the Argyre basin (Figure 5.12), with a diameter of 700 km. Both basins are surrounded by mountains formed from impact explosions.

Figure 5.12 The Argyre impact basin of Mars, about 700 km in diameter, photographed by the Viking orbiter. *(NASA/JPL)*

In addition to the basic division of the planet into old uplands and younger volcanic plains, Mars displays a continent the size of North America that rises nearly 10 km above its surroundings. This is the Tharsis continent, a volcanically active region crowned by four great volcanoes that rise another 15 km into the pink martian sky. This large volcanic continent has no direct counterpart on either Venus or Earth.

Volcanoes on Mars

The rolling, volcanic plains of the northern hemisphere of Mars look very much like the lunar maria, and they have about the same numbers of impact craters. Like the lunar maria, they probably formed between 3 and 4 billion years ago. Apparently Mars experienced extensive volcanic activity at about the same time that the Moon did, producing similar basaltic lavas.

The largest volcanic mountains of Mars are found in the Tharsis area, although many smaller volcanoes dot much of the surface of the younger northern half of the planet. The most dramatic is Olympus Mons (Mount Olympus), on the northwest slope of Tharsis (Figure 5.13). It is more than 500 km in diameter, with a summit 25 km above the surrounding plains. The volume of this immense volcano is nearly 100 times greater than that of the largest terrestrial volcano, Mauna Loa in Hawaii.

Figure 5.13 The largest volcano in the solar system is Olympus Mons. This Viking oblique perspective shows the 25 km high volcano surrounded by clouds. *(NASA/JPL)*

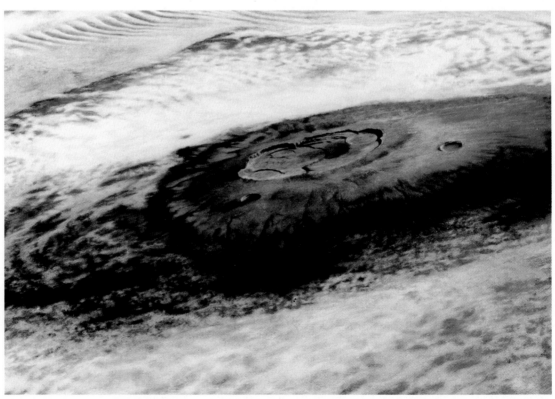

The Viking imagery permits a detailed examination of the slopes of these volcanoes to search for impact craters and estimate surface age. Many of the volcanoes show fair numbers of such craters, suggesting that they ceased activity a billion years or more ago. However, Olympus Mons has very few impact craters. Its present surface cannot be more than about 100 million years old, and it could even be much younger. Some of the fresh-looking lava flows we see might have been formed a hundred, or a thousand, or a million years ago, but geologically speaking they are young. It is probable that Olympus Mons remains intermittently active today.

Cracks and Canyons

The Tharsis region consists of more than a collection of huge volcanoes. In this part of the planet the surface itself has bulged upward, forced by great pressures from below, resulting in tectonic cracking of the crust. None of these cracks, however, shows any evidence of sliding motion such as that associated with faults on Earth. Apparently Mars never reached the stage of plate tectonics, perhaps because of its smaller size. Instead, there is just the single feature of Tharsis, forced upward but not induced to shift sideways. It is as if crustal forces began to act but then subsided before full-scale plate tectonics could begin.

How does the geology of Mars compare with that of Venus and Earth?

Among the most spectacular features on Mars are the great canyons called the Valles Marineris, which extend for about 5000 km (nearly a quarter of the way around Mars) along the slopes of the Tharsis bulge (Figure 5.14). The main canyon is about 7 km deep and up to 100 km wide. It is so large that the Grand Canyon of the Colorado River would fit comfortably into one of its side canyons.

The term canyon is somewhat misleading, because the Valles Marineris canyons were not cut by running water. They have no outlets. They are basically tectonic cracks, produced by the same crustal tensions that caused the Tharsis uplift. However, water is believed to have played a later role in shaping the canyons, primarily through undercutting of the cliffs by seepage from deep springs. This undercutting led to landslides, gradually widening the original cracks into the great valleys we see today.

The View from the Surface

Viking 1 landed at a latitude of 22 N, on a 3-billion-year-old windswept plain near the lowest point of a broad basin. Its desolate but strangely beautiful surroundings included numerous angular rocks, some more than a meter across, interspersed with dune-like deposits of fine-grained soil (Figure 5.15a). At the Viking 2 site in Utopia, at latitude 48 N, the surface was somewhat similar, but with substantially greater numbers of rocks (Figure 5.15b). At both sites, it seems that winds, blowing sometimes up to 100 km/hr or more, have stripped the surface of loose, fine material to leave the rocks exposed.

Each lander peered at its surroundings through color stereo cameras, sniffed the atmosphere with a variety of analytical instruments,

Figure 5.14 High-resolution Viking view of intricate cliffs and mesas in the martian canyonlands. Image width is approximately 200 km. *(USGS/NASA, courtesy of Alfred McEwen)*

(a)

(b)

Figure 5.15   The martian surface as photographed by Viking. (a) The V-1 landing site in Chryse. (b) The V-2 landing site in Utopia.   *(NASA/JPL)*

and poked at nearby rocks and soil with its mechanical arm. As part of its primary mission of searching for martian life, each lander collected soil samples and brought them on board for analysis, as described in more detail below. The soil was found to consist of clays and iron oxides, as had long been expected from the red color of the planet.

Each lander also carried a weather station to measure temperature, pressure, and wind. As expected, temperatures vary much more on Mars than on Earth, because of the absence of moderating oceans and clouds. Typically, the summer maximum was 240 K (−33° C), dropping to 190 K (−83° C) at the same location just before dawn. The lowest air temperatures, measured farther north by Viking 2, were about 173 K

What is it like on the surface of Mars?

$(-100°\ \text{C})$. During the winter, Viking 2 also photographed H_2O frost deposits on the ground (Figure 5.16).

Most of the winds measured at the Viking sites were low to moderate, only a few km/hr. However, Mars is capable of great windstorms, which can shroud the entire planet in dust. At such times the Sun was greatly dimmed at the Viking sites, and the sky turned a dark red color.

Martian Samples

Much of what we know of the Moon, including the circumstances of its origin, comes from studies of lunar samples, but spacecraft have not yet returned martian samples to Earth for laboratory analysis. It is with great interest, therefore, that scientists have recently concluded that there are samples of martian material already available for study on Earth.

There are about a dozen of these martian rocks, all members of a rare class of meteorites called SNC meteorites (Figure 5.17). The most obvious special characteristic of these meteorites is that they are volcanic *basalts*, and they are relatively *young*, about 1.3 billion years. We know from details of their composition that they are not from the Moon, and in any case there was no lunar volcanic activity as recently as 1.3 billion

Figure 5.17 One of the SNC meteorites, believed to be fragments of basalt ejected from Mars. *(NASA)*

years ago. By process of elimination, the only reasonable origin seems to be Mars, where the Tharsis volcanoes were certainly active at that time.

The martian origin of the SNC meteorites has been confirmed by analysis of tiny bubbles of gas trapped in several of these meteorites, which match the atmospheric properties of Mars as measured directly by Viking. Apparently the atmospheric gas was emplaced in the rock by the shock of the impact that ejected it from Mars and started it on its way toward Earth. As to helping us understand Mars, the work on these meteorites has just begun, and we do not know how much they will tell us. This is an intriguing chapter in martian science that is only beginning to be written.

5.4
MARTIAN POLAR CAPS, ATMOSPHERE, AND CLIMATE

The atmosphere of Mars is particularly interesting because it is related to the presence of water in the atmosphere and polar caps. Water is the key to the past; if Mars once had liquid water on its surface, than we might expect life to have evolved there also.

The Martian Atmosphere

The atmosphere of Mars has an average surface pressure of only 0.007 bar, less than 1 percent that of the Earth. It is composed primarily of carbon dioxide (95 percent), with about 3 percent nitrogen and 2 percent argon. The predominance of carbon dioxide over nitrogen is not surpris-

Figure 5.18   Morning fog in the martian canyonlands.   *(NASA/JPL)*

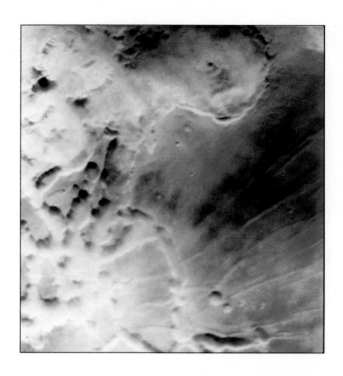

What is the evidence for water on Mars, past or present?

ing when you remember (Section 3.3) that the Earth's atmosphere would also be mostly carbon dioxide if the CO_2 were not locked up in marine sediments. Water vapor is not expected to be a major component of the martian atmosphere, since temperatures are almost always well below the freezing point.

Several types of clouds can form in the martian atmosphere. First there are dust clouds, raised by winds, which can sometimes grow to cover a large fraction of the surface. Second are water ice clouds similar to those on Earth. These often form around mountains, just as happens on our planet (Figure 5.7). Finally, the CO_2 of the atmosphere can itself condense at high altitudes to form hazes of dry ice crystals. The CO_2 clouds have no counterpart on Earth, since on our planet the temperature never drops low enough (about 150 K) for this gas to condense.

Although there is water vapor in the atmosphere and occasional clouds of water ice can form (Figure 5.18), *liquid* water is not stable under present conditions on Mars. Part of the problem is the low temperatures on the planet. But even if the temperature on a sunny summer day rises above the freezing point, liquid water cannot exist. At pressure of less than 0.006 bars, only the solid and vapor forms are possible. In effect, the boiling point is as low or lower than the freezing point, and water changes directly from solid to vapor without an intermediate liquid state. This condition applies over the entire upland regions of Mars.

The Polar Caps

Through a telescope, the most prominent surface features on Mars are the bright polar caps, which change with the seasons. These seasonal

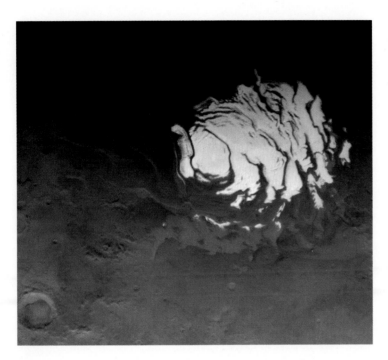

Figure 5.19   The residual south polar cap of Mars, composed of frozen H_2O and CO_2. The diameter of the cap is 350 km.   *(NASA/USGS, courtesy of Alfred McEwen)*

caps are similar to the seasonal snow cover on Earth. We do not usually think of the winter snow as a part of our polar caps. But seen from space, the thin snow would blend with the thick permanent ice caps to create a situation much like that seen on Mars.

The seasonal caps on Mars are composed not of ordinary snow, but of frozen CO_2 (dry ice). These deposits condense directly from the atmosphere when the surface temperature drops below about 150 K. The caps develop during the cold martian winters, extending down to a latitude of about 50 degrees by the start of spring.

Distinct from these thin seasonal caps of CO_2 are the permanent, or residual, caps that are always present near the poles (Figure 5.19). As the seasonal cap retreats during spring and early summer, it reveals a brighter, thicker cap beneath. The southern permanent cap has a diameter of 350 km and is composed of thick deposits of frozen CO_2, presumably mixed with water ice. Throughout the southern summer, it remains at the freezing point of CO_2, 150 K, and this cold reservoir is thick enough to survive the summer heat intact. The northern permanent cap is different. It is much larger, never shrinking below a diameter of 1000 km, and it is composed of ordinary H_2O. The two caps are different because the seasons are complicated by the substantial variation in distance from the Sun that Mars experiences during its year.

These two permanent caps represent a huge reservoir of water, in comparison with the very small amounts of water vapor in the atmosphere. But they are only the tip of the iceberg, for we think Mars contains a great deal of ice beneath its surface in the form of permafrost. Only at the two poles does this subsurface ice emerge where it can be studied directly.

Figure 5.20 Detail of the southern polar cap showing terracing on exposed slopes. *(NASA/JPL)*

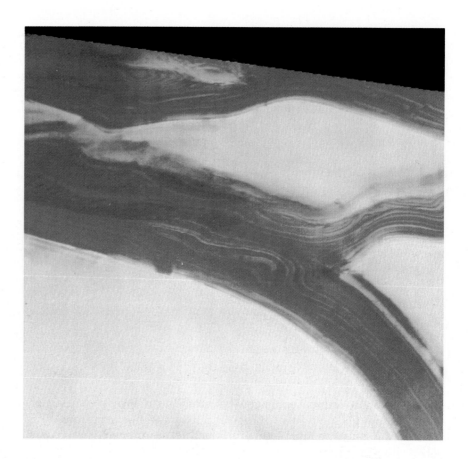

The terrain surrounding the permanent polar caps and seen in ice-free areas within the caps is remarkable (Figure 5.20). At latitudes above 80 degrees in both hemispheres, the surface consists of recent, layered sedimentary deposits that entirely cover the older cratered ground below. Individual layers are typically a few tens of meters in thickness, marked by alternating light and dark bands of sediment. Probably the material in the polar deposits is dust carried by wind from equatorial regions. Calculations suggest that each major layer corresponds to an interval of tens of thousands of years.

Channels and Floods

Although there is no liquid water today on Mars, there is fascinating evidence that rain once fell and rivers once flowed. Two kinds of geological features appear to be the remnants of ancient water-courses: the runoff channels of the old uplands and the larger outflow channels that lead from the uplands down to the great northern basins.

In the upland equatorial plains are multitudes of small sinuous (twisting) channels typically a few meters deep, some tens of meters wide, and perhaps 10 or 20 km long. They are called runoff channels, be-

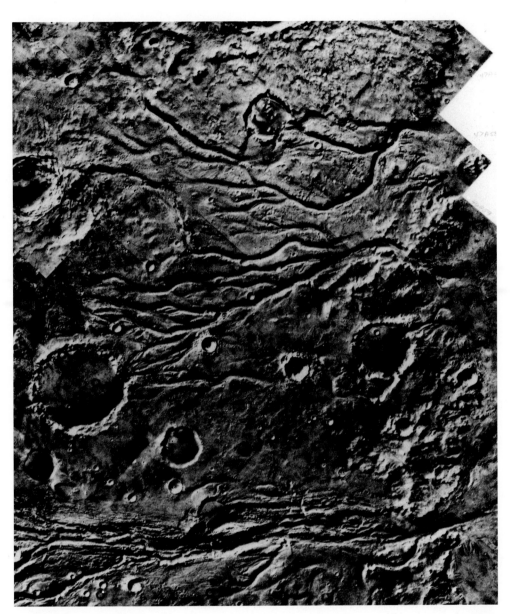

Figure 5.21 Large outflow channels photographed by Viking. These features appear to have been formed in the distant past from massive floods of water. Image width is approximately 400 km.   *(NASA/JPL)*

cause they appear to have carried the surface runoff of ancient rainstorms. These channels seem to be telling us that the planet had a very different climate long ago. How can we tell when rain might have last fallen on Mars? Crater counts show that this part of Mars is more cratered than the lunar maria but less so than the lunar highlands. Thus the runoff channels are older than the maria, presumably at least 3.9 billion years.

The outflow channels (Figure 5.21) are much larger than the older runoff channels. The largest of these, which drain into the Chryse basin where Viking 1 landed, are 10 km or more in width and hundreds of kilometers long. Many features of these outflow channels have convinced geologists that they were carved by huge volumes of running water, far too great to be produced by ordinary rainfall. Where did this flood-water come from? As far as we can tell, the source regions contained abundant water that is now frozen in the soil as permafrost. Some local source of heating must have released this water, leading to catastrophic flooding. Perhaps this heating was associated with the formation of the volcanic plains, which are contemporaneous with the channels.

Climate Change

The evidence suggests that the climate of Mars has varied on at least two different time scales. Billions of years ago temperatures were warmer, rain fell, and the atmosphere must have been much more substantial than it is today. And much more recently there has been a cyclic variation as recorded in the layered deposits of the martian polar regions.

The long-term cooling of Mars and the loss of its atmosphere are results of its small size relative to the Earth and of its greater distance from the Sun. Mars presumably formed with a much thicker atmosphere, and the atmosphere maintained a higher surface temperature because of the greenhouse effect. Escape of the atmosphere to space, however, gradually lowered the temperature. Because its surface gravity is only one-third that of the Earth, this atmospheric escape proceeds much faster than on our own planet or on Venus. Eventually it became so cold on Mars that the water froze out of the atmosphere, further reducing its ability to retain heat. The result is the cold, dry planet we see today. Probably this loss of atmosphere took place within a few hundred million years; from the absence of runoff channels in the northern plains, it seems that rain has not fallen on Mars for at least 3 billion years.

The climate variations recorded in the polar regions are much more recent. The time scales represented by the polar layers are tens of thousands of years. Apparently the martian climate experiences periodic changes with this frequency, which is similar to the intervals between ice ages on the Earth. Calculations indicate that the causes are probably also similar: variations in the orbit and tilt of the planet induced by gravitational perturbations from other planets.

The Search for Life on Mars

If there was running water on Mars in the past, perhaps there was life as well; perhaps life, in some form, remains in the martian soil. Testing this possibility, however unlikely, was one of the primary objectives of the Viking landers (Figure 5.22).

The Viking landers, in addition to the instruments we have already discussed, carried miniature biological laboratories to test for micro-organisms in the martian soil. They looked for evidence of *respiration* by living animals, *absorption of nutrients* offered to organisms that might be

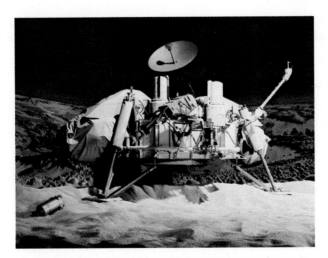

Figure 5.22 Engineering model of the Viking lander in a Mars simulation laboratory. The boom picks up soil and rocks for analysis. The two cylinders at the top of the lander are survey cameras. Below the right-hand camera is one of three rocket engines used during the final soft landing on Mars. *(NASA/JPL)*

present, and *exchange of gases* between the soil and its surrounding area for any reason whatsoever. In various tests the martian soil was scooped up by the spacecraft's long arm and provided to the experimental chambers, where it was isolated and incubated in contact with a variety of gases, radioactive isotopes, and nutrients to see what would happen. A fourth instrument pulverized the soil and analyzed it carefully to see what organic (carbon-bearing) material it contained.

The Viking experiments were sensitive enough that had one of the spacecraft landed anywhere on Earth, with the possible exception of Antarctica, it would easily have detected life. Those experiments that tested martian soil for absorption of nutrients and gas exchange did show activity, but this could have been caused by inorganic chemical reactions. In fact, these experiments showed that the martian soil seems to be much more chemically active than terrestrial soils, as a result of its exposure to solar ultraviolet radiation. The organic chemistry experiment showed no trace of organic material, which is apparently destroyed by the sterilizing effect of this ultraviolet light. Although the possibility of martian life has not been eliminated, most experts consider the chance of life on that planet to be negligible. Although Mars has the most Earthlike environment of any planet, nobody seems to be home.

Was there ever life on Mars?

5.5
THE MASSIVE ATMOSPHERE OF VENUS

In some ways the atmosphere of Venus resembles that of Mars; in other ways it is very different. In this section we will look at possible ways that the atmosphere of Venus could have evolved to its present hellish state. Perhaps if we understood what happened to Venus, we could evaluate whether there is any danger of a similar climatic catastrophe on the Earth.

TABLE 5.2 Atmospheric Compositions of Earth, Venus, and Mars (%)

	Venus	Mars	Earth
Carbon dioxide (CO_2)	96.5	95.3	0.03
Nitrogen (N_2)	3.5	2.7	78.1
Argon (Ar)	0.006	1.6	0.93
Oxygen (O_2)	0.003	0.15	21.0
Neon (Ne)	0.001	0.0003	0.002

Composition and Structure

How does the atmosphere of Venus compare with that of the Earth?

The most abundant gas in the atmosphere of Venus is carbon dioxide, which accounts for 96 percent of the atmosphere. The second most abundant gas is nitrogen. Table 5.2 compares the compositions of the atmospheres of Venus, Mars, and the Earth. Expressed as they are in the table, as percentages, the proportions of major gases are very similar for Venus and Mars, but in other respects their atmospheres are dramatically different. With its surface pressure of 90 bars, the venerian atmosphere is more than 10,000 times more massive than its martian counterpart.

In addition to these gases, there are measurements of sulfur dioxide (SO_2) in the middle atmosphere of Venus, a subject we will return to when we look at the clouds. The atmosphere is very dry; unlike Mars, there is no possibility that Venus' water is frozen in polar caps or beneath the surface. The absence of water on Venus is one of the important characteristics that distinguishes Venus from Earth.

The venerian atmosphere (Figure 5.23) has a huge troposphere that extends up to at least 50 km above the surface. Within the troposphere, the gas is heated from below and circulates slowly, rising near the equator and descending over the poles. With no rapid rotation to break up this flow, the atmospheric circulation is highly stable. In addition, the very size of the atmosphere maintains stability. Being at the base of the atmosphere of Venus is something like being a kilometer below the ocean surface on the Earth. There, also, the mass of water evens out temperature variations and results in a uniform environment.

The thick clouds of Venus are composed primarily of sulfuric acid droplets. Sulfuric acid (H_2SO_4) is formed from the chemical combination of sulfur dioxide (SO_2) and water (H_2O). In the atmosphere of the Earth, sulfur dioxide is one of the primary gases emitted by volcanoes, but it is quickly diluted and washed out of the atmosphere by rainfall. In the dry atmosphere of Venus, this unpleasant substance is apparently stable. If there are active volcanoes on the surface, the source of sulfur dioxide is readily understood.

The clouds lie in the upper troposphere, between 30 and 60 km above the surface. Below 30 km, the air is clear. In the middle of the cloud layer, at the 53-km altitude where the VEGA balloons floated for 46 hours each in 1985, the conditions are almost Earth-like. The pressure

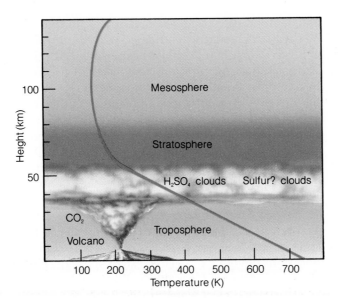

Figure 5.23 Structure of the massive atmosphere of Venus.

here is 0.5 bar and the temperature a comfortable 305 K, just a little warmer than the room in which you are reading this book. Were it not for the absence of oxygen and the nasty sulfuric acid clouds, this would not be a bad place to visit. Certainly it beats conditions on the *surface* of our "twin" planet.

Surface Temperature

The high surface temperature of Venus was discovered by radio astronomers in the late 1950s and confirmed by Mariner 2 observations and by the early Venera probes. It was not easy to understand, however, how this planet could be so much hotter than would be calculated from solar heating. A major question therefore became: What is heating the surface of Venus to a temperature above 700 K? The answer is to be found in the greenhouse effect.

Why is Venus so hot?

The greenhouse effect works on Venus just as it does for the Earth. But since Venus has so much more CO_2—almost a million times more—the effect is much stronger. Sunlight that diffuses through the atmosphere of Venus heats the surface, but the CO_2 acts as a blanket, making it very difficult for the energy to leak back to space. In consequence, the surface heats up until eventually it is emitting so much heat that an energy balance is reached with incoming sunlight.

Has Venus always had such a massive atmosphere and high surface temperature, or might it have evolved to such conditions from a climate that was once more nearly Earth-like? The answer to this question is of particular interest to us when we look at the increasing levels of CO_2 in the Earth's atmosphere, as discussed in Section 3.5. Are we in any danger of transforming our own planet into a hellish place like Venus?

Let us try to reconstruct the possible evolution of Venus from an Earth-like beginning to its present state. Imagine that it began with moderate temperatures, water oceans, and with much of its CO_2 dissolved in the ocean or chemically combined with the surface rocks, as is the case on Earth today. In our thought experiment, we then allow for modest additional heating, for example by a small rise in the energy output of the Sun or by an increase in atmospheric CO_2. One consequence is a further increase in atmospheric CO_2 and H_2O, as a result of increased evaporation from the oceans and release of gas from surface rocks. These two gases would in turn produce a stronger greenhouse effect, further raising the temperature and leading to still more CO_2 and H_2O in the atmosphere. Unless some other processes intervene, the temperature will continue to rise. Such a situation is called the **runaway greenhouse effect.**

The runaway greenhouse is not just a larger greenhouse effect; it is a process whereby an atmosphere can evolve from a state in which the greenhouse effect is small, such as on the Earth, to one with a much larger effect, such as we see today on Venus. Once the larger greenhouse conditions develop, the planet establishes equilibrium, and reversing the situation is difficult if not impossible.

If large bodies of water are available, the runaway greenhouse leads to their evaporation, creating an atmosphere of hot water vapor, which itself is a major contributor to the greenhouse effect. Water vapor is not stable in the presence of solar ultraviolet light, however, which tends to dissociate, or break apart, the molecules of H_2O into their constituent parts, oxygen and hydrogen. As we have seen, hydrogen can escape from the atmospheres of the terrestrial planets, leaving the oxygen to combine chemically with surface rock. The loss of water is therefore an *irreversible* process; once the water is gone, it cannot be restored. There is evidence that this is exactly what happened to the water once present on Venus.

5.6
COMPARISONS OF EARTH, MOON, MERCURY, MARS, AND VENUS

Much of the information in Chapters 3 through 5 has been presented in comparative form. By comparing the processes at work on the various terrestrial planets, we can understand them better and relate what we have learned to condition on the Earth. This final section is devoted entirely to such comparative planetology.

Geological Activity

What causes the differences in the geology of the terrestrial planets?

The geological activity we see manifested on the surface of a planet is driven by heat escaping from the interior. This heat can be either left over from the formation of a planet or produced from radioactive elements in the interior. The larger the planet, the more likely it is to retain its internal heat, and therefore the more we expect to see evidence on the surface of continuing geological activity, both tectonic and volcanic in nature.

Accretion, heating,
  differentiation

Formation of solid
crust, heavy cratering

Widespread mare-like
volcanism

Reduced volcanism,
possible plate tectonics

Mantle solidification,
end of tectonic activity

Cool interior, no
activity

Geologic history of the terrestrial planets

Venus   Earth

Mars

Mercury

Moon

Figure 5.24 Stages in the geological history of a terrestrial planet. The smaller the planet, the more quickly it will pass through these stages and reach "old age."

For the most part, the geology of the terrestrial planets confirms this expectation. The Moon, the smallest of these objects, was internally active until about 3.3 billion years ago, when volcanism ceased. Since that time the mantle has cooled and become solid, and today even internal seismic activity has declined almost to zero. The Moon is a geologically dead world. Mercury is almost dead, but it retains a partially liquid core.

Mars represents an intermediate case, and it has been much more active than the Moon or Mercury. The southern hemisphere crust had formed by 4 billion years ago, and the northern hemisphere volcanic plains seem to be contemporary with the lunar maria. However, the Tharsis bulge formed somewhat later in response to tectonic forces, and volcanic activity in the large Tharsis volcanoes has apparently continued intermittently to the present. Although there has been no large-scale plate tectonic activity, hot spots in the interior must still be carrying significant quantities of heat to the surface.

The Earth is the largest and the most active terrestrial planet, with its global plate tectonics driven by mantle convection. As a result, our surface is continually reworked, and over most of the planet the age of the surface material is less than 100 million years. Venus, with its similar size, should have similar levels of mantle convection carrying heat to the surface. The Magellan data show that the level of volcanic activity of Venus is comparable to that on the Earth and that the crust has been subject to great tectonic forces; yet there is no plate tectonics. Earth and Venus appear to differ more in their *style* of geologic activity than in the *magnitude* of this activity.

Figure 5.24 summarizes the geological stages through which a planet should evolve, and it indicates where the five terrestrial objects lie along this evolutionary sequence.

Figure 5.25 Comparison of the highest mountains on Earth, Venus, and Mars. The vertical scale is three times greater than the horizontal.

Elevation Differences

The mountains on the terrestrial planets have very different origins. On the Moon the major mountains are ejecta thrown up by the large basin-forming impacts that took place billions of years ago. On Mars the largest mountains are huge volcanoes, produced by repeated eruptions of lava from the same vents. Finally, the highest mountains on Earth and Venus are tectonic in origin, the result of pressure and buckling within the crust. On Earth this pressure comes from one continental plate pressing against another; on Venus the mechanism is less clear. Large volcanoes up to 10 km high are also present on Earth and up to 5 km high on Venus.

It is interesting to compare the maximum heights of the mountains on Earth, Venus, and Mars (Figure 5.25). On both of the larger planets, the maximum elevation differences between these mountains and their surroundings are about 10 km. Olympus Mons, in contrast, towers 26 km above its surroundings, and nearly 30 km above the lowest areas on Mars.

One reason that Olympus Mons is so much larger than its terrestrial counterparts is that the crustal plates on Earth never stop long enough to let a really large volcano grow. Instead, the moving plate creates a row of volcanoes, like the Hawaiian islands. On Mars, the crust remains stationary with respect to the underlying hot-spot, and the volcano can continue to grow for hundreds of millions of years.

A second difference relates to the force of gravity on the planets. The surface gravity on Mars is only about one-third that of the Earth (or Venus). In order for a mountain to survive, its internal strength must be great enough to support its weight against the force of gravity. Volcanic rocks have known strengths, and it is apparent that 10 km is about the limit on Earth; for instance, when new lava is added to the top of Mauna Loa, the mountain slumps downward under its own weight. The same height limit applies on Venus, where the force of gravity is the same. On Mars, however, with its lesser surface gravity, much greater elevation differences can be supported, and it should be no surprise that Olympus Mons is more than twice as high as the mountains of Venus or Earth.

Oceans and Atmospheres

How have the atmospheres of Venus, Earth, and Mars evolved to their present state?

The predominant atmospheric gas on Venus and Mars is carbon dioxide. On Earth, in contrast, the presence of life has eliminated most atmospheric CO_2 and introduced abundant free oxygen into its atmosphere. The Earth is also the only planet with liquid water on its surface.

Venus, Earth, and Mars may each have started with abundant water, but the fate of this water was different for each of these three planets, depending on its size and distance from the Sun. Early in its history, Mars apparently had a thick atmosphere and oceans, but it could not retain these conditions. The carbon dioxide necessary for a substantial greenhouse effect was lost, the temperature dropped, and eventually the water froze. On Venus the reverse process took place, with a runaway greenhouse effect leading to the permanent loss of water. Only on Earth was the delicate balance maintained that permits liquid water to persist.

With the water gone, Venus and Mars each ended up with an atmosphere that is about 96 percent carbon dioxide and a few percent nitrogen. On Earth, the presence first of water and then of life led to a very different kind of atmosphere. The CO_2 was removed to be deposited in marine sediment, whereas proliferation of photosynthetic life eventually led to the release of enough oxygen to overcome natural chemical reactions that are eliminating this gas from the atmosphere. As a result, we find ourselves on Earth with a great deficiency of carbon dioxide and with the only planetary atmosphere containing molecular oxygen.

▶ SUMMARY

5.1 Venus, the nearest planet, is a great disappointment through the telescope because of its impenetrable cloud cover. Mars is more tantalizing, with dark markings and polar caps. Mars has only 11 percent the mass of the Earth, but Venus is nearly our twin in size and mass. Mars completes a rotation every 24 hrs and has seasons like the Earth; Venus has a retrograde rotation period of 243 days. Both planets have been extensively explored by spacecraft.

5.2 Venus has been mapped by the Magellan radar orbiter, revealing a lowland surface with two continents (Aphrodite and Ishtar) of folded mountains but no plate tectonics. The geology is dominated by volcanic and **tectonic** processes, with very little erosion. Distinctive features include pancake domes, long lava rivers, tectonic grid patterns, and coronae, the latter related to "blob tectonics." The surface is extraordinarily inhospitable, with pressure of 90 bars and temperature of 730 K, yet several Russian Venera landers have investigated it successfully.

5.3 Mars has heavily cratered uplands in its southern hemisphere but younger and lower volcanic plains over much of its northern half. The Tharsis continent, as big as North America, includes several huge volcanoes; Olympus Mons is 25 km high and 500 km in diameter. The Valles Marineris canyons are tectonic features widened by erosion. The Viking landers revealed barren, windswept plains at Chryse and Utopia. Currently there is great interest in the SNC meteorites, which appear to be samples of martian basalts.

5.4 The martian atmosphere has less than 0.01 bar surface pressure and is made up primarily (95 percent) of CO_2. There are dust clouds, water clouds, and carbon dioxide (dry ice) clouds. Liquid water is not possible, but there may be subsurface permafrost. Seasonal polar caps are made of dry ice, but the northern residual cap is H_2O ice. Evidence of a very different climate in the past is found in water erosion features, both runoff channels and outflow channels, the latter carved by catastrophic floods. The Viking landers searched for martian life in 1976, with negative results, but life might have flourished long ago.

5.5 The atmosphere of Venus is primarily CO_2 (96 percent). Thick clouds at altitudes of 30–60 km are made of sulfuric acid, and a CO_2 greenhouse maintains the high surface temperature. Venus presumably reached its current state from initially more Earth-like conditions as a result of a **runaway greenhouse effect,** which include the loss of large quantities of water.

5.6 Comparisons of the terrestrial planets can be made to understand their common evolution. Mass is the primary determinant of geological activity. The mountains have different origins on different planets, but limitations on their height are set by the surface gravity. Venus, Earth, and Mars are thought to have begun with comparable atmospheres and oceans, but Venus suffered a runaway greenhouse effect and a loss of water, whereas Mars froze into a terminal ice age. Only Earth can support life today.

▶ REVIEW QUESTIONS

1. Compare the basic data on orbits, mass, size, density, and rotation for Venus, Earth, and Mars.

2. Describe the main topographic and geological features on Venus, and compare them with those of Mars and the Earth.

3. Describe the main topographic and geological features on Mars, and compare them with those of the Earth and the Moon.

4. Describe the current state of the martian atmosphere, and indicate how it must have been different in the past, judging from the evidence that there was once abundant water on its surface.

5. Describe the current state of the venerian atmosphere, and indicate how it might have evolved to this condition through a runaway greenhouse effect.

6. Summarize the main reasons for differences among the terrestrial planets in their: (1) geological activity; (2) mountain heights; (2) atmospheric pressure and composition; and (4) amount and state of water.

▶ THOUGHT QUESTIONS

7. What are the advantages of using radar imaging rather than ordinary cameras to study the topography of Venus? What are the relative advantages of these two approaches to mapping the Earth or Mars?

8. Contrast the tectonic and volcanic activity on Venus with that on Earth.

9. Why is Mars red? Why aren't the Moon or the Earth the same color?

10. Explain how the theory that all of the terrestrial planets had similar impact histories can be used to date the formation of the martian uplands, the martian basins, and the Tharsis volcanoes. How certain are the ages derived for these features?

11. List the main differences among the martian canals, runoff channels, and outflow channels.

12. Explain the major divisions of Earth, Venus, Mars, and the Moon each into two distinct types of topography. In each case, what is the origin of these two types of topography?

13. How do the mountains on Mars compare with those on Earth and the Moon?

14. Explain some of the problems that would be encountered in trying to build a spacecraft that could operate on the surface of Venus for a full venerian year.

15. Why is there so much more carbon dioxide in the atmosphere of Venus than in that of the Earth? Why so much more than in the martian atmosphere?

16. Estimate the maximum height of the mountains on a hypothetical planet similar to the Earth but with twice the surface gravity of our planet.

▶ PROBLEMS

17. At its nearest, Venus comes within about 40 million km of the Earth. How distant is it at its farthest?

18. Calculate the relative *land* areas of Earth, Moon, Venus, and Mars. (*Note:* 70 percent of the Earth is covered with water.)

***19.** Mariner 2 to Venus was the first successful interplanetary spacecraft. It traveled to Venus from the Earth on an elliptical orbit in which the Earth was at aphelion and Venus at perihelion. Calculate the total period of revolution about the Sun for a spacecraft on such an orbit. Then calculate the one-way travel time from Earth to Venus on this same trajectory.

***20.** A similar trajectory from Earth to Mars requires a longer trip than to Venus. Calculate the one-way trip time to Mars using the same approach as in Problem 19. Explain in general why it takes longer to go to Mars than it does to go to Venus (*Hint:* The reason is *not* just that Mars is slightly farther away.)

***21.** Where is the water on Mars? Try to estimate how much water might be present in various forms, such as in the polar caps (using the dimensions given in the text) or in subsurface permafrost (assuming various thicknesses for the permafrost, from 1 to 10 km, and a concentration of ice in the permafrost of 10 percent by volume).

▶ SUGGESTIONS FOR ADDITIONAL READING

Books

On Mars and Venus

Morrison, D. *Exploring Planetary Worlds.* W. H. Freeman, 1993. Up-to-date overview of the solar system, with comparative descriptions of Mars, Earth, and Venus.

On Mars

Cooper, H. *The Search for Life on Mars.* Harper & Row, 1980. Eyewitness journalist's account of the Viking program.

Washburn, M. *Mars at Last.* Putnam, 1977. Well-written journalist's survey of the Viking project and its discoveries.

Carr, M. *The Surface of Mars.* Yale U. Press, 1981. Moderately technical account of the Viking mission results, beautifully illustrated.

On Venus

Cooper, H. *The Evening Star: Venus Observed.* Farrar, Straus & Giroux, 1993. A journalist recounts the history of Venus exploration and the results of the Magellan mission.

Burgess, E. *Venus: An Errant Twin.* Columbia U. Press, 1985. A science writer's summary of our knowledge before the Magellan mission.

Articles

On Mars

Carroll, M. "The Changing Face of Mars" in *Astronomy,* Mar. 1987, p. 6.

Hartmann, W. "What's New on Mars" in *Sky & Telescope,* May 1989, p. 471.

Edgett, K., *et al.* "The Sands of Mars" in *Astronomy,* June 1993, p. 26.

Carr, M. "The Surface of Mars: A Post-Viking View" in *Mercury,* Jan./Feb. 1983, p. 2.

Gore, R. "Sifting for Life in the Sands of Mars" in *National Geographic,* Jan. 1977. On the results of the Viking missions.

Albin, E. "Observing the New Mars" in *Astronomy,* Nov. 1992, p. 74. A guide for beginning observers.

On Venus

Stofan, E. "The New Face of Venus" in *Sky & Telescope,* Aug. 1993, p. 22. Excellent summary, together with 3-D photos (glasses supplied).

Saunders, S. "The Exploration of Venus: A Magellan Progress Report" in *Mercury,* Sept./Oct. 1991, p. 130.

Burnham, R. "What Makes Venus Go?" in *Astronomy,* Jan. 1993, p. 40.

THE REALM OF THE GIANTS

The four giant planets—Jupiter, Saturn, Uranus, and Neptune—all shown to the same scale, with the Earth for comparison. *(NASA/JPL)*

*T*he Earth and other terrestrial planets discussed in the preceding three chapters are becoming familiar worlds. They are places we can relate to. Humans have already visited the Moon, and within the next 25 years they are likely to set foot on Mars as well. The outer solar system is a very different place. Most obviously, the scale of the system is greatly expanded. The planets themselves are much larger, distances between them are vastly increased, and they are accompanied by extensive systems of satellites and rings.

The giant planets are so massive that they have been able to retain even the light gases hydrogen and helium, producing a chemistry different from that of the rocky inner planets. As a consequence of their large size and hydrogen-based chemistry, these planets have no solid surfaces. When we look at them, we

129

see the upper layers of their deep cloudy atmospheres. These are strange worlds, yet we should remember that they represent most of the material in the planetary system. Indeed, the solar system has sometimes been described simply as "one star, four giant planets and some debris."

Most of what we know about the giant planets and their systems of satellites and rings has been obtained from two Voyager spacecraft launched in 1977. These robotic explorers visited each of the giant planets between 1979 and 1989, returning spectacular photos and other data that are still being analyzed by planetary scientists. Both Voyagers are still alive as of late 1993, each speeding away from the solar system and into the unknown of interstellar space.

6.1
JUPITER, THE KING OF THE PLANETS

Jupiter (Figure 6.1) is the true giant of the planetary system, containing more matter than all of the other eight planets combined. Its retinue of 16 satellites makes Jupiter the center of a sort of miniature solar system. We begin our look at the outer solar system with this giant, which will serve as the basis for a comparative study of the other outer planets and their systems of satellites and rings.

Figure 6.1 Jupiter, as photographed by Voyager. The banded structure on the clouds represents strong east-west winds in the atmosphere. *(NASA/JPL)*

Basic Properties of a Giant

Jupiter's diameter of 143,000 km is more than 10 times greater than that of the Earth or Venus. The volume of a planet is proportional to its diameter cubed, so the volume of Jupiter must be more than 1000 times that of the Earth. Thus Jupiter is big enough to hold more than 1000 Earths.

The mass of Jupiter is 318 times that of the Earth. Since its volume is 1000 times greater and its mass only 318 times greater, the density of Jupiter must be much less than that of the Earth. An accurate calculation gives a density for Jupiter of 1.3 g/cm^3, as compared with 5.5 g/cm^3 for the Earth. This low density proves that Jupiter is not composed of the same rocks and metals that dominate the interior of our own planet. It is made up mostly of hydrogen and helium, the lightest and most abundant gases in the universe.

What are the size and composition of Jupiter?

The average distance of Jupiter from the Sun is 778 million km, 5.2 times that of the Earth; its period of revolution is 12 years. Despite its great size, Jupiter rotates in only 9 hours and 56 minutes, less than half as long as the day on Earth.

It is interesting to note that Jupiter has very nearly the maximum possible diameter for a body of "cold" hydrogen, that is, one that is not generating energy by nuclear reactions as the Sun and stars do. Less massive bodies than Jupiter would occupy a smaller volume. Saturn is an example, being both less massive and smaller than Jupiter. But more massive bodies, if they are "cold," would also be compressed to a smaller volume than Jupiter's, because their internal pressure cannot support the weight of their upper layers. Such an object, with a mass larger than Jupiter's but not large enough to support nuclear reactions, is called a brown dwarf. We discuss brown dwarfs, objects that are intermediate between stars and planets, in Chapter 10.

Composition and Structure

From its density alone, astronomers long ago concluded that Jupiter was composed mostly of hydrogen. This conclusion did not depend on direct observations of the atmosphere of the planet. As we should expect, the compounds subsequently detected in its atmosphere are chemically reduced gases, such as methane (CH_4) and ammonia (NH_3), or more complex hydrocarbons such as ethane (C_2H_6) and acetylene (C_2H_2). The second most abundant element in Jupiter is helium, which accounts for about 25 percent of its mass. All the other elements combined make up only about 1 percent of the material in Jupiter, a situation that mimics that of the Sun itself.

The internal structure of Jupiter also differs from those of the rocky inner planets (Figure 6.2). When we look in from the outside, we see the swirling, colorful clouds of the upper atmosphere. Sometimes Jupiter has been called incorrectly a "gas" planet. Most of Jupiter, however, is not a gas but a liquid. At depths of only a few thousand kilometers below the visible clouds, pressures become so large that hydrogen changes

Figure 6.2 Internal structure of Jupiter. The planet is composed mostly of liquid hydrogen.

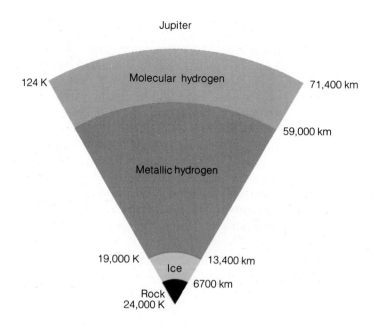

from a gaseous to a liquid state. The equivalent of a mantle for Jupiter is thus a huge region composed primarily of liquid hydrogen. Still deeper, this liquid hydrogen becomes electrically conductive, like a metal. This hot, conductive, spinning interior generates a magnetic field, the largest of any planet.

Beneath the liquid hydrogen mantle of Jupiter lies a denser core with 10 to 15 times the mass of the Earth. This core is thought to be composed of rock and metal with the addition of large quantities of water ice. This dense core is probably the original mass that later was able to attract hydrogen and other gases to it as the planet formed.

Jupiter was heated during its formation by the collapse of surrounding nebular gas onto this core. Its center remains extremely hot, although it has been cooling for more than 4 billion years. This primordial energy, still leaking slowly into space, provides as much heat to the atmosphere as does the Sun shining down from above. The atmosphere of Jupiter is therefore somewhat of a cross between a normal planetary atmosphere, which obtains most of its energy from the Sun, and the atmosphere of a star, which is entirely heated from within. The internal heat leaking away from Jupiter provides an energy source of 4×10^{17} watts. The planet is glowing with the equivalent of 4 million billion hundred-watt light bulbs.

Poisonous Atmosphere and Colorful Clouds

Thick, colorful clouds envelop Jupiter, twisting in a changing kaleidoscope of patterns (Figure 6.3). These clouds must be composed of some material other than hydrogen and helium, both of which are transparent

Figure 6.3 Jupiter's complex and colorful clouds. Also shown in this Voyager photo are two of Jupiter's satellites: Io (left) and Europa (right). (*NASA/JPL*)

gases. What are the clouds made of? The gas ammonia (NH_3), which has been detected in small quantities in the atmosphere, provides the answer. At the low temperatures in Jupiter's upper atmosphere, ammonia gas can condense as a liquid, just as water vapor condenses in the Earth's atmosphere; these tiny droplets of condensed ammonia thus become visible as clouds. When we look down on Jupiter, what we see primarily are the tops of the ammonia clouds.

Below the visible ammonia clouds are several additional layers of clouds. Let us imagine riding down into the atmosphere on an instrumented entry probe, such as the Galileo Probe to be deployed into the atmosphere of Jupiter in 1995 (Figure 6.4). We would reach the ammonia clouds at a level where the atmospheric pressure is about 0.1 bar (a little less than at the summit of Mt. Everest) and the temperature is 140 K (about the same as on the polar caps of Mars). After passing quickly through these clouds we enter a somewhat warmer clear region, but at a pressure of about 3 bars we expect the probe to reach another thick deck of condensation clouds, this time composed of ammonium hydrosulfide (NH_4SH). These clouds have never been detected directly, but there is strong circumstantial evidence for their existence.

As we descend to a pressure of 10 bars and ever-higher temperatures, we should pass next into a region of frozen water clouds, and below that into clouds of liquid water droplets perhaps similar to the common clouds of the terrestrial troposphere. This region corresponds almost to a "shirt-sleeve" environment, in which astronauts could work comfortably if they carried scuba gear for breathing. But with no solid surface to stop it, our probe continues to descend, penetrating into dark regions of higher and higher pressure and temperature. No matter how strongly it was built, eventually the probe will be crushed and swal-

Figure 6.4 Artist's impression of the Galileo Probe descending through the jovian clouds in 1995. *(NASA/ARC)*

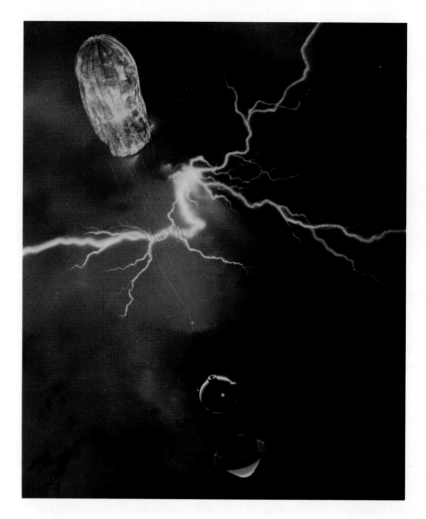

lowed in the black depths, where the great pressures transform the atmospheric hydrogen into a hot, dense liquid. In the case of the Galileo Probe, it has been designed to withstand a pressure of 20 bars; thus we can expect it to radio back information to this depth but no further.

Circulation of the Atmosphere

How do the atmosphere and weather of Jupiter compare with the Earth?

The weather patterns on Jupiter differ fundamentally from those of the terrestrial planets. There are three primary reasons for these differences: (1) the atmosphere is much deeper, with no solid lower boundary; (2) the planet spins rapidly, suppressing north-south circulation and accentuating east-west gas flow; and (3) the internal heat source contributes about as much energy as sunlight, forcing the atmosphere into deep convection to carry the internal heat outward.

The main features of the visible clouds of Jupiter are alternating dark and light bands that stretch around the planet parallel to the equator. These bands are related to underlying east-west wind patterns in the atmosphere. The light zones are regions of upwelling gas, capped by

Figure 6.5 The Great Red Spot of Jupiter, large enough to contain two Earths. Below it and to the right is one of the "white ovals," which are similar but smaller high-pressure storms. *(NASA/JPL)*

white ammonia cirrus clouds. They apparently represent the tops of up-ward-moving convection currents. The darker bands are regions where the cooler atmosphere moves downward, completing the convective cycle. They are darker because there are fewer ammonia clouds, and it is possible to see deeper in the atmosphere, perhaps down to the ammonium hydrosulfide clouds.

Storms on Jupiter

Superimposed on the regular pattern described above are many local disturbances—weather systems or storms, to borrow terrestrial terminology. The most prominent of these are oval high-pressure regions, some of which can be seen from the Earth.

The largest and most famous storm on Jupiter is the Great Red Spot, a reddish oval in the southern hemisphere that is almost 30,000 km long—big enough to hold two Earths side by side (Figure 6.5). First seen 300 years ago, the Red Spot is much longer lived than are storms in our own atmosphere. Unlike terrestrial cyclones or hurricanes, it is a high-pressure region, with a counterclockwise rotation period of six days. Three similar but smaller disturbances on Jupiter formed about 1940, called the white ovals; these are only about 10,000 km across.

We do not know what causes these jovian storms, but it is possible to understand how they can last so long once they do form. On Earth, a large oceanic hurricane or typhoon typically has a lifetime of a few weeks, or even less when it moves over the continents and encounters friction with the land. On Jupiter, there is no solid surface to slow down an atmospheric disturbance, and furthermore the sheer size of these features lends them stability. It is possible to calculate that on a planet with no solid surface, the lifetime of anything as large as the Great Red Spot should be measured in centuries. The same calculations show that the lifetime of features like the white ovals should be measured in decades.

ESSAY The Big Splash

In July 1994, Jupiter is at the center of a celestial event without precedent: the collision of a comet with a planet. Although astronomers have known for a long time that comet collisions must happen occasionally, and they have suspected a comet as the cause of the extinction of the dinosaurs on Earth 65 million years ago, no one had expected actually to witness an example of such celestial pyrotechnics.

The comet that strikes Jupiter in 1994 was discovered in March 1993 by astronomers Carolyn Shoemaker and David Levy as part of a photographic patrol for faint asteroids and comets (see discussion in Chapter 8). As their ninth comet discovery, it is called Comet Shoemaker-Levy 9. From the beginning it was recognized as an odd object, apparently consisting of about 20 separate lumps of material spread along the orbit like beads on a string (Figure 1). Tracing the orbit backwards, astronomers found

that this object had made an extremely close brush by Jupiter on July 8, 1992, swinging just 40,000 km above the cloud tops. Tidal forces at that time split the comet, while the powerful gravity of Jupiter captured the fragments into an elongated 2-year orbit about the planet.

Jupiter thus suddenly gained about 20 new small satellites, but they turned out to be temporary additions to the jovian system. As a consequence of their very close passage by Jupiter and subsequent perturbations caused by the gravity of the Sun, these fragments did not follow stable Keplerian orbits. Instead, their paths were distorted so that on the next return to Jupiter, about July 21, 1994, they would hit the planet. Astronomers predicted that within a few hours most of the cometary fragments would smash at high speed into the jovian atmosphere (Figure 2). There is some dispute about the sizes of the

Ground Based Wide Angle View   HST View Region Containing the Nuclei   HST View Closeup Near Brightest Nucleus

Figure 1 Hubble Space Telescope image of approximately 20 objects that comprise comet Shoemaker-Levy 9, giving it the resemblance of a "string of pearls." The individual objects began to separate in July 1992, and by the time of impact in July 1994 they spanned a width of more than 100,000 km. (*H. A. Weaver and T. E. Smith, STScI, NASA*)

fragments, but if any of them are as large as 5–10 km in diameter, the energy of their impact amounts to hundreds of million of megatons—as powerful as the impact of 65 million years ago that led to the extinction of more than half the species on Earth.

What happens when objects plunge into the jovian atmosphere at a speed of 60 km/s? No one knows, because we have never before witnessed such an event. Calculations suggest, however, that the larger bodies will penetrate far below the visible clouds before they explode in the atmosphere. Each explosion is expected to form an intense fireball hundreds of kilometers in diameter, which will rise through the clouds carrying with it material (including plentiful water vapor) from deeper re-

gions of the atmosphere. The result could be new layers of clouds or even major new systems near the point of impact.

Unfortunately it will be difficult to determine exactly what happens, since the impacts all take place on the far side of Jupiter, out of sight from Earth. If the Galileo spacecraft were in orbit about Jupiter, it would provide a wonderful vantage point, but it is still 16 months away from Jupiter when the collisions occur in July 1994. Nevertheless, astronomers from all over the world expect to be watching the planet near the time of the collision. The impact point rotates into view from Earth just a few hours after the explosions, and no one knows what we may see in the aftermath of such a remarkable celestial event.

Figure 2   The two jovian encounters of Comet Shoemaker-Levy 9, as viewed from the direction of the Sun. In July 1992, the comet was disrupted as it passed just 40,000 km above the jovian clouds, while at its return 2 years later the fragments crash into the back side of Jupiter.

These time scales are consistent with the observed lifetimes of jovian storms.

There is one other mystery of the jovian clouds that should be mentioned, and that concerns their colors. The widespread ammonia clouds identified on the planet should be white, like water clouds on Earth, yet we see beautiful and complex patterns of red, orange, and brown. Some additional chemical or chemicals must be present to color the clouds, but we do not know what they are. Various organic compounds have been suggested, produced by the action of sunlight on the upper atmosphere. Elemental sulfur and red phosphorus have also been proposed as coloring agents. But there are no firm identifications nor any immediate prospects of solving this mystery.

6.2
FOUR GIANTS COMPARED

There are five planets in the outer solar system: Jupiter, Saturn, Uranus, Neptune, and Pluto. The first four of these are the giant or jovian planets, whereas Pluto is a small world, by far the smallest in the planetary system. In this section we compare Saturn, Uranus, and Neptune with their big brother, Jupiter.

Basic Properties

Saturn (Figure 6.6) is about twice as far from the Sun as Jupiter, orbiting at a distance of nearly 10 AU, with a period of 30 years. Uranus and Neptune are even more distant, with orbital periods of 84 and 165 years, respectively.

Saturn, Uranus, and Neptune all share with Jupiter a hydrogen-rich composition and a reducing chemistry. In many other respects, however, they are quite different. For one thing, they are smaller and less massive. Saturn has only about one-fourth the mass of Jupiter, and Uranus and Neptune are each about five times smaller yet. In fact, Uranus and Neptune are almost exactly intermediate in mass between giant Jupiter and the terrestrial planets, being 1/20 the mass of Jupiter and about 15 times the mass of the Earth.

Saturn has a day that is almost as short as Jupiter's: 11 hours. Uranus and Neptune each rotate more slowly, with periods of about 16 hours. Uranus also has a peculiar angle of rotation. Unlike the other planets we have discussed, it rotates about an axis that is tilted on its side. This produces very exaggerated seasons on Uranus. During part of its orbit, one pole is pointed almost directly at the Sun, and the other pole experiences constant darkness. About 40 years later, Uranus has traveled half way around its orbit, and the formerly dark pole experiences continuous sunlight. Table 6.1 compares some of the basic properties of the four giant planets.

How do Saturn, Uranus, and Neptune compare with Jupiter?

Figure 6.6 Saturn and its rings, photographed by Voyager. The clouds of Saturn are less colorful than those of Jupiter, but the structure and dynamics of the atmosphere are similar. (*NASA/JPL*)

TABLE 6.1 Basic Properties of the Giant Planets

	Jupiter	Saturn	Uranus	Neptune
Semimajor axis (AU)	5.2	9.6	19.2	30.1
Period (years)	11.9	29.5	84	165
Mass (Earth = 1)	318	94	15	17
Diameter (km)	142,800	120,660	50,800	50,500
Density (g/cm^3)	1.3	0.7	1.3	1.5
Rotation (hours)	10	11	−15	17

Composition and Internal Structure

As far as is known, Saturn has a nearly identical composition to Jupiter, with more than 95 percent hydrogen and helium. Both planets have solid or liquid cores of rocky and metallic material with masses of about ten times the Earth's mass. Like Jupiter, Saturn has a mantle of liquid hydrogen and an extensive reducing atmosphere.

Uranus and Neptune are different, being much more like scaled-up terrestrial planets. Their interiors are composed mostly of rock, metal, and water (either ice or liquid, depending on the temperature and pressure) (Figure 6.7). Probably these solid or liquid cores are similar in size to the cores of Jupiter and Saturn. Above these cores are deep atmospheres composed primarily of hydrogen and helium gas. Although these atmospheres resemble the atmospheres of Jupiter and Saturn, Uranus and Neptune lack the liquid hydrogen mantles on the inside that are the dominant features of the larger giants.

Each of the giant planets was heated during its formation. However, because of their smaller sizes, the energy deposited in Saturn, Uranus, and Neptune was less than that for Jupiter. In addition, being smaller than Jupiter they have had more opportunity to cool and shed this primordial energy.

Do Saturn, Uranus, and Neptune still retain detectable internal sources of heat? Saturn certainly does. Its internal energy source is about

Figure 6.7   Internal structures of the four jovian planets, drawn to scale. Jupiter and Saturn are composed primarily of hydrogen and helium, but the mass of Uranus and Neptune consists in large part of compounds of carbon, nitrogen, and oxygen. (*NASA/JPL*)

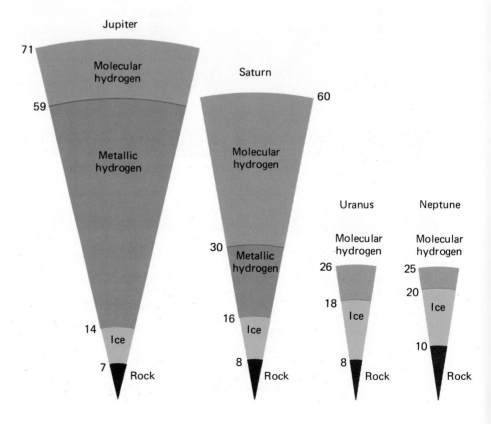

half as powerful as that of Jupiter, but its mass is only about one-quarter as great, which means that Saturn is producing twice as much energy per ton of material as Jupiter. Because Saturn is expected to have less *primordial* heat, there must be another source at work generating most of this 2×10^{17} watts of power. This source is thought to be the separation of helium from hydrogen in the interior. In the liquid hydrogen mantle, the heavier helium forms drops that sink toward the core, releasing gravitational energy. In effect, Saturn is still differentiating. This precipitation of helium is possible in Saturn because it is cooler than Jupiter; at the temperatures in Jupiter's interior, hydrogen and helium remain well mixed.

Again, Uranus and Neptune are different. Neptune has a small internal energy source, whereas Uranus does not emit enough internal heat for it to have been measured. No one knows why these two planets differ in this respect.

Atmospheres, Clouds, and Weather

Despite differing interiors, all four of the giant planets have similar atmospheric composition and structure. The distinctions we do see in their atmospheres are due primarily to their different distances from the Sun and thus their different temperatures. The absence of an internal energy source on Uranus also contributes to a distinctive kind of circulation on that planet.

Each of these planets is expected to have clouds of ammonia, ammonium hydrosulfide, and water, but because atmospheric temperatures and pressures differ, the clouds are spaced differently on each planet. Because Saturn is cooler and its atmosphere is more extended than Jupiter's, the ammonia cirrus clouds form deeper and are more difficult to see (Figure 6.8). As a consequence, the cloud patterns appear much

Figure 6.8   The structure of the atmospheres and clouds for the giant planets.   (*NASA/JPL*)

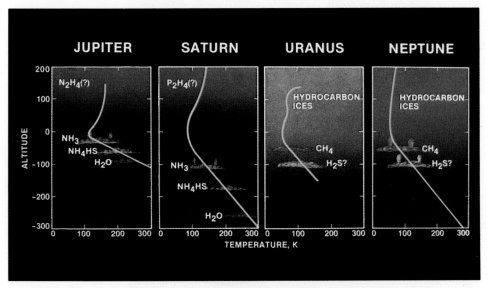

Figure 6.9 The planet Uranus as photographed by Voyager in 1986, when its rotation pole was tipped toward the Sun. (*NASA/JPL*)

How do Uranus and Neptune differ from the true giants, Jupiter and Saturn?

subdued. In addition, Saturn seems to lack the mysterious coloring agents that produce the bright colors seen on Jupiter. The planet has a rather bland, butterscotch color, and the individual cloud patterns and storms are difficult to see.

The clouds on Uranus and Neptune lie even deeper, with the addition of an upper cloud deck of methane. Uranus has practically no visi-

Figure 6.10 The planet Neptune as photographed in 1989 by Voyager. (*NASA/JPL*)

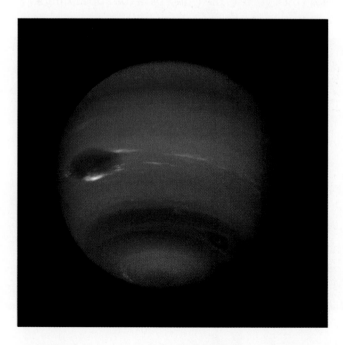

ble cloud features; without an internal heat source, there are no convection currents to bring up gases from depth, and the atmosphere remains stable and dull (Figure 6.9). Neptune also has a deep, cold atmosphere, but convection stirs the atmosphere and produces weather patterns that are nearly as visible as those on Jupiter, at least when seen up close with spacecraft cameras (Figure 6.10). Plumes of methane gas rise into the cold stratosphere to condense as bright white clouds. A large tropospheric storm system, called the Great Dark Spot, appears to mimic the Great Red Spot of Jupiter. However, Neptune is so far away that these features are impossible to study with telescopes on Earth; we have only the "snapshot" from the Voyager encounter in 1989.

6.3
THE SATELLITES: WORLDS OF FIRE AND ICE

There are 61 known satellites in the solar system, and all but four of them orbit the four giant planets. Table 6.2 summarizes these satellite systems. Although small in comparison to the planets they accompany, a number of these satellites are larger than the planet Mercury, and many of them show evidence of surprising degrees of geological and atmospheric evolution.

Temperature provides the main distinction between the terrestrial planets of the inner solar system and the larger satellites of the outer solar system. Because it was cooler where these satellites formed, water ice and other volatile compounds were available as planetary building blocks. The fact that most outer planet satellites are largely composed of ice is apparent from their densities, which are generally less than 2.0 g/cm^3, in contrast to the densities of the terrestrial planets, which are in the range from 3.3 to 5.6 g/cm^3. Detailed calculations show that most of these satellites are composed of nearly equal quantities of water ice and rocky minerals.

There are far too many satellites in the outer solar system for us to discuss them all. In this chapter we make no effort to cover most of them, but instead we focus attention on just four of the more interesting satellites and on Pluto.

TABLE 6.2   Satellite Systems

Planet	No. of Satellites	Large Satellites
Jupiter	16	Ganymede, Callisto, Io, Europa
Saturn	19	Titan
Uranus	15	none
Neptune	8	Triton

Callisto, A Garden-Variety Icy Satellite

Jupiter has four large satellites, discovered by Galileo in 1610 and called the Galilean satellites. These are, counting outward from the planet, Io, Europa, Ganymede, and Callisto (all named for lovers of Zeus/Jupiter in classical mythology). We begin our discussion with Callisto, not because it is remarkable, but because it is not. Callisto is a simple large icy satellite, an object with which other more peculiar objects can be compared.

What is the range of environments on the galilean satellites of Jupiter?

Callisto is the outermost galilean satellite. Its distance from Jupiter is about 2 million kilometers, and it orbits the planet in 17 days. Like our own Moon, Callisto rotates in the same period as it revolves, so that it always keeps the same face toward Jupiter. Callisto's day thus equals its year: 17 days. Its noontime surface temperature is only 130 K (about 140° C below freezing).

Callisto has a diameter of 4820 km, almost the same as the planet Mercury. Its mass (and therefore its density) are only one-third as great, however, immediately telling us that Callisto is an icy body. This satellite can help us answer the question: How does the geology of an icy object compare with one made primarily of rock?

Like other icy bodies of the outer solar system, Callisto has differentiated. Actually, it is much easier for an icy body to differentiate than for a rocky one, because the melting temperature of ice is so low. Only a little heating will soften the ice and get the process started, as the rock and metal sink to the center and the slushy ice floats to the surface. Callisto thus consists of a rocky and metallic core about the size of the Moon, surrounded by a mantle and crust of water ice.

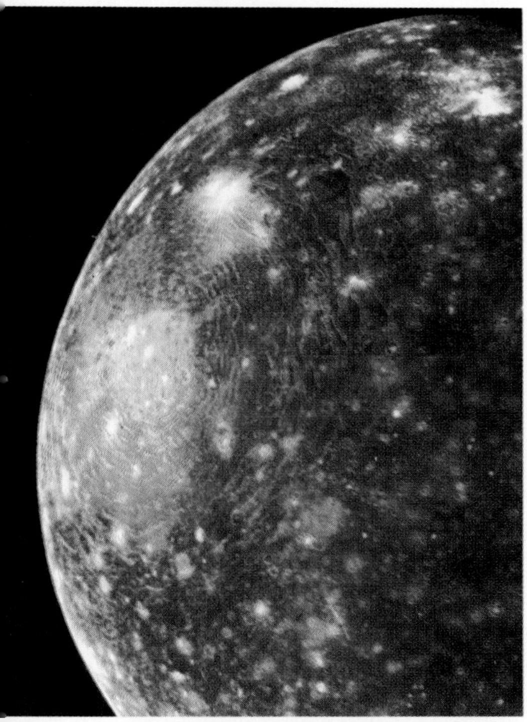

The surface of Callisto is covered with impact craters, like the lunar highlands (Figure 6.11). The survival of these craters tells us that an icy object can form and retain impact craters in its surface, and that Callisto, like both the Moon and Mercury, has experienced little if any geological activity for billions of years. The craters on Callisto also demonstrate that there was a period of heavy bombardment in the outer solar system as well as nearer the Sun.

In form, the smaller icy craters of Callisto look very much like the rocky craters of the Moon and Mercury. At the temperatures of the outer solar system, ice is nearly as hard as rock, and it behaves similarly. Ice on Callisto does not deform or flow like ice in glaciers on the Earth. Thus we see, upon reflection, that it is quite reasonable that the surface of Callisto should look like that of the Moon, despite their different compositions.

The jovian system contains a near-twin of Callisto in the form of Ganymede, the next galilean satellite inward toward the planet. With a diameter 5262 km, Ganymede is the biggest of all the planetary satellites, yet it is actually only a few percent larger than Callisto. Like Callisto, it is composed of about half rock and half ice. Surprisingly, however, the Voyager spacecraft found that Ganymede has experienced a much more complex geologic history than Callisto, with evidence of both surface melting and tectonic activity in the distant past. Like the Earth and Venus, these two objects don't turn out to be close twins after all. One of

Figure 6.11 The surface of Callisto is heavily cratered like that of the Moon or Mercury. (*NASA/JPL*)

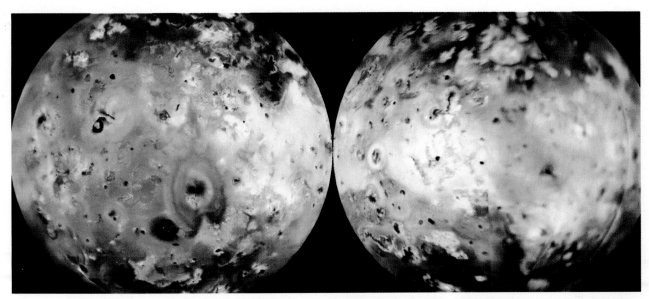

Figure 6.12 Io, showing the colorful volcanic features that dominate its highly active surface and make it unlike any other planet or satellite. (*NASA/USGS*)

the important objectives of the Galileo spacecraft will be to obtain high-resolution images that allow geologists to understand the geological differences between Callisto and Ganymede.

Io, a Volcanic Satellite

Io, the innermost of Jupiter's four Galilean satellites, is in many ways a close twin of the Moon, with nearly the same size and density. Therefore, we might expect it to have experienced a similar history. Its appearance, as photographed from space, tells us another story, however (Figure 6.12). Instead of being a dead, cratered world, Io turns out to have the highest level of volcanism in the solar system, exceeding even that of the Earth.

Io's active volcanism was discovered by the Voyager spacecraft. Eight volcanoes were actually seen erupting when Voyager 1 passed in March 1979, and six of these were still active 4 months later when Voyager 2 passed (Figure 6.13). These eruptions consisted of graceful plumes that extended to heights of hundreds of kilometers into space. The material erupted is not lava or steam or carbon dioxide, all of which are vented by terrestrial volcanoes, but sulfur and sulfur dioxide. As the rising plumes cool, the sulfur and sulfur dioxide recondense as solid particles, which fall back to the surface in colorful "snowfalls" that extend as much as 1000 km from the vent. Io also has a variety of volcanic hot spots, one of them a sort of "lava lake" 200 km in diameter, with "lava" composed of liquid sulfur.

How can Io maintain this remarkable level of volcanism, despite its small size? The answer lies in gravitational heating of the satellite by Jupiter. Io is about the same distance from Jupiter as our Moon is from

(a)

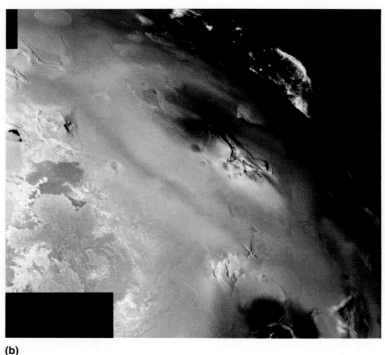

(b)

Figure 6.13 Two views of erupting volcanoes on Io. (a) Crescent view with two eruptions near the edge of the image. (b) A large plume rising above the volcano called Pele. (*NASA/JPL*)

the Earth. Yet Jupiter is more than 300 times more massive than Earth, causing forces that pull the satellite into an elongated shape, with a several-kilometer-high bulge extending toward Jupiter. If Io always kept exactly the same face turned toward Jupiter, this bulge would not generate heat. However, Io's orbit is not exactly circular, because of gravitational perturbations from Europa and Ganymede. In its slightly eccentric orbit, Io twists back and forth with respect to Jupiter, at the same time moving nearer and farther from the planet on each revolution. The twisting and flexing heats Io, much as repeated flexing of a wire coathanger heats the wire.

After billions of years, constant flexing and heating have taken their toll on Io, driving away water and carbon dioxide and other gases, until now sulfur and sulfur compounds are the most volatile materials remaining. The inside is entirely melted, and the crust itself is constantly recycled by volcanic activity. Although Io was well mapped by Voyager, we expect that its surface will wear a different face (at least in part) when the Galileo spacecraft reaches it in 1995, after a 16-year interval.

The Galileo spacecraft will also focus its cameras and other instruments on Europa, the smallest of the four galilean satellites. Like Io, Eupora has a relatively high density and is thought to be composed mostly of rock. However, it lacks Io's tidally driven volcanism, and so it has retained water ice on its surface. Some theoretical work indicates that the icy crust of Europa is only a few kilometers thick. If this is the case, Europa might have an extensive ocean of liquid water beneath the ice—the only other place in the solar system besides Earth where liquid water might be plentiful. Voyager did not obtain very close views of Europa, so this satellite will remain an enigma until the 1995 arrival of Galileo.

Titan: Satellite with an Atmosphere

Titan, found in 1655 by the Dutch astronomer Christian Huygens, is the largest satellite of Saturn. It was the first satellite discovered since Galileo had seen the four moons of Jupiter that bear his name. Titan has about the same diameter, mass, and density as Callisto. Presumably it also has a similar composition: about half ice and half rock. What is remarkable about this satellite, however, is the presence of a substantial atmosphere, discovered telescopically in 1944 (Figure 6.14).

The Voyager flyby of Titan in 1980 determined that the atmospheric surface pressure on this satellite is 1.6 bar, higher than that on any other satellite, and even higher than that of the terrestrial planets Mars and Earth. The atmospheric composition is primarily nitrogen, an important respect in which Titan's atmosphere resembles Earth's.

A variety of additional compounds have been detected in Titan's upper atmosphere, including carbon monoxide (CO), hydrocarbons such as methane (CH_4), ethane (C_2H_6), and propane (C_3H_8), and nitrogen compounds such as hydrogen cyanide (HCN), cyanogen (C_2N_2), and cyanoacetylene (HC_3N). These indicate an active chemistry, in

What would it be like on the surfaces of Titan, Triton, and Pluto?

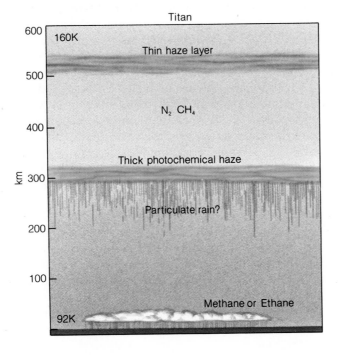

Figure 6.14 Atmospheric structure of Titan. The satellite has the most Earth-like atmosphere in the solar system, although it is much colder than our planet. *(NASA/JPL)*

Figure 6.15   Enhanced color photo of the upper atmosphere of Titan, showing multiple haze layers. (*NASA/JPL*)

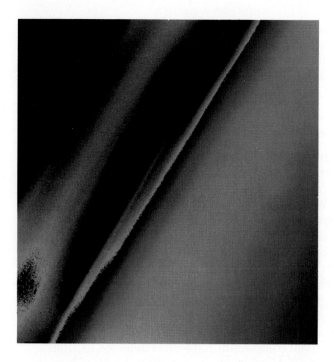

which sunlight interacts with atmospheric nitrogen and methane to create a rich organic mix, in much the way we believe organic compounds were formed on the Earth when it still had a reducing atmosphere. The discovery of HCN is particularly interesting, since this molecule is the starting point for formation of some of the components of DNA, the fundamental genetic molecule essential to life on Earth.

There are multiple cloud layers on Titan (Figure 6.15). The lowest clouds are in the troposphere, within the bottom 10 km of the atmosphere; these are condensation clouds composed of methane. Methane plays the same role in Titan's atmosphere that water does on Earth; the gas is only a minor constituent of the atmosphere, but it condenses to form the major tropospheric clouds. Much higher, there is a dark reddish haze or smog consisting of complex organic chemicals. Formed at an altitude of several hundred kilometers, this aerosol slowly settles downward, where it presumably has built up a deep layer of tar-like organic chemicals on the surface of Titan.

Titan's surface temperature is about 90 K, held uniform by the blanketing atmosphere. At such a low temperature, there may be seas of liquid methane and ethane. Organic compounds are chemically stable, unlike the situation on the warmer, oxidizing Earth; therefore, Titan's surface probably records a chemical history that goes back billions of years. Many scientists believe that this satellite will provide more insights into the early history of Earth's atmosphere, and even into the origin of life, than any other object in the solar system.

Triton, the Coldest Satellite

Triton is the only major satellite in a retrograde orbit, and we have no idea how it came to circle its planet in a backward direction. Beyond this fact, we knew very little about this satellite before the 1989 Voyager encounter.

Triton has a density of 2.1 g/cm^3, indicating a primarily rocky composition with the addition of about 25 percent water ice. Its surface is very bright, reflecting 80 percent of incident sunlight. Consequently, the surface is very cold—at below 40 K, the coldest place known in the solar system. At such low temperatures, almost all known gases freeze, including methane (CH_4), carbon monoxide (CO), carbon dioxide (CO_2), and nitrogen (N_2), and ices of all of these materials probably coat the surface of Triton. A very small amount of nitrogen remains gaseous, however, to form a tenuous atmosphere with a surface pressure of only 10^{-5} bar, nearly a million times lower than on Titan.

Despite its low surface temperature, Triton experiences a kind of low-temperature volcanism. Its surface is geologically young, with few impact craters and numerous other features caused by surface flooding or collapse (Figure 6.16). The volcanically active materials are probably combinations of methane, ammonia, and nitrogen, which can melt at very low temperatures. Voyager even photographed examples of ongoing eruptions, in which small plumes of unknown composition shot up about 10 km above the surface. All in all, this distant satellite turned out to be far more dynamic and interesting than anyone could have guessed before the spacecraft encounter.

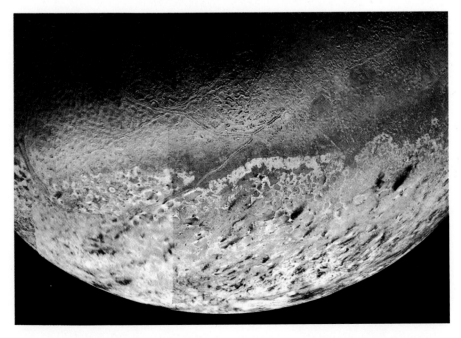

Figure 6.16 Global Voyager view of Triton, showing a wide variety of surface features. The large polar cap dominates the lower part of this image. (*NASA/JPL*)

Figure 6.17 The appearance of Pluto and Charon as computer modeled by Marc Buie and David Tholen, based on observations of mutual occultations. On the left, Charon moves in front of Pluto; on the right it moves behind Pluto.   (*Space Telescope Science Institute, courtesy Marc Buie*)

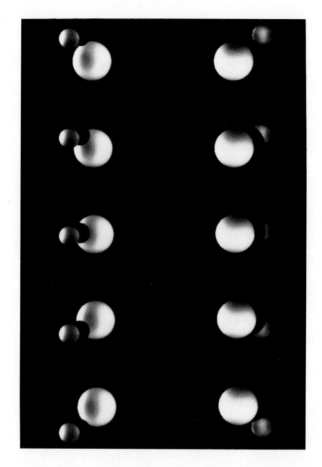

The Pluto-Charon System

We discuss Pluto and its satellite Charon along with the icy satellites of the outer solar system, which they resemble. Pluto is one of the curiosities of the solar system. This little planet, discovered by the American astronomer Clyde Tombaugh in 1930, has never been visited by a spacecraft mission. Even through the largest telescopes, it is faint and unresolved, appearing as a star-like, featureless point of light. Yet astronomers have managed to learn a great deal about Pluto during the past few years, largely because of the 1978 discovery of its satellite, Charon.

The presence of a satellite has permitted us to calculate the mass of Pluto, using Newton's formulation of Kepler's third law. Equally important, the satellite has provided a tool to probe the size and surface composition of both Pluto and Charon, owing to a fortunate alignment of the Earth with the orbit of Charon that occurred between 1985 and 1990. During these few years, Pluto and Charon passed alternately in front of each other, as seen from the Earth, and the changing patterns of reflected light revealed much that could never have been determined otherwise for such small and distant objects (Figure 6.17).

Pluto turns out to be a very small planet indeed, with a diameter of only 2300 km, about 60 percent the diameter of the Moon. Its mass is only 1/20 that of the next smallest planet, Mercury. Pluto is thus very similar in diameter, mass, and density of Neptune's satellite Triton. Charon, in contrast, is relatively large for a satellite, with about half the diameter of Pluto. These two thus make up a sort of double planet, with the two objects orbiting each other in a period of about 6 days.

Pluto and Charon have provided some surprises in terms of their chemistry, too. Pluto has a density of 2.1 g/cm^3, implying a larger fraction of rock and metal (and a smaller fraction of water and other ices) than had been expected this far from the warmth of the Sun, where temperatures are always below 50 K. The surface of Pluto is covered by frozen methane (CH_4), and like Triton it has a thin atmosphere composed of nitrogen with a trace of methane added. Charon, in contrast, has a lower density and a surface of ordinary water ice.

We may never see a spacecraft photo of Pluto or its satellite within our lifetimes. But thanks to careful observations made during the 1980s, we know far more about these distant worlds than had seemed possible before the discovery of Charon.

6.4
PLANETARY RINGS

All four of the giant planets have rings, with each ring system consisting of billions of small particles or moonlets orbiting close to their planet. Each of these rings displays a complex structure apparently related to interactions between the ring particles and the larger satellites. However, the ring systems are very different from each other in mass, structure, and composition (Table 6.3).

Saturn's main ring system, which is by far the largest, is made up of icy particles spread out into several vast flat rings with a great deal of fine structure. Both the Uranus and Neptune ring systems are nearly the reverse, consisting of dark particles confined to a few narrow rings, with broad empty gaps in between. The Jupiter ring and at least one ring of Saturn are merely transient dust bands, constantly renewed by erosion

What causes planetary rings?

TABLE 6.3 Properties of Ring Systems				
Planet	**Outer Radius (km)**	**(R_{planet})**	**Mass (kg)**	**Reflectivity (%)**
Jupiter	128,000	1.8	10^{10} (?)	?
Saturn	140,000	2.3	10^{19}	60
Uranus	51,000	2.2	10^{14}	5
Neptune	63,000	2.5	10^{12}	5

of dust grains from small satellites. In this section, we focus on the two largest ring systems, those of Saturn and Uranus.

What Causes Rings?

A ring is a collection of vast numbers of particles, each obeying Kepler's laws as it follows its own orbit around the planet. Thus the inner particles revolve faster than those farther out, and the ring as a whole does not rotate as a solid body. In fact, it is better not to think of a ring *rotating* at all, but rather to consider the *revolution* of its individual moonlets.

If the ring particles were widely spaced, they would move independently, like separate small satellites. However, in the main rings of Saturn and Uranus the particles are close enough to each other to exert mutual gravitational influence, and occasionally even to rub together or bounce off each other in low-speed collisions. Because of these interactions, phenomena such as waves can be produced that move across the rings, like water waves moving over the surface of the ocean.

There are two basic theories of ring origin. First is the breakup theory, which suggests that the rings are the remains of a shattered satellite. The second theory, which takes the reverse perspective, suggests that the rings are made of particles that were unable to come together to form a satellite in the first place. In either theory, an important role is played by the gravitation of the planet. Close to the planet, gravitational forces can tear a small satellite apart or can inhibit loose particles from

Figure 6.18 Schematic diagram of the ring systems of Jupiter, Saturn, Uranus, and Neptune. All four ring drawings are scaled to the diameters of their respective planets. The dots represent the inner satellites of each planet.

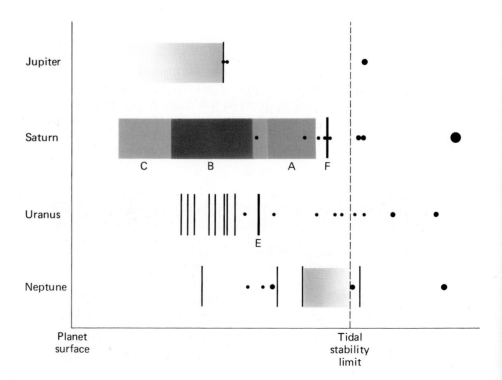

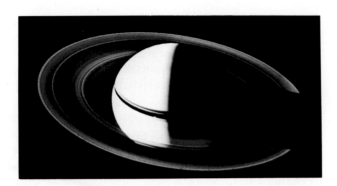

Figure 6.19   Voyager image of the rings of Saturn seen from below; note that except for the B Ring, these rings transmit enough sunlight to be clearly visible.   *(NASA/JPL)*

coming together to form a satellite. The rings of both Saturn and Uranus lie within the region in which large satellites are probably not stable against such gravitational forces (Figure 6.18).

In the breakup theory of ring formation, we can imagine a satellite or even a passing comet coming too close and being torn apart. The fragments then remain in orbit as one or more rings. A more likely variant of this idea suggests that a small satellite close to the planet might be broken apart in a collision, with the fragments dispersing into a disk. We do not know which explanation holds for the rings, although many scientists have concluded that at least a few of the rings are relatively young and must therefore be the result of break-up.

Rings of Saturn

The rings of Saturn are one of the most beautiful sights in the solar system (Figure 6.19). The three brightest rings are labeled (from outer to inner) the A, B, and C Rings. In Table 6.4 the dimensions of the rings are given in both kilometers and in units of the radius of Saturn, R_s. The B Ring is the brightest and has the most closely packed particles, whereas the A and C rings are translucent. The total mass of the B Ring, which is probably close to the mass of the entire ring system, is about equal to that of a 250-km diameter icy satellite.

TABLE 6.4   Rings of Saturn

Ring Name	Outer Edge (R_s)	Outer Edge (km)	Width (km)
F	2.324	140,180	90
A	2.267	136,780	14,600
Cassini Division	2.025	122,170	4,590
B	1.949	117,580	25,580
C	1.525	92,000	17,490

The rings of Saturn are very broad and very thin. The width of the main rings is 70,000 km, yet their thickness is only about 20 m. If we made a scale model of the rings out of paper the thickness of the sheets in this book, we would have to make the rings a kilometer across—about eight city blocks. On this scale, Saturn itself would loom as high as an 80-story building.

The ring particles are composed primarily of water ice, and they span a range of sizes from grains of sand up to house-sized boulders. An insider's view of the rings would probably resemble a bright cloud of floating snowflakes and hailstones, with a few snowballs and larger objects, many of these loose aggregates of smaller particles.

In addition to the broad A, B, and C Rings, Saturn has a handful of very narrow rings, no more than 100 km wide. The largest of these, which lies just outside the A Ring, is called the F Ring. These narrow rings resemble the rings of Uranus and Neptune.

Rings of Uranus

The rings of Uranus are narrow and black, making them almost invisible from the Earth. The nine main rings were discovered in 1977 from observations made of a star as Uranus passed in front of it. We call the passage of one astronomical object in front of another an **occultation.** During the 1977 occultation, the star dimmed briefly as each narrow ring passed between the star and the telescope; thus the rings were mapped out in detail even though they could not be seen or photographed directly. When Voyager approached Uranus in 1986, it was able to study the rings at close range, and the spacecraft also photographed two new rings (Figure 6.20).

Figure 6.20 The rings of Uranus: A dozen narrow ribbons of dark particles, much different from the broad bright rings of Saturn. (*NASA/JPL*)

The outermost and most massive of the rings of Uranus is called the Epsilon Ring. It is only about 100 km wide, and its thickness is probably no more than 100 m (similar to the F Ring of Saturn). The Epsilon Ring circles Uranus at a distance of 51,000 km, or 2.0 Uranus radii. This ring probably contains as much mass as all of the other ten rings combined. With one exception, all of the other rings of Uranus are narrow ribbons less than 10 km in width—just the reverse of the broad rings of Saturn.

The individual particles in the uranian rings are nearly as black as lumps of coal. Although astronomers do not understand the composition of this material in detail, it seems largely to consist of black carbon and hydrocarbon compounds. Organic material of this sort is rather common in the outer solar system. Many of the asteroids and comets (discussed in the next chapter) also are composed of dark tar-like materials. In the case of Uranus, its ten small inner satellites have a similar composition, suggesting that one or more satellites might have broken up to make the rings.

The rings of Neptune are generally similar to those of Uranus but even more tenuous (Figure 6.21). There are just four rings, and the particles are not uniformly distributed along their length. Because these rings are so difficult to investigate from the Earth, it will probably be a long time before we understand them well.

How do the rings of Saturn and Uranus differ?

Figure 6.21 The rings of Neptune: Similar to the narrow dark rings of Uranus. (*NASA/JPL*)

Satellite-Ring Interactions

Much of our current fascination with planetary rings is a consequence of the intricate structures discovered by the Voyager spacecraft. Most of this structure owes its existence to the gravitational effect of satellites. If there were no satellites, the rings would be flat and featureless. Indeed, it there were no satellites, there would probably be no rings at all, because left to themselves, thin disks of matter gradually spread and dissipate. The sharp edges as well as the fine structure of rings are due to gravitational effects of satellites.

Most of the gaps in Saturn's rings, and the location of the outer edge of the A Ring, result from gravitational resonances with small inner satellites. A **resonance** takes place when two objects have orbital periods that are exact ratios of each other, such as one-half or one-third. For example, any particle in the gap at the inner side of the Cassini Division of Saturn's rings would have a period equal to one-half that of Saturn's satellite Mimas. Such a particle would be nearest Mimas in the same part of its orbit every second revolution. The repeated gravitational tugs of Mimas, acting always in the same direction, would perturb it, forcing it into a new orbit that is outside the gap and thus no longer represents a resonance.

One of the most interesting rings of Saturn is the narrow F Ring, which contains several apparent ringlets within its 90-km width. In places, the F Ring breaks up into two or three parallel strands, which sometimes show bends or kinks. Most of the rings of Uranus and Neptune are narrow ribbons like the F Ring of Saturn. What defines these narrow rings and keeps their particles from spreading out?

The best theory is that the rings are controlled gravitationally by small satellites orbiting very close to them. This certainly seems to be the

What produces the intricate structures seen in rings?

Figure 6.22   Voyager close-up of Saturn's shepherd satellite Pandora with the thin F Ring, which the satellite helps to control.   (*NASA/JPL*)

Figure 6.23 Saturn's B Ring, showing intricate structure, much of it the result of waves propagating through billions of orbiting, self-gravitating particles that comprise the rings.   (*NASA/JPL*)

case for Saturn's F Ring, which is bounded by the orbits of satellites Pandora and Prometheus (Figure 6.22). These two small objects (having diameters of about 100 km) are referred to as "shepherd satellites," since their gravitation serves to "shepherd" the ring particles and keep them confined to a narrow ribbon. A similar situation applies to the Epsilon Ring of Uranus, which is shepherded by the satellites Cordelia and Ophelia. These two shepherd satellites, each about 50 km in diameter, orbit about 2000 km inside and outside the ring.

Theoretical calculations suggest that the other narrow rings in the uranian and neptunian systems should also be controlled by shepherd satellites, but none has been located. The calculated diameter for such shepherds—about 10 km—was just at the limit of detectability for the Voyager cameras, so it is impossible to say if they are present or not. Given the multiplicity of such narrow rings, many scientists hope that we will find a more satisfactory alternative explanation.

Making Waves in Rings

Saturn's major B Ring has no gaps, but photos show intricate structure, partly in the form of waves (Figure 6.23). Each wave corresponds to alternating ringlets in which the ring particles are bunched together or spread more thinly. These waves, which are typically separated by 100 km or so, look like the grooves in an old-fashioned phonograph record.

The A Ring has even more of this wave-like structure, and smaller waves appear to be present even in narrow rings, such as those in the saturnian F Ring and the uranian Epsilon Ring.

Many of the observed waves take the form of tightly wound spirals. These are called spiral density waves, produced at distances from Saturn corresponding to resonances, mostly with the inner satellites. Astronomers are especially interested in these waves since their cause seems to be similar to that which generates the spiral arms of galaxies, with the role of individual stars being played here by the ring particles. We discuss the structure of galaxies in Chapter 13.

The main rings of Saturn include several narrow gaps that are clearly the result of satellites, which clear lanes in a manner similar to that of the shepherd satellites. As each small satellite moves through its gap, it produces waves in the surrounding ring material like the wake left by a moving ship. One of these satellites, called Pan, was discovered in 1991 from its wake; calculations based on the waves it produced pinpointed its location, and when the appropriate ten-year-old Voyager picture was examined, there it was!

Studies of planetary rings and of the gravitational interactions between rings and small satellites have come a long way in the past few years, but many problems in understanding these complex phenomena remain unsolved. Without additional spacecraft missions to the outer planets, however, it may be difficult to obtain the data needed to test the new dynamical theories now being developed. NASA and ESA (the European Space Agency) are planning a new mission to Saturn called Cassini, which is tentatively scheduled to begin orbiting the planet in 2002. If successful, the Cassini mission (which will also drop a probe into the atmosphere of Titan), should provide a vast amount of new information on the rings.

▶ SUMMARY

6.1 Jupiter, with a mass more than 300 times that of the Earth, contains more matter than the rest of the planetary system combined. Its diameter (143,000 km) is about the largest possible for a planetary object. Jupiter has a *cosmic* composition (mostly hydrogen and helium), and most of its interior is liquid. Heated at birth, it still radiates as much heat from its interior as it receives from the Sun. Clouds composed of ammonia, ammonium hydrosulfide, and water float in the atmosphere, which displays a variety of colorful forms, including such long-lived storms as the Great Red Spot, which is twice the size of the Earth.

6.2 Saturn, Uranus, and Neptune are also giant planets, each with atmospheres composed of hydrogen and helium. Saturn (about one-fourth the mass of Jupiter) resembles Jupiter most, although its clouds are less colorful and its internal heat has a different source. Uranus and Neptune are much less massive and lack the large liquid hydrogen interior of Jupiter and Saturn. Uranus has a peculiar rotation and no internal heat source; its clouds lie low and are invisible from above. Neptune, in contrast, has highly visible clouds in its upper atmosphere.

6.3 All four giant planets have satellite systems, and most of these satellites are mainly composed of H_2O ice. The largest satellites are comparable in size to the Moon and Mercury. Callisto is a prototypical, large icy satellite, with a differentiated interior and a heavily cratered surface. Io is the most volcanically active object in the solar system, with constant eruptions of sulfur and SO_2. Titan has an N_2 atmosphere greater than that of the Earth and supports a complex organic chemistry. Triton is the coldest satellite, yet it has experienced some kind of volcanic activity. Finally, Pluto and its satellite Charon are distant icy worlds, with Pluto perhaps resembling Triton.

6.4 All of the jovian planets have rings, but the largest and best studied are the rings of Saturn and Uranus, using both spacecraft imaging and **occultation** observations. The ring systems of both of these planets lie close to the planet and may be the result of break-up of one or more inner satellites. The Saturn rings are broad, bright, and have several gaps. The rings of Uranus are narrow, black, and separated by wide gaps. The rings of both planets have intricate structures (such as gaps and waves) that result from gravitational interactions with satellites, especially at positions corresponding to **resonances.**

▶ REVIEW QUESTIONS

1. Describe the fundamental ways in which Jupiter differs from the terrestrial planets (i.e., size, composition, structure, internal energy, etc.).

2. Compare the atmospheres of the four jovian planets in terms of composition, structure, clouds, and weather.

3. List the main ways in which Uranus and Neptune differ from Jupiter and Saturn.

4. Contrast the following objects, all of which have similar sizes: the Moon, Callisto, Io, and Titan. Why are they so different from each other?

5. Why was the mass of Pluto not known accurately before 1978?

6. Compare and contrast the rings of Saturn and Uranus.

▶ THOUGHT QUESTIONS

7. What would you expect a "planet" with five times the mass of Jupiter to be like?

8. Jupiter is more dense than water, yet it is composed for the most part of two light gases, hydrogen and helium. How can it be so dense?

9. Explain why the clouds (and their associated storms) are less easily observed on Saturn and Uranus than they are on Jupiter.

10. Would you expect to find oxygen gas in the atmospheres of the jovian planets? Why or why not?

11. The water clouds believed to be present on Jupiter and Saturn exist at pressures and temperatures similar to those at the locations of the clouds in the terrestrial troposphere. What would it be like to visit such a location on Jupiter or Saturn? In what ways would the environment differ from that in the clouds of Earth?

12. Would you expect to find more impact craters on Callisto or on Io? Why?

13. Io is almost the same size and mass as our Moon. Why is it so much more active volcanically? How does the level of volcanism on Io today compare with that on the Moon 3.5 billion years ago, when the maria were formed?

14. Do you think there are many impact craters on the surface of Titan? Why or why not?

15. Why do you suppose the rings of Saturn are made of bright particles, whereas the particles in the rings of Uranus are black?

▶ PROBLEMS

16. Jupiter's Great Red Spot rotates completely in six days. If the spot is circular (not quite true) and 20,000 km in diameter (approximately correct), what are the wind speeds at the outer edges of this storm? How do these speeds compare with the winds in terrestrial hurricanes?

17. Astronomers estimate that the volcanic eruptions of Io add about 12 mm of new material to the surface each year. At this rate, how long will it take to recycle the entire crust of Io if it is 20 km thick?

18. The rings of Saturn are about 70,000 km across but only 20 m thick. Suppose you wanted to build a model of Saturn's rings out of 2-mm thick plastic. To keep the proportions right, how broad would the rings in your model be?

*19. Use the data in Table 6.4 to calculate the ratio of the period of revolution around Saturn for particles at the inner edge of the rings compared to those in the F Ring.

*20. The mass of Saturn is about six times greater than the mass of Uranus. Use Newton's modification of Kepler's laws to calculate which revolves more quickly: the F Ring of Saturn or the Epsilon Ring of Uranus.

*21. The one-way trip time between two planets is just half the period of revolution for the class of orbits described in Problem 2.18. Calculate the one-way trip time from Earth to (a) Jupiter, (b) Saturn, (c) Uranus, and (d) Neptune. Compare these trip times with the voyage of Voyager 2, which was launched in 1977 and reached these four planets in 1979, 1981, and 1989, respectively. Do you know the reason for the differences?

▶ SUGGESTIONS FOR ADDITIONAL READING

Books

Morrison, D. *Exploring Planetary Worlds.* W. H. Freeman, 1993. Up-to-date overview of the solar system.

Littmann, M. *Planets Beyond: Discovering the Outer Solar System,* rev. ed. John Wiley, 1990. An overview of Uranus, Neptune, and Pluto by a science educator.

Morrison, D. *Voyage to Saturn*. NASA Special Publication 451, available from the U. S. Government Printing Office, 1981. Story of the Voyager encounters with Jupiter and Saturn, from the perspective of the scientist involved.

Washburn, M. *Distant Encounters: The Exploration of Jupiter and Saturn*. Harcourt, Brace, Jovanovich, 1983. A well-written journalist's account of the Voyager Program and results.

Miner, E. *Uranus*. Ellis Horwood/Simon & Schuster, 1990. A review of the Voyager mission results by one of the project scientists.

Davis, J. *Flyby*. Atheneum, 1987. Journalist's account of the Voyager flyby of Uranus.

Rothery, C. *Satellites of the Outer Planets*. Oxford U. Press, 1992. A geologist examines the moons and their properties.

Articles

Jupiter

Gore, R. "Voyager Views Jupiter" in *National Geographic*, Jan. 1980.

Kaufmann, W. "Jupiter: Lord of the Planets" in *Mercury*, Nov./Dec. 1984, p. 169.

Morrison, D. "Four New Worlds" in *Astronomy*, Sept. 1980, p. 6. On the Galilean satellites.

Talcott, R. "The Violent Volcanoes of Io" in *Astronomy*, May 1993, p. 41.

Elliott, J. & Kerr, R. "How Jupiter's Ring Was Discovered" in *Mercury*, Nov./Dec. 1985, p. 162.

Saturn

Gore, R. "Saturn: Lord of the Rings" in *National Geographic*, July 1981.

Overbye, D. "Lord of the Rings" in *Discover*, Jan. 1981, p. 24.

Murrill, M. "Voyager: The Grandest Tour" in *Mercury*, May/June 1993, p. 67. Good introduction to all the Voyager destinations.

Sanchez-Lavega, A. "Saturn's Great White Spot" in *Sky & Telescope*, Aug. 1989, p. 141.

Uranus

Gore, R. "Uranus: Voyager Visits a Dark Planet" in *National Geographic*, Aug. 1986.

Overbye, D. "Voyager Was on Target Again" in *Discover*, Apr. 1986, p. 70.

Detailed coverage of the Voyager Uranus flyby can be found in the April and October 1986 issues of *Sky & Telescope;* and the April and May 1986 issues of *Astronomy.*

Elliott, J., et al. "Discovering the Rings of Uranus" in *Sky & Telescope*, June 1977, p. 412.

Neptune

Gore, R. "Voyager's Last Picture Show" in *National Geographic,* Aug. 1990.

Limaye, S. "Neptune's Weather Forecast: Cloudy, Windy, and Cold" in *Astronomy*, Aug. 1991, p. 38.

Kaufmann, W. "Voyager at Neptune" in *Mercury*, Nov./Dec. 1989, p. 174.

Coverage of the Voyager Neptune encounter can be found in the October 1989 and February 1990 issues of *Sky & Telescope*; and the November and December 1989 issues of *Astronomy*.

Croswell, K. "The Titan/Triton Connection" in *Astronomy*, Apr. 1993, p. 26.

Pluto

Croswell, K. "Pluto: Enigma at the Edge of the Solar System" in *Astronomy*, July 1986, p. 6.

Beatty, J. & Killian, A. "Discovering Pluto's Atmosphere" in *Sky & Telescope*, Dec. 1988, p. 624.

Harrington, R. & B. "The Discovery of Pluto's Moon" in *Mercury*, Jan./Feb. 1979, p. 1.

ORIGIN OF THE SOLAR SYSTEM

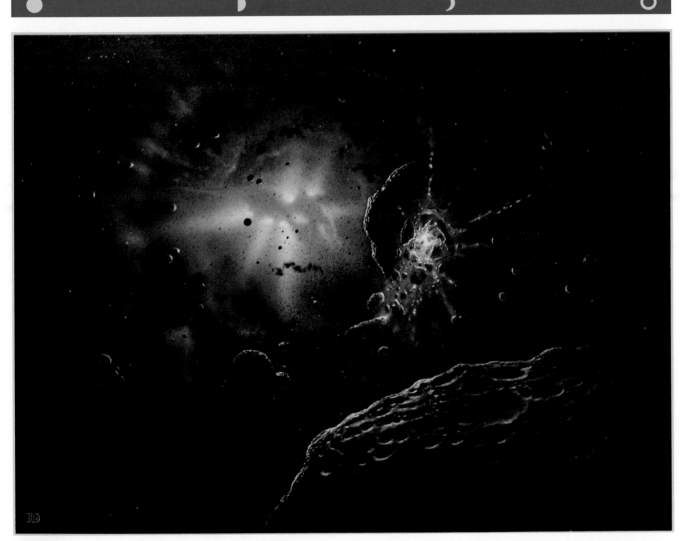

Where did we come from? This is one of the most basic questions asked by astronomers (or anyone else). In previous chapters we have traced much of the history of the Earth and the other planets since they were formed, but we have not addressed the more basic problem of the origin of the solar system itself.

On questions of origins, the planets themselves are largely mute. Melted, battered by giant impacts, repeatedly resurfaced, they retain little evidence of their births. Rather, we must turn to clues provided by the surviving remnants of the creation process: the comets, asteroids, meteors, and meteorites.

Collision between planetesimals during the formative stage of the solar system. *(Painting by Don Dixon)*

163

Asteroids are small objects, differing from the planets primarily in size. Sometimes they are called minor planets. A comet is also small, but it is defined as an object with a visible transient atmosphere and extended tail of gas and dust. This definition reflects primarily a difference in composition between comets and asteroids: comets contain water ice and other volatiles that vaporize when heated by the Sun, whereas asteroids are rocky objects with little volatile material.

All comets and most asteroids are primitive bodies, *which have been relatively little altered chemically or physically since the formation of the solar system. They thus provide a window on the period before the planets formed. The meteorites and other bits of cosmic debris are fragments of comets and asteroids that we can study directly in our laboratories. These laboratory studies, in combination with astronomical observations of the comets and asteroids, provide much of the information that is needed to probe into the birth of the planets.*

7.1
ASTEROIDS

The asteroids are rocky, metallic objects that probably represent the population of small bodies that formed the building blocks of the terrestrial planets. Most of the asteroids have orbits between those of Mars and Jupiter, in a region called the **asteroid belt.**

Discovery of the Asteroids

The asteroids are too small to be seen without a telescope, and the first of them was not discovered until the beginning of the 19th century. The largest asteroid is Ceres, with a diameter just under 1000 km. Two (Pallas and Vesta) have diameters near 500 km, about 15 are larger than 250 km (Table 7.1), and perhaps 100,000 are larger than 1 km. The number of asteroids increases rapidly with decreasing size; there are about 100 more objects 10 km across than 100 km across. The total mass, which probably represents just a tiny fraction of the original asteroid population, is less than that of the Moon.

The asteroids all revolve about the Sun in the same direction as the planets (from west to east), and most of them have orbits that lie near the plane of the Earth's orbit. The asteroid belt is defined to contain all asteroids with semimajor axes in the range 2.2 to 3.3 AU, with corresponding periods of orbital revolution about the Sun from 3.3 to 6 years (Figure 7.1). Although more than 75 percent of the asteroids are in the main belt, they are not closely spaced. The volume of the belt is actually very large, and the typical spacing between objects (down to 1 km size) is several million kilometers.

Composition and Classification

Where are the asteroids and what are they made of?

Asteroids are as different as black and white. The majority are very dark, with reflectivities of only 3 to 4 percent, like a lump of coal. However, there is another large group with typical reflectivities of about

TABLE 7.1   The Largest Asteroids

Name	Discovery	Semimajor Axis (AU)	Diameter (km)	Type	Surface Composition
Ceres	1801	2.77	940	C	Carbonaceous
Pallas	1802	2.77	540	C	Carbonaceous
Vesta	1807	2.36	510	U	Basaltic
Hygeia	1849	3.14	410	C	Carbonaceous
Interamnia	1910	3.06	310	C	Carbonaceous
Davida	1903	3.18	310	C	Carbonaceous
Cybele	1861	3.43	280	C	Carbonaceous
Europa	1868	3.10	280	C	Carbonaceous
Sylvia	1866	3.48	275	C	Carbonaceous
Juno	1804	2.67	265	S	Silicate
Psyche	1852	2.92	265	M	Metallic
Patientia	1899	3.07	260	C	Carbonaceous
Euphrosyne	1854	3.15	250	C	Carbonaceous
Eunomia	1851	2.64	250	S	Silicate

16 percent, a little greater than that of the Moon, and still others have reflectivities as high as 60 percent.

The dark asteroids are revealed from telescopic analysis to be primitive bodies, composed of silicates mixed with dark organic carbon compounds. These are usually called C asteroids, where the C stands for "carbonaceous." Most of the large asteroids (including Ceres) are primitive, as are almost all of the objects in the outer third of the belt. Many of the lighter colored asteroids with silicate surfaces are probably also primitive objects, but their exact composition is a matter of dispute, since no definitive measurements exist.

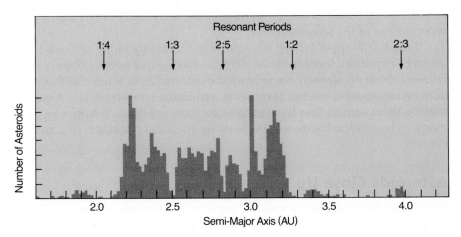

Figure 7.1   The main asteroid belt: The distribution of the number of asteroids at various distances from the Sun.

ALHA78132, 2

Figure 7.2 Photo of one of the basaltic meteorites that are believed to be fragments from the crust of asteroid Vesta. *(NASA/JSC)*

Why are we interested in the Earth-approaching asteroids?

Observations have identified a few asteroids that are either differentiated objects or the fragments of differentiated objects. These include a class of asteroids composed primarily of metal, called the M asteroids. A metal asteroid, like an airplane or a ship, is a much better reflector of radar than is a stony object. Thus, the largest of the M asteroids (Psyche) has been clearly identified by its high radar reflectivity. There is enough metal in even a 1-km M asteroid to supply the world's needs for iron and most other industrial metals for the foreseeable future, if we could only bring one to Earth. There are also a few known asteroids that have basaltic surfaces like the volcanic plains of the Moon and Mars. The large asteroid Vesta is in the latter category (Figure 7.2). Apparently some of the asteroids were heated and experienced surface volcanic activity early in the history of the solar system, but why such a small percentage of the total number were heated, we do not know.

Earth-Approaching Asteroids

Not all of the asteroids are in the main asteroid belt. Many lie beyond the belt, all of them dark, primitive objects. Of greater interest, because they are easier to study, are the asteroids with orbits that come close to the orbit of the Earth. Some of these are the nearest approaching celestial objects, excluding the Moon and meteorites. In 1993, one passed within 150,000 km. Some have collided with the Earth in the past, and others will continue to do so, as we saw in Chapter 3. These are known collectively as **Earth-approaching asteroids.**

At present, fewer than 200 Earth-approaching asteroids have been located. The total population of such objects is calculated at approximately 2000 asteroids that are larger than 1 km across, and at least 1 million asteroids larger than 10 m. The largest is about 20 km across. Searches for additional Earth-approaching asteroids result in the discovery of two or three new objects each month.

The orbits of Earth-approaching asteroids are unstable. These objects will meet one of two fates: Either they will impact one of the terrestrial planets, or they will be ejected gravitationally from the inner solar system as the result of a near-encounter with a planet. The probabilities of these two outcomes are about the same. The timescale for impact or ejection is only a few hundred million years, very short in comparison with the age of the solar system.

If the Earth-approaching asteroids are constantly being removed by impact or ejection, there must be a continuing source of new objects. Some of these apparently come from the asteroid belt, where collisions between asteroids can eject fragments into Earth-crossing orbits. Others may be dead comets that have exhausted their volatiles. Possibly as many as half of the Earth-approachers are the solid remnants of former comets.

Asteroids Close Up

The asteroids were the last major component of the solar system to be visited by spacecraft. In late 1991, the Galileo spacecraft flew within

1600 km of the small, main-belt asteroid, Gaspra, on its way toward Jupiter. A second flyby of a larger asteroid called Ida took place in the summer of 1993. Both Gaspra and Ida are in the S or stony class, and each is thought to be a fragment produced in an ancient collision within the asteroid belt.

Two other objects that are closely related to the asteroids are the martian moons Phobos and Deimos. Both are irregular, dark objects about 20 km across, quite probably captured asteroids. For us, their importance is amplified because they were photographed in detail by the Viking orbiters, and we are likely to obtain additional detailed information on them as a byproduct of future Mars missions.

Radar has been used to obtain images of a few Earth-approaching asteroids that pass very close to the Earth. The best radar images are of Toutatis, which flew by in December of 1992. Figure 7.3 illustrates a series of these radar images as the asteroid rotates. Toutatis is a double asteroid, consisting of two roughly spherical components with diameters of about 4 and 3 km. Radar data on another small asteroid, Castalia, show that it is double also, and from some angles Gaspra also appears to have formed by joining two independent objects. Apparently many small asteroids are double—a surprising result that had not been anticipated before we actually looked with sufficient resolution to see what these asteroids really are like.

Spacecraft photos of Gaspra, Phobos, and Deimos are illustrated in Figure 7.4 (a), (b), (c). Each is irregular and cratered, but there are significant differences between them. Gaspra has the most small craters, telling us that small debris is more plentiful within the asteroid belt than had been expected. Phobos has dozens of long grooves or cracks that suggest the satellite was once nearly broken apart in a collision; Gaspra has a few smaller grooves, but none have been seen on Deimos. Deimos is characterized by a relatively smooth surface that looks as if its craters have been buried under a thick blanket of debris. In contrast, the craters on both Gaspra and Phobos are mostly sharp and fresh looking. Asteroid Ida is shown in Figure 7.4 (d).

7.2
THE LONG-TAILED COMETS

Comets have been observed from the earliest times; accounts of spectacular comets are found in the histories of virtually all ancient civilizations. A typical comet has the appearance of a rather faint, diffuse spot of light, somewhat smaller than the Moon and many times less brilliant (Figure 7.5). There may be a very faint nebulous **tail,** extending several degrees away from the main body of the comet.

Today we recognize comets as the best preserved, most primitive material available in the solar system. Stored in the deep freeze of space, these icy objects are messengers from the distant past, providing us unique access to the initial material from which the planets formed 4.5 billion years ago.

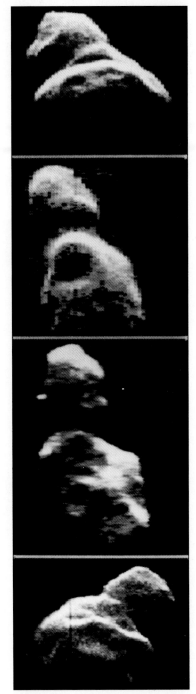

Figure 7.3 Radar images of the Earth-approaching asteroid Toutatis, obtained during its close pass by the Earth in December 1992. The asteroid appears to consist of two irregular, roughly spherical objects with diameters of about 3 and 2 km, respectively, slowly rotating in contact with each other.   (*NASA/JPL, courtesy of Steve Ostro*)

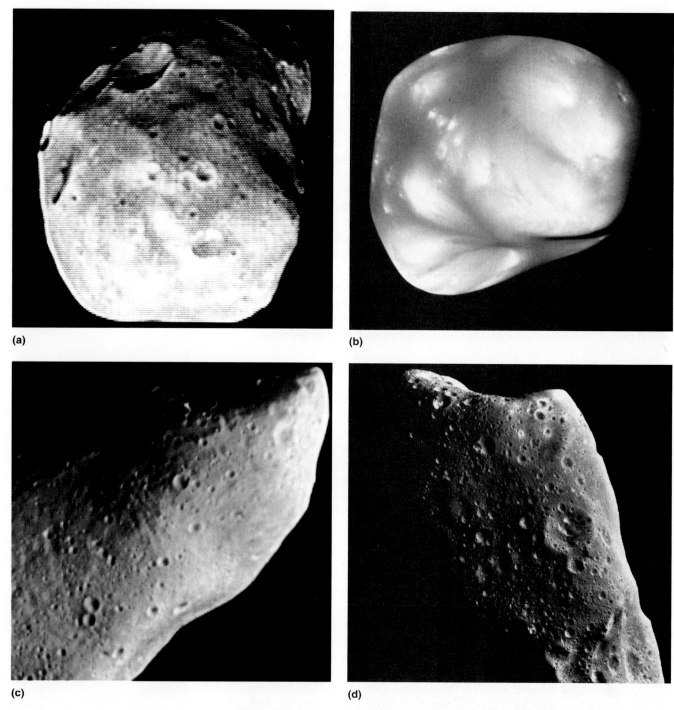

(a)

(b)

(c)

(d)

Figure 7.4 Comparison of the two martian satellites, Phobos (a) and Deimos (b), with the main-belt asteroids Gaspra (c) and Ida (d). All have irregular shapes, reflecting a history of catastrophic collisions; however, the number and freshness of their craters differ greatly. The images of Phobos and Deimos were obtained by Viking, and those of Gaspra and Ida by Galileo; all four images have a resolution of about 100 m. (*NASA/JPL*)

Comet Orbits

The study of comets as members of the solar system dates from the time of Newton, who first suggested that their orbits were extremely elongated ellipses. Newton's colleague Edmund Halley developed these ideas, and in 1705 he published calculations of 24 cometary orbits. In particular, he noted that the orbits of the bright comets of 1531, 1607, and 1682 were so similar that the three could well be the same comet, returning to perihelion at average intervals of 76 years. If so, he predicted that the object should return about 1758. When the comet did return as predicted, it was given the name Comet Halley, in honor of the man who first recognized it to be a permanent member of the solar system.

Comet Halley has been observed and recorded on every passage near the Sun at intervals from 74 to 79 years since 239 B.C. The period varies somewhat because of changes in its orbit produced by the jovian planets. In 1910, the Earth was brushed by the comet's tail, and Comet Halley last appeared in 1986 (Figure 7.6).

Observational records exist for about 1000 comets. Today, new comets are discovered at an average rate of five to ten per year. Most never become conspicuous and are visible only on photographs made with large telescopes. Every few years, however, a comet may appear

Figure 7.5 Comet Halley in the spring of 1986 had the appearance typical of a moderately bright comet with a tail a few degrees long. It is shown here rising above Mauna Kea Observatory in Hawaii. (*William Golisch, University of Hawaii*)

Figure 7.6 Comet Halley photographed from Australia in December 1985. This color picture is produced from three separate exposures taken through individual filters, which is why the stars appear as individual colored streaks. (*Anglo-Australian Observatory*)

that is bright enough to be seen easily with the unaided eye. The brightest comet of recent years was Comet West in 1976. Comet Halley, in 1986, was less bright as a result of its poor placement relative to the Earth, although it put on a good show, when seen from the southern hemisphere.

The Comet's Nucleus

How do comets differ from asteroids?

When we look at a comet we see its transient atmosphere of gas and dust illuminated by sunlight. Because the escape velocity from such small bodies is very low, the rapidly escaping atmosphere we see must be coming from the heart of the comet's head, the small solid **nucleus** hidden by the glow of the atmosphere. The nucleus is the *real* comet, the fragment of primitive material preserved since the beginning of the solar system.

No comet nucleus had been seen or measured until this past decade, and the very existence of the nucleus as a single solid object was speculative. Now most of the studies of comets are directed toward a better understanding of the nucleus.

The modern theory of the physical and chemical nature of comets was first proposed by Harvard astronomer Fred L. Whipple in 1950.

Figure 7.7   A particle that is believed to be a tiny fragment of cometary dust, collected in the upper atmosphere of the Earth. It is composed of a mixture of silicate and carbonaceous (organic) material. (*Donald Brownlee, University of Washington*)

Before Whipple's work, many astronomers had thought that the nucleus of a comet might be a loose aggregation of solids of meteoritic nature—a sort of orbiting gravel bank. Whipple proposed instead that the nucleus is a solid object a few kilometers across, composed of water ice mixed with silicate grains and dust. This proposal became known as the "dirty snowball" model for the nucleus of a comet.

The water vapor and the other volatiles that escape from the nucleus when it is heated can be detected telescopically in the comet's head and tail. We are somewhat less certain of the non-icy component of the nucleus, however. No large fragments of solid matter from a comet have ever survived passage through the Earth's atmosphere to be studied as meteorites. Some very fine, microscopic grains of comet dust have been collected in the Earth's upper atmosphere, however, and have been studied in the laboratory (Figure 7.7). The spacecraft that encountered Comet Halley in March 1986 also carried dust detectors. From these various investigations it seems that much of the "dirt" in the dirty snowball is in the form of tiny bits of dark, primitive hydrocarbons and silicates, somewhat like the material thought to be present on dark primitive asteroids.

Because the comet nucleus is small and dark, it is a difficult object for astronomers to study. Even measuring its diameter has been a problem. However, recent radar observations of several faint comets have indicated nucleus diameters of 5 km to 10 km. Comet Halley's nucleus, as measured in 1986, has dimensions of about 8 × 12 km.

Activity of the Nucleus

The spectacular activity of a comet that gives rise to its atmosphere and tail results from the evaporation of cometary ices as they are heated by sunlight. Beyond the asteroid belt, where a comet spends most of its

What would it be like on a comet as it orbits the Sun?

Figure 7.8 Composite photograph of the black, irregularly shaped nucleus of Comet Halley. This image, obtained by the Giotto spacecraft at a distance of about 1000 km, has a resolution of better than 1 km. *(Max Planck Institut für Aeronomie and Ball Aerospace Corporation, courtesy Harold Reitsema)*

time, the ices are solidly frozen. But as a comet approaches the Sun, it begins to warm up. If water is the dominant ice, a significant quantity vaporizes as temperatures rise toward 200 K, somewhat beyond the orbit of Mars. The evaporating water in turn releases any dust that was mixed with the ice. Because the comet nucleus is so small, its gravity cannot hold back either the vapor or the dust, both of which flow away into space at speeds of about 1 km/s.

The comet continues to absorb energy as it approaches the Sun. A great deal of this energy goes into the evaporation of its ice as well as into the heating of its surface. However, observations of many comets indicate that the evaporation is not uniform, and that most of the gas is released in sudden spurts, perhaps confined to only a few areas of the surface. Such jets were observed directly on the surface of Comet Halley by the spacecraft that photographed it in 1986, which showed that the jets resembled volcanic plumes or geysers (Figure 7.8). Most of Comet Halley's surface is apparently inactive, with the ice buried under a layer of black silicates and carbon compounds.

The Comet's Atmosphere

The atmosphere of a comet is composed of the gas released from the nucleus together with the dust and other solid material being carried along with it. Expanding at a speed of about 1 km/s, the atmosphere can reach an enormous size. The diameter of the comet's head is usually as large as Jupiter, and it often approaches 1 million km. The composition of the gas is primarily H_2O (about 80 percent in the case of Comet Halley), a

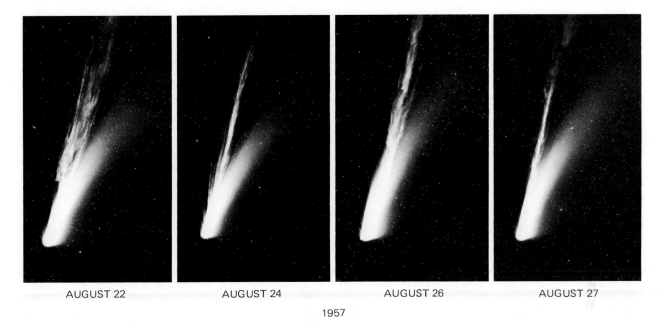

AUGUST 22   AUGUST 24   AUGUST 26   AUGUST 27

1957

Figure 7.9 Comet Mrkos, photographed with the Schmidt telescope at Palomar Observatory. Note the features in the tail that are swept back by the solar wind. (*CalTech/Palomar Observatory*)

few percent CO_2, and small quantities of many additional gases, including hydrocarbons. Hydrogen is produced when solar ultraviolet light breaks up the molecules of H_2O. Huge hydrogen clouds, up to tens of millions of kilometers across, are formed around comets.

Many comets develop tails as they approach the Sun (Figure 7.9). The tail of a comet is an extension of its atmosphere, consisting of the same gas and dust that make up the head. As early as the 16th century, observers realized that comet tails point away from the Sun (Figure 7.10), not back along the comet's orbit. Newton attempted to account for

Figure 7.10 Orientation of a typical comet tail as the comet passes perihelion.

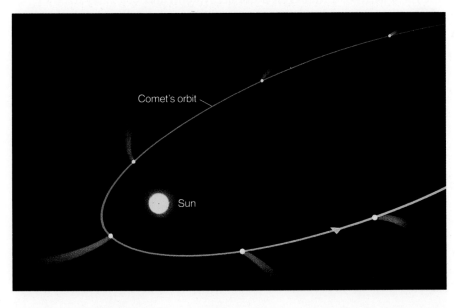

TABLE 7.2   Some Famous Comets

Name	Period	Significance
Great Comet of 1811	Long	Largest head observed (2 million km across)
Great Comet of 1843	Long	Brightest ever; visible in daylight
Daylight Comet of 1910	Long	Brightest of 20th century
West (1976)	Long	Best recent comet; nucleus split
Swift–Tuttle	133 years	Parent of Perseid meteor shower
Halley	76 years	Studied in detail by spacecraft in 1986
Biela	6.7 years	Broke up in 1846 and disappeared
Giacobini–Zinner	6.5 years	First spacecraft encounter in 1985
Encke	3.3 years	Shortest known period
Shoemaker–Levy 9	Unknown	Collides with Jupiter in 1994

comet tails by a repulsive force of sunlight driving particles away from the comet's head, an idea that is close to our modern view. In addition to sunlight, however, we now know that cometary gas is also repulsed by streams of charged atomic particles that are emitted by the Sun.

Table 7.2 provides information on several famous comets. Included are long-period comets, which have been seen only once, as well as those that return regularly.

7.3
ORIGIN AND EVOLUTION OF COMETS

Comets are messengers from the cold depths of space. They represent the most primitive matter in the solar system, and they may even include material that has been preserved from interstellar clouds that predate the formation of the solar system.

The Oort Comet Cloud

Where do comets come from?

Although comets are part of the solar system, observations show that they come initially from very great distances. Observationally, the aphelia of new comets typically have values near 50,000 AU. This clustering of aphelion distances was first noted by Dutch astronomer Jan Oort, who in 1950 proposed a scheme for the origin of the comets that is still accepted today.

It is possible to calculate that the gravitational sphere of influence of a star—the distance within which it can exert sufficient gravitation to hold onto orbiting objects—is about one-third of the distance to the nearest other stars. In the vicinity of the Sun, stars are spaced such that the Sun's sphere of influence extends only a little beyond 50,000 AU. At

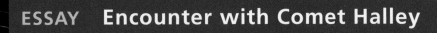

ESSAY   Encounter with Comet Halley

Exploration of the small, primitive members of the solar system began in 1986, when an international flotilla of six spacecraft reached Comet Halley as it crossed the plane of the Earth's orbit a month after perihelion. Two of these—the U.S. ICE spacecraft and a small Japanese craft named Sakagaki—served primarily to monitor the comet from a distance of several million kilometers, while a second Japanese spacecraft—Suisei—passed about 1 million km from the comet. The primary exploration tasks, however, were undertaken by three craft targeted for the nucleus itself.

Because of its retrograde (backward) orbit, Comet Halley can only be approached head-on, at speeds of about 70 km/s. The encounter period is, therefore, very short—on the order of an hour or two—and the nucleus itself is passed in less than a second. The high speed of the spacecraft relative to the comet also created extra risks, since at 70 km/s even a tiny grain of dust could have pierced the spacecraft like a bullet.

The Soviet VEGA 1 and VEGA 2 were the first craft to arrive, on March 6 and 9, 1986. Each plunged deeply into the inner atmosphere and dust cloud of the comet, passing within about 8000 km of the nucleus. In addition to making many direct measurements of the gas and dust, they photographed the dust-shrouded nucleus. However, they could see very little beyond the bright plumes of material jet-ting out from the two most active regions on its surface. Both VEGA craft were severely damaged by dust impacts, losing most of their solar cells and suffering the loss of several instruments at the time of closest approach.

The trajectory data for the VEGA craft were provided to the European Space Agency to allow them to target their Giotto spacecraft for an even closer encounter on March 14, 1986—just 605 km from the comet's nucleus. Giotto also carried out many measurements of the near environment of the nucleus, confirming and extending the Soviet results. The same two prominent jets were photographed, as was the dark, lumpy surface of the nucleus, which was considerably larger (about 8 × 12 km) than had been estimated in advance.

Among the most exciting discoveries was the fact that much of the dust is in the form of very small particles consisting largely of carbon and hydrocarbon compounds, rather than silicates. Thanks to the pathfinder data from VEGA, Giotto was able to image the nucleus at resolutions as high as 100 m before its camera was knocked off target by impacts a few seconds before closest approach.

Although the Giotto camera was destroyed by the impacts of Halley dust, the spacecraft itself recovered, and ESA controllers were able to redirect it for a later, non-imaging encounter with another small comet.

such distances, objects in orbit about the Sun are perturbed by the gravitation of passing stars. Oort suggested, therefore, that the new comets were objects orbiting the Sun with aphelia near the edge of its sphere of influence, and that the perturbing effects of other nearby stars modified their orbits to bring them close to the Sun where we can see them. This region of space from which the new comets are derived is called the **Oort comet cloud.**

There are thought to be about one thousand billion (10^{12}) comets in the Oort cloud. In addition, it is hypothesized that about ten times this number of potential comets are orbiting the Sun in the volume of space between the orbit of Neptune and the Oort cloud at 50,000 AU. The orbits of these objects are too stable to permit any of them to be deflected inward close to the Sun. The first member of this group, with an orbit just beyond that of Pluto, was discovered in 1992. The total number of cometary objects is thus of the order of 10^{13}. If these objects are similar in size to the few comets measured (typical diameters of 10 km or so), the entire cometary mass in the solar system is equivalent to at least 100 Earths. Thus, the mass of material in the form of comets may be comparable to the total mass of other solid matter in the solar system: planets, satellites, and asteroids combined.

The Fate of Comets

Any comet we see today will have spent nearly its entire existence in the Oort cloud, at a temperature near absolute zero. But once a comet enters the inner solar system, its previously uneventful life history begins to accelerate. It may, of course, survive its initial passage near the Sun and return to the cold reaches of space where it spent the previous 4.5 billion years. At the other extreme, it may impact the Sun or pass so close that it is destroyed on its first perihelion passage. Observations from space indicate that at least one comet collides with the Sun every year. Frequently, however, the new comet does not come that close to the Sun, but instead it interacts with one or more of the planets.

A comet coming within the gravitational influence of a planet has three possible fates: (1) it can impact the planet, ending the story at once; (2) it can be speeded up and ejected, leaving the solar system forever; or (3) it can be perturbed into a shorter period. In the last case, its fate is sealed. Each time it approaches the Sun, it will lose part of its material, and it also has a significant chance of collision with a planet. Once in a short-period orbit, the comet's lifetime is measured in thousands, not billions, of years.

Measurements of the amount of gas and dust in the atmosphere of a comet permit an estimate of the total losses during one orbit. Typical loss rates are up to 1 million tons per day from an active comet near the Sun, adding up to some tens of millions of tons per orbit. At that rate, the comet will be gone after a few thousand orbits.

Whether or not the comet evaporates completely is not known. If the gas and dust are well mixed, we would expect the nucleus to shrink each time the comet orbits around the Sun until it has entirely disappeared. However, there remains the suggestion that many of the Earth-approaching asteroids are extinct comets. If there is a silicate core in the comet, or if the dirty snowball includes large blocks of nonvolatile material that are held gravitationally to the surface, then there could be a substantial solid residue after the ices are gone. We simply do not know which of these alternatives is correct.

Comet Dust and Meteors

Whatever the fate of the remnants of the nucleus after the volatiles are exhausted, we do know what happens to the dust that is carried away from a comet by the evaporation of the nucleus. This dust fills the inner part of the solar system. The Earth is surrounded by it. And each of the

Figure 7.11 A meteor results when a small fragment of comet dust encounters the Earth's atmosphere and burns up.

larger dust particles that reaches the Earth creates a shooting star or **meteor** (Figure 7.11).

Although the layperson often confuses comets and meteors, these two phenomena are very different. Comets can be seen when they are many millions of miles away from the Earth and may be visible in the sky for weeks or even months, slowly shifting their positions from day to day. They rise and set with the stars, and during a single night they appear motionless to the casual glance. Meteors, on the other hand, are small solid particles that enter the Earth's atmosphere from interplanetary space. Since they move at speeds of many kilometers per second, the high friction they encounter in the air vaporizes them. The light caused by the luminous vapors formed in such an encounter appears like a star moving rapidly across the sky, fading out within a few seconds.

On a dark, moonless night an alert observer can see half a dozen meteors per hour. To be visible, a meteor must be within 200 km of the observer. Over the entire Earth, the total number of meteors bright enough to be visible must total about 25 million per day.

The typical bright meteor is produced by a particle with a mass less than 1 gram—no larger in size than a pea. Of course, the light you see comes from the much larger region of glowing gas surrounding this little grain of interplanetary material. A particle the size of a marble produces a bright fireball when it strikes the atmosphere, and one as big as a fist has a fair chance of surviving its fiery entry to become a meteorite, if its approach speed is not too high. The total mass of meteoritic material entering the Earth's atmosphere is estimated to be about 100 tons per day.

Meteor Showers

What is the fate of comet dust?

Many—perhaps most—of the meteors that strike the Earth can be associated with specific comets. These interplanetary dust particles retain approximately the orbit of their parent comet, and the particles travel together through space. When the Earth crosses such a dust stream, we see a sudden burst of meteor activity, usually lasting several hours. These events are called **meteor showers.**

The meteoric dust is not always evenly distributed along the orbit of the comet. Therefore, sometimes more meteors are seen when the Earth intersects the dust stream and sometimes fewer meteors are seen. A very clumpy distribution is associated with the Leonid meteors, which in 1833 and again in 1866 (after an interval of 33 years—the period of the comet) yielded the most spectacular showers ever recorded. The last good Leonid shower was on November 17, 1966, when in some southwestern states up to 100 meteors could be observed per second.

The best meteor shower that can be depended on at present is the Perseid shower, which appears for about three nights near August 11 each year. In the absence of bright moonlight, meteors can be seen with a frequency of about one per minute during a typical Perseid shower. It is estimated that the total combined mass of the particles in the Perseid swarm is nearly 5×10^8 tons; this gives at least a lower limit for the original mass of its associated comet, called Swift-Tuttle. The mass of Comet Halley, however, is nearly 1000 times greater, suggesting that only a very small fraction of the original material survives in the meteor stream.

How are meteors and meteor showers related to comets?

The characteristics of some of the more famous meteor showers are summarized in Table 7.3. Other spectacular meteor showers can occur, however, at almost any time, just as some bright comets appear unexpectedly.

No shower meteor has ever survived its flight through the atmosphere and been recovered for laboratory analysis. However, there are other ways to investigate the nature of these particles and thereby to gain additional insight into the comets from which they are derived. Analysis of the photographic tracks of meteors show that most of them are very light or porous material with densities typically less than 1.0 g/cm^3. If you placed a fist-sized lump of such a meteor on a table, it would fall apart under its own weight. Such particles break up very easily in the atmosphere, accounting for the failure of even relatively large shower meteors to produce meteorites. Comet dust is apparently fluffy, rather inconsequential stuff. This fluff, by its very nature, does not reach the Earth's surface intact. However, the more substantial fragments from asteroids do make it into our laboratories, as we will see in the next section. These meteorites are a major source of information on conditions in the solar system at the time of its formation.

TABLE 7.3 Major Annual Meteor Showers		
Shower Name	**Date**	**Associated Comet**
Quadrantid	January 2	unknown
Lyrid	April 21	Thatcher
Eta Aquarid	May 4	Halley
Delta Aquarid	July 30	unknown
Perseid	August 11	Swift-Tuttle
Draconid	October 9	Giacobini-Zinner
Orionid	October 20	Halley
Taurid	October 31	Encke
Leonid	November 16	Tempel-Tuttle
Geminid	December 13	Phaethon

7.4
METEORITES: STONES FROM HEAVEN

Any fragment of interplanetary debris that survives its fiery plunge through the Earth's atmosphere is called a meteorite. Meteorites fall only very rarely in any one locality, but over the entire Earth hundreds of meteorites fall each year. These rocks from the sky carry a remarkable record of the formation and early history of the solar system.

Meteorite Falls and Finds

Meteorites sometimes fall in groups or showers. Such a fall may result when a group of particles were moving together in space before they collided with the Earth, but more likely the different stones are fragments of a single particle that broke up during its violent passage through the atmosphere. It is important to remember that such a *shower of meteorites* has nothing to do with a *meteor shower*. No meteorites have ever been recovered in association with meteor showers. Whatever the ultimate source of these objects, they do not appear to come from the comets or their associated particle streams.

Meteorites are found in two ways. First, sometimes bright meteors (fireballs) are observed to penetrate the atmosphere to very low altitudes. A search of the area beneath the point where the fireball was observed to burn out may reveal one or more remnants of the particle. Observed **meteorite falls,** in other words, may lead to the recovery of fallen meteorites.

Second, unusual-looking rocks are occasionally discovered that turn out to be meteoritic. These are termed **meteorite finds.** Now that the public has become meteorite conscious, many suspected meteorites are sent to experts each year. Some scientists refer to these objects as "mete-

Figure 7.12   Harold C. Urey (1893–1981) was the father of modern cosmochemistry. Working at the University of California in the 1960s, Urey first drew attention to the role of primitive meteorites as remnants of the birth of the solar system.
(University of California at San Diego)

Figure 7.13 Collecting a meteorite on the Antarctic ice. *(NASA/JSC)*

orites" and "meteorwrongs." Outside of Antarctica (see below), genuine meteorites are turned up at an average rate of approximately 25 per year. Most of these end up in natural history museums or specialized meteoritical laboratories throughout the world.

Recently, a new source of meteorites from the Antarctic has dramatically increased the rate of discovery of meteorites and is greatly enriching our knowledge of these objects. Thousands of meteorites have been recovered from the Antarctic ice because of the low precipitation and the peculiar motion of the ice in some parts of that continent (Figure 7.13). Meteorites that fall in regions where ice accumulates are buried and then carried slowly, with the motion of the ice, to other areas where the ice is gradually worn away. After thousands of years the rock again finds itself on the surface, along with other meteorites carried to these same locations. The ice thus concentrates the meteorites from both a large area and a long period of time.

Meteorite Classification

How are meteorites classified?

The meteorites in our collections include a wide range of compositions and histories, but traditionally they have been placed into three broad classes. First, there are the **irons,** which are composed of nearly pure metallic nickel-iron. Second are the **stones,** which is the term used for any silicate or rocky meteorite. Third are the much rarer **stony-irons,** which are (as the name implies) made of mixtures of stony and metallic iron materials.

Of these three types, the irons (Figure 7.14) and the stony-irons (Figure 7.15) are the most obviously extraterrestrial in origin because of

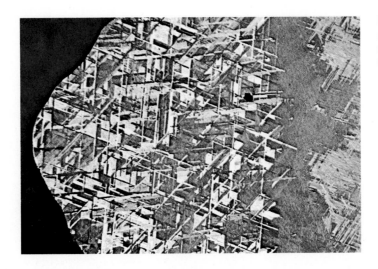

Figure 7.14   Slice of the Kamkas iron meteorite, polished and etched to show the crystal pattern in the metal. *(Ivan Dryer)*

their metallic content. Native, or unoxidized, iron almost never occurs naturally on Earth. This metal is always found here as an oxide or other mineral ore. Therefore, if you ever come across a chunk of metallic iron, it is sure to be either man-made or a meteorite.

The stones (Figure 7.16) are much more common than the irons, but they are more difficult to recognize. Often a laboratory analysis is required to demonstrate that a particular sample is really of extraterrestrial origin, especially if it has lain on the ground for some time and been subject to weathering. The most scientifically valuable stones are those that are collected immediately after they fall or the Antarctic samples that have been preserved in a nearly pristine state by the ice.

Figure 7.15   Polished slice of the Albin stony-iron meteorite. This type of meteorite, called a pallasite, consists of nickel iron mixed with crystals of the green mineral olivine. *(Ivan Dryer)*

Figure 7.16 Stony meteorite of the type called ordinary chondrites. To the layperson, such a meteorite looks very much like a terrestrial rock. *(NASA/JSC)*

Ages and Compositions of Meteorites

It was not until the ages of meteorites were measured and techniques were developed for the detailed analysis of their compositions that their true significance, as the oldest and most primitive materials available for direct study in the laboratory, was appreciated. The ages of stony meteorites can be determined from the careful measurement of radioactive isotopes and their decay products (Section 4.2). Almost all meteorites have radiometric ages between 4.48 and 4.55 billion years. The few exceptions are mostly igneous rocks—basalts—which are believed to be either fragments of asteroid Vesta or ejecta from cratering events on the Moon or Mars.

What is meant by the age of the solar system?

The average age for all of the old meteorites, calculated using the best data and the most accurate values now available for the radioactive half-lives, is 4.52 billion years, with an uncertainty of less than 0.1 billion years. This value is taken to represent the *age of the solar system*—the time since the first solids condensed and began to form into larger bodies.

The traditional classification of meteorites into irons, stones, and stony-irons is easy to use because it is obvious from inspection which category a meteorite falls into (although it may be much more difficult to distinguish a meteoritic stone from a terrestrial rock). Much more significant, however, is the distinction between *primitive* and *differentiated* meteorites.

The differentiated meteorites are, as the name implies, fragments of differentiated parent bodies. They can tell us much about the asteroids that are their source. To learn about the earliest history of the solar sys-

Figure 7.17   The Allende carbonaceous meteorite that fell in Mexico in 1969.   *(NASA/JSC)*

tem, however, we turn to the primitive meteorites. A primitive meteorite is one that is made of materials that have never been subject to great heat or pressure since their formation. The fiery passage of the meteorite through the air takes place so rapidly that the interior (below a burned crust a few millimeters thick) never even becomes hot, so a fragment of primitive interplanetary debris is still primitive (in the sense in which we use the word) after it lands on the Earth.

The great majority of the meteorites that reach the Earth are primitive stones. Many of them are composed of light-colored gray silicates with some metallic grains mixed in, but there is also an important group of darker stones called **carbonaceous meteorites** (Figure 7.17). As their name suggests, these meteorites contain carbon, various complex organic compounds, and often chemically bound water; they are also depleted in metallic iron. The carbonaceous meteorites are presumably related to the dark, carbonaceous asteroids, which are concentrated in the outer part of the asteroid belt.

The Allende and Murchison Primitive Meteorites

The carbonaceous meteorites are the most primitive materials available for laboratory study, excepting the tiny dust particles from comets. Two large carbonaceous meteorites that fell within a few months of each other have proved particularly valuable in probing the birth of the solar system.

The Allende meteorite fell in Mexico and the Murchison meteorite fell in Canada, both in 1969. Like other meteorites, Allende and Murchison are named for the towns near where they fell.

Murchison is best known for the variety of organic, or carbon-bearing, chemicals that it has yielded. Most of the carbon compounds in these meteorites are complex, tar-like substances that defy exact analysis. However, Murchison also contains 16 amino acids, 11 of which are rare on Earth. Unlike terrestrial amino acids, which are formed by living things, the Murchison chemicals include equal numbers with right-handed and left-handed molecular symmetry. The presence of these naturally occurring amino acids and other complex organic compounds in Murchison demonstrates that a great deal of interesting chemistry must have taken place when the solar system formed. Perhaps some of the molecular building blocks of life on Earth were derived directly from the primitive meteorites and comets.

The Allende meteorite (see Figure 7.17) is a rich source of information on the formation of the solar system because it contains many individual grains with varied chemical histories. As much as 10 percent of the material in Allende has been estimated to be of pre-solar system origin—interstellar dust grains that were not destroyed in the processes that gave rise to our own system.

7.5
FORMATION OF THE SOLAR SYSTEM

The comets, asteroids, and meteorites are surviving remnants from the origin of the solar system. The planets and the Sun, of course, are also the products of the formation process. We are now ready to put together the information from the past five chapters in order to discuss what is known of the origin of the solar system.

Observational Constraints

There are certain basic properties of the planetary system that any theory of formation should explain. These may be summarized under three categories: dynamical constraints, chemical constraints, and age constraints.

Dynamical Constraints. The planets all move around the Sun in the same direction and approximately in the plane of the Sun's own rotation. In addition, most of the planets share this same sense of rotation, and most of the satellites also move in counterclockwise orbits. With the exception of the comets, the members of the system define a disk shape. On the other hand, exceptions are possible in the form of retrograde rotation, like that of Venus.

Chemical Constraints. The planets Jupiter and Saturn have approximately the same composition as the Sun and the stars, dominated by hydrogen and helium. Each of the other members is to some degree lacking in the light elements. A careful examination of the composition of solar system objects shows a striking progression from the metal-rich inner planets through predominantly rocky materials out to ice-dominated composition in the outer solar system. The comets are also icy objects, whereas the asteroids represent a transitional rocky composition with abundant dark, carbon-rich material. This general chemical pattern can be interpreted as a temperature sequence, with the inner parts of the system strongly depleted in materials that could not condense at the high temperatures found near the Sun. Again, however, there are important exceptions to the general pattern. In particular, it is difficult to explain the presence of water on Earth and Mars if these planets had formed in a region where the temperature was too hot for ice to condense, unless the ice or water was brought in later from cooler regions. There are also problems with the composition of the Moon and Mercury, as we discussed in Chapter 4.

Age Constraints. Radioactive dating demonstrates that there are rocks on the surface of the Earth that have been present for at least 3.8 billion years and lunar samples that are 4.4 billion years old. In addition, the primitive meteorites all have radioactive ages near 4.5 billion years. The age of these unaltered building blocks is considered the age of the planetary system. The similarity of the measured ages tells us that planets formed and their crusts cooled within a few hundred million years, at most, of the beginning of the solar system. Further, detailed examination of primitive meteorites indicates that they are made primarily from material that condensed or coagulated out of a hot gas; few identifiable fragments or grains survived from before this hot vapor stage 4.5 billion years ago.

The Solar Nebula

All of the above constraints lead to the conclusion that the solar system formed 4.5 billion years ago out of a rotating cloud of hot vapor called the **solar nebula.** The initial composition of the solar nebula was similar to that of the Sun today. As the cloud cooled, grains of solid material coagulated from it, like raindrops condensing from moist air as it rises over a mountain. The meteorites and comet dust are remnants of this original condensate.

How did the solar system originate?

Initially, the solar nebula was large—much larger than the present dimensions of the solar system—and cool—perhaps only a few tens of degrees Kelvin. Slowly the pull of gravity caused this cool, diffuse mass of gas and dust to collapse. As the solar nebula shrank, it was heated by its own gravitational energy, and its rotation speed increased. To understand why its rotation speeded up, we need to introduce the idea of **angular momentum.** Angular momentum is a property of any object that rotates or revolves about some fixed point. It depends on the mass of the moving object, its speed, and its size. Whenever we study spinning or rotating objects, from planets to galaxies, we must consider angular momentum.

Angular momentum is a property that is *conserved,* or stays constant, for any rotating system in which no external forces act. This means that if one part of the system changes—for example, if its size becomes smaller—then in order for the total angular momentum to remain the same, some other property must also adjust itself. In this example, the rotation speed must increase to compensate for the smaller size. Thus a shrinking cloud of dust or a stream of matter falling into a black hole increases its spin rate as it contracts. The same thing is accomplished by figure skaters, who draw in their arms and legs to spin more rapidly and extend their arms and legs to slow down. A diver does the same thing by taking a tuck to spin and then stretching out for a smooth entry into the water.

Increasing temperatures in the shrinking nebula vaporized most of the solid material that was originally present. The nebula eventually collapsed into a disk shape, with most of the material confined to a thin spinning sheet. At the center, continuing collapse ultimately led to the birth of the Sun. The existence of this disk-shaped rotating nebula explains the primary dynamical properties of the solar system as described above.

Condensation and Accretion

Picture the solar nebula at the end of the collapse phase, when it is at its hottest. With no more gravitational energy to heat it, most of the nebula begins to cool. In the center, however, the newly formed Sun keeps the temperatures up. The temperature within the nebula therefore decreases with increasing distance from the Sun, much as the planetary temperatures vary with position today. As the nebula cools, the gases interact chemically to produce compounds, and eventually these compounds condense into liquid droplets or solid grains.

The chemical condensation sequence (Figure 7.18) in the cooling nebula was calculated by geochemists in the 1970s. The first materials to

Figure 7.18   The chemical condensation sequence in the solar nebula, showing the primary chemical species that would be expected to form in a cooling gas cloud of solar composition under equilibrium conditions.   *(Adapted from diagrams published by John Lewis, University of Arizona)*

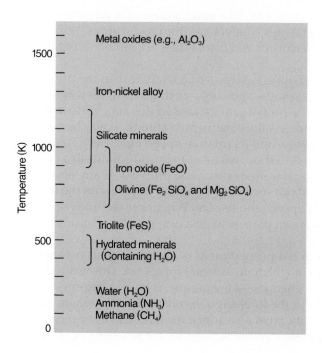

form grains are the metals and various rock-forming silicates. As the temperature dropped, these were joined throughout much of the solar nebula by sulfur compounds and by carbon- and water-rich silicates such as those now found abundantly among the asteroids. However, in the inner parts of the nebula the temperature never dropped low enough for such materials as ice or carbonaceous organic compounds to condense, so they are lacking on the innermost planets. Far from the Sun, where temperatures continued to decline, the oxygen combined with hydrogen and condensed in the form of water ice (H_2O). Beyond the orbit of Saturn, carbon and nitrogen combined with hydrogen to condense as additional ices such as methane (CH_4) and ammonia (NH_3). This chemical condensation sequence explains the basic chemistry of the solar system, and it also tells us why the oldest materials (the primitive meteorites) all have about the same age—the age corresponding to the time when all of these solid grains formed.

How did planets form from the raw materials in the solar nebula?

It is thought that the grains that condensed in the solar nebula rather quickly formed into larger and larger aggregates, until most of the solid material was in the form of **planetesimals** a few kilometers to a few tens of kilometers in diameter. Some planetesimals were large enough to attract their neighbors gravitationally and thus to grow by the process called **accretion**—the gradual addition of infalling material. Although the intermediate steps are not well understood, there seem to have developed several dozen centers of accretion in the inner solar system, the largest of which became the planets Mercury, Venus, Earth, and Mars. In the outer solar system, where the building blocks included ices as well as silicates, much larger bodies grew, with masses 10 to 15 times greater than the mass of the Earth.

Formation of the Giant Planets

In the inner solar system, the rocky objects continued to grow by accretion, but they had little interaction with the residual gas of the solar nebula. In contrast, the proto-planets of the outer solar system became so large that they were able to attract the surrounding gas. As the hydrogen and helium rapidly collapsed onto their cores, the giant planets were heated by the energy of contraction, just as the contraction of the solar nebula had ignited the nuclear fires of the Sun. But these giant planets were far too small to achieve the central temperatures and pressures necessary to initiate self-sustaining nuclear reactions. After glowing dull red for a few thousand years, they gradually cooled to their present state.

How did the early history of the giant planets differ from that of the terrestrial planets?

The collapse of nebular gas onto the cores of the giant planets explains how these objects came to have about the same hydrogen-rich composition as the Sun itself. The process was most efficient for Jupiter and Saturn, so that their composition is most nearly "cosmic." Much less gas was captured by Uranus and Neptune, which is why these two planets have compositions dominated by the icy and rocky building blocks that made up their large cores, rather than by hydrogen and helium.

Dynamical Evolution of the System

All of the processes described above, from the collapse of the solar nebula to the formation of protoplanets, took place within at most a few million years, and possibly even less time. However, the story of the formation of the solar system is not complete at this stage—there remains the fate of the planetesimals and other debris that did not initially accumulate to form the planets.

The Oort comet cloud was probably formed from icy planetesimals in the outer solar system. The gravitational influence of the giant planets is thought to have ejected the comets from their initial orbits in the disk, probably near the present orbits of Uranus and Neptune. If this idea is correct, then the comets are leftovers from the building blocks of the outer planets, preserved in the Oort cloud.

In the inner parts of the system, remnant planetesimals and perhaps several dozen large objects (up to the size of Mercury or Mars) continued to whiz about and to interact gravitationally with the proto-terrestrial planets. Collisions between these objects were inevitable. Giant impacts at this stage probably stripped Mercury of part of its mantle and crust, reversed the rotation of Venus, and splashed out material from the Earth's mantle to create the Moon.

Smaller-scale impacts also added mass to the inner proto-planets. This impacting material could have come from almost anywhere within the solar system. Unlike the previous stage of accretion, therefore, this new material did not represent just a narrow range of compositions as specified by the initial temperatures in the solar nebula. Much of the debris striking the inner planets, for example, was ice-rich material that had condensed in the outer part of the solar nebula. As this comet-like bombardment progressed, the Earth accumulated the water and various organic compounds that would later be critical to the formation of life. Mars and Venus should also have acquired water and organic materials from the same source.

Gradually, as the planets swept up the remaining debris, most of the leftover planetesimals disappeared. In the region between Mars and Jupiter, however, there exist stable orbits where small bodies can avoid impacting the planets or being ejected from the system. The remnant of objects that survives in this special location is what we call the asteroids.

The foregoing account of the solar system's beginnings is probably close to what actually took place, but future research can be expected to fill in many details and show others to be incorrect. The problem of origins remains an area of active research, and some of our questions will probably be answered only when we get information on the formation of other planetary systems in addition to our own. We will return to some of these ideas in Chapter 11, where we discuss the births of stars.

▶ SUMMARY

7.1 The **asteroid belt** (2.2–3.3 AU) includes 100,000 asteroids down to 1-km diameter. Ceres is the largest. Most asteroids are dark, primitive objects (Ceres,

Pallas); others are differentiated (basaltic Vesta, metallic Psyche). There are also **Earth-approaching asteroids;** their orbits are unstable, and they will either impact a planet or be ejected from the solar system on a time scale of 100 million years. We have pictures of three asteroids; Gaspra and Toutatis both appear to be double.

7.2 Cometary orbits are elongated ellipses; as the comet nears the Sun, its ices evaporate to produce the visible head and **tail.** Comet Halley has the relatively short period of 76 years. The solid part of a comet is its **nucleus,** typically about 10 km in diameter and composed (like Whipple's "dirty snowball") of ices and dark, primitive material. The comet's extensive but tenuous atmosphere is produced by gas jets and other activity on the nucleus, stimulated by solar heating. Comet Halley was studied by spacecraft in 1986.

7.3 Comets come from the **Oort comet cloud,** a halo of more than 10^{12} volatile-rich bodies orbiting the Sun at distances up to 50,000 AU. Perturbations by passing stars deflect a few into the inner solar system, where they impact a planet, are ejected, or evaporate from solar heating. Comet dust is visible when it encounters the Earth as meteors; **meteor showers** represent the dust streams generated by active comets. Meteors are made of fluffy material that never survives its passage through the Earth's atmosphere.

7.4 Meteorites are the debris (mostly fragments from asteroids) that survives to reach the surface of the Earth. Meteorites are called **finds** or **falls** according to how they are found; the most productive source today is the Antarctic ice cap. Compositional classification includes **irons, stony-irons,** and **stones.** Most stones are primitive objects, dated to the origin of the solar system 4.5 billion years ago. The most primitive are the **carbonaceous meteorites,** such as Murchison and Allende.

7.5 Meteorites, comets, and asteroids are survivors from the formation of the solar system out of the **solar nebula.** The solar nebula formed by the collapse of an interstellar cloud of gas and dust, which contracted (conserving its **angular momentum**) to form the proto-sun, which was surrounded by a thin disk of dust and hot vapor. Condensation and **accretion** in the disk led to **planetesimals,** which formed into the larger planets and satellites. The giant planets were also able to attract and hold gas from the solar nebula. After a few million years of violent impacts, most of the debris was swept up or ejected, leaving only the asteroids and cometary remnants surviving to the present.

▶ REVIEW QUESTIONS

1. Compare the asteroids of the asteroid belt with the Earth-approaching asteroids. What are the main differences between the two groups?

2. Describe the nucleus of a typical comet, and compare it with an asteroid of similar size.

3. Describe the origin and the eventual fate of comets.

4. In what ways are meteors different from meteorites? What is the probable origin of each?

5. Describe the solar nebula, and outline the sequence of events within the nebula that gave rise to the planets.

6. Comets and asteroids are considered to be relics from the origin of the solar system. Why do astronomers think they are older and more primitive than the planets and their satellites?

▶ THOUGHT QUESTIONS

7. There is a great deal of interest today in the discovery of additional Earth-approaching asteroids. Can you think of several reasons for this high interest?

8. Comets are considered to be the "most primitive" solid bodies in the solar system. What does this statement mean?

9. Meteors apparently come primarily from comets, while the meteorites are thought to be fragments of asteroids. This may seem contradictory. Explain why we do not believe meteorites come from comets, or meteors from asteroids.

10. Which meteorites are the most useful for defining the age of the solar system? Why?

11. Suppose a new primitive meteorite is discovered and analysis shows that it contains a trace of amino acids, all of which show the same rotational symmetry (unlike the Murchison meteorite). What might you conclude from this finding?

12. Give some everyday examples of the conservation of angular momentum.

13. Describe the chemical building blocks that are thought to have been available in the grains that condensed from the solar nebula. If each planet formed in place from these grains, what would be the chemical composition of objects at 0.4 AU, 1.0 AU, 5.0 AU, and 25 AU from the Sun?

14. Suppose a comet is discovered approaching the Sun, and it is found to be on an orbit that will cause it to collide with the Earth 20 months later, after perihelion passage. (This is approximately the situation described in the popular science fiction novel *Lucifer's Hammer*, by Larry Niven and Jerry Pournelle.) What could we do? Is there any way to protect ourselves from a catastrophe?

▶ PROBLEMS

15. What is the period of revolution about the Sun for an asteroid in the middle of the asteroid belt (semimajor axis 3.0 AU)?

16. What is the period of revolution for a comet with aphelion at 5 AU and perihelion at the orbit of the Earth?

17. Suppose that the Oort comet cloud contains 10^{12} comets, with an average comet diameter of 10 km. Calculate the mass of a 10-km diameter comet, assuming that is it composed mostly of water ice (density of 1 g/cm^3). Next calculate the total mass of the comet cloud. Finally, compare this mass with that of the Earth and Jupiter.

18. The calculation in Problem 17 refers to the known Oort cloud, the source for the comets we see. If, as some astronomers suspect, there are ten times this many cometary objects in the solar system, how does the total mass of cometary matter compare with the total mass of the planets?

19. If the Oort comet cloud contains 10^{12} comets and ten new comets are discovered each year, what percentage of the comets have been used up since the beginning of the solar system?

*20. The angular momentum of an object is proportional to the square of its size divided by the period of rotation (D^2/P). If angular momentum is conserved, then any change in size must be compensated by a proportional change in period, so as to keep D^2 divided by P a constant. Suppose that the solar nebula began with a diameter of 10,000 AU and a rotation period of 1 million years. What would be its rotation period when it had shrunk to the size of the orbit of Pluto? To the orbit of Jupiter? To the orbit of the Earth?

▶ SUGGESTIONS FOR ADDITIONAL READING

Books

Dodd, R. *Thunderstones and Shooting Stars.* Cambridge U. Press, 1986. An outstanding primer on meteorites and their significance.

Hutchinson, R. *The Search for Our Beginnings.* Oxford U. Press, 1983. Readable account of meteoritics and the early history of our solar system.

Whipple, F. *The Mystery of Comets.* Smithsonian Institution Press, 1985. Summary of our knowledge just before the last Halley encounter, by one of the world's leading cometary experts.

Kowal, C. *Asteroids: Their Nature and Utilization.* Ellis Horwood/J. Wiley, 1988. A nice introduction by an astronomer.

Yeomans, D. *Comets: A Chronological History of Observations, Science, Myth, and Folklore.* John Wiley, 1991. An up-to-date survey of comets through history.

Sagan, C. & Druyan, A. *Comet.* Random House, 1986. A colorful popular introduction.

Burke, J. *Cosmic Debris: Meteorites in History.* U. of California Press, 1986. A historical survey.

Articles

On the Formation of the Solar System

Barnes-Swarney, P. "The Chronology of Planetary Bombardments" in *Astronomy,* July 1988, p. 21.

Reeves, H. "The Origin of the Solar System" in *Mercury,* Mar./Apr. 1977, p. 7.

On Asteroids and the Moons of Mars

Beatty, J. "Gaspra: A Picture Perfect Asteroid" in *Sky & Telescope,* Feb. 1992, p. 135.

Hartmann, W. "Vesta: A World of Its Own" in *Astronomy,* Feb. 1983, p. 6.

Morrison, D. "Asteroids" in *Astronomy,* June 1976, p. 6.

Binzel, R., *et al.* "The Origin of the Asteroids" in *Scientific American,* Oct. 1991, p. 88.

Dick, S. "Discovering the Moons of Mars" in *Sky & Telescope,* Sept. 1988, p. 242.

Stooke, P. "Sizing Up Phobos" in *Sky & Telescope*, May 1989, p. 477.

On Comets

The March 1987 issue of *Sky & Telescope* was devoted to what we learned from the 1986 pass of Halley's Comet.

Weissman, P. "Comets at the Solar System's Edge" in *Sky & Telescope*, Jan. 1993, p. 26.

Benningfield, D. "Where Do Comets Come From?" in *Astronomy*, Sept. 1990, p. 28.

Berry, R. "Search for the Primitive" in *Astronomy*, June 1987, p. 6. Halley's Comet results. (See also the Sept. 1986 issue, p. 6.)

Delsemme, A. "Whence Come Comets" in *Sky & Telescope*, Mar. 1989, p. 260.

Gingerich, O. "Newton, Halley, and the Comet" in *Sky & Telescope*, Mar. 1986, p. 230.

Levy, D. "How To Discover a Comet" in *Astronomy*, Dec. 1987, p. 74.

On Meteors and Meteorites

Kronk, G. "Meteor Showers" in *Mercury*, Nov./Dec. 1988, p. 162.

MacRobert, A. "Meteor Observing" in *Sky & Telescope*, Aug. 1988, p. 131; Oct. 1988, p. 363.

Spratt, C. & Stephens, S. "Against All Odds: Meteorites that Have Struck Home" in *Mercury*, Mar./Apr. 1992, p. 50.

Wagner, J. "The Sources of Meteorites" in *Astronomy*, Feb. 1984, p. 6.

Lewis, R. & Anders, E. "Interstellar Matter in Meteorites" in *Scientific American*, Aug. 1983.

LIGHT: MESSENGER FROM SPACE

The McMath Telescope, which is located on Kitt Peak in Arizona, is the largest solar telescope in the world. Much of the observing time at this telescope is used to study the spectrum of the Sun. *(National Optical Astronomy Observatories)*

It is easy to see how we know so much about the planets in our solar system. Except for distant Pluto, we have actually sent spacecraft to each planet in order to obtain photographs or to land on their surfaces. The planets, however, occupy only a tiny portion of the vast universe in which we live. If we want to learn more about the Sun and the stars, we must devise techniques that will allow us to analyze them from a distance. The temperature of the Sun is so high that a spacecraft would be destroyed by its heat long before reaching the solar surface, and the stars are too far away to visit in our lifetimes with the technology now available. Even light, which travels at a speed of 300,000 km/s, takes more than four years to reach us from the nearest star. To learn more about the Sun and the stars, we must rely on the messages contained in the light, x rays, radio waves, and other radiation that they emit.

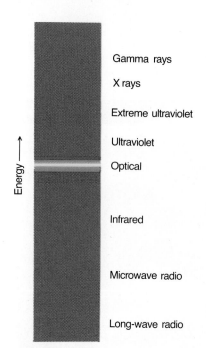

Gamma rays

X rays

Extreme ultraviolet

Ultraviolet

Optical

Infrared

Microwave radio

Long-wave radio

Energy ⟶

Figure 8.1 The electromagnetic spectrum.

What is electromagnetic radiation?

8.1
THE NATURE OF LIGHT

Astronomy is basically the study of *light radiation* from celestial objects. This light radiation comes in many forms. Most familiar is visible light—the rays (with all the colors of the rainbow) to which our eyes are sensitive. Until a few decades ago, astronomy consisted entirely of the study of visible light, as seen directly with the eye or recorded on photographic film. But we know there are many other forms of radiation, differing from visible light in the energy of the individual packets of energy called **photons**. Although these other forms of light cannot be seen with the eye, we have detectors that are sensitive to them. Taken all together, the different kinds of light make up what is called the **electromagnetic spectrum** (Figure 8.1).

Maxwell's Theory of Electromagnetism

With one simple equation, Newton's theory of gravitation accounts for the motions of the planets as well as for the motions of objects on the Earth. Application of this theory to a variety of problems dominated the work of scientists for nearly two centuries. In the 19th century, many physicists turned to the study of electricity and magnetism. The scientist who played a role analogous to that of Newton was physicist James Clerk Maxwell (1831–1879), who was born and educated in Scotland. Maxwell developed a single theory that describes in a small number of elegant equations both electricity and magnetism. It is this theory that allows us to understand the behavior of light.

It was known in the 19th century as a result of laboratory experiments that changing magnetic fields could produce electric currents. Maxwell showed through theoretical calculations that changing electric currents could also produce magnetic fields, a result that was subsequently confirmed experimentally. The word *field* is a technical term used in physics to describe the consequences of forces that act on distant objects. For example, the Sun produces a gravitational field that controls the Earth's orbit, even though the Sun and the Earth do not come directly into contact. Stationary electric charges produce electric fields, and as Maxwell showed, moving electric charges produce magnetic fields.

Maxwell found that electric and magnetic fields propagate through space. Maxwell was able to calculate the speed at which an electromagnetic disturbance moves through space and found that it was equal to the speed of light, which had been measured experimentally. On that basis, he speculated that light was one form of electromagnetic radiation, a conclusion that was again confirmed in the laboratory. When light enters a human eye, its changing electric and magnetic fields stimulate nerve endings, which then transmit the information contained in these changing fields to the brain.

Because the word "radiation" will be used frequently in this book, it is important to understand what it means. In the modern world, radiation is commonly used to describe certain kinds of dangerous high-energy subatomic particles released by radioactive materials in our environment, but this is not what we mean when we speak of radiation in an astronomy text. Radiation as used in this book is a general term for light, x rays, and other photons. This radiation provides almost our only link with the universe beyond our own solar system.

The Wave-like Characteristics of Light

Light radiation can be thought of as consisting of photons of different energies. However, it also has the properties of waves, because it is produced by regularly repeating changes in electric and magnetic fields. Some of the characteristics of waves are shown in Figure 8.2. Ocean waves, for example, have alternating crests and troughs. The distance between successive crests is called the **wavelength.** In the Pacific Ocean, waves generated by storms are typically about 200 m long.

The wavelength of visible light lies in the range of 0.0000004 to 0.0000007 m. Rather than write all of these zeros, it is customary to express the wavelength of light in nanometers (nm). One nm is one-billionth of a meter (10^{-9} m). Visible light has wavelengths that range from about 400 to 700 nm. The exact wavelength of visible light determines its color. Radiation with a wavelength of 400–450 nm is perceived by the retina of the eye as the color violet. Radiations of successively longer wavelengths are perceived as the colors blue, green, yellow, orange, and red.

The **frequency** of a wave is defined as the number of wave crests (or troughs) that pass a specific point in 1 s. Wind-driven waves break against the coastline at intervals of 5 to 20 seconds. Their frequency is thus 12 to 3 cycles per minute. The frequency of visible light is much higher—4.3×10^{14} cycles per second for red light to 7.5×10^{14} cycles per second for blue light.

The relationship between the velocity with which a wave propagates, its wavelength, and its frequency are illustrated in Figure 8.2. Imagine a long train of waves moving to the right, past point O, at the speed of light c. If we measure to the left of O a distance of c km, we arrive at the point P along the wave train that will just reach point O after a period of 1 s. The frequency f of the wave train—that is, the number of waves between P and O—times the length of each, λ, is equal to the distance c. Thus, we see that for any wave motion, the speed of propagation equals the frequency times the wavelength. That is,

$$c = \lambda f$$

The Greek letter λ (lambda) is almost always used to denote wavelength.

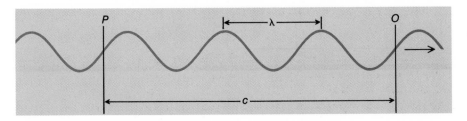

Figure 8.2 Electromagnetic radiation has wave-like characteristics. The relationships between the length of the wave (λ), the frequency of the wave, and the speed (c) with which it moves are shown.

OBSERVING YOUR WORLD   REFRACTION AND THE RAINBOW

Telescopes with lenses form images because light is *refracted* or bent and brought to a focus when it passes from air into the glass lens. In fact, light will be bent anytime that it passes from one transparent medium, such as air, into any other transparent medium that has a different density, such as glass or water. Rainbows are an excellent illustration of refraction and also of what happens when white sunlight is spread out into a spectrum of colors.

You have a good chance of seeing a rainbow anytime you are between the Sun and a rain shower (Figure 8A). Remember that white sunlight is composed of all the colors from violet to red. Now suppose that a ray of sunlight encounters a raindrop and passes into it. The light changes direction, or to use the proper technical term, the light is refracted when it passes from air to water (Figure 8B). Measurements show that blue light changes direction more than red light. The light is then reflected at the backside of the

drop and re-emerges from the front, where it is again refracted. Now the white light has been spread out into a rainbow of colors.

ACTIVITIES

1. Look at a rainbow. Which color lies higher in the sky—blue or red? To understand why, note that in Figure 8B, violet light lies above the red light after it emerges from the raindrop. When you look at a rainbow, however, it is the red light that is higher in the sky. This result may seem surprising, but look again at Figure 8A. If the observer looks at a raindrop that is high in the sky, the violet light passes over her head, while the red light enters her eye. Similarly, if the observer looks at a raindrop that is low in the sky, the violet light reaches her eye and

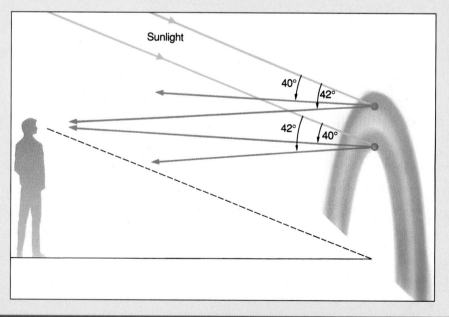

Figure 8A   This diagram shows how light from the Sun, which is located behind the observer, can be refracted by raindrops to produce a rainbow.

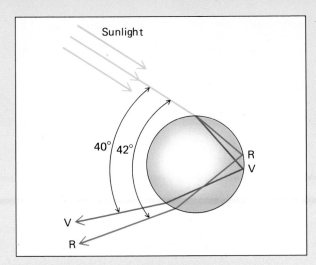

Figure 8B A diagram showing the path of light passing through a raindrop. Refraction separates white light into its component colors.

the drop appears violet. The red light from that same drop strikes the ground and is not seen. Colors of intermediate wavelengths are refracted to the eye by drops that are intermediate in altitude between the drops that appear violet and the drops that appear red. Thus, a single rainbow always has red on the outside and violet on the inside.

2. How does the altitude of a rainbow vary as the altitude of the Sun changes? Make a prediction using the information in Figure 8A, which shows that sunlight always emerges from a raindrop at an angle that is about 40° different from the angle at which the sunlight entered the raindrop. Verify your prediction by noting the altitude of the top of the rainbow and the altitude of the Sun when you see rainbows at different times of day.

3. Could you see a rainbow if the Sun were at the zenith? Could you ever see a rainbow if the rain were between you and the Sun? Again, refer to Figure 8A.

4. For an even simpler example of refraction, put a pencil at a slanted angle in a clear glass filled with water. Look at the pencil from above the surface of the water and to one side of it (do not look directly down into the water). Does the pencil look straight? Where does the tip of the pencil appear to be relative to its actual position?

The explanation for this effect is that when light passes from water to air (or from any transparent medium to one that is less dense), it is bent *away* from the line perpendicular to the water. Figure 8C shows rays of light leaving the tip of the pencil and entering the air. The brain assumes that the tip of the pencil is located at the position from which the light rays that enter the eye appear to come.

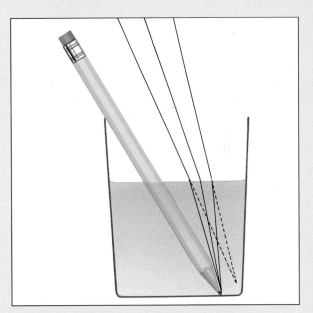

Figure 8C A pencil in a glass of water. The solid lines show the direction of light rays. The tip of the pencil appears to lie at the origin of the light rays. This origin is found by projecting the rays back along the direction that they are following as they enter the eye. The projections of two rays are shown as dashed lines.

Figure 8.3   The inverse-square law for light. As light energy radiates away from its source, it spreads out, so that the energy passing through a unit area decreases as the square of the distance from its source.

How rapidly does the apparent brightness of a source of electromagnetic radiation decrease as the distance from the source increases?

Propagation of Light

As electromagnetic radiation moves away from its source, it spreads out and covers an ever-widening area. The increase in area is proportional to the square of the distance that the radiation has traveled (Figure 8.3). For example, when light from the Sun reaches the Earth, it is spread out over a sphere 1 AU in radius. When it has gone twice as far, to 2 AU from the Sun, that same light is spread over an area four times as great, because the surface area of a sphere is proportional to the square of its radius. When the Sun's radiation reaches Saturn, 10 AU from the Sun, it is spread over an area 100 times that at the Earth's distance. This decrease in energy with increasing distance from the source is called the **inverse-square law,** and it also applies to other kinds of energy, including sound. It explains why the inner planets are hot and the outer planets are cold; the farther each planet is from the Sun, the less solar energy it receives on a given area of its surface.

The **apparent brightness** of a light source depends on how much of its energy (that is, light) enters the pupil of our eye or a telescope. Because the collecting area of the eye or telescope is constant, the larger the area over which light is spread, the smaller is the fraction observed. Thus, the apparent brightness of a light source varies inversely with the square of its distance. At 2 AU from the Sun, the Sun would appear to be only one-fourth as bright as it would be at the surface of the Earth. A typical star emits about the same total energy as does the Sun, but even the nearest star is about 270,000 times further away, and so it appears about 73 billion times fainter ($73 \times 10^9 = 270,000 \times 270,000$).

8.2
THE ELECTROMAGNETIC SPECTRUM

Now that we have described the wave-like nature of light, let us look again at the electromagnetic spectrum and the ways that radiation of different wavelength (or energy) is produced. We need to understand how the radiation is produced if we are to use astronomical observations to determine the nature of celestial sources of radiation.

Radiation of Different Wavelengths

Light has wavelengths within the range of 400–700 nm. The wavelengths of other types of electromagnetic radiation are shown in Figure 8.4. **Gamma rays,** which have the shortest wavelengths and the highest energy, are often emitted in the course of nuclear reactions and by radioactive elements. Gamma radiation is generated in the deep interiors of stars. Radiation with wavelengths between 0.01 nm and 20 nm is referred to as **x rays,** whereas **ultraviolet** radiation has wavelengths intermediate in length between x rays and visible light.

Between visible light and radio waves are the wavelengths of **infrared** or heat radiation. Radio radiation has wavelengths from about 1 mm to several kilometers. The **microwaves** used in shortwave communication and in television transmission have wavelengths ranging from a few centimeters to a few meters. The types of radiation with longer wavelengths also have the lowest energy photons.

In 1672, in the first paper that he submitted to the Royal Society, Newton described an experiment in which he permitted sunlight to pass through a small hole and then through a prism. Newton found that sunlight, which gives the impression of being white, is made up of a mixture of all the colors of the rainbow—a spectrum (Figure 8.5). Table 8.1 summarizes the types of electromagnetic radiation and gives examples of astronomical objects that emit electromagnetic radiation in specific ranges of wavelength.

Radiation Laws

Some astronomical objects emit mostly infrared radiation, others emit mostly visible light, and still others emit mostly ultraviolet radiation. What determines the type of electromagnetic radiation emitted by the Sun, stars, and other astronomical objects? The answer is temperature.

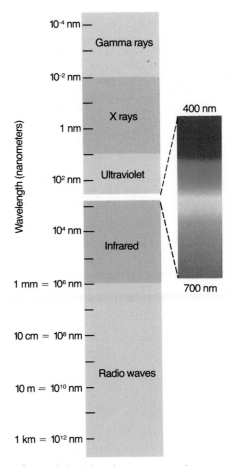

Figure 8.4 The electromagnetic spectrum showing wavelengths.

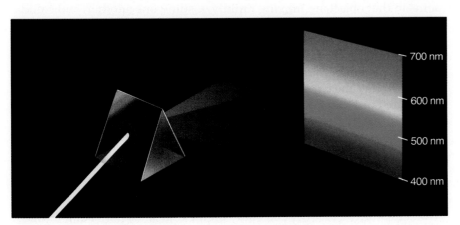

Figure 8.5 When Newton passed a beam of white sunlight through a prism, he saw a rainbow-colored band of light that we now call a continuous spectrum.

TABLE 8.1 Electromagnetic Radiation

Type of Radiation	Wavelength Range (nm)	Radiated by Objects at this Temperature	Typical Sources
Gamma rays	Less than 0.01	More than 10^8 K	No astronomical sources this hot; some gamma rays produced in nuclear reactions
X rays	0.01–20	10^6–10^7 K	Gas in clusters of galaxies; supernova remnants; solar corona
Ultraviolet	20–400	10^5–10^6 K	Supernova remnants, very hot stars
Visible	400–700	10^3–10^5 K	Stars
Infrared	10^3–10^6	10–10^3 K	Cool clouds of dust and gas, planets, satellites
Radio	More than 10^6	Less than 1 K	No astronomical objects this cold; radio emission produced by electrons moving in magnetic fields (synchrotron radiation)

What is the relationship between the temperature of an object and the electromagnetic radiation that it emits?

Electromagnetic radiation is emitted when electric charges change either the speed or the direction of their motion. If an object is hot, the atoms in that object are moving rapidly, jostling one another, and colliding with electrons, changing their motions at each collision. Each collision therefore results in the emission of electromagnetic radiation— radio, infrared, visible, ultraviolet, or x ray. How much of each type depends on the temperature of the object producing the radiation.

The higher the temperature of the source, the higher the energy of the emitted radiation. To understand in more quantitative detail the relationship between temperature and electromagnetic radiation, it is useful to resort to a common tactic used by physicists. Consider an idealized object that absorbs all the electromagnetic energy that impinges on it. Such an object is called a **blackbody.** A blackbody absorbs all of the energy incident upon it and heats up until it is emitting energy at the same rate that it is being absorbed. The blackbody reaches an equilibrium temperature that depends only on the total energy striking it each

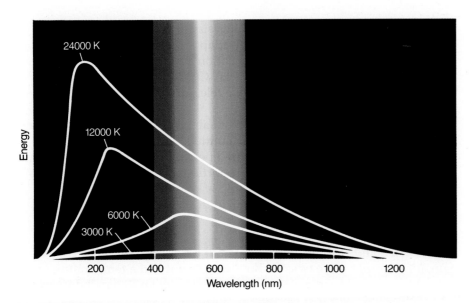

Figure 8.6 Energy emitted at different wavelengths for black bodies at four different temperatures. At hotter temperatures, more energy is emitted at all wavelengths. The peak amount of energy is radiated at shorter wavelengths for higher temperatures (Wien's law).

second. The emitted energy from blackbodies was measured experimentally in the 19th century, and the results for three different temperatures are shown in Figure 8.6.

The radiation from a blackbody has several characteristics that are illustrated in Figure 8.6. First, a blackbody with a temperature higher than absolute zero emits some energy at all wavelengths. Second, a blackbody at higher temperature emits more energy at all wavelengths than does a cooler one. Third, the higher the temperature, the shorter the wavelength (and higher the energy) at which the maximum radiation is emitted.

This third characteristic is one that we have all observed in everyday life. For example, when a burner on an electric stove is turned on low, it emits heat, which is infrared radiation, but not light. If the burner is set to a higher temperature, it glows a dull red. At a still higher setting, it glows a brighter orange-red. At still higher temperatures, which cannot be reached with ordinary stoves, metal can appear brilliant yellow or even blue-white hot.

The Sun and the stars emit energy that approximates the energy from a blackbody, and so it is possible to estimate temperatures by measuring the energy that they emit as a function of wavelength—that is, by measuring their colors. The temperature at the surface of the Sun, which is where the radiation that we see is emitted, turns out to be 5800 K.

The wavelength at which a blackbody emits its maximum energy can be calculated according to the equation

$$\lambda_{max} = 3{,}000{,}000/T$$

in which the wavelength is measured in nanometers and the temperature is measured in Kelvins. This relationship is called **Wien's law.** For the Sun, the wavelength at which the maximum energy is emitted is 520 nm, which is near the middle of that portion of the electromagnetic spectrum that is called visible light. It is surely no coincidence, but rather a consequence of evolutionary adaptation, that human eyes are most sensitive to electromagnetic radiation at those wavelengths at which the Sun puts out the most energy. Characteristic temperatures of other astronomical objects, and the wavelengths at which they emit most of their energy, are listed in Table 8.1.

If we sum up the contributions from all parts of the electromagnetic spectrum, we obtain the total energy emitted by a blackbody over all wavelengths. That total energy, emitted per second per square meter by a blackbody at a temperature T, is proportional to the *fourth power* of its absolute temperature. This relationship is known as the **Stefan-Boltzmann law,** which can be written in the form of an equation as

$$E = \sigma T^4$$

in which E stands for the total energy and σ is a constant number. If the Sun, for example, were twice as hot—that is, if it had a temperature of 11,600 K—it would radiate 2^4, or 16 times more energy than it does now.

8.3
STRUCTURE OF THE ATOM

As Figure 8.6 shows, a blackbody emits some radiation at all wavelengths. The visible spectrum of a blackbody is therefore a continuous rainbow of colors. If the Sun behaves like a blackbody, then when sunlight passes through a prism, we would expect to see radiation of all the possible colors. In fact, this is what Newton, who used very simple equipment, did see. Later, however, it was discovered that some colors are missing from the solar spectrum, and this discovery led to our

Figure 8.7 The visible spectrum of the Sun. The spectrum is crossed by dark lines produced by atoms in the solar atmosphere that absorb light at certain wavelengths. *(National Solar Observatory/National Optical Astronomy Observatories)*

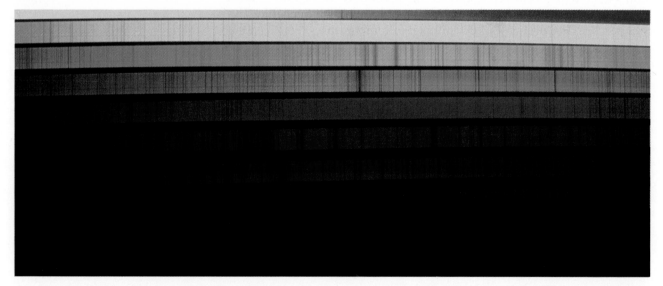

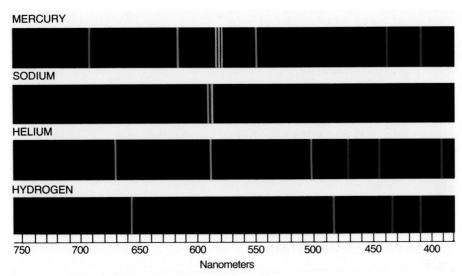

MERCURY

SODIUM

HELIUM

HYDROGEN

750   700   650   600   550   500   450   400

Nanometers

Figure 8.8   The line spectra produced by several different kinds of hot gas. Each gas produces its own unique pattern of lines, and so the composition of a gas can be identified from observations of its spectrum.

understanding of the nature of atoms and their interaction with light radiation.

Spectral Lines

In 1802, the English chemist William Wollaston (1766–1828) repeated Newton's experiment with the solar spectrum, but he used a slit instead of a round hole. After the sunlight passed through the prism, Wollaston used a lens to form an image of the slit at every wavelength—and saw narrow lines in the spectrum where some of the colors were missing! We now know that there are thousands of places where light is missing. Sunlight has thousands of dark **spectral lines** (Figure 8.7).

During the 19th century, scientists learned that hot gases produce bright spectral lines. Each gas—sodium, sulfur, carbon, etc.—produces its own unique pattern of spectral lines (Figure 8.8). Just as we can identify an individual person from fingerprints, so it is possible to identify a gas purely by examination of its spectrum. In 1860, Gustav Kirchhoff concluded from his analysis of the solar spectrum that sodium is present in the Sun's atmosphere. (Later in this section, there is an explanation of why spectral lines are sometimes bright and sometimes dark.)

Analysis of spectral lines is the key to modern astrophysics. In 1835, the French philosopher Auguste Comte concluded that it would be possible to measure the sizes, distances, and motions of stars but that it would never be possible by any means to learn their chemical composition. Only 25 years later, Kirchhoff proved him resoundingly wrong. In the years following Kirchhoff's discovery of sodium in the Sun, astronomers identified many other chemical elements in the Sun and stars, but it was only in the 20th century, with the development of a model for the atom, that scientists learned how spectral lines are formed.

What is the structure of an atom?

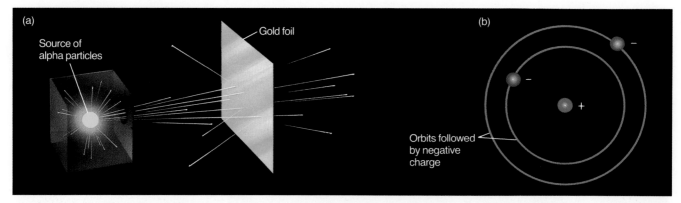

Figure 8.9 (a) When Rutherford allowed alpha particles from a radioactive source to strike a target of gold foil, he found that some of the alpha particles rebounded back in the direction from which they came. (b) From this experiment, Rutherford concluded that the atom must be constructed like a miniature solar system, with the positive charge concentrated in the nucleus. The negative charge was assumed to orbit in the large volume around the nucleus.

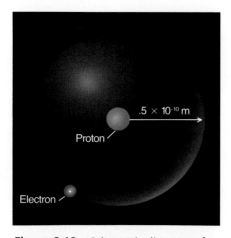

Figure 8.10 Schematic diagram of a hydrogen atom in its lowest energy state, which is also called the ground state. Although the orbit of the electron is drawn as if it were well defined, in fact the orbit shown is only the most probable one. The electron may move in a variety of orbits all relatively close to the one shown. The proton and electron have equal but opposite charges, which exert an electromagnetic force that binds the hydrogen atom together.

Structure of the Atom

The idea that matter is composed of tiny particles—atoms—massed together is at least 25 centuries old. It was not until the 20th century, however, that scientists invented instruments that permitted them to probe inside an atom and discover that it was not, as Newton thought, hard and indivisible. Instead, it is a complex structure composed of still smaller particles.

The first of these smaller particles was discovered by J.J. Thomson in 1897. Named the **electron,** this particle is negatively charged. Because an atom in its normal state is electrically neutral, each electron in an atom must be balanced by the same amount of positive charge. The obvious next problem was to determine where in the atom the positive and negative charges are located. In 1911, British physicist Ernest Rutherford devised an experiment that provided part of the answer to this question. He bombarded a piece of gold foil, which was about 400 atoms thick, with a beam of alpha particles emitted from a radioactive material (Figure 8.9). We now know that alpha particles are helium atoms that have lost all of their electrons. Most of the alpha particles passed through the gold foil just as if it, and the atoms composing it, were nearly empty space. About 1 in 8000 of the alpha particles, however, completely reversed direction and bounced backward from the foil. Rutherford wrote, "It was quite the most incredible event that has ever happened to me in my life. It was almost as incredible as if you fired a 15-inch shell at a piece of tissue paper and it came back and hit you."

The only way to account for the alpha particles that reversed direction when they hit the gold foil is to assume that nearly all of the mass and all of one type of charge, either positive or negative, in each individual gold atom is concentrated in a tiny **atomic nucleus.** We now know that it is the positive charge that is located in the nucleus. When an alpha particle strikes a nucleus it reverses direction, much as a cue ball reverses direction when it hits another billiard ball. Rutherford's model placed the other type of charge—the electrons as we now know—in orbit around this nucleus. This model of an atom resembles a miniature solar system with the electrons revolving around the nucleus, just as planets revolve around the Sun. The electrons must be in motion.

Because positive and negative charges attract each other, stationary electrons would simply fall into the nucleus.

Because most of the atom is empty, just as most of the solar system is empty space, nearly all of Rutherford's alpha particles were able to pass right through the gold foil without colliding with anything.

The Atomic Nucleus

The simplest atom of all is hydrogen, and we now know that the nucleus of ordinary hydrogen contains a single positively charged particle called a **proton.** In orbit around this proton is a single electron. The mass of an electron is nearly 2000 times smaller than the mass of a proton, but the electron carries an amount of charge that is exactly equal in magnitude but opposite in sign to that of the proton. The electron's charge is negative instead of positive (Figure 8.10). Opposite charges attract one another, and so it is the electromagnetic force that binds the proton and electron together, just as gravity is the force that keeps the planets in orbit around the Sun.

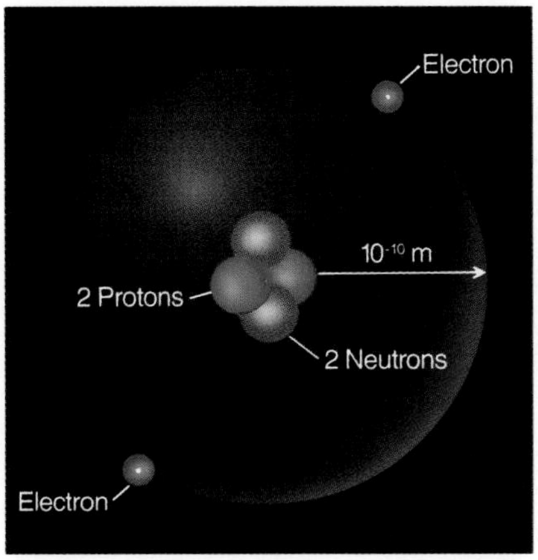

There are, of course, other types of atoms. Helium, for example, is the second most abundant element in the Sun. Helium has two protons in its nucleus, instead of the single proton that characterizes hydrogen. In addition, the helium nucleus contains two **neutrons,** particles with a mass slightly larger than that of the proton but with no electric charge. Orbiting around this nucleus are two electrons, so that the total net charge of the helium atom is zero (Figure 8.11).

From this description of hydrogen and helium, perhaps you have guessed the pattern for building up all of the elements that we find in the universe. The specific element is determined by the number of protons in the nucleus: carbon has 6, oxygen has 8, iron has 26, and uranium has 92. In its normal state, each atom has the same number of electrons as protons, and these electrons follow complex orbital patterns around the nucleus.

Although the number of neutrons in the nucleus is usually approximately equal to the number of protons, the number of neutrons is not necessarily the same for all atoms of a given element. For example, most atoms of hydrogen contain no neutron at all. There are, however, hydrogen atoms that contain one proton and one neutron, and others that contain one proton and two neutrons. The various types of nuclei of hydrogen are called **isotopes** of hydrogen (Figure 8.12).

Other elements have isotopes as well. There is a form of helium with two protons and one neutron, rather than the more common form with two neutrons. Normal oxygen has eight protons and eight neutrons, but there are also rare isotopes of oxygen with nine and ten neutrons.

Figure 8.11 Schematic diagram of a helium atom in its lowest energy state. Two protons occur in the nucleus of all helium atoms. In the most common variety of helium, the nucleus also contains two neutrons, which have nearly the same mass as the proton but carry no charge.

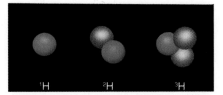

Figure 8.12 Schematic diagram of the nuclei of the isotopes of hydrogen. A single proton in the nucleus defines the atom to be hydrogen, but there may be zero, one, or two neutrons. By far the most common isotope is the one with only a single proton. A hydrogen nucleus with one neutron is called deuterium; one with two neutrons is called tritium.

8.4
SPECTROSCOPY

There is one serious problem with Rutherford's model for atoms. As we have already seen, Maxwell's theory states that when electrons change either their speed or their direction of motion, they emit energy in the form of electromagnetic radiation. Thus, because orbiting electrons constantly change their direction of motion, they should emit a constant stream of energy. Earth-orbiting satellites spiral back toward Earth as they lose energy through friction with the Earth's atmosphere. So too should electrons spiral into the nucleus of the atom as they radiate electromagnetic energy. But they don't. The solution to this problem opened the door to understanding the structure of the atom.

The Bohr Model

Danish physicist Niels Bohr (1885–1962) solved the mystery of electrons. He suggested that the spectrum of hydrogen can be understood if it is assumed that only orbits of certain sizes are possible for the electron. Bohr further assumed that so long as the electron moves in only one of these allowed orbits, it radiates no energy. If the electron moves from one orbit to another orbit closer to the atomic nucleus, then it must give up some energy in the form of electromagnetic radiation, just as a satellite gives up energy when it spirals back into the Earth's atmosphere. Conversely, energy is required to boost the electron from a smaller orbit to one further from the nucleus, and one way to obtain the necessary energy is to absorb electromagnetic radiation if some is streaming past the atom from an outside source.

According to Bohr's model, one fundamental difference between an electron orbiting an atomic nucleus and a satellite orbiting the Earth is that only certain orbits are allowed. The amount of energy that the electron must either absorb or emit to move from one orbit to another is therefore fixed and definite.

When an electron moves from a larger orbit to a smaller orbit, it emits a discrete packet of electromagnetic energy, or photon. The energy of this photon is proportional to the frequency of the electromagnetic radiation. Higher frequency corresponds to higher energy. The highest energy photons of all are gamma rays; those of lowest energy are radio waves. Photons have no mass.

How does the structure of an atom determine the spectral lines that it produces?

Particles and Waves

In this description of how the hydrogen atom absorbs and emits energy, we have talked about light as though it were made up of little energetic particles—photons. But we have already seen that electromagnetic energy also propagates as waves. So what is light really—is it made up of particles or waves? The answer is neither or both. In some situations, it is easier to describe the way light interacts with the material world by treating it as if it were a wave. In other situations, the description is easier to understand if we assume that light is composed of particles. What is really happening is that words that are adequate for describing ocean waves and billiard balls and other aspects of our everyday world simply fail when we try to apply them to the microscopic world of the atom and its interactions with light. The mathematical equations that describe the behavior of light can rigorously predict what will happen in a given situation, and those equations include both wave-like and particle-like behavior. Indeed, the case of light is a good example of why scientists use mathematics. Equations can be devised that will predict the outcome of any particular experiment or set of events, not only in the world that is familiar, but also on the much larger scales that apply to stars, galaxies, and the entire universe and in the smaller world of the atom and the particles within it. Words alone, which are based on the direct experiences of human beings, are merely descriptive and may become misleading when applied to events on vastly different scales.

Nevertheless, based on the mathematical equations that describe the behavior of light, we can provide a somewhat better description in words of its dual nature. We can think of electromagnetic waves simply as a mathematical formula that tells us the probability that a photon is in any particular place. Thus, the wave crests are the places where the photon is more likely to be found. A photon is less likely to be found where there is a wave trough. However, we cannot know where the photon actually is until we observe it. It also turns out that material particles—like protons and electrons and even billiard balls—can similarly be thought of as waves. We cannot say precisely where an electron, for example, is located at any given instant. The best we can do is solve an equation that gives us the probability of its having various locations. The equation turns out to be the equation for a system of waves. The situation is exactly the same as with photons, and it is appropriate to think of the electron, too, as a wave.

Spectral Lines

In order to understand how spectral lines are formed, suppose a beam of white light (which consists of photons of all wavelengths) shines through a gas of atomic hydrogen. Because a photon of wavelength 656 nm has the energy necessary to raise an electron in a hydrogen atom from the second to the third orbit, it can be absorbed by those hydrogen atoms that are in their second-to-lowest energy states. Other photons will have the right energies to raise electrons from the second to the fourth orbit, or the first to the fifth orbit, and so on, and only photons that have exactly these correct energies can be absorbed. All of the other photons stream past the atom untouched. Thus, the hydrogen atoms absorb light only at certain wavelengths and produce spectral absorption lines. Conversely, hydrogen atoms in which electrons move from larger to smaller orbits emit light — but again only light of those energies or wavelengths that correspond to the energy difference between permissible orbits. The changes in orbits of electrons giving rise to spectral lines are shown in Figure 8.13.

Similar pictures can be drawn for atoms other than hydrogen. However, because these other atoms ordinarily have more than one electron each, the energies of the orbits of their electrons are much more complicated, and the spectra are more complex as well. Each type of atom has its own unique pattern of orbits and, therefore, its own unique

Figure 8.13 Emission and absorption of light by a hydrogen atom according to the Bohr model. Several different series of spectral lines are shown, corresponding to transitions of electrons from or to certain allowed energy levels. Each series of lines that terminates on a specific inner orbit is named for the physicist who studied it.

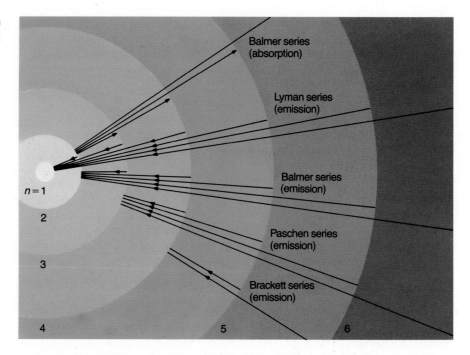

set of spectral lines. Just as fingerprints can be used to identify a person, spectral lines can be used to identify the type of atom that produced them.

Excitation of Atoms

Bohr's model of the hydrogen atom was a great step forward in our understanding of the atom. However, we know today that atoms cannot be represented by quite so simple a picture as the Bohr model. Even the concept of sharply defined orbits of electrons is not really correct. According to modern theory, it is impossible at any instant to measure simultaneously the exact position and the exact velocity of an electron in an atom. Because an electron has wave-like characteristics, we can only estimate the *probability* that it will follow a particular orbit. Nevertheless, because the most likely orbits are fairly narrow with respect to the size of an atom, we still retain the concept that only certain discrete energies are allowable for an atom. These energies, called **energy levels,** can be thought of as representing certain average distances of the most likely of the electron's possible orbits around the atomic nucleus.

Ordinarily, an atom is in the state of lowest possible energy, its **ground state.** In the Bohr model, the ground state corresponds to the electron being in the innermost orbit. However, an atom can absorb energy, which raises it to a higher energy level (corresponding, in the Bohr picture, to the movement of an electron to a larger orbit). The atom is then said to be in an **excited state.** Generally, an atom remains excited for only a very brief time. After a short interval, typically only a hun-

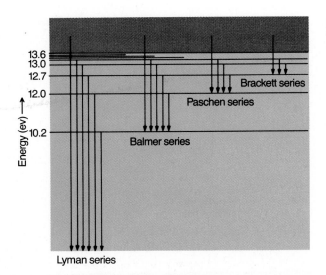

Figure 8.14   Energy level diagram for hydrogen. The shaded region represents energies at which the atom is ionized.

dred-millionth of a second or so, it drops back down to its ground state, with the simultaneous emission of light. (In the Bohr model, this corresponds to a jump by the electron back to the innermost orbit.) The atom may return to its lowest state in one jump, or it may make the transition in steps of two or more jumps, stopping at intermediate levels on the way down. With each jump, it emits a photon of the wavelength that corresponds to the energy difference between the levels at both the beginning and the end of that jump. An energy-level diagram for a hydrogen atom and several possible atomic transitions are shown in Figure 8.14; compare this figure with the Bohr model in Figure 8.13.

Because atoms that have absorbed light and have thus become excited generally de-excite themselves and emit that light again, we might wonder why dark spectral lines are ever produced. In other words, why doesn't this re-emitted light "fill in" the absorption lines? Some of the re-emitted light actually is received by us, but this light fills in the absorption lines only to a slight extent. The reason is that the atoms re-emit light in random directions, and only a small fraction of the re-emitted light is directed toward the observer. We can observe the re-emitted light as emission lines if and only if we can view the absorbing atoms from a direction from which no background light is coming—as we do, for example, when we look at clouds of hot gas located in the space between the stars. Figure 8.15 illustrates the situation.

Atoms in a gas are moving at high speeds and continually collide with each other as well as with electrons. They can be excited and de-excited by these collisions as well as by absorbing and emitting light. The velocity of atoms in a gas depends on its temperature, and so the higher the temperature, the higher the velocity and the higher the energy of the collisions. Therefore, the hotter the gas, the more likely it is that electrons will occupy the outermost orbits, which correspond to the highest energy levels.

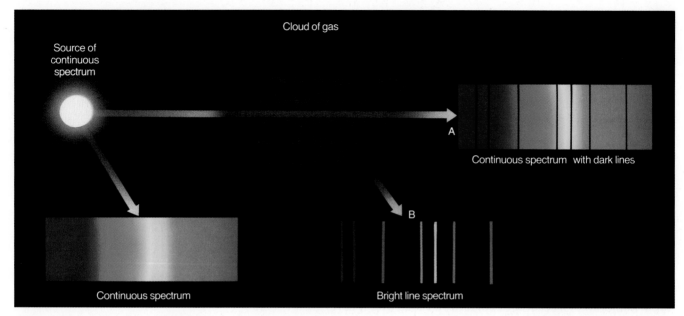

Figure 8.15 Production of bright and dark spectral lines. The atoms in the gas cloud produce absorption lines in the continuous spectrum of the white light source when viewed from direction A, but they produce emission lines (of the light they re-emit) when viewed from direction B.

Ionization

An atom can acquire energy through a collision with another atom, with an electron (most likely), or through absorption of electromagnetic radiation. If enough energy is transferred to the atom, one or more of its electrons can be removed completely. The atom is then said to be **ionized** (it is called an **ion**). An atom that has become ionized has lost a negative charge—that carried away by the electron—and thus is left with a net positive charge. It therefore has a strong affinity for a free electron (because opposite electrical charges attract each other). Eventually it will capture one and become neutral (or ionized to one less degree) again.

The energy levels of an ionized atom are different from those of the same atom when it is neutral. In each degree of ionization, the energy levels of the ion, and thus the wavelengths of the spectral lines it can produce, have their own characteristic values. Ionized hydrogen has no electron and can produce no absorption lines.

Doppler Effect

One very important thing that can be learned from spectral lines is whether or not the source of light is moving closer to or farther away from us. In 1842, the Austrian physicist and mathematician Christian Doppler pointed out that if a light source is approaching an observer, the light waves are crowded closer together. If the light source is receding, the light waves are spread out. This principle, known as the Doppler principle or **Doppler effect,** is illustrated in Figure 8.16. In (a),

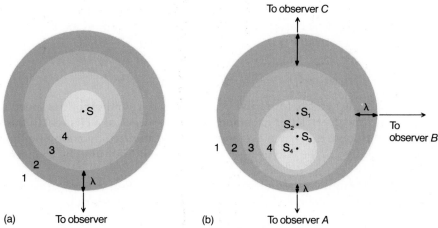

Figure 8.16 The Doppler effect. See text for explanation.

the light source is stationary with respect to the observer. As successive wave crests 1, 2, 3, and 4 are emitted, they spread out evenly in all directions, like the ripples from a splash in a pond. They approach the observer at a distance λ behind each other, where λ is the wavelength of the light. Conversely, if the source is moving with respect to the observer, as in (b), the successive wave crests are emitted with the source at different positions, S1, S2, S3, and S4, respectively. Thus, to observer A, the waves seem to follow each other by a distance less than λ, whereas to observer C they follow each other by a distance greater than λ. The wavelength of the radiation received by A is shortened; the wavelength of the radiation received by C is lengthened. Because the wave crests arrive at A following each other more closely than they do from a stationary source, they are observed by A at a higher than normal frequency, whereas C receives the light at a diminished frequency. To observer B, in a direction at a right angle to the motion of the source, no effect is observed. The effect is produced only by a motion toward or away from the observer, a motion called **radial velocity.** Observers between A and B and between B and C would observe some shortening or lengthening of the light waves, respectively, because a component (or part) of the motion of the source is either toward them or away from them.

The Doppler effect is also observed in sound. Most of us have heard the higher-than-normal pitch of a siren on an approaching emergency vehicle and the lower-than-normal pitch of a receding one. Standing by the highway, you can even hear the characteristic "whoosh," or declining pitch, of each car or truck as it speeds past.

The precise amount by which the wavelengths of light from a source appear lengthened or shortened (the Doppler shift) provides us a means of determining how fast that source is moving in the line of sight—that is, toward or away from us. That velocity is its **radial velocity.** If its speed is small compared to the speed of light, the change in wavelength

How can the measurement of the wavelength of a spectral line be used to determine whether a source of electromagnetic radiation is moving toward or away from an observer?

of its light divided by the wavelength itself is proportional to its radial velocity divided by the speed of light. This relationship is expressed in the following formula:

$$\frac{\Delta\lambda}{\lambda} = \frac{v}{c}$$

where v is the symbol for radial velocity, c is the symbol for the speed of light, and the greek letter (Δ) stands for the concept of a change. In this case, the change is in the wavelength of a line emitted by a moving light source relative to the wavelength of that same line emitted by a stationary source.

Even though we cannot travel directly to the Sun or the stars, we can learn a great deal about them from measurements made at a distance. Spectroscopy provides the tools we need to determine what kinds of gas are present, how hot that gas is, and even something about what is going on at the turbulent, boiling surface of the Sun. The nucleus of the atom holds the secret of energy generation deep in the interior of the Sun—energy that has been warming the Earth for 4.5 billion years. In the next chapter, we use the messages contained in sunlight to embark on a journey first to the surface and then to the center of the Sun.

8.5
ASTRONOMICAL OBSERVATIONS

Before we can analyze the light or other forms of radiation from a celestial source, we must collect and focus that radiation. This is the function of a telescope. All telescopes work in the same general way, whether they are collecting gamma rays, visible light, or radio waves.

Telescopes

A telescope is the basic tool of the observational astronomer. The term telescope refers to any device that collects photons, of whatever energy. Until the 20th century, all telescopes were built to operate in the visible part of the spectrum. About 50 years ago, astronomers also began to build radio telescopes. Today, astronomical telescopes (both optical and radio) tend to be huge devices, constructed at costs of tens or even hundreds of millions of dollars.

The most important functions of a telescope are to *collect* the faint light from an astronomical source and to *focus* it into an image. A large curved lens or mirror to accomplish these tasks is the essential optical element of a telescope. For the astronomer, the magnification of this image is largely incidental, and it is meaningless to ask what the "power" of an astronomical telescope is. Instead, the measure of size and capability for an astronomical telescope is its light-gathering ability, which is determined by the diameter, or **aperture,** of its primary lens or mirror. Just as a broader bucket standing in the rain collects more raindrops, so a larger aperture telescope collects more photons.

Why are telescopes used to observe astronomical objects?

Figure 8.17   The 40-in. (1-m) refracting telescope of Yerkes Observatory was the largest telescope in the world when completed in the 1890s.   *(Yerkes Observatory)*

Most people, when they think of a telescope, picture a long tube with a large glass lens at one end. This design is called a **refracting telescope** (Figure 8.17). Galileo's telescopes were refracting, as are common binoculars or opera glasses today. But refracting telescopes are not very good for most astronomical applications. They cannot be built with apertures larger than about 1 m, and lenses do not provide images that are as sharp as those formed by concave polished mirrors. For these and other reasons, most astronomical telescopes (both amateur and professional) use a mirror rather than a lens as their primary optical part; these are called **reflecting telescopes** (Figure 8.18).

Because the main purpose of a telescope is to collect light from faint sources, astronomers usually want a telescope with as large an aperture as possible. They begin with a large disk of glass, which is laboriously ground and polished to produce a concave mirror of high optical quality. This mirror is supported mechanically so that it does not distort or sag by as much as a millionth of an inch as it is pointed in various directions. One or more additional mirrors are used to direct the image into auxiliary instruments, where it is recorded for later analysis.

The Astronomical Observatory

An observatory is a location where astronomical telescopes are operated. It should be in a place where the skies are frequently clear and conditions for astronomical observation are good. The best such sites are on

Figure 8.18 Various focus arrangements for reflecting telescopes. These are called: (a) prime focus; (b) Newtonian focus; (c) Cassegrain focus.

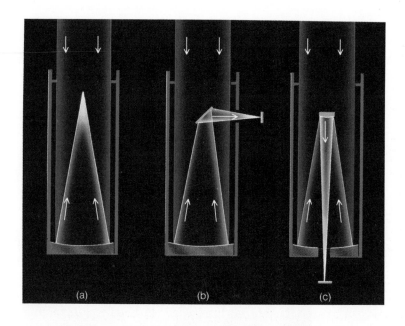

(a)   (b)   (c)

mountains far from the lights and pollution of cities. Although a number of old observatories remain in cities, especially in Europe, they have become administrative centers or museums. The real action takes place far away, often on desert mountains or on isolated peaks in the Atlantic and Pacific oceans (Figure 8.19).

Typically, the astronomer travels for hours by plane or car in order to reach the observatory. Sometimes the trip is half-way around the world, as for the British and French astronomers who operate large tele-

Figure 8.19 Several telescope domes are shown in this view of Mauna Kea Observatory located at an altitude of 4200 m on an extinct volcano in Hawaii. *(University of Hawaii photo by David Morrison)*

scopes in Hawaii. At the observatory are living quarters, computers, electronic and machine shops, and of course the telescopes themselves. A large observatory today requires a support staff of 20 to 100 persons, in addition to the astronomers.

The popular view of an astronomer as a person in a cold observatory peering through a telescope all night is no longer valid. Many astronomers work only with radio telescopes or with space experiments, whereas others work at purely theoretical problems and never observe at a telescope of any kind. Even optical astronomers seldom inspect telescopic images visually except to center the telescope on a desired region of the sky or to make adjustments. On the contrary, electronic detectors are used to record the data permanently for detailed analysis after the observations are completed. Typically, one successful night at the telescope yields enough data to keep an astronomer busy for weeks of analysis and interpretation.

Major Optical Observatories

The world's largest optical telescope is the Keck Telescope of Mauna Kea Observatory in Hawaii (Figure 8.20). Its light-collecting mirror, 10 m in aperture, consists of 36 individual mirrors all mounted and polished so that they work together like a single large piece of glass. A twin 10-m telescope, Keck II, is under construction on an adjacent site, and

Figure 8.20   The 10-m aperture segmented mirror of the Keck Telescope on Mauna Kea consists of 36 individually ground and polished pieces of glass.   *(CARA)*

TABLE 8.2 Large Optical Telescopes

Aperture (M)	Telescope Name	Location
16.0	VLT (four 8-m telescopes)	Cerro Paranal, Chile
10.0	Keck I	Mauna Kea, HI, USA
10.0	Keck II	Mauna Kea, HI, USA
8.2	Subaru (Pleiades)	Mauna Kea, HI, USA
8.0	Gemini	Mauna Kea, HI, USA
8.0	Gemini	Cerro Puchon, Chile
6.5	MMT	Mount Hopkins, AZ, USA
6.5	Magellan	Las Campanas, Chile
6.0	Large Alt-Azimuth	Mount Pastukhov, Russia
5.0	Hale	Mount Palomar, CA, USA
4.2	William Herschel	Canary Islands, Spain
4.0	NOAO Cerro Tololo	Cerro Tololo, Chile*
3.9	Anglo-Australian	Siding Spring, Australia
3.8	NOAO Mayall	Kitt Peak, AZ, USA*
3.8	United Kingdom Infrared	Mauna Kea, HI, USA
3.6	Canada-France-Hawaii	Mauna Kea, HI, USA
3.6	ESO	Cerro La Silla, Chile**
3.6	ESO New Technology	Cerro La Silla, Chile**
3.5	Max Planck Institut	Calar Alto, Spain
3.5	WIYN	Kitt Peak, AZ, USA
3.0	Shane (Lick Observatory)	Mount Hamilton, CA, USA
3.0	NASA IRTF	Mauna Kea, HI, USA

* U.S. National Optical Astronomy Observatories
** European Southern Observatory

eventually the two telescopes will be used together. At an altitude of 4200 m, Mauna Kea is the premier observing site in the world, and two other large telescopes, with apertures of about 8 m, are under construction there by Japan and the U.S. National Optical Astronomy Observatory.

Palomar Mountain in California is the home of the Hale 5-m (200-in.) telescope, perhaps the most famous astronomical instrument in the world. Completed in 1948, the Hale telescope has pioneered in many areas of astronomy, but especially in studies of distant galaxies. The U.S. National Observatory, open to astronomers from all over the nation, operates several telescopes at Kitt Peak near Tucson, Arizona. A number of major observatories have been built in South America, in the foothills of the Andes Mountains of Chile. Here the Europeans are building the

Figure 8.21 Grote Reber's original radio telescope. *(National Radio Astronomy Observatory)*

largest optical telescope in the world, which will consist of four telescopes, each with an 8-m mirror, that can be used either independently or as a single telescope, with a collecting area equivalent to that of a single 16-m mirror. Table 8.2 lists the world's largest optical telescopes, including those currently under construction.

Radio Astronomy

In the latter half of the 20th century, astronomy expanded rapidly beyond the narrow spectral window available to the traditional optical astronomer. Radio astronomy began in 1931, as a result of observations made by electrical engineer Karl G. Jansky of the Bell Telephone Laboratories. Jansky was experimenting with antennas for long-range radio communication when he encountered interference in the form of radio radiation coming from an unknown source. He discovered that this radiation was not of terrestrial or even solar origin, but was coming from the Milky Way.

In 1936, Grote Reber, an American amateur astronomer and radio ham, built from galvanized iron and wood the first antenna specifically designed to receive these cosmic radio waves. Over the years, Reber built several such antennas and used them to carry out pioneering surveys of the sky for celestial radio sources. During the first decade he worked practically alone, for professional astronomers had not yet recognized the vast potential of radio astronomy. Many of the objects that Reber discovered became subjects of intensive investigation years later. His original radio telescope is on display at the National Radio Astronomy Observatory (Figure 8.21).

How do telescopes that operate at radio frequencies or in space differ from conventional optical telescopes?

Figure 8.22 The 1000-ft (305-m) radio dish at the National Astronomy and Ionosphere Center, Arecibo, Puerto Rico, operated by Cornell University and sponsored by the National Science Foundation. *(Cornell University)*

It is important to understand that radio waves are not "heard." They are not the same as sound waves. Although radio waves are modulated or coded to carry sound information in commercial radio broadcasting, the sound itself is not transmitted. The radio waves merely carry the information that a radio receiver must decode and convert into sound by means of a speaker or earphones. Sound is a physical vibration of mat-

Figure 8.23 The Very Large Array (VLA) near Socorro, New Mexico. This telescope consists of 27 movable antennas spread over a total span of 36 km. *(National Radio Astronomy Observatory)*

ter; radio waves, like light, are made up of photons. We can also code visible light to carry sound information, as is done, for example, by the sound track on a movie film or in the fiber optics used today for telecommunications.

The world's largest single radio antenna is a 1000-ft. (305-m) dish at Arecibo, Puerto Rico. Although this immense, bowl-shaped antenna is fixed in the ground, tracking of astronomical sources is possible by moving the receiving equipment at the focus of the dish (Figure 8.22). The largest fully steerable radio antennas, with apertures of 100 m, are a German instrument located near Bonn and an American instrument at Green Bank, West Virginia.

The most impressive radio facilities consist of many antennas coupled together. The largest such telescope, operated by the U.S. National Radio Astronomy Observatory in New Mexico (Figure 8.23), is the Very Large Array (VLA). The VLA consists of 27 individual antennas, each 25 m in diameter, all operated together as a single instrument with an effective diameter of 36 km. Completed in 1980, the VLA can generate images of the radio sky at several wavelengths that compete favorably with the best optical images obtained with more conventional telescopes. Substantially higher resolution (more detail) is obtained with the recently completed Very Long Baseline Array (VLBA), also operated by the National Radio Astronomy Observatory. These antennas are located at widely spaced intervals, from the Virgin Islands to Hawaii, yet they all work together as one giant telescope of almost planetary dimensions. Table 8.3 lists the major radio observatories of the world.

TABLE 8.3   Major Radio Observatories of the World

Observatory	Location	Description (Main Instruments)
National Astronomy and Ionospheric Center	Arecibo, Puerto Rico	305-m fixed dish
National Radio Astronomy Observatory (NRAO)	Green Bank, WV	100-m steerable dish and 43-m steerable dish
Max Planck Institute for Radioastronomy	Bonn, Germany	100-m steerable dish
Jodrell Bank Radio Observatory	Manchester, England	76-m steerable dish and 66-m steerable dish
Very Large Array (NRAO)	Socorro, NM	27-element array of 25-m dishes (36 km length)
Very Long Baseline Array (NRAO)	Many U.S. sites from the Caribbean to Hawaii	14-element array of 25-m dishes (9,000 km length)
Westerbork Radio Observatory	Westerbork, The Netherlands	12-element array of 25-m dishes (1.6 km length)

Observatories in Space

Most astronomical objects emit in all parts of the electromagnetic spectrum, i.e., radio photons, light, infrared and ultraviolet radiation, etc. The radio photons we receive from space are those that can penetrate the Earth's atmosphere. Other radio photons, as well as most of the photons in the infrared, ultraviolet, x-ray, and gamma-ray regions of the electromagnetic spectrum, are absorbed by the atmosphere before they can reach our telescopes.

Ultraviolet, x-ray, and gamma-ray observations can be made only from space. The first all-sky survey for celestial x-ray sources was carried out in 1970 by the U.S. built satellite Uhuru (Swahili for "freedom"), which was launched from a European facility in Kenya. A similar initial infrared survey was made in 1983 by the Infrared Astronomy Satellite (IRAS) built jointly by the United States, Britain, and The Netherlands (Figure 8.24). Another small infrared observatory called COBE has contributed greatly to our knowledge of the early history of the universe, which we will describe in Chapter 16.

In addition to a continuing series of small satellite observatories, several large space telescopes have either been already launched or are in the planning stages. The first of these new telescopes is the Hubble

Figure 8.24 The Infrared Astronomy Satellite (IRAS), a joint project of NASA, The Netherlands, and the United Kingdom. Here the spacecraft is being tested before its 1983 launch. *(NASA/JPL)*

ESSAY   How to Choose a Telescope

How do you choose a telescope? This question is one of the first asked by most people who become interested in astronomy. Unfortunately, telescopes are expensive. Even the small telescopes used by amateur astronomers cost several hundred dollars. It is certainly possible to purchase telescopes at lower cost, but cheaper models are unlikely to produce very good images or to be much fun to use.

The principal purpose of a telescope is to make it possible to see objects too faint to be seen with the unaided eye. A telescope is a light bucket. It gathers together all of the light that passes through it and focuses that light into a small image that can be viewed with the human eye. The bigger the telescope, the more light that is captured and the fainter the object that can be seen. The human eye works the same way. In dim illumination, the pupil of the eye dilates to a maximum diameter of about 1/3 in. in order to intercept more light. Compare this size with the typical size of amateur telescopes, which are usually several inches in diameter, or with the Palomer telescope, which has a diameter of 200 in.

Amateur telescopes, like the large telescopes used by professional astronomers, come in two types—refractors and reflectors. To be useful for astronomy, the lens of a refracting telescope should be at least 3 in. in diameter. For a given size, refractors cost more than reflectors. The higher cost comes about because both sides of the lens must be polished to high accuracy. Furthermore, because the light passes through the lens, the lens must be made of high quality glass throughout. In contrast, only the front surface of the mirror in a reflector must be accurately polished, and there can be bubbles and other flaws in the glass behind the polished surface.

The mount of a telescope is one of its critical elements. Telescopes operate at much higher magnification than binoculars, and even tiny vibrations are readily seen. A sturdy and stable mount is essential.

One commonly asked question concerns the magnification of a telescope. Magnification is not one of the criteria on which to base the choice of a telescope. The magnification of a telescope is determined not by the diameter of the primary lens or mirror but by the eyepiece used for viewing. As a result, the magnification of a telescope can be anything the user wants it to be, and most telescopes come with a selection of eyepieces that provide varying amounts of magnification.

A telescope magnifies not only the astronomical image but also the turbulence in the Earth's atmosphere. If the chosen magnification is too high, then blurring effects caused by the Earth's atmosphere make the image shimmer and shake. It should also be noted that all stars appear as points of light no matter what magnification is chosen.

The best way to choose a telescope is to try it out first. Most communities have amateur astronomy clubs that sponsor star parties open to the public. Before spending a substantial sum of money on a telescope, one should get to know what type of telescope would give the most satisfaction.

Space Telescope (HST), an optical reflector with an aperture of 2.4 m (Figure 8.25). Operating above the atmosphere, this telescope has access to the entire ultraviolet part of the spectrum. Free from the distortions imposed by the Earth's atmosphere even in the visible part of the electromagnetic spectrum, the HST is also designed to probe deeper into space than can any ground-based telescope. Unfortunately, however, an error in the manufacture of the HST primary mirror has restricted its usefulness since its launch in 1989. Late in 1993, astronauts installed corrective optics in HST, with the objective of restoring it to its originally planned capability.

Figure 8.25   The 2.4-m Hubble Space Telescope (HST) at the time of its launch from the Space Shuttle. *(NASA)*

The second large, new space observatory is a more specialized facility called the Compton Gamma Ray Observatory (GRO), launched in 1990. The third new observatory is the Advanced X-Ray Astrophysics Facility (AXAF), with an anticipated launch sometime in 1997. Finally, the proposed U.S. infrared observatory, SIRTF (the Space Infrared Telescope Facility), is a 1-m telescope that could be in operation early in the next century. This observatory will cover the fourth spectral region from space, the infared. Meanwhile, the European Space Agency is launching a slightly smaller Infrared Space Observatory (ISO) in 1995.

All of the space observatories have unique capabilities that cannot be duplicated from the ground. Most of them operate in spectral regions where ground-based astronomy is simply impossible. However, these capabilities are bought at a great price. The cost of HST, including the 1993 refurbishment, exceeds $3 billion, and each of the other three large telescopes costs considerably more than $1 billion. For comparison, even the largest ground-based optical or radio telescope costs only about $100 million. Astronomers must establish their priorities carefully to balance the capabilities, and costs, of ground-based and space facilities.

▶ SUMMARY

8.1 Light is one form of electromagnetic radiation. Light can be thought of as **photons,** but it also has wave characteristics. Photons of a wide range of **wavelength** and **frequency** make up the **electromagnetic spectrum.** Wavelength (λ) times frequency (f) equals the speed of light (c). The **apparent brightness** of a source of electromagnetic energy decreases with increasing distance from that source. Quantitatively, the amount of the decrease in apparent brightness is proportional to the square of the distance. The mathematical equation describing this relationship is known as the **inverse-square law.**

8.2 Gamma rays, x rays, and **ultraviolet** radiation are forms of electromagnetic radiation with wavelengths shorter than that of visible light. **Infrared, microwave,** and radio radiation have wavelengths longer than that of light. The higher the temperature of a **blackbody,** the shorter the wavelength at which the maximum amount of electromagnetic radiation is emitted. The mathematical equation describing this relationship ($\lambda_{max} = 3{,}000{,}000/T$) is known as **Wien's law.** The total energy emitted per sq m increases with increasing temperature. The relationship between emitted energy and temperature ($E = \sigma T^4$) is known as the **Stefan-Boltzmann Law.**

8.3 Atoms consist of a nucleus containing one or more positively charged particles known as **protons.** All atoms except hydrogen also contain one or more **neutrons** in the **atomic nucleus.** Neutrons have no charge but are about the same mass as the proton. Negatively charged **electrons** orbit around the nucleus, and the number of electrons normally equals the number of protons. The number of protons defines the type (hydrogen, helium, etc.) of atom, while atoms that differ only in the number of neutrons are **isotopes** of an element.

8.4 When an electron moves from one orbit or **energy level** to another orbit closer to the atomic nucleus, a photon is emitted, and a spectral emission line is formed. Absorption lines are formed when an electron moves to an orbit farther away from the nucleus. Because each atom has its own characteristic set of energy levels, each atom possesses a unique pattern of **spectral lines.** The lowest energy level is the **ground state;** an atom in a higher energy level is **excited.** If an atom contains fewer electrons than protons, it is said to be **ionized.** If an atom is moving toward an observer when an electron changes orbits and produces a spectral line, that line is shifted slightly to the blue of its normal position. If the atom is moving away, the line is shifted to the red. This shift is known as the **Doppler effect** and can be used to measure the **radial velocities** of distant objects.

8.5 Astronomical telescopes, whether they are **refracting** or **reflecting,** collect the light and focus it. The greater the **aperture** of a telescope, the more powerful it is. Most astronomical data in the visible spectrum come from remote mountain observatories such as Mauna Kea, the home of the world's largest telescope. Radio astronomy, exploiting the radio spectrum, was the first major technique to extend astronomical observations into other parts of the electromagnetic spectrum. Major radio observatories include Arecibo, VLA, and VLBA. To pursue other opportunities, it is necessary to get above the Earth's atmosphere. Space observatories launched by the United States include HST and GRO, with AXAF and SIRTF planned for the future.

▶ REVIEW QUESTIONS

1. What is a wave? Use the terms "wavelength" and "frequency" in your definition.

2. What distinguishes one type of electromagnetic radiation from another? Identify types of electromagnetic radiation that you make use of, or are exposed to, in everyday life.

3. What is a blackbody? How does the energy emitted by a blackbody depend on its temperature? (As we shall see in later chapters, the energy emitted by the Sun and the stars is related to temperature in nearly the same way.)

4. Where in an atom would you expect to find electrons? protons? neutrons?

5. For astronomical telescopes, bigger is better. Why?

6. What are the main reasons for building observatories in space?

▶ THOUGHT QUESTIONS

7. What type of electromagnetic radiation is best suited to observing a star with a temperature of 5800 K? A gas heated to a temperature of 1,000,000 K? A human being on a dark night?

8. Go outside at night and look carefully at the brightest stars. Some should look red and some should look blue. The primary factor determining the color of a star is its temperature. Which is hotter—a blue star or a red star? Explain.

9. Water faucets are often labeled with a red dot for hot water and a blue dot for cold water. Given Wien's law, does this labeling make sense?

10. Suppose that you are standing at the exact center of a park surrounded by a circular road. Suppose an ambulance drives completely around this circular road. How will the pitch of the ambulance's siren change as the ambulance circles around you?

11. Why is it dangerous to be exposed to x rays but not (or at least very much less) dangerous to be exposed to radio waves? Are gamma rays likely to be more dangerous than x rays?

12. The planet Jupiter appears yellowish, whereas Mars appears red. Does this observation mean that Mars is cooler than Jupiter? Explain.

13. How could you measure the Earth's orbital speed by photographing the spectrum of a star at various times throughout the year? (Hint: Suppose the star lies in the plane of the Earth's orbit.)

14. The space between stars contains clouds of cold gas that are transparent to light. Suppose you observe a distant star. How might the presence of a cloud between you and the star be detected?

15. The largest observatory complex in the world is on Mauna Kea in Hawaii at an altitude of 4200 m. Consider other potential sites on high mountains and discuss their advantages and disadvantages. Should astronomers, for example, consider building an observatory on Mt. McKinley (Denali) or Mt. Everest?

16. Another potential site sometimes considered for astronomy is the Antarctic plateau. Discuss its advantages and disadvantages.

▌ PROBLEMS

17. Galileo suggested a technique for measuring the speed of light with two experimenters separated by a mile and each equipped with a lantern that can be covered. The first experimenter opens his lantern, and the second experimenter, on seeing the light from the first, uncovers his. The time that elapses between the time that the first experimenter opens his lantern and the time that he sees the light of his associate's lantern, after correction for the human reaction time, is how long light spends making the round trip. Would you expect this technique to work?

18. From the fact that it takes light 16.5 minutes to cross the orbit of the Earth, show that the speed of light is about 10,000 times that of the Earth.

19. "Tidal waves," or tsunamis, are waves of seismic origin that travel rapidly through the ocean. If tsunamis travel at the speed of 600 km/h and approach a shore at the rate of one wave crest every 15 minutes, what would be the distance between those wave crests at sea?

20. How many times brighter or fainter would a star appear if it were moved to (a) twice its present distance? (b) ten times its present distance? (c) half its present distance?

21. Two stars are at the same distance from the earth. The two stars have identical diameters. One has a temperature of 5800 K; the other has a temperature of 2900 K. Which is brighter? How much brighter is it?

22. A typical large telescope today requires about $2 million per year to operate. Suppose you wish to observe with that telescope. What is the cost per night for the telescope time? The cost per hour? The cost per minute? Can you think of any other activities that have such a high associated cost, in dollars per hour?

23. The Hubble Space Telescope cost about $1.4 billion for construction, $600 million for its Shuttle launch, and $100 million per year for operations (not counting the flight for repair and refurbishment). If the telescope lasts a total of 10 years, what is the cost per year? The cost per day? If the telescope can be used just 30 percent of the time for actual observations, what is the cost per hour and per minute for the astronomer's observing time on this instrument?

▌ SUGGESTIONS FOR ADDITIONAL READING

Books

On Light and the Atom

Bova, B. *The Beauty of Light.* J. Wiley, 1988. A readable introduction by a science writer.

Sobel, M. *Light.* U. of Chicago Press, 1987. A nontechnical primer by a physicist.

Rowan-Robinson, M. *Cosmic Landscape: Voyage Back Along the Photon's Track.* Oxford U. Press, 1980. Examining the cosmos in each region of the electromagnetic spectrum.

Hearnshaw, J. *The Analysis of Starlight.* Cambridge U. Press, 1986. A history of spectroscopy.

On Telescopes and Observatories

Cohen, M. *In Quest of Telescopes.* Cambridge U. Press, 1980. An astronomer describes what it's like to observe with the world's largest instruments.

Tucker, W. & K. *Cosmic Inquirers.* Harvard U. Press, 1986. The story of some of the most important astronomical telescopes, on the ground and in space.

Field, G. & Goldsmith, D. *Space Telescope: Eyes Above the Atmosphere.* Contemporary Books, 1990. Two well-known astronomers tell the story of the Hubble; written before the mirror flaw was discovered.

Krisciunas, K. *Astronomical Centers of the World.* Cambridge U. Press, 1988. Interesting history of the major observatories around the world.

Learner, R. *Astronomy Through the Telescope.* Van Nostrand Reinhold, 1981. Illustrated history of telescopes and the discoveries they made possible.

Articles

On Light and the Atom

Augensen, H. & Woodbury, J. "The Electromagnetic Spectrum" in *Astronomy,* June 1982, p. 6.

Darling, D. "Spectral Visions: The Long Wavelengths" in *Astronomy,* Aug. 1984, p. 16; "The Short Wavelengths" Sept. 1984, p. 14.

Gingerich, O. "Unlocking the Chemical Secrets of the Cosmos" in *Sky & Telescope,* July 1981, p. 13.

On Telescopes and Observatories

Eicher, D. "A New Era in Space: Space Astronomy for the 1990's" in *Astronomy,* Jan. 1990, p. 22.

Robinson, L. "Monster Telescopes for the 1990's" in *Sky & Telescope,* May 1987, p. 495.

Wolff, S. "The Search for Aperture: A Selective History of Telescopes" in *Mercury,* Sept./Oct. 1985, p. 139.

Fienberg, R. "Space Telescope: Picking Up the Pieces" in *Sky & Telescope,* Oct. 1990, p. 352. An excellent summary of the mirror problem; updated Jan. 1991, p. 14; Sept. 1991, p. 242.

Sinnott, R. "The Keck Telescope's Giant Eye" in *Sky & Telescope,* July 1990, p. 15.

Keel, W. "Galaxies Through a Red Giant" in *Sky & Telescope,* June 1992, p. 626.

Shore, L. "VLA: The Telescope that Never Sleeps" in *Astronomy,* Aug. 1987, p. 15.

Kniffen, D. "The Gamma-Ray Observatory" in *Sky & Telescope,* May 1991, p. 488.

Robinson, L. "The Frigid World of IRAS" in *Sky & Telescope,* Jan. 1984, p. 4.

Tucker, W. & Giacconi, R. "The Birth of X-Ray Astronomy" in *Mercury,* Nov./Dec. 1985, p. 178; Jan./Feb. 1986, p. 13.

THE SUN: A NUCLEAR POWERHOUSE

 he Sun is a star.

Grade school children memorize this fact, but what does it mean? It is far from obvious that this "fact" is even correct. In appearance and in its influence on the Earth, the Sun is very different from a star. The Sun appears much brighter. It also seems to be much larger than stars, which look like mere points of light. The Sun is the source of heat that sustains life on Earth and controls its climate. There is evidence that tiny decreases in the output of energy from the Sun can cause ice ages. The Sun heats some parts of the Earth more than others. This uneven heating creates winds that determine weather patterns. The Sun bombards the Earth with charged atomic particles that can disrupt radio communications. Given its influence, it is little wonder that the Sun was worshipped as a god by many ancient civilizations.

The Sun, just rising above the horizon, with the silhouette of the mirror of McMath Solar Telescope at Kitt Peak National Observatory in the foreground. This mirror tracks the Sun across the sky and directs sunlight downward into the rest of the optics of the telescope. *(National Optical Astronomy Observatories)*

227

The idea that the stars might be distant suns, perhaps surrounded by their own families of planets, is relatively modern. The Italian philosopher Giordano Bruno (1548–1600) was one of the first to speculate that the universe is infinite, with an uncountable number of suns.

Modern astronomers have found that the Sun is a rather ordinary star— not unusually hot or cold, old or young, or large or small. It is fortunate that we have such a garden-variety star so near at hand (astronomically speaking!), because by observing the Sun closely, we can learn a great deal about the physical processes that determine the structure and evolution of other stars. The Sun plays a role in stellar astronomy similar to that of the Earth in planetary studies: It is the one well-studied example that serves as a benchmark for all of the much fainter stars for which we cannot obtain the same enormously detailed information.

Table 9.1 lists some of the basic facts about the Sun.

9.1
HOW FAR? HOW BIG? HOW MASSIVE? HOW DENSE?

Distance to the Sun

"How far away is the Sun?"

"93,000,000 miles (150,000,000 km)."

"Prove it!"

Stymied by that challenge? So were astronomers for a very long time. One of the great scientific achievements of the 17th century was the mea-

TABLE 9.1 Solar Data

Mean distance	1 AU
	149,597,892 km
Mass	333,400 Earth masses
	1.99×10^{30} kg
Diameter at surface (photosphere)	109 Earth diameters
	1.39×10^6 km
Density	1.41 g/cm³
Luminosity	3.8×10^{26} watts
Rotation period at equator	24 days, 16 hours
Escape velocity	618 km/s
Surface temperature	5800 K
Age	4.5×10^9 yrs

surement of the size of the solar system, including the distance to the Sun. At the beginning of that century, the best estimates of the distances to the Sun and Moon were still those derived by Ptolemy, who worked as both an astronomer and a geographer in Alexandria, Egypt, between the years 125 and 150 A.D. Based on the motions of the Sun and the Moon, and on a direct measurement of the distance to the Moon, Ptolemy calculated that the diameter of the Sun is about 5.5 times the diameter of the Earth—in fact it is about 100 times the diameter of the Earth—and that the distance to the Sun is about 8 million km, a value that is nearly 20 times too small.

By the beginning of the 18th century, astronomers had made new measurements that agreed in showing that the distance to the Sun is about 130,000,000 km, an estimate that is within 15 percent of the modern value. Astronomers now use the technique of radar to measure distances within the solar system (Section 2.2). It is not possible to use radar to measure the distance to the Sun directly. The Sun does not reflect radar efficiently, and the Sun itself is a source of radio waves that are much stronger than any radar signal we could send. Radar does tell us precisely the size of the orbits of the planets around the Sun, and from that information it is possible to calculate the distance to the Sun. Modern measurements give a value of 149,597,892 km for the astronomical unit (AU), which is the average distance from the Earth to the Sun.

Diameter of the Sun

Once we know the distance to the Sun, it is very easy to calculate its diameter. Figure 9.1 illustrates the method. The first step is to measure the angular diameter of the Sun. The angle between one edge of the Sun and the other is its angular diameter. Angles are usually measured in a system in which a complete circle contains 360°. Viewed from the Earth, the average angular diameter of the Sun is about 0.53°. Because the Sun is (on average) about 150,000,000 km away, the diameter of the Sun in kilometers must be in the same proportion to the circumference of a circle with a radius of 150,000,000 km as 0.53° is to 360°. That is, if D is the diameter of the Sun, then

$$\frac{0.53°}{360°} = \frac{D(\text{km})}{2\pi \times 150,000,000 \text{ km}}$$

How do astronomers measure the distance, diameter, and mass of the Sun?

Figure 9.1 The angular diameter of the Sun as viewed from the Earth is 0.53°. The size of this angle plus the distance of the Sun provides enough information to calculate the linear diameter of the Sun in kilometers.

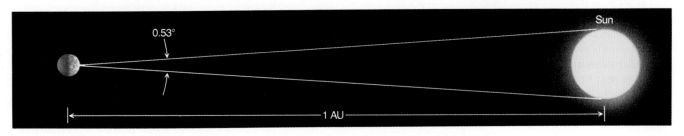

ESSAY Spectroscopy of the Sun: How We Know What We Know

What is the Sun made of? How fast does the Sun rotate? Is its visible surface smooth and motionless or in constant motion like the surface of the seas on Earth? Does it have magnetic fields? Spectroscopy—measurement of spectral lines—is the most important tool astronomers have for answering these questions. Perhaps you can already guess how spectroscopy might be used to answer these questions about the Sun.

Composition of the Sun

Each element is associated with a unique set of spectral lines. By matching the spectral lines in the light of the Sun with spectra produced in the laboratory, scientists can identify the elements that are present in the Sun. More than 60 of the elements known on the Earth have now been identified in the solar spectrum. The remaining elements are also thought to be present in the Sun, but in quantities so small that their spectral lines are too weak to measure.

The same elements are present in both the Sun and the Earth, but in very different proportions. About three-quarters of the Sun (by weight) is hydrogen, and about 98 percent is hydrogen and helium. The next most abundant elements are carbon, nitrogen, and oxygen. These three elements, along with all of the other elements that are so essential to life here on Earth, make up less than 2 percent of the material in the Sun.

Velocities

Measurements of the positions of spectral lines can be used to determine the velocity of parts of the Sun. For example, it is possible to measure the rotation of the Sun at various latitudes. The technique is to observe light at a given latitude first at the edge of the Sun that is rotating toward us, and then at the opposite side of the Sun at the same latitude, where the rotation is away from us. The results show that the Sun rotates more slowly at higher latitudes than it does near its equator. Material at the solar equator

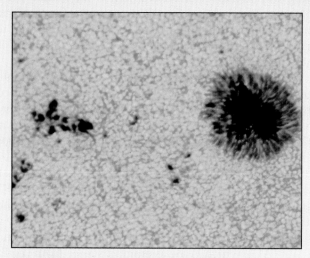

Figure 9A The solar granulation. Each bright spot is a rising cell of hotter gas about 1000 km across; cooler gas descends in the darker regions between the granules. *(National Solar Observatory)*

completes one full revolution in about 25 days. The rotation period near the poles is nearly 37 days. On the Earth, which is a solid body, all places along a fixed line of longitude remain forever lined up along that longitude as the Earth rotates. Regions on the Sun, however, change their longitudes relative to one another.

Pictures show that the surface of the Sun has a mottled appearance (Figure 9A), which is called solar granulation. Measurements of the Doppler shift indicate that the bright regions are columns of hot gas bubbling up to the surface from the interior. As the rising gas reaches the surface, it spreads out, cools, and sinks back down again. The sinking gas appears darker in Figure 9A because it is cooler than the hotter, brighter rising gas. The temperature difference between the bright and dark regions is 50 to 100 K. The speed with which the gas moves is 2 to 3 km/s. Each rising column lasts for about 8 minutes. The surface of the Sun is a seething, bubbling cauldron of hot gas.

Magnetic Fields

Observations of the Sun show that very strong magnetic fields are present in dark, cool regions called sunspots (Figure 9B). If a strong magnetic field is present, then single spectral lines split into multiple components (Figure 9C). These magnetic fields hold the key to explaining why sunspots are cooler and darker than the regions without strong magnetic fields. The forces produced by the magnetic field resist the motions of the bubbling columns of rising hot gases. Because these rising columns of hot gas carry most of the heat from inside the Sun to the surface, there is less heating where there are strong magnetic fields. As a result, darker cooler sunspots appear in regions where magnetic forces are strong.

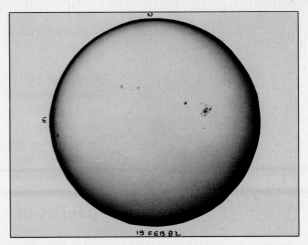

Figure 9B Photograph of the surface of the Sun showing several sunspots. *(National Solar Observatory/National Optical Astronomy Observatories)*

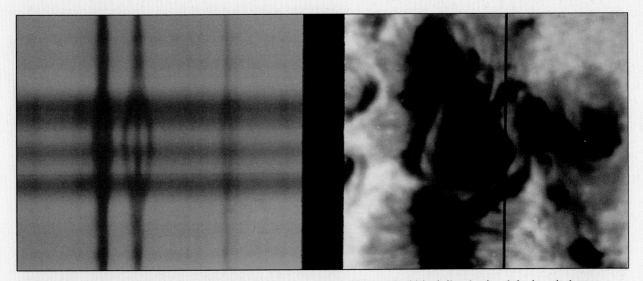

Figure 9C Measurement of the magnetic field of a sunspot. The vertical black line in the right-hand picture indicates the position of the spectrograph slit through which light passed in order to obtain the spectrum in the left-hand picture. Note that the strongest spectral line in the left-hand picture is split into three components. *(National Solar Observatory)*

If you work out the arithmetic, you will find that the diameter of the Sun is approximately equal to 1,400,000 km, or about 109 times the diameter of the Earth.

Mass of the Sun

The mass of the Sun can be determined from Kepler's third law as modified by Newton (Section 2.3). This equation can be written in the form

$$(M_1 + M_2)P^2 = \text{(constant factor)} \times D^3$$

Suppose M_1 is the mass of the Sun, M_2 is the mass of the Earth, P is the length of time it takes for the Earth to move completely around the Sun, and D is the distance between the Sun and the Earth. Because P is equal to one year, D is equal to 1 AU, and M_2, the mass of the Earth, is already known, it is possible to calculate the mass of the Sun.

The mass of the Sun turns out to be about 333,000 times greater than the mass of the Earth.

Density of the Sun

Now that we know the mass and the diameter of the Sun, we can also calculate its density, which is about 1.4 times the density of water. For comparison, the density of gold is about 19 times the density of water, ordinary rock is 2 to 3 times the density of water, and typical ices have densities comparable to water (see Section 2.3). From the density alone, we know then that the Sun is not solid gold. Unless it is hollow, and if it were hollow there would be no way to generate the enormous amount of energy it pours forth, it cannot be composed of rock or minerals. It cannot even be composed of ice. The Sun produces enough heat to melt ice at the surface of the Earth, which is 150,000,000 km away. It is obvious that ice could not survive the intense heat at the surface of the Sun. Therefore, from the density we know that the Sun must be a gas.

9.2
THE SOLAR ATMOSPHERE

The only parts of the Sun that can be observed directly are its outer layers, collectively known as the Sun's *atmosphere.* There are three general regions in the Sun's atmosphere—the photosphere, the chromosphere, and the corona.

The Solar Photosphere

The Sun is so hot that it is a gas throughout. There is no solid surface. The outer parts of the solar atmosphere are transparent to visible light, and we can look through them to deeper layers of the Sun. The **photosphere** is that depth in the Sun past which we cannot see. The photo-

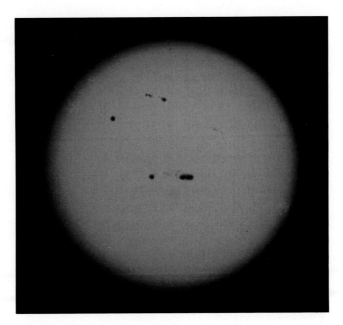

Figure 9.2   The visible surface of the Sun is called the photosphere. This photograph of the photosphere shows several sunspots.   *(National Solar Observatory/National Optical Astronomy Observatories)*

sphere, like a cloud seen from an airplane window, looks solid even though it really is not.

The distance over which the gases in the solar atmosphere change from almost perfectly transparent to almost completely opaque is only 200 km. This distance is very small relative to the diameter of the Sun. In effect, therefore, the photosphere defines a very sharp boundary. When we speak of the "size" of the Sun, we generally mean the size of the region contained within the photosphere (Figure 9.2).

At a typical point in the photosphere, the pressure is only a few hundredths of sea-level pressure on the Earth. The temperature increases from a low of 4500 K at the outer boundary of the photosphere to about 6800 K at the point where the gases become totally opaque.

The chemical composition of the ten most abundant gases in the solar photosphere is summarized in Table 9.2. The relative abundances of the chemical elements in the Sun are similar to the relative abundances found for other stars. Approximately 73 percent of the Sun (by mass) is hydrogen, and another 25 percent is helium. The remaining chemical elements make up only about 2 percent of the total mass of the Sun.

What is the atmosphere of the Sun like?

The Chromosphere

The Sun's outer gases extend far beyond the photosphere, but they are transparent to most visible radiation. The region of the Sun's atmosphere that lies immediately above the photosphere is the **chromosphere.** The chromosphere is about 2500 km thick. The temperature *increases* from 4500 K at the photosphere to 100,000 K or so at the upper chromospheric levels.

TABLE 9.2 Abundance of Elements in the Sun		
Element	Percentage by Number of Atoms	Percentage by Mass
Hydrogen	92.0	73.4
Helium	7.8	25.0
Carbon	0.03	0.3
Nitrogen	0.008	0.1
Oxygen	0.06	0.8
Neon	0.008	0.05
Magnesium	0.002	0.06
Silicon	0.003	0.07
Sulfur	0.002	0.04
Iron	0.004	0.2

The Corona

The chromosphere merges into the outermost part of the Sun's atmosphere, the **corona.** The corona extends millions of kilometers above the photosphere and emits half as much light as the full moon. Under ordinary circumstances, we cannot see the corona because of the overpowering brilliance of the photosphere. Although the best way to see the corona remains a total solar eclipse (Figure 9.3), the brighter parts of the corona can now be photographed with special instruments.

The temperature of the corona is in the range 1,000,000 K to 2,000,000 K. Because of its high temperature, the corona is very bright at x-ray wavelengths (Figure 9.4).

Figure 9.3   An image of the Sun taken at the time of the solar eclipse on February 16, 1980. Since the light from the brilliant surface (photosphere) of the Sun is blocked by the Moon, it is possible to see the tenuous outer atmosphere of the Sun, which is called the corona.   *(High Altitude Observatory/NCAR)*

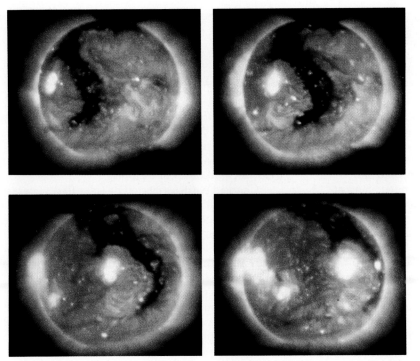

Figure 9.4 X-ray images of the Sun taken from the Skylab satellite show hot coronal gas. The corona is patchy, with bright spots indicating regions where hot gas is concentrated. The long dark area where there is no x-ray emission is called a coronal hole. In these regions, hot gas streams away from the solar surface out through the solar system. This stream of particles is called the solar wind. The four successive images of the Sun clearly show how the positions of solar features change as the Sun rotates. *(Harvard College Observatory/NASA)*

The high temperatures of the chromosphere and the corona should seem surprising. The surface (photospheric) temperature of the Sun is only about 6000 K. Just as the air grows cooler the farther away you move from a bonfire, so too we might have expected the outer atmosphere of the Sun to be cooler than the photosphere. Why then are the outer layers so much hotter? Observations indicate that magnetic fields play a major role. Magnetic fields apparently store energy and carry it to the chromosphere and corona, where the magnetic energy is converted to kinetic energy or electrical currents, which in turn heat the gases of the Sun's outer atmosphere. The precise way in which magnetic energy is converted to heat energy is not yet understood. Explaining this process is one of the major challenges facing solar astronomers.

The Solar Wind

The **solar wind** is a stream of charged particles, mainly protons, flowing outward from the Sun at a speed of about 400 km/s. This solar wind exists because the gases in the corona are too hot to be confined by solar gravity. In just the same way, an atmosphere of light gas would quickly escape from the Moon.

In optical photographs, the solar corona appears to be fairly uniform and smooth. In x-ray and radio pictures, it is very patchy (see Figure 9.4). Hot gas is present mainly where magnetic fields have trapped and concentrated it. The regions between these concentrations of gas, where

ESSAY Solar Flares

The most awesome event on the surface of the Sun is a solar flare. A typical flare lasts for 5 to 10 minutes and releases a total amount of energy equivalent to that of perhaps a million hydrogen bombs. The largest flares last for several hours and emit enough energy to power the entire United States at its current rate of consumption for 100,000 years.

The detailed process that leads to a solar flare is not well understood but apparently involves the liberation of energy stored in magnetic fields high in the solar corona. Solar flares are most likely to occur near times of maximum solar activity. Whatever the mechanism may be, the total amount of energy involved is astounding, and the effects on the Earth are profound.

Flares are often observed in the red light of hydrogen (Figure 9D), but the visible emission is only a tiny fraction of what happens when a solar flare explodes. At the moment of the explosion, the matter associated with the flare is heated to temperatures as high as 10,000,000 K. At such high temperatures, a flood of x-ray and extreme ultraviolet radiation is emitted, along with energetic particles, mainly protons and electrons, that stream outward into the solar system.

The most obvious effect of solar flares on the Earth is the appearance of the aurora borealis, or northern lights. In March 1989, a gigantic flare, which occurred as the Sun approached maximum activity, produced an aurora visible as far south as Arizona. Auroras occur when electrons and protons ejected during a solar flare strike atoms high in the Earth's atmosphere. The atoms are excited or ionized, and when electrons then rejoin the atoms and return to lower energy states, characteristic emission lines are produced. Auroras occur preferentially near the magnetic poles of the Earth because the charged particles from the Sun tend to flow down into the Earth's atmosphere along the magnetic field and to penetrate to lowest altitudes near the poles.

Charged particles ejected by solar flares interact with the Earth's magnetic field and cause it to fluctuate. The changes in the Earth's magnetic field in turn generate changing electric currents. The effects are most noticeable in long power lines, and solar flares can even cause components to burn out in power stations. As a result of the flare in March 1989, parts of Montreal and Quebec Province were without power for up to 9 hours. Other effects occurred as well. For example, because of the electrical interference, people found their automatic garage doors opening and closing for no apparent reason.

The ultraviolet and x-ray emission from flares can affect the ability of the atmosphere to reflect radio waves and can disrupt shortwave radio transmissions.

The short wavelength radiation produced during solar flares heats the outer atmosphere of the Earth. In 1981, a very large solar flare occurred while the space shuttle Columbia was in orbit. The astronauts aboard made measurements that showed that the flare, which lasted for 3 hours, increased the temperature of the atmosphere at an altitude of 260 km from its normal value of 1200 K to 2200 K. When the outer atmosphere is heated, it also expands. As a consequence, friction between the atmosphere and spacecraft increases and drags satellites to lower altitude. During solar maximum, many satellites are brought to such a low altitude that they are destroyed by friction with the atmosphere.

x-ray emission is weak because of low gas density, are called **coronal holes.** The solar wind comes predominantly from these coronal holes, streaming through them into space unhindered by magnetic fields.

At the surface of the Earth, we are protected from the solar wind by the atmosphere and by the Earth's magnetic field. The solar wind does, however, disturb the ionized layers of gas in the Earth's ionosphere.

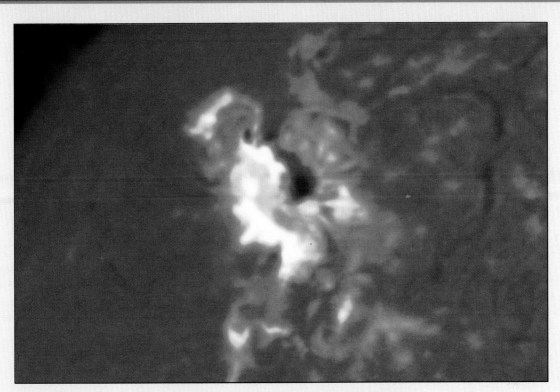

Figure 9D   A giant solar flare is shown as seen in the red light of hydrogen. This flare occurred on March 10, 1989.   *(National Optical Astronomy Observatories)*

The level of solar activity is a critical factor in calculating the lifetimes and orbits of shuttles and satellites in near-Earth orbit. Obviously, it would be extremely valuable to be able to predict both the overall level of solar activity and the occurrence of individual flares. Solar astronomers are working very hard to learn how to make reliable predictions, but accurate forecasts of solar "weather" are proving to be as elusive as reliable forecasts of the weather on Earth.

Unusually intense streams of particles, which can be produced by solar flares (see Essay), are responsible for auroras—the northern and southern lights. As the particles strike atoms and molecules in the upper atmosphere, they excite them. Radiation from the ions and atoms in the atmosphere gives rise to the auroral emission of light. The most spectacular auroras occur at altitudes of 75 to 150 km.

Figure 9.5 Photograph of a large sunspot group. Note the dark central regions (umbra) surrounded by less dark regions (penumbra). *(National Solar Observatory/National Optical Astronomy Observatories)*

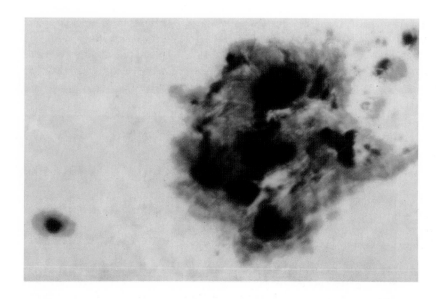

Sunspots

The most conspicuous of the features observed in the photosphere are the **sunspots** (Figures 9.5 and 9.6). Sunspots are darker than the surrounding photosphere because the gases in sunspots are as much as 1500 K cooler than the surrounding gases. Sunspots are nevertheless hotter than the surfaces of many stars. If they could be removed from the Sun, they would shine brightly. They appear dark only by contrast with the hotter, brighter photosphere that surrounds them.

Figure 9.6 Photographs of the surface of the Sun showing a large group of sunspots. The series of exposures follows the rotation of sunspots across the visible hemisphere of the Sun. The top sequence shows the Sun in ordinary light; the bottom sequence is taken through a special filter that shows chromospheric emission. One exposure was taken each day from March 7 through March 17, 1989, except for March 11 and 12, when it was cloudy. *(National Solar Observatory/National Optical Astronomy Observatories)*

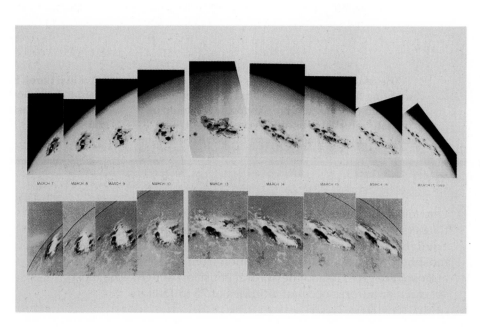

Individual sunspots have lifetimes that range from a few hours to a few months. Many spots are much larger in diameter than the Earth. A few have reached diameters of 50,000 km. Frequently spots occur in groups of 2 to 20 or more. The largest groups are very complex and may have over 100 spots. Like storms on the Earth, sunspots move on the surface of the Sun, but their individual motions are slow. Solar rotation carries the sunspots across the disk of the Sun (see Figure 9.6).

The Sunspot Cycle

In 1851, a German apothecary and amateur astronomer, Heinrich Schwabe, found that the total number of spots visible on the Sun at any one time is likely to be much greater during certain time periods than during others. The periods when the most sunspots are visible are called sunspot maxima. Sunspot maxima have occurred at an average interval of 11.1 years. Individual cycles have been as long as 16 years and as short as 8 years. During sunspot maxima, more than 100 spots can often be seen on the Sun at once. During sunspot minima, the Sun sometimes has no visible spots (Figure 9.7).

Sunspots have strong magnetic fields. Whenever sunspots are observed in pairs or in groups containing two principal spots, one of the spots usually has the magnetic polarity of a north-seeking magnetic pole and the other has the opposite polarity. During a given sunspot cycle, that is, during the period of time from one sunspot maximum to the next, the leading spots in the northern hemisphere all tend to have the same polarity. Those in the southern hemisphere all tend to have the opposite polarity. During the next sunspot cycle, the polarity of the leading spots is reversed in each hemisphere. For example, if during one cycle

How does the number of sunspots change with time?

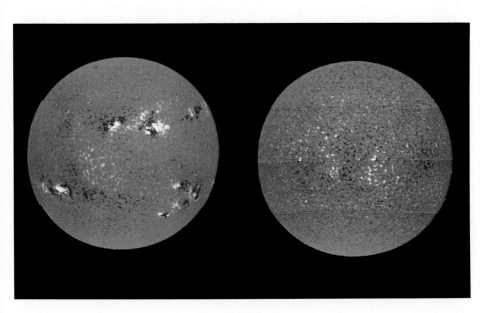

Figure 9.7 A comparison of the number of sunspots and magnetic activity on the active (left) and quiet Sun. The computer-generated images use yellow to indicate positive or north polarity and blue for negative or south polarity. In the image of the active Sun, note that pairs of sunspots have opposite polarity. Note also that the polarity of the leading spot is different in the upper and lower hemispheres. At solar minimum (right) there are no large sunspots and the magnetic fields are weak.   *(National Solar Observatory/National Optical Astronomy Observatories)*

the leading spots in the northern hemisphere all had the polarity of a north-seeking pole, the leading spots in the southern hemisphere would have the polarity of a south-seeking pole. During the next cycle, the leading spots in the northern hemisphere would have south-seeking polarity and those of the southern hemisphere would have north-seeking polarity. We see, therefore, that the sunspot cycle does not repeat itself with respect to magnetic polarity until two 11-year maxima have passed.

Many aspects of the Sun besides sunspots vary in an 11-year cycle. For example, the solar wind is stronger when there are more sunspots. Even the corona changes with the cycle. It is large and roughly spherical during sunspot maximum. At sunspot minima, it is compressed along the Sun's rotational axis but extended in the equatorial direction. Solar flares are much more common near and shortly after sunspot maxima (see essay on solar flares). All of the changes that occur during the sunspot cycle are referred to as **solar activity.** When the number of sunspots is large, the activity of the Sun is described as high.

The solar cycle is closely related to magnetism in the Sun, and it is the changing magnetic field of the Sun that provides the driving force for many aspects of solar activity. Although astronomers have developed some understanding of the solar magnetic field, many mysteries remain. Solar magnetism is an exceedingly complex subject, and our understanding of why the number of sunspots varies is very limited.

9.3
THE SOURCE OF THE SUN'S ENERGY

What makes the Sun shine? The source of energy that is most familiar to us here on Earth is burning (the chemical term is oxidation) of wood, coal, gasoline, or other fuel. Even if the immense mass of the Sun consisted of a burnable material like coal or wood, it could not produce energy at its present rate for more than a few thousand years. Geologists have found fossils in rocks that are 3.5 billion years old. We know, therefore, that the Sun must have been heating the Earth to nearly its current temperature for at least that long.

In the 19th century, scientists used the *law of conservation of energy* to look for a source of energy for the Sun. The law of conservation of energy simply says that energy cannot be created or destroyed, but it can be transformed. The steam engine, which was the key to industrial development during the 19th century, relies on the transformation of heat energy to mechanical energy. The steam from a boiler drives the motion of a piston.

The reverse is also true. Mechanical motion can be transformed into heat. If you clap your hands vigorously, your palms become hotter. If you rub ice on the surface of a table, the heat produced by friction melts the ice. In the 19th century, scientists considered the possibility that the motion of meteorites falling into the Sun might provide an adequate source of heat. Calculations show, however, that in order to produce the total amount of energy—heat and light—emitted by the Sun, the mass in meteorites that would have to fall into it every 100 years would equal the mass of the Earth. The increase in the mass of the Sun would, according to Kepler's third law, change the period of the Earth's orbit by 2 s per year. Such a change would be easily measurable and has not been detected.

As an alternative, German scientist Hermann von Helmholtz and British physicist Lord Kelvin, in about the middle of the 19th century, proposed that the outer layers of the Sun might "fall" inward and thereby produce heat energy. The outer layer of the Sun is a gas made up of individual atoms, all moving about in random directions; temperature is simply a measure of the speed of their motion. Now imagine that this outer layer starts to fall inward. The atoms acquire an additional velocity as a result of this falling motion. As the outer layer falls inward, it also contracts and the atoms move closer together. Collisions become more likely. Some collisions serve to transfer the velocity associated with the falling motion to other atoms and increase their velocities, and so increase the temperature of the Sun. Other collisions may actually excite electrons within the atoms to higher energy orbits. When these electrons return to their normal orbits, they emit photons, which can then escape from the Sun as heat or light.

Kelvin and Helmholtz calculated that a contraction of the Sun at a rate of only about 40 m per year would be enough to provide for its total energy output. Over the time span of human history, the decrease in the Sun's size from such a slow contraction would be undetectable. The total energy available from a contraction that started at the time that the Sun was formed is enough to keep the Sun shining for about 100 million years.

In the 19th century, this length of time seemed adequate, but 20th-century geologists have shown that the Earth, and therefore the Sun, have an age of several billion years. Contraction of the Sun cannot account for the energy it has generated over this long a period of time.

Even as geologists were ruling out one hypothesis about the source of the Sun's energy, physicists were developing a new one. The key lies in the nucleus of the atom and in Einstein's special theory of relativity.

Mass, Energy, and the Special Theory of Relativity

According to the law of conservation of energy, energy cannot be created or destroyed but only converted from one form to another. One of the remarkable results of Einstein's special theory of relativity is that mass and energy are equivalent. This equivalence is expressed in one of the most famous equations in all of science:

$$E = mc^2$$

In this equation, E is the symbol for energy, m is the symbol for mass, and c, the constant that relates the two in a precise mathematical way, is the speed of light. What this equation says is that mass can be converted to energy, and energy can be converted to mass. Although we do not see this conversion taking place in everyday life, it is seen in high-energy physics experiments involving the neutrons, protons, and electrons that make up individual atoms.

For example, the electron has a twin called a **positron,** which has a mass identical to that of the electron but carries an opposite, positive charge. When a positron and an electron come into contact, they annihilate each other, turning into photons of energy equal to the combined mass of the positron and electron times the square of the speed of light. The mass in the positron and electron has been converted to pure energy. Energetic photons can also combine to produce a positron and an electron—that is, interaction of photons, which are pure energy, can produce mass, in this case in the form of an electron and a positron.

Because c^2, the speed of light squared, is a very large quantity, the conversion of even a small amount of mass results in a great amount of energy. For example, the annihilation of 1 g of electrons by 1 g of positrons (about 1/14 ounce) would produce as much energy as 30,000 barrels of oil.

The application to the Sun is obvious. If we can find a set of interactions of atoms that leads to the destruction of some of the Sun's most abundant element hydrogen and the conversion of that lost mass into energy, then we will have identified a source of energy for the Sun that can last for billions of years. With Einstein's equation $E = mc^2$, it is possible to calculate that the amount of energy radiated by the Sun could be produced by the conversion of about 4 million tons of hydrogen to energy each second. This sounds like a lot of hydrogen, but in fact the Sun contains enough hydrogen to continue shining at its present rate for about 10 billion years before it exhausts its supply of hydrogen fuel.

> How is matter converted to energy in the interior of the Sun?

Nuclear Reactions in the Sun's Interior

The Sun taps the energy contained in the nucleus of atoms through a process called **nuclear fusion.** Four hydrogen atoms combine or fuse together to form a helium atom. A helium atom is about 0.71 percent less massive than the four hydrogen atoms that combine to form it, and that lost mass is converted to energy.

The steps required to form one helium nucleus from four hydrogen nuclei are shown in Figure 9.8. First, two protons combine to make a deuterium nucleus, which contains one proton and one neutron. In effect, one of the original protons has been converted to a neutron. Electric charge is conserved in nuclear reactions, and so the positive charge originally associated with one of the protons is carried away by a positron.

This positron instantly collides with an electron and both are annihilated, producing pure electromagnetic energy in the form of gamma rays. After about 10 million years, this electromagnetic energy makes its way to the surface of the Sun, being constantly absorbed and re-emitted by atoms along the way and converted to photons of longer wavelength and lower energy in the process. The photons that we observe directly are only those that are emitted so close to the surface of the Sun that they can escape without being absorbed again.

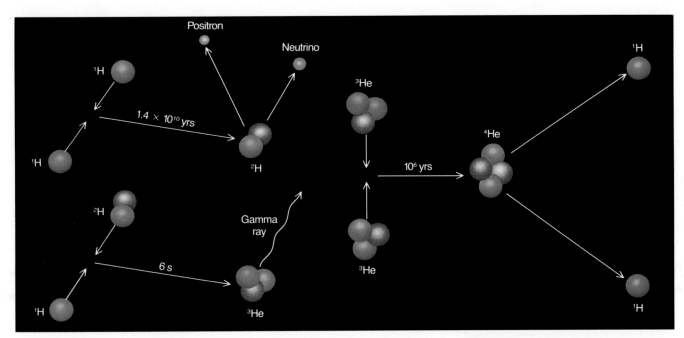

Figure 9.8 The Sun generates its energy by fusing four hydrogen nuclei to form helium. The steps involved in the process are shown.

In addition to the positron, the fusion of two hydrogen atoms to form deuterium results in the emission of a **neutrino.** The neutrino is a massless, or nearly massless, particle that rarely interacts with other matter. Neutrinos produced by fusion reactions near the center of the Sun travel directly to the Sun's surface and then on toward the Earth without interacting with other atoms along the way.

The next step in forming helium from hydrogen is the addition of a proton to the deuterium nucleus to form a helium nucleus that contains two protons and one neutron. In the process, more gamma radiation is emitted. Finally, this helium nucleus combines with another just like it to form normal helium, which has two protons and two neutrons in its nucleus. The two protons that are left over can participate in still more fusion reactions.

This series of reactions can be described succinctly through the following equations:

$$^1H + {}^1H \rightarrow {}^2H + e^+ + \nu$$

$$^2H + {}^1H \rightarrow {}^3He + \gamma$$

$$^3He + {}^3He \rightarrow {}^4He + 2\,{}^1H,$$

in which the superscripts indicate the total number of neutrons and protons in the nucleus, e^+ is the symbol for the positron, ν is the symbol for neutrino, and γ indicates that gamma rays are emitted.

Why do these reactions take place? Protons are positively charged, and positive charges repel each other. These reactions can occur only in regions of very high temperature where the velocities of the protons are high enough to overcome the electrical forces that try to keep protons apart. In the Sun, hydrogen fusion takes place only in regions where the temperature is greater than about 10 million K and the velocities of the protons average 1000 km/s or more. Such extreme temperatures are reached only in the regions surrounding the center of the Sun, which has a temperature of 15 million K. Calculations show that nearly all of the Sun's energy is generated within about 150,000 km of its core, or within about one-quarter of its total radius.

Even at these high temperatures, it is exceedingly difficult to force two protons to combine. On average, a proton will rebound from other protons for about 14 billion years, at the rate of 100 million collisions per second, before it fuses with a second proton. Of course, some protons are lucky and take only a few collisions to achieve a fusion reaction. It is those protons that are responsible for producing the energy radiated by the Sun. Because the Sun is only about 4.5 billion years old, most of its protons have not yet been involved in fusion reactions. The low probability of the interaction of protons is fortunate for us, because it means that the Sun's fuel lasts for a long time—long enough to permit the slow biological processes on Earth to produce complex forms of life.

After the deuterium nucleus is formed, the remaining reactions happen very quickly. After about 6 s on average, the deuterium nucleus is converted to ^{3}He. About 1 million years after that, the ^{3}He nucleus combines with another to form ^{4}He.

9.4
THE INTERIOR OF THE SUN: THEORY

Fusion of protons occurs in the center of the Sun only if the temperature exceeds 10 million K. How do we know whether the Sun is actually this hot? In order to determine what the interior of the Sun is like, it is necessary to resort to mathematical calculations. In effect, astronomers teach a computer everything they know about the physical processes that are going on in the interior of the Sun. The computer then calculates the temperature and pressure at every point inside the Sun and determines what nuclear reactions, if any, are going on. The computer can also calculate how the Sun will change with time.

What do we know about the structure of the solar interior?

The Sun must change. In its center the Sun is slowly depleting its supply of hydrogen and creating helium instead. Will this change in composition have measurable effects? Will the Sun get hotter? cooler? larger? smaller? brighter? fainter? Ultimately, the changes must be catastrophic, because the hydrogen fuel will eventually be exhausted. Either a new source of energy must be found or the Sun will cease to shine. What will happen to the Sun is described in Chapters 11 and 12. For now, let us look at what we need to teach the computer about the Sun in order to carry out the calculations.

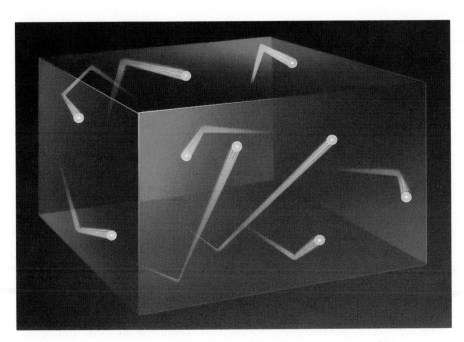

Figure 9.9   Gas pressure. The particles in a gas are in rapid motion and produce pressure through collisions with the surrounding material. Here particles are shown bombarding the sides of a container.

The Sun Is a Gas

The Sun is so hot that the material in it is gaseous from the center to the surface. The particles that constitute a gas are in rapid motion, frequently colliding with each other. This constant bombardment is what causes the gas to exert *pressure* on the material surrounding it (Figure 9.9). The greater the number of particles within a given volume of the gas, the greater the pressure of the gas, because the combined impact of the moving particles increases with their number. The pressure also increases if the velocity of the gas particles increases. Because the velocity is determined by the temperature of the gas, higher temperature corresponds to higher pressure.

These two ideas combine to give us the *perfect gas law,* which states that the pressure of a gas is proportional to the product of its density and temperature.

The Sun Is Stable

Apart from some very low-amplitude pulsations (see Section 9.5), the Sun is *stable.* It is neither expanding nor contracting. The Sun is in *equilibrium.* All the forces are balanced, so that at each point within the star the temperature, pressure, density, and so on are maintained at constant values.

The mutual gravitational attraction between the masses of various regions within the Sun produces tremendous forces that tend to collapse the Sun toward its center. Yet the Sun has been emitting approximately the same amount of energy for billions of years and thus has managed to

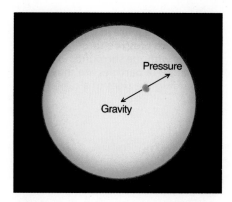

Figure 9.10 Hydrostatic equilibrium. In the interior of a star, the inward force of gravity is exactly balanced at each point by the outward force of gas pressure.

resist collapse for a very long time. The gravitational forces must therefore be counterbalanced by some other force, and that force is the pressure of the gases within the Sun. In order to exert enough pressure to prevent the collapse of the Sun due to the force of gravity, the gases at the center of the Sun must be at a temperature of 15 million K. Thus, temperatures high enough to fuse protons are *required* by the fact that the Sun is not contracting.

What if pressure and gravity did not exactly balance each other? If the internal pressure in a star were not great enough to balance the weight of its outer parts, the star would shrink a little. The pressure would build up. Alternatively, if the pressure were greater than the weight of the overlying layers, the star would expand, thus decreasing the internal pressure. Expansion would stop when the pressure at every internal point again equaled the weight of the stellar layers above that point. An analogy is an inflated balloon, which will expand or contract until an equilibrium is reached between the excess pressure of the air inside over that of the air outside and the tension of the rubber. This condition, where everything is in balance, is called **hydrostatic equilibrium** (Figure 9.10). Stable stars are all in hydrostatic equilibrium; so is the Earth's atmosphere. The pressure of the air keeps the air from falling to the ground.

The Temperature of the Sun Is Not Changing

From observation we know that electromagnetic energy flows from the surface of the Sun. Furthermore, heat always flows from hotter to cooler regions. Because energy is radiating away from the surface of the Sun, it must be flowing to the solar surface from inner hotter regions. The temperature cannot ordinarily decrease inward in a star, or energy would flow in and heat up those regions until they were at least as hot as the outer ones.

The highest temperature occurs at the center of a star, and temperatures drop to successively lower values toward the stellar surface. The outward flow of energy through a star robs it of its internal heat and would cool the interior if that energy were not replaced. Therefore, there must be a source of energy within each star, and in the Sun that source of energy is the fusion of hydrogen to form helium.

The Sun is in a steady state—neither contracting nor expanding, and emitting energy at a constant rate. Therefore, the temperature and pressure at each point within it must remain approximately constant. If the temperature were to change suddenly at some point, the pressure would similarly change, causing the star to contract suddenly, to expand, or otherwise to deviate from hydrostatic equilibrium. Energy must be supplied, therefore, to each layer in the Sun at just the right rate to balance the loss of heat in that layer as it passes energy outward toward the surface. Moreover, the rate at which energy is supplied to the Sun as a whole must, at least on the average, exactly balance the rate at which the Sun loses energy by radiating it into space. We call this balance of heat gain and heat loss for the Sun as a whole, and at each point within it, the condition of *thermal equilibrium.*

Transport of Energy

Convection and radiation are the two ways in which heat is transported from the interior of the Sun to its surface.

Convection occurs in the Sun when hot gases become buoyant and rise, carrying their heat outward (see Section 3.4). Convection also occurs in a pot of boiling water. Convection currents in the Sun travel at moderate speeds and do not upset the condition of hydrostatic equilibrium or result in a net transfer of mass either inward or outward. However, they carry heat very efficiently (Figure 9.11).

The other method for transporting energy out of the Sun is **radiation.** Photons initially produced by nuclear fusion travel only a short distance before they are absorbed by the hot solar gases. They are then immediately re-emitted by the hot material that absorbed them, but they may be emitted in any direction. Although the net flow of energy is outward, from the hot interior toward the cooler surface, there are almost as many steps backward as forward, with the consequence that it takes about 10 million years for energy generated in the solar interior to reach the surface. If the photons were not absorbed and re-emitted along the way, they would travel at the speed of light and could reach the surface in a little over 2 s, just as the neutrinos do (Figure 9.12).

These, then, are the facts that theorists use to calculate a model of the Sun. The Sun is a gas. It is neither expanding nor contracting. Its temperature does not vary. Heat flows outward.

How is energy transported from the interior of the Sun to its atmosphere?

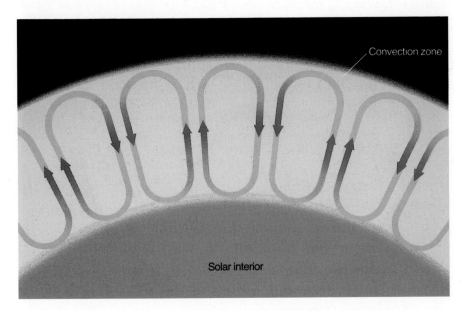

Figure 9.11 Rising convection currents carry heat from the interior of the Sun to its surface. Cooler material sinks downward.

Convection zone

Solar interior

Figure 9.12 (a) Photons generated by fusion reactions in the solar interior travel only a short distance before they are absorbed. The re-emitted photons usually have lower energy and may travel in any direction. As a consequence, it takes about 10 million years for energy to make its way from the center of the Sun to its surface. (b) In contrast, neutrinos do not interact with matter but traverse the Sun at the speed of light, reaching the surface in only a little more than 2 s.

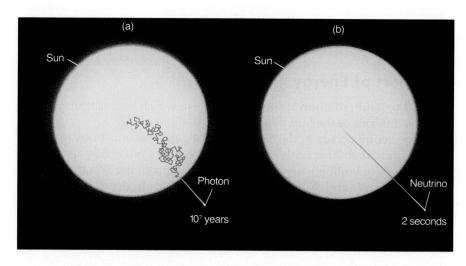

A Model of the Interior of the Sun

Figure 9.13 illustrates schematically what calculations show the interior of the Sun to be like. Energy is generated through fusion of hydrogen in the core of the Sun, which extends only about one-quarter of the way to the surface. This core contains about one-third of the total mass of the Sun. At the center the temperature reaches a maximum of about 15 million K, and the density is nearly 150 times the density of water. The

Figure 9.13 The interior structure of the Sun. Energy is generated in the core by the fusion of hydrogen to form helium. This energy is transmitted outward by radiation, i.e., by the absorption and re-emission of photons. In the outermost layers, energy is transported mainly by convection.

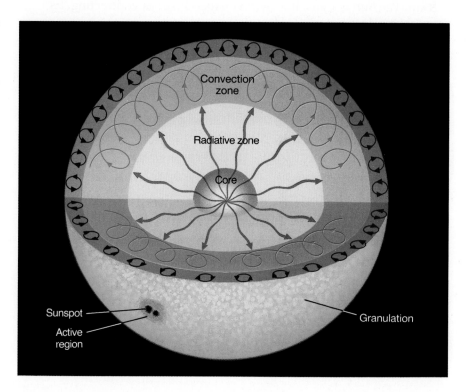

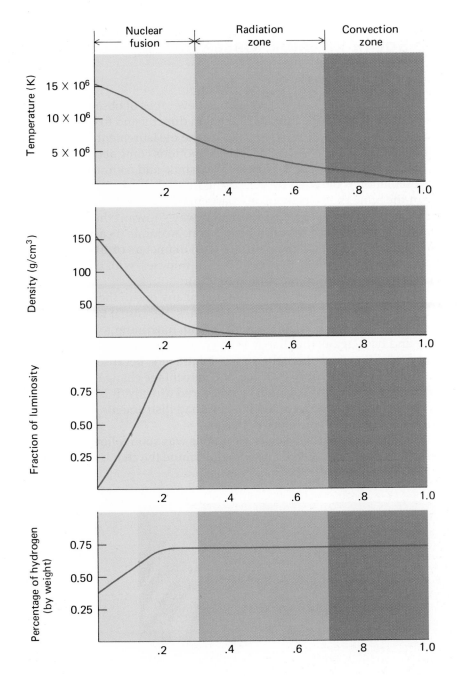

Figure 9.14 Diagrams showing how temperature, density, rate of energy generation, and percentage (by mass) abundance of hydrogen vary inside the Sun.

energy generated is transported toward the surface by radiation until it reaches a point about 70 percent of the distance from the center to the surface. At this point convection begins, and energy is transported the rest of the way primarily by rising columns of hot gas.

Figure 9.14 shows how the temperature, density, composition, and rate of energy generation vary from the center of the Sun to its surface.

THE SOLAR INTERIOR: OBSERVATIONS

Solar Oscillations

Recently, astronomers have devised two new types of observations that can be used to study the interior of the Sun directly and to check their calculations. The first technique relies on the measurement of very small motions of the surface of the Sun. These motions come about because the Sun pulsates—that is, it alternately expands and contracts slightly. This oscillation is detected by measuring the radial velocity of the surface of the Sun. The velocity of any particular small region on the solar surface is observed to change in a regular way—toward the Earth, away from the Earth, toward the Earth, and so on. Accurate measurements show that regions on the solar surface with diameters of 4000 to 15,000 km fluctuate back and forth every 2.5 to 11 minutes, with the dominant periods being about 5 minutes (Figure 9.15).

The velocity of approach or recession of any one of these small regions is very small—only a few hundred m/s. Because it takes typically about 5 minutes to complete a full cycle from maximum to minimum velocity and back again, the change in the radius measured at any given point on the Sun is no more than a few kilometers. Because the total radius of the Sun is about 700,000 km, the percentage change in the size of the Sun is so small that it cannot be measured directly. It is only in the past 20 years or so that astronomers have had instruments that permit them to detect the small velocity changes.

The discovery of the velocity variations was soon followed by the realization that they could be used to determine the characteristics of the

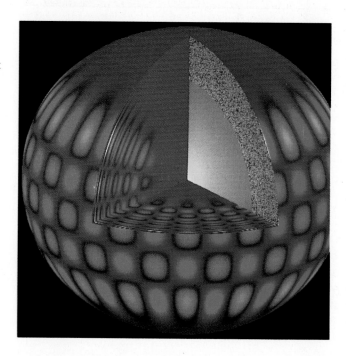

Figure 9.15 New observational techniques permit astronomers to measure small differences in velocity at the surface of the Sun to infer what the deep interior of the Sun is like. In this computer simulation, red denotes regions of the surface that are moving away from the observer, and blue denotes regions moving toward the observer. Note that the velocity changes penetrate deeply into the interior of the Sun. *(National Optical Astronomy Observatories)*

interior of the Sun. The motion of the Sun's surface is caused by waves generated deep within it. Study of the amplitude and period of the velocity changes produced by this motion can yield information about the temperature, density, and composition beneath the Sun's surface. The situation is somewhat analogous to the use of earthquakes to infer the properties of the interior of the Earth. For this reason, studies of solar oscillations are referred to as **solar seismology.** It takes about an hour for these waves to traverse the Sun, and so they provide information about what the solar interior is like at the present time. As we have already noted, this statement is not true for visible light. Because photons take about 10 million years to travel from the center of the Sun to its surface, the luminosity that we measure *now* was generated millions of years ago.

Solar seismology has already yielded some important results. Calculations suggested that the convection zone extended only 15 to 20 percent of the distance from the surface of the Sun to its center. Measurements of solar pulsation indicate that convection extends inward from the surface 30 percent of the way to the center, and this result is shown in Figure 9.13. The observed oscillations indicate that the abundance of helium inside the Sun, except in the center where nuclear reactions have converted hydrogen to helium, is about the same as at its surface. That result means that it is correct to use the abundances measured in the solar atmosphere to construct models of the solar interior.

Recent measurements of solar pulsations show that the outer 30 percent of the Sun rotates at about the same rate as does the surface, with the rotation rate slower at the poles than at the equator (Figure 9.16).

Why do solar oscillations allow us to "observe" the optically invisible interior of the Sun?

Figure 9.16 The period of solar rotation in days as a function of position within the Sun. Note that the rotation period increases from 25 days at the equator to 36 days at the pole. These same periods persist inward through the Sun's convective zone to a depth of 200,000 km. Deeper inside, the Sun appears to rotate as a solid body, with a period of 27 days. *(Kenneth Libbrecht/Caltech)*

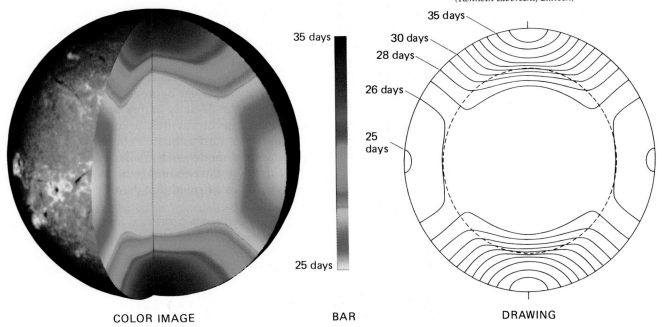

COLOR IMAGE   BAR   DRAWING

Deeper in the interior, the Sun appears to rotate like a solid body, with a period of about 27 days.

The next step is to make better measurements so that solar pulsations can be used to probe the structure of the Sun still closer to its center, where nuclear reactions are taking place. Astronomers are building a network of stations around the Earth to monitor solar velocities continuously. As it is now, observations from any single observatory are interrupted by sunset. The name given to this program to build a worldwide network of telescopes is called the Global Oscillations Network Group (GONG) project. Observations will begin in about 1995, and analysis of the velocity measurements will take perhaps 10 years.

Solar Neutrinos

The second technique for obtaining information about the solar interior involves the measurement of neutrinos. About 3 percent of the total energy generated by nuclear fusion in the Sun is carried away by neutrinos. Neutrinos only rarely interact with matter, and the neutrinos created in the center of the Sun make their way directly out of the Sun and to the Earth at the speed of light. So far as neutrinos are concerned, the Sun is transparent. If we can devise a way to detect some of the 300 billion billion (3×10^{20}) solar neutrinos that pass through each square meter of the Earth's surface every second, then we can obtain information directly about what is going on in the center of the Sun.

Unfortunately, the very property that makes neutrinos an interesting source of information about the interior of the Sun makes them very difficult to detect. Neutrinos pass through material on the Earth as readily as they escape from the Sun. On very, very rare occasions, however, a neutrino will interact with another atom. Several experiments have been devised to detect these interactions.

The first of these experiments was designed to detect the interaction of solar neutrinos with chlorine. In order to detect these rare events, Raymond Davis, Jr., and his colleagues at Brookhaven National Laboratory placed a tank containing nearly 400,000 L of cleaning fluid (C_2Cl_4) 1.5 km beneath the surface of the Earth in a gold mine at Lead, South Dakota. Calculations show that solar neutrinos should produce about one radioactive atom of argon-37 daily in this tank.

Amazingly, it is possible to detect individual argon-37 atoms, and the results are that only about one-third as many neutrinos reach the Earth as are predicted by standard models of the solar interior. A second experiment carried out by Japanese astronomers, who looked for the interaction of neutrinos with water, confirmed that there are too few high-energy neutrinos.

Fewer than 1 percent of the neutrinos produced in the Sun have energies high enough to be detected by the chlorine, and the water experiment is also sensitive only to the neutrinos with very high energies. A small decrease in the central temperature of the Sun relative to what is

What do observations of neutrinos from the Sun tell us about the interior of the Sun?

assumed in standard solar models might be able to account for the deficiency of high-energy neutrinos.

A small change in central temperature could not, however, account for a change in the number of low-energy neutrinos. Nearly all of the solar energy as well as nearly all of the solar neutrinos are produced when two hydrogen atoms combine to form deuterium. Because we know accurately how much total energy is emitted by the Sun, we also know accurately how many times each second two protons combine to form deuterium, and so we also know how many times each second the associated neutrinos are produced by this interaction. Once it became clear that standard solar models predicted far more high-energy neutrinos than are detected here on Earth, scientists began to design experiments to look for the low-energy neutrinos associated with the proton-proton reaction.

It turns out that these low-energy neutrinos will interact with gallium. In the early 1990s, the first results of two gallium experiments, one in Russia and the other in Italy, were reported. Preliminary observations indicated that the number of low-energy neutrinos is about two-thirds the value predicted by standard solar models. Measurements over a longer time period are required to confirm this early result. Again, we find fewer neutrinos than standard solar models predict.

Scientists are beginning to think that the problem may turn out to be not with our models of the Sun, but with our ideas about neutrinos. There are three types of neutrinos, and standard theory assumes that all three types have no mass. There is actually no convincing laboratory evidence that the mass of a neutrino is exactly zero. If neutrinos have even a tiny mass, then it is possible for one type of neutrino to change to another type on its journey from the center to the photosphere of the Sun, and then on to the Earth. The Sun produces only one type of neutrino, the so-called electron neutrino, and the chlorine, water, and gallium experiments are sensitive only to this one type of neutrino. If some of the electron neutrinos change to another type of neutrino on their way from the center of the Sun to the Earth, then the experiments performed so far would have missed those neutrinos.

To test this idea, we need a new experiment that can measure all types of neutrinos. If the total number of neutrinos emitted by the Sun is found to be greater than the number of electron neutrinos, then we would have evidence that the mass of the neutrino is not zero. Such an experiment, which involves heavy water (water with the hydrogen atoms replaced by deuterium), is now being built in Canada, and the results should be available in the late 1990s.

So it will be another ten years before we have the results of the new experiments on solar pulsations and neutrinos. Until then, we will not know whether our models of the Sun are wrong or whether the assumption that the neutrino has no mass is wrong. Although it may seem discouraging that there are questions for which definitive answers are not yet available, science often works this way. Observations lead to the de-

OBSERVING YOUR WORLD   THE SUN

Looking directly at the Sun can damage your eyes, and so the Sun is not a good object for direct visual observations. You can, however, estimate the diameter of the Sun by using a pinhole camera.

Activities

1. A pinhole camera is a very simple device that can be used to project an image onto a screen. The way a pinhole camera works is illustrated in Figure 9E. A very small hole is cut in an opaque piece of material. Light travels from a particular part of an object through the pinhole and strikes the screen in a specific location. Light rays traveling in other directions are blocked by the opaque material surrounding the pinhole. The result is that an upside-down image is formed on the screen.

Try viewing the Sun with a pinhole camera. In a piece of cardboard or a stiff sheet of paper, cut a smooth circle about one-half inch in diameter. Let the Sun shine through the pinhole onto a white sheet of paper. This experiment will work best if the white sheet of paper is at least 6 feet away from the pinhole. How does the shape of the solar image change if you make the pinhole a different shape—a square or a diamond? How does the size of the image depend on the size of the hole?

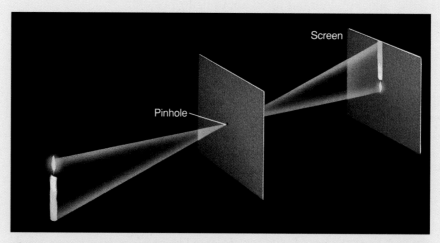

Figure 9E   A pinhole camera forms an image because light travels in straight lines. Only those rays that pass through the pinhole reach the screen, where an upside-down image is formed.

velopment of a model. This model then suggests a number of other measurements that could be made and predicts the outcome of those measurements. Frequently, the predictions are incorrect, and the models must be modified to take into account the new measurements. And so science moves forward by successive approximations, each step provid-

2. Try measuring the diameter of the Sun. The geometry is shown in Figure 9F. Because the two triangles are similar triangles, we have the following:

$$\frac{\text{Diameter of Sun}}{\text{Distance to Sun}} = \frac{\text{Diameter of solar image}}{\substack{\text{Distance between pinhole} \\ \text{and screen}}}$$

Use the known distance to the Sun to calculate the diameter of the Sun. How does your answer compare with the actual value of the diameter? How does your measurement of the diameter vary depending on both the size of the pinhole and the distance between the pinhole and the screen?

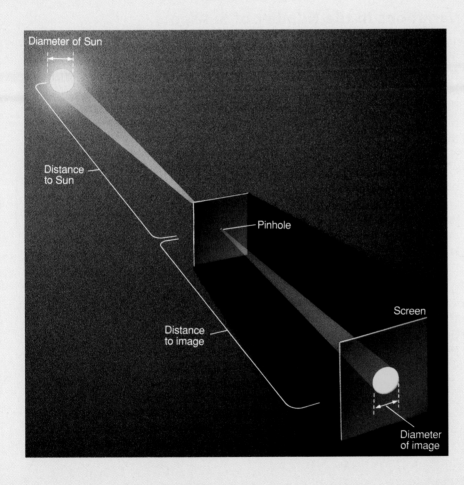

Figure 9F The two triangles on either side of the pinhole are similar triangles. Therefore the ratio of base to altitude is the same for both triangles.

ing a better and more complete description of what is actually occurring. Rather than finding the lack of final answers discouraging or frustrating, scientists find this situation a source of never-ending challenge. The possibility of learning something never before known by anyone is what attracts many scientists to research.

9.6
IS THE SUN A VARIABLE STAR?

The Sun pulsates with so small an amplitude that the effects on the Earth are undetectable. Does the Sun ever vary by larger amounts? Do these variations ever become large enough to affect the Earth or its climate? We already know that solar variations, if any do exist, will be subtle. The existence of life on Earth demonstrates that there have been no major recent changes in the climate of the Earth. There is, however, growing evidence that long-term changes in the energy output of the Sun do have measurable effects on the Earth.

Variations in the Number of Sunspots

There is considerable evidence that the number of sunspots was much lower from 1645 to 1715 than it is now. This interval of low activity was first noted by Gustav Spörer in 1887, and by E.W. Maunder in 1890, and is now called the Maunder Minimum. The incidence of sunspots over the past four centuries is shown in Figure 9.17. According to the data presented in this figure, sunspot numbers were also somewhat lower than they are now during the first part of the 19th century, and this period of time is called the Little Maunder Minimum.

When the number of sunspots is high, the Sun is active in a number of other ways as well, and this activity affects the Earth directly. For example, auroras are caused by the impact of charged particles from the Sun on the Earth's magnetosphere. Energetic charged particles are much more likely to be ejected by the Sun when the Sun is active and when the sunspot number is high. There is a strong correlation between sunspot

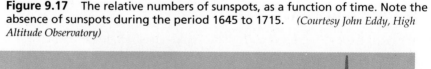

Figure 9.17　The relative numbers of sunspots, as a function of time. Note the absence of sunspots during the period 1645 to 1715.　*(Courtesy John Eddy, High Altitude Observatory)*

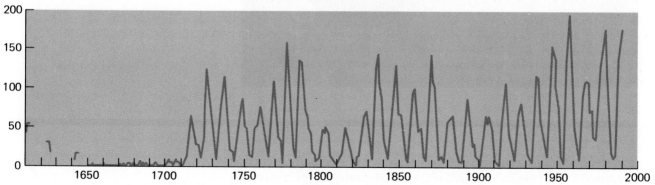

number and the frequency of auroral displays. Historical accounts indicate that auroral activity was low during the Maunder Minimum.

The best quantitative evidence of variations in the level of solar activity comes from studies of the radioactive isotope carbon-14. The Earth is constantly bombarded by cosmic rays, which are high-energy, charged particles, including protons and nuclei of heavier elements. The rate at which cosmic rays from sources outside the solar system reach the upper atmosphere depends on the level of solar activity. When the Sun is active, charged particles streaming away from the Sun out into the solar system carry the Sun's strong magnetic field with them. This magnetic field shields the Earth from incoming cosmic rays. At times of low activity, when the Sun's magnetic field is weak, cosmic rays reach the Earth in larger numbers.

When the energetic cosmic-ray particles strike the upper atmosphere, they produce several different radioactive isotopes. One such isotope is carbon-14, which is produced when nitrogen is struck by high-energy cosmic rays. The rate of production of carbon-14 is higher when the activity of the Sun is lower and the solar magnetic field does not shield the Earth from bombardment by cosmic rays.

Some of the radioactive carbon is contained in carbon dioxide molecules, which are ultimately incorporated into trees through photosynthesis. By measuring the amount of radioactive carbon in tree rings, we can estimate the historical levels of solar activity back about 8000 years into the past.

These measurements show that the Sun has been, at different times, both more and less active than it is now. They also confirm that the amount of carbon-14 was unusually high, and that solar activity was correspondingly low, during both the Maunder Minimum and the Little Maunder Minimum. During the past 1000 years, activity was also low during the periods 1410–1530 and 1280–1340. Between the years 1100 and 1250, the level of solar activity may have been higher than it is now.

> Does the number of sunspots change on time scales of hundreds of years?

Solar Variability and the Earth's Climate

Variations in the overall level of the Sun's activity seem to be well established. Do these variations have any direct impact on the Earth or its climate? It has long been known that the period of the Maunder Minimum was a time of exceptionally low temperatures in Europe—so low that this period is described as the Little Ice Age. The global climate also appears to have been unusually cool from 1400 to 1510, and this period was one of low solar activity as well.

The most obvious way in which the Sun and the Earth's climate might be linked is through variations in the luminosity of the Sun. If the Sun puts out less energy, then logically one might expect the Earth to become colder. The relationship between solar luminosity and activity level was determined only recently from measurements made by a satel-

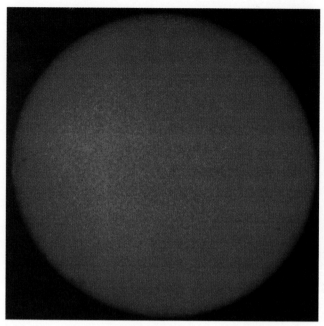

MINIMUM

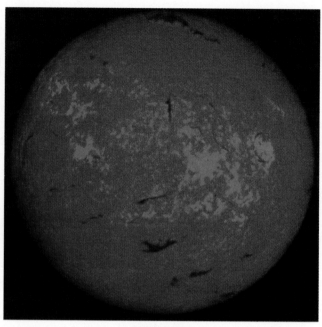

MAXIMUM

Figure 9.18 These images show the appearance of the Sun in the red light of hydrogen at the times of a minimum (left) and maximum (right) in the number of sunspots. The bright areas are called plages, and more plages are present at the time of sunspot maximum. *(National Solar Observatory/National Optical Astronomy Observatories)*

lite orbiting the Earth. Measurements indicate that there is a gradual overall increase in luminosity that accompanies increasing solar activity. At the time of solar maximum, large numbers of bright, hot regions called *plages* appear and produce excess radiation (Figure 9.18).

These observations support the idea that the Maunder Minimum was indeed associated with the Little Ice Age. The unusually cold temperatures at that time imply a drop in the solar luminosity of about 1 percent. It seems possible that such a drop in luminosity might have accompanied a long period of reduced solar activity.

There are, however, a wide variety of other phenomena that also affect the global climate. For example, the eruption of the Philippine volcano Pinatubo in June 1991 launched 25 to 30 million tons of sulfur dioxide gas into the Earth's upper atmosphere. The gas then turned into sulfuric acid droplets, which blocked the Sun's radiation. The Earth's surface temperature, averaged around the globe throughout 1992, was about 0.6°C cooler than it was the year before. This cooling is only temporary, and Pinatubo's effects should disappear by the end of 1993. Climate is also affected by the increasing amounts of greenhouse gases in the atmosphere, which are the result of the burning of fossil fuels. Both of these effects appear to have a greater impact on the Earth's climate than do changes in solar luminosity, at least in the present era.

Do changes in the average worldwide temperature correlate with changes in the number of sunspots?

▶ SUMMARY

9.1. The distance, mass, and diameter of the Sun can all be determined by measuring the orbits of the planets and the apparent size of the Sun. The density of the Sun is about 1.4 times the density of water. This low density indicates that the Sun is gaseous.

9.2. The outer atmosphere of the Sun consists of three regions. In order of increasing distance from the center of the Sun and increasing temperature, these three regions are the **photosphere,** the **chromosphere,** and the **corona.** Charged particles, mainly protons, stream away from the Sun into the solar system. These outward-streaming particles, which come primarily from regions of low gas density called **coronal holes,** are referred to as the **solar wind.** Sunspots are dark regions of lower temperature that appear in the photosphere. The number of visible spots varies on an 11-year **solar cycle.** Sunspots, flares, and other phenomena that vary with the solar cycle are referred to as **solar activity.**

9.3 The source of the Sun's energy is the **fusion** of hydrogen to form helium. A helium atom is about 0.71 percent less massive than the four hydrogen atoms that combine to form it, and that lost mass is converted to energy.

9.4 Even though we cannot see inside the Sun, it is possible to calculate what the solar interior must be like. As input for these calculations, we use what we know about the Sun. It is a gas. Apart from some very tiny changes, the Sun is neither expanding nor contracting (it is in **hydrostatic equilibrium**) and puts out energy at a constant rate. Fusion of hydrogen occurs in the core of the Sun, which contains about one-third of the Sun's total mass. Energy is transported outward from the core by **radiation** until it reaches a layer that is about 70 percent of the distance from the center to the surface. Energy is transported the rest of the way by **convection.**

9.5 Two types of measurements permit us to "observe" what the interior of the Sun is like. **Solar seismology** makes use of measurements of tiny velocity changes at the surface of the Sun to infer such quantities as the depth of the solar convection zone, the rate of rotation in the interior of the Sun, and the composition of the solar interior. **Neutrinos** emerge directly from the center of the Sun, where the conversion of hydrogen to helium is taking place. The standard solar model predicts that we should detect more neutrinos than we actually do. It may be that the solar model is wrong. Alternatively, it may be that these observations are the first evidence that neutrinos have a small amount of mass and that on their journey from the center of the Sun to the Earth, some solar neutrinos change from one type of neutrino to another type, thereby becoming undetectable by the experiments performed thus far.

9.6 There are long-term (100 years or more) variations in the level of solar activity and in the number of sunspots. The lull in solar activity that lasted from the period 1645 to 1715 is called the **Maunder Minimum.** It appears that there is a tendency for the Earth to be cooler when the number of sunspots is unusually low for several decades.

▶ REVIEW QUESTIONS

1. Describe what information is required to calculate the density of the Sun. What is the density of the Sun? How does this density compare with various types of matter on Earth?

2. Describe the main differences between the composition of the Earth and the composition of the Sun.

3. Make a sketch of the Sun that shows the location of the photosphere, the chromosphere, and the corona. What is the approximate temperature of each of these regions?

4. Why are sunspots dark?

5. What makes the Sun shine?

6. In what respect (or respects) is a neutrino very different from a neutron?

7. Summarize the evidence that over a period of several decades or more there have been variations in the level of solar activity.

▶ THOUGHT QUESTIONS

8. The astronomer William Herschel (1738–1822) proposed that the Sun has a cool interior and is inhabited. Give at least three good arguments against this idea.

9. Stars that are hotter than the Sun also derive their energy by fusing hydrogen to form helium, but a different set of reactions is involved. This set of reactions is called the carbon-nitrogen cycle, and the individual steps in the process are given in Appendix 6. Draw a picture (like Figure 9.8) that shows what happens in each step of the carbon-nitrogen cycle.

10. In the carbon-nitrogen cycle, carbon and nitrogen are referred to as catalysts, because the carbon and nitrogen nuclei are required to make the reactions proceed; however, the total number of carbon and nitrogen atoms does not change when all of the steps are completed. From your drawing for the previous problem, show that this statement is true.

11. Why is a higher temperature required to fuse hydrogen to helium by means of the carbon-nitrogen cycle than by the process that occurs in the Sun, which involves only isotopes of hydrogen and helium?

12. The Earth's atmosphere is in hydrostatic equilibrium. Explain what this means. Would you expect the pressure in the Earth's atmosphere to increase or decrease as you climb from the bottom of a mountain to its summit? Why?

13. Give some everyday examples of the transport of heat by convection and by radiation.

14. Why do you suppose that such a large fraction of the Sun's energy comes from its central regions? Within what fraction of the Sun's radius does practically all of the Sun's luminosity originate? (See Figure 9.14.) Within what radius of the Sun has its original hydrogen been partially used up? Discuss how the answers to these questions are interrelated.

15. The Sun obtains its energy by fusing four hydrogen nuclei to form a helium nucleus. It is also possible to obtain energy by breaking up atomic nuclei of such heavy elements as uranium and plutonium to form lighter nuclei. This process is called nuclear fission. Why is fission not an important source of energy in the Sun?

▶ PROBLEMS

16. In 1630, astronomers for the first time observed Mercury as it passed directly between the Earth and the Sun. They expected (erroneously) that the apparent diameter of Mercury would be about one-tenth the diameter of the Sun. How big would Mercury be in kilometers if its apparent diameter were this large? (The radius of Mercury's orbit is about 58×10^6 km.) In fact, Mercury appeared to be a black dot against the Sun's bright surface. How big (in fractions of a degree) does Mercury actually appear to be? (Mercury's diameter is approximately 5000 km.)

17. Use Kepler's third law in the form $(M_1 + M_2)P^2 = D^3$ to calculate the ratio of the mass of the Sun plus the mass of the Earth to the mass of the Earth plus the mass of the Moon. Show that this ratio is nearly the same as the ratio of the mass of the Sun to the mass of the Earth.

18. Use the data in Table 9.1 to confirm the result that the density of the Sun is 1.4 g/cm³.

19. If an observed oscillation of the solar surface has a period of 10 minutes and the average radial velocity is 1 m/s in and out, calculate the total distance moved by the surface. What percent is this of the total radius of the Sun?

20. The text states that a sunspot actually emits light but appears dark in contrast with the photosphere. Use the Stefan-Boltzmann law and the data in Chapter 9 to estimate how much more energy per square meter is emitted by the photosphere than by a sunspot.

▶ SUGGESTIONS FOR ADDITIONAL READING

Books

Wentzel, D. *The Restless Sun.* Smithsonian Inst. Press, 1989. Excellent introduction by an astronomer and educator.

Frazier, K. *Our Turbulent Sun.* Prentice-Hall, 1983. A journalist explains our current understanding of the Sun.

Friedman, H. *Sun and Earth.* W. H. Freeman, 1986. Attractive introduction by an astronomer.

Goodwin, J., *et al. Fire of Life.* Smithsonian, 1981. Coffee table book about solar science and legend.

Hufbauer, K. *Exploring the Sun: Solar Science Since Galileo.* Johns Hopkins Press, 1991.

Sutton, C. *Spaceship Neutrino.* Cambridge U. Press, 1992. A science writer introduces neutrinos in physics and astronomy.

Articles

On the Sun

Pasachoff, J. "The Sun: A Star Close-Up" in *Mercury,* May/June 1991, p. 66.

Gibson, E. "The Sun as Never Seen Before" in *National Geographic,* Oct. 1974. Observations from a space laboratory.

LoPresto, J. "Looking Inside the Sun" in *Astronomy*, Mar. 1989, p. 20. On helio-seismology.

MacRobert, A. "Close-up of a Star" in *Sky & Telescope*, May 1985, p. 397.

Golub, L. "Heating the Sun's Million Degree Corona" in *Astronomy*, May 1993, p. 27.

Talcott, R. "Seeing the Unseen Sun: The Ulysses Mission" in *Astronomy*, Jan. 1990, p. 30.

Overbye, D. "John Eddy: The Solar Detective" in *Discover*, Aug. 1982, p. 68.

Bartusiak, M. "The Sunspot Syndrome" in *Discover*, Nov. 1989, p. 44.

Schaefer, B. "The Astrophysics of Suntanning" in *Sky & Telescope*, June 1988, p. 595.

Verschuur, G. "The Day The Sun Cut Loose" in *Astronomy*, Aug. 1989, p. 48. About a huge flare.

Verschuur, G. "The Many Faces of the Sun" in *Astronomy*, Mar. 1989, p. 46.

Kanipe, J. "The Rise and Fall of the Sun's Activity" in *Astronomy*, Oct. 1988, p. 22.

Robinson, L. "The Disquieting Sun: How Big, How Steady?" in *Sky & Telescope*, Apr. 1982, p. 354.

Robinson, L. "The Sunspot Cycle: Tip of the Iceberg" in *Sky & Telescope*, June 1987, p. 589.

Taylor, P. "Watching the Sun" in *Sky & Telescope*, Feb. 1989, p. 220. Observing instructions for amateurs.

On the Neutrino Problem

Bahcall, J. "Where are the Solar Neutrinos" in *Astronomy*, Mar. 1990, p. 40.

Fischer, D. "Closing In on the Solar Neutrino Problem" in *Sky & Telescope*, Oct. 1992, p. 378.

Bahcall, J. "The Solar Neutrino Problem" in *Scientific American*, May 1990.

DISCOVERY OF THE STARS

The Jewel Box (NGC 4755). This famous cluster of young bright stars is about 7800 LY from the Sun. It was named the Jewel Box from its description by Sir John Herschel as "a casket of variously colored precious stones."   *(Copyright Anglo-Australian Telescope Board, 1977)*

The stars are suns.

To the astronomer, this statement means that instead of reflecting light from some external source, stars generate the energy that they emit. Most stars produce energy in the same way that the Sun does. Deep in stellar interiors, hydrogen nuclei fuse to form helium, and the small amounts of mass lost in the process are converted to energy. Late in their lives, when all available hydrogen has been depleted, stars are forced to utilize new energy sources. The manner in which stars evolve and eventually die as they exhaust their sources of energy is described in Chapter 12.

We can understand stellar evolution only after we know what stars are like. Simple observations are enough to prove that not all stars are identical. For example, if you look at the sky you will see that different stars appear to have different degrees of brightness. Do some stars emit more energy than others, or do stars merely appear to have different degrees of brightness because some are nearby and others are far away? Some stars appear to be blue-white; others are obviously red. In Chapter 8 we showed that color is a good indication of temperature, so stars can be either hotter or cooler than the Sun. The Sun, of course, is brightest at the wavelengths that the eye sees as yellow.

Other questions about stars can be answered only by means of careful observations with large telescopes. Are stars larger or smaller than the Sun? Are they made of the same material as the Sun? How massive are they? How do the ages of stars compare with the age of the Sun?

Because the stars are suns, it should not surprise you that the same techniques used to answer questions about the Sun can also be used to study the stars. Spectroscopy is the most powerful technique of all. There are, however, two problems encountered when studying stars. First, because they are so far away, stars appear to our eyes and to our telescopes to be much fainter than the Sun. Thus astronomers cannot study stars in the same detail. Second, measuring the long distances to the stars is a difficult task, whereas with modern radar the measurement of distances within the solar system has become easy.

10.1
STELLAR DISTANCES

Triangulation and Parallaxes

The distance from the planets to the stars is enormous. The distance from the Earth to the nearest stars is more than 200,000 times greater than the distance from the Earth to the Sun. Yet in principle, we survey distances to the nearest stars by the same technique that a civil engineer uses on Earth to survey the distance to an inaccessible mountain or tree—the technique of *triangulation*.

Triangulation works as follows: Two observing stations are set up some distance apart. That distance (AB in Figure 10.1) is called the baseline. The direction to the remote object (C in the figure) in relation to the baseline is observed from each station, A and B. Note that C appears in different directions from the two stations. The apparent change in direction of the remote object that results from the observer's change of vantage point is called **parallax.** A knowledge of the angles at A and B and the length of the baseline, AB, allows the triangle ABC to be solved for any of its dimensions, including the distance AC or BC. The solution can be accomplished either by constructing a scale drawing or by numerical calculation with trigonometry.

Depth perception is an example of the same principle. Our eyes are separated by a baseline of a few inches. Therefore, our two eyes see an object in front of us from slightly different directions. The brain, like an electronic computer, solves the triangle and gives us an impression of

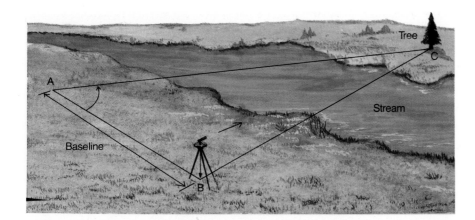

Figure 10.1 Triangulation of an inaccessible object.

the distance of the object. The greater the parallax, the nearer the object. Hold a pencil a few inches in front of your face and look at it first with one eye and then with the other. Note the large apparent shift in direction against the more distant wall across the room. Now hold the pencil at arm's length and note how the parallax is less. If an object is fairly distant, the shift—the parallax—is too small to notice with the eyes. Depth perception fails for objects more than a few tens of meters away.

Because astronomical objects are so far away, a very large baseline must be found or highly precise angular measurements must be made, or both. The Moon is the only object close enough to the Earth that its distance can be found fairly accurately with measurements made without a telescope. Hipparchus, a Greek astronomer who worked in the second century B.C., determined the distance to the Moon within about 20 percent of the correct value. He used measurements of the shift of position or parallax of the Moon relative to the Sun from two different spots on the Earth at the time of a total solar eclipse. Hipparchus assumed that the Sun is so far away that its parallax is zero.

With the aid of telescopes, subsequent astronomers were able to measure the distances to the nearer planets using the Earth's diameter as a baseline. To reach for the stars, however, requires a much longer baseline for triangulation.

Distances to Nearby Stars

As the Earth revolves around the Sun, we should observe the parallax of the nearer stars against the background of more distant objects. When the sky is viewed first from one side of the Sun and then six months later from the opposite side, nearby stars should be seen to shift their positions relative to very distant stars (Figure 10.2). The amount of shift or parallax provides a quantitative measurement of distance. In the fourth century B.C., the Greek philosopher Aristotle looked for this effect, failed to find it, and concluded that the Earth must be stationary.

Figure 10.2 As the Earth revolves around the Sun, the direction in which we see a nearby star varies with respect to distant stars. The parallax of the nearby star is defined to be one-half of the total change in direction, and is measured in fractions of a degree.

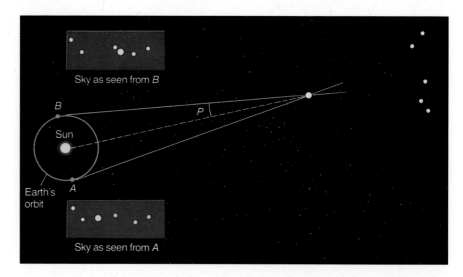

We now know that the stars are so distant that the parallaxes of even the nearest ones are far too small to be measured with the techniques available to Aristotle. Successful measurements of stellar parallax were not made until 1838. However, even the nearest star, Alpha Centauri, shows a total displacement of only about 0.0004° during the year. Just how small this displacement is can be illustrated by a comparison. A coin the size of a quarter would appear to have a diameter of 0.0004° if it were at a distance of about 3 km (nearly 2 miles). (Astronomers actually measure such tiny changes in direction in seconds of arc. Because 60 seconds of arc equal 1 minute of arc, and 60 minutes equal 1 degree, 0.0004° is equal to about 1.5 seconds of arc. The parallax of Alpha Centauri is equal to one-half of the total shift seen during a year or about 0.75 seconds of arc.

Units of Stellar Distance

How do astronomers measure the distances to stars?

The baseline for the measurement of stellar distance is the diameter of the Earth's orbit, which is 2 AU (astronomical units) or about 300 million km. The parallax angle p is defined as one-half of the total shift in angle measured from opposite sides of the Earth's orbit (Figure 10.2). The same argument that was used to derive the diameter of the Sun (see Section 9.1) can be used to show that the distance to a star in AU is related to the parallax in the following way:

$$\text{Distance} = (57.3/p) \text{ AU}$$

where p is in degrees.

Suppose that the total displacement in the position of a star, as seen from opposite sides of the Earth's orbit, is 1/3600 of a degree. The parallax p is half this amount, or 1/7200 of a degree. The distance to the star can be calculated to be 413,000 AU or 6.2×10^{13} km—62 million million km! And all but the *nearest* stars are farther away than even this huge distance.

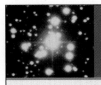

ESSAY   The Scale of Stellar Distances

In Chapter 2 we introduced a scale model for the solar system in which all dimensions were reduced by a factor of 1 billion (10^9). In this model, the Earth was the size of a grape and the Sun was 1.5 m in diameter, with the distance from the Sun to the most distant planets equal to about 5 km. On this same scale, the nearest stars are tens of thousands of kilometers away, making our model rather too large to visualize.

Let us therefore shrink our model by another factor of a thousand, so that all dimensions are reduced by 1 trillion (10^{12}). The Sun is now 1.5 mm in diameter, about the size of a mustard seed, and the entire solar system (except the Oort comet cloud) will fit inside a closet. But the nearest star is still 10 km away, on the other side of town.

The space between the stars is extraordinarily empty. Separations between individual stars amount to 1 million times their diameters. The stars in our model are like mustard seeds scattered many miles apart. Even in star clusters where the stars are much more tightly packed, these seeds are still separated by tens or hundreds of meters. The stellar universe is a big place, and it is a wonderful accomplishment, when you think about it, that we are able to learn so much about the stars by the analyses of the feeble light we collect with our astronomical telescopes.

Clearly we need a more convenient unit than the kilometer or the AU to express such large distances. One commonly used unit is the **light year (LY),** which is defined as the distance that light travels in one year in a vacuum. We can calculate the length of the LY easily by noting that the number of seconds in one year is about 3.15×10^7; thus one LY = 1.0 10^{13} km (approximately). In this text, most distances to stars, galaxies, and quasars will be given in LY.

One advantage of the LY as a unit is that it emphasizes the fact that as we look out into space, we are also looking back into time. The light that we see from a star 100 LY from us left that star 100 years ago. What we study is not the star as it is now but as it was in the past. The light from the most distant galaxies that we observe today started its journey before the Earth even existed.

The Nearest Stars

The stars nearest to the Sun are members of a system of three stars. To the unaided eye the system appears as a single bright star, which is named Alpha Centauri. Because Alpha Centauri is only 30° from the south celestial pole, it is not visible from most of the mainland United States, Canada, and Europe. The nearest of the three stars in the Alpha Centauri system is a faint star known as Proxima Centauri, which is 4.3 LY distant from the Sun. The remaining two stars in the triple system, including Alpha Centauri itself, are 4.4 LY from the Sun.

The nearest star visible to the unaided eye from most parts of the United States is Sirius, which also appears to be the brightest star in the night sky. Sirius is at a distance of 8 LY.

How far away are the stars?

ESSAY Stellar Magnitudes

In this text we describe the energy output of stars by comparing their luminosities with that of the Sun. Sirius, for example, emits 23 times as much energy as the Sun, and so we say that it has a luminosity of 23 L_s. Astronomers, however, have been looking at the sky for thousands of years—long before they understood much about energy or how to measure it or even that stars are not all at the same distance from the Earth. All those early astronomers could estimate was the *apparent* brightness of stars, and they couldn't even measure that quantity very precisely. The system of brightness measurements developed by early astronomers is still used today in some scientific papers and in star maps used by stargazers and amateur astronomers. It is helpful to know something about this traditional system of measurement if you want to learn to identify the constellations.

Around 150 B.C., Hipparchus erected an observatory on the island of Rhodes in the Mediterranean. Here he made the measurements necessary to prepare a catalog of nearly 1000 stars. This catalog included not only the positions of stars but also estimates of their apparent brightness. Obviously, Hipparchus did not have photographic plates or any instruments that could measure brightness accurately, so he simply made estimates with his eye. He sorted the stars according to brightness into six categories, which he called *magnitudes*. He referred to the

TABLE 10.A Magnitude Differences and Light Ratios

Difference in Magnitude	Ratio of Brightness
0.0	1:1
0.5	1.6:1
0.75	2:1
1.0	2.5:1
1.5	4:1
2.0	6.3:1
2.5	10:1
3.0	16:1
4.0	40:1
5.0	100:1
6.0	251:1
10.0	10,000:1
15.0	1,000,000:1
20.0	100,000,000:1

To date, parallaxes have been measured for nearly 10,000 stars. Only for a fraction of them, however, are the parallaxes large enough (about 1.5×10^{-5} degrees or more) to be measured with a precision of 10 percent or better. When accurate observations from space become available, it will be possible to extend parallax measurements from a distance of about 60 LY, which is the best that groundbased astronomy can do, to distances of nearly 200 LY.

Although satellites can provide a major step forward in our ability to measure distances of stars in the Sun's neighborhood, direct parallax measurements can be used to explore only a tiny fraction of the universe in which we live. The diameter of our own Milky Way Galaxy is about 100,000 LY, and the most distant galaxies and quasars are billions of light years away. In order to determine distances to these objects, astronomers must use a variety of methods that are based on the physical properties of stars. Therefore, we will put aside the question of distances

brightest stars in his catalog as being of the first magnitude. Stars so faint that he could just barely see them were called sixth magnitude stars.

This system of brightness estimates has remained in use for the past 2000 years. During the last century, an effort was made to make the scale more precise by establishing just exactly how much the brightness of a sixth magnitude star differs from that of a first magnitude star. Measurements show that we receive about 100 times more light from a star of the first magnitude as from one of the sixth. Therefore, astronomers adopted a precise scale in which a difference of five magnitudes corresponds exactly to a ratio of brightness of 100. Table 10A gives the approximate ratios of brightness corresponding to several selected magnitude differences. Figure 10A shows the range of observed magnitudes from the brightest to the faintest along with the actual magnitudes of several well-known objects.

The important fact to remember when using star charts is that numerically *smaller* magnitudes are associated with *brighter* stars. A numerically large magnitude, therefore, refers to a faint object. The stars that Hipparchus referred to as first magnitude stars actually have a range of apparent brightness of about a factor of 10. On the modern magnitude scale, the brightest of these stars have been assigned negative magnitudes.

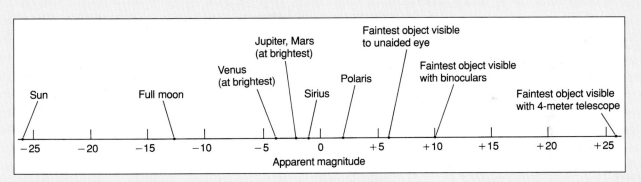

Figure 10A

to stars to discuss what has been learned from the study of nearby stars whose parallaxes can be measured. In Chapters 13 and 14, we will return to the problem of determining the distances of stars and galaxies for which parallax measurements are not possible.

10.2
LUMINOSITIES OF STARS

Even if all stars were identical, they would still not all appear to have identical brightnesses because they are located at various distances from the Earth. The light energy that we receive from a star is inversely proportional to the square of its distance (see Section 8.1). Therefore, the apparent brightnesses of stars do not provide a basis for comparing the amounts of light that they actually emit into space.

The Sun, for example, gives us billions of times more light than do any of the other stars, but it is hundreds of thousands of times closer to us than is any other star. To compare the intrinsic luminous outputs of other stars, we need to determine their true energy output, or **luminosity.** This is one of the reasons that the measurement of the distances to the nearest stars is so important. Until we know the distance, we cannot use the inverse-square law to determine luminosity.

Once the distance to a star is known, it is easy for the astronomer to use the observed brightness (watts per square meter reaching the Earth) to calculate true luminosity—the total energy output in, for example, watts. The luminosity of the Sun is about 4×10^{26} watts, a very large number indeed. Often it is convenient to express the luminosity of other stars in terms of the Sun's luminosity. For example, the luminosity of Sirius is 23 times that of the Sun. If we use the symbol L_s to denote the Sun's luminosity, the luminosity of Sirius is thus 23 L_s. We will frequently use this notation in the following chapters of this book.

The luminosity of stars ranges from less than 0.0001 L_s to more than 10^6 L_s.

The Luminosity Function

Once we measure the luminosities of a large number of stars, we can also determine how many stars in a given volume of space are intrinsically very luminous and how many stars are intrinsically faint. Figure 10.3 shows how many stars in the solar neighborhood fall within each successive interval of luminosity. This relationship is called the **luminosity function.**

Figure 10.3 shows that the Sun is more luminous than are the vast majority of stars. Only a tiny fraction of the stars are brighter than the Sun.

How does the luminosity of the Sun compare with the luminosity of nearby stars?

Figure 10.3   A luminosity function in astronomy indicates how many stars there are in different ranges of intrinsic luminosity. Shown here is the luminosity function of stars in the solar neighborhood. Note that there are many more faint stars than luminous stars in this given volume of space.

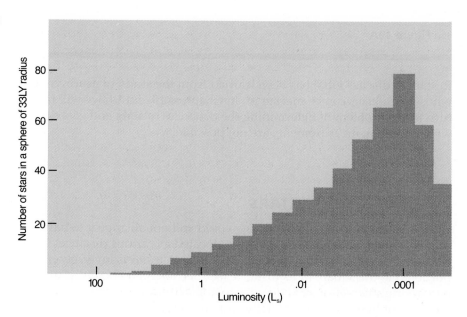

Figure 10.4 Sir William Herschel (1738–1822), a German musician, emigrated to England to avoid service in the Seven Years' War. While composing and giving music lessons, he built the first large reflecting telescopes, which he used to survey the skies. *(Yerkes Observatory)*

STELLAR MASSES

Binary Stars

Many stars are members of double-star systems. Masses for these stars can be calculated from measurements of their orbits, just as the mass of the Sun was derived by measuring the orbits of the planets around it.

The first binary star was discovered in 1650, less than half a century after Galileo first observed the sky with a telescope. John Baptiste Riccioli, an Italian astronomer, observed that the star Mizar, in the middle of the handle of the Big Dipper, appeared through his telescope as two stars. Since that discovery, thousands of double stars have been catalogued. Among the 59 nearest stars listed in Appendix 10, nearly 50 percent (28) are members of systems containing more than one star. Although stars most commonly come in pairs, there are also triple and quadruple systems. All such multiple star systems are usually classed together under the general term *binary stars.*

One well-known double star is Castor, which is located in the constellation of Gemini. By 1804, astronomer William Herschel (Figure 10.4), who also discovered the planet Uranus, had noted that the fainter component of Castor had slightly changed its direction relative to the

Figure 10.5 Revolution of a double star. Three photographs covering a period of about 12 years show the mutual revolution of the components of the nearby star Kruger 60. *(Yerkes Observatory)* →

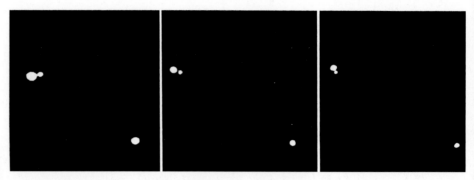

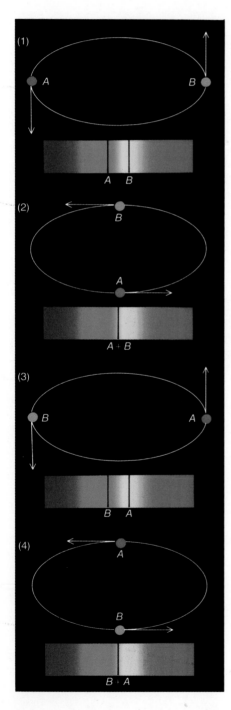

brighter component. Here was evidence that one star was moving about another; it was the first evidence that gravitational influences exist outside the solar system. The orbital motion of another binary star, Kruger 60, is shown in Figure 10.5.

Another class of double stars was discovered by E.C. Pickering at Harvard in 1889. He found that the dark absorption lines in the spectrum of the brighter component of Mizar (the first double star to be discovered) are usually double, but that the spacing of the components of the lines varies periodically. At times the lines even become single. He correctly deduced that the brighter component of Mizar, which is called Mizar A, is really two stars that revolve about each other over a period of 104 days.

It would be incorrect to describe the motion of a double-star system by saying that one star orbits around the other. Gravitation is a mutual attraction. Each star exerts a gravitational force on the other, with the result that both stars orbit about a point between them called the **center of mass.** Imagine that the two stars are seated at opposite ends of a seesaw. The point at which one would have to place the support to make the seesaw balance is the center of mass.

When one star is approaching the Earth, relative to the center of mass, the other star is receding from the Earth. The radial velocities of the two stars and, therefore, the Doppler shifts of their spectral lines, are different, so that when the composite spectrum of the two stars is observed, each line appears double. When the two stars are both moving across our line of sight, however, they both have the same radial velocity (that of the center of mass of the pair), and the spectral lines of the two stars coalesce (Figure 10.6). A plot showing how the velocities of the stars change with time is called a radial velocity curve (Figure 10.7).

Stars such as Mizar A, which appear to be single stars when photographed or observed visually through the telescope, but which really

←

Figure 10.6 A schematic drawing of the motions of two stars orbiting around each other. The Doppler effect causes the spectral lines to move to longer and shorter wavelengths.

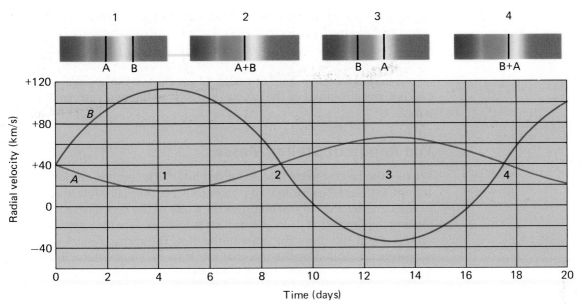

Figure 10.7 Radial velocity curves for a spectroscopic binary system showing how the two components alternately approach and recede from the Earth. The positions on the curve corresponding to the illustrations in Figure 10.6 are marked.

are double stars, as shown by spectroscopy, are called **spectroscopic binaries.** Systems that can be observed visually as double stars are called **visual binaries.**

Masses from the Orbits of Binary Stars

We can estimate the masses of double-star systems by using Newton's reformulation of Kepler's third law. The method is similar to that used to calculate the mass of the Sun (see Section 9.1). If two objects are in mutual revolution, the square of the period with which they orbit around each other is proportional to the cube of the semimajor axis of the orbit of one object with respect to the other divided by their combined mass ($M_1 + M_2$). Thus, if we can observe the size of the orbit and the period of mutual revolution of the stars in a binary system, we can calculate the sum of their masses.

Most spectroscopic binaries have periods ranging from a few days to a few months, with separations of their member stars usually less than 1 AU. If the two stars of a spectroscopic binary are not too different in luminosity, the spectrum of the system displays the lines of both stars, and each set of lines oscillates in the period of mutual revolution. More often, one star is much fainter than the other, and only the lines of the brighter star can be observed.

The actual analysis of a radial velocity curve (Figure 10.7) to determine the masses of the stars in a spectroscopic binary is complex, although in principle the idea is simple. The speeds of the stars are measured from the Doppler effect. The speed together with the period (determined from the velocity curve), allows us to calculate the circumference of the orbit, and hence the separation of the stars in kilometers or AU. Knowing the period and the separation, we can calculate the sum of the masses of the stars from Newton's reformulation of Kepler's third

How can the binary stars be used to measure stellar masses?

law. In addition, the relative orbital speeds of the two stars allow us to calculate how much of the total mass each star has. The more massive star, which is closer to the center of mass, has a smaller orbit and thus moves more slowly to get around in the same time. If the binary system is properly oriented and all of the steps are carried out carefully, the result is a measurement of the masses for each of the two stars in the system.

The Range of Stellar Masses

What is the largest mass that a star can have? The limit is not known for certain, but searches for massive stars indicate that very few stars have masses greater than about 60 times the mass of the Sun. There is no convincing evidence that there are any stars with masses that significantly exceed about 100 times the mass of the Sun. The rarity of stars with large masses is illustrated by the fact that there are no stars within 30 LY of the Sun that have masses greater than four times the mass of the Sun.

According to theoretical calculations, objects with masses less than 1/12 the mass of the Sun never become hot enough to ignite nuclear reactions and so cannot become true stars. Objects with masses between 1/65 and 1/12 the mass of the Sun may produce energy for a brief period of time by means of nuclear reactions involving deuterium, but do not become hot enough to fuse protons to form helium. Such objects are called **brown dwarfs.** Still smaller objects, with masses less than 1/65 the mass of the Sun, are true planets. They may radiate energy produced by the radioactive elements that they contain. They may also radiate heat generated by slow gravitational contraction, but their interiors will never reach temperatures high enough for nuclear reactions to occur.

The Mass-Luminosity Relation

What is the relationship between the mass of a star and its luminosity?

When we compare the masses and luminosities of stars, it is generally found that the more massive stars are also the more luminous stars. This phenomenon, known as the **mass-luminosity relation,** is shown graphically in Figure 10.8. Each point represents a star of known mass and luminosity. Its horizontal position indicates its mass, given in units of the Sun's mass, and its vertical position indicates its luminosity, given in units of the Sun's luminosity.

Most stars fall along a line running from the lower-left (low mass, low luminosity) corner of the diagram to the upper-right (high mass, high luminosity) corner of the diagram. It is estimated that approximately 90 percent of all stars obey the mass-luminosity relation illustrated in Figure 10.8.

The range of stellar luminosities is much greater than the range of stellar masses. Luminosities of stars are roughly proportional to their masses raised to the 3.5 power. Stars have masses between 1/12 and 100 times that of the Sun. According to the mass-luminosity relation, however, the corresponding luminosities of stars at either end of the range are, respectively, less than 0.001 L_s and greater than 1 million L_s. If two

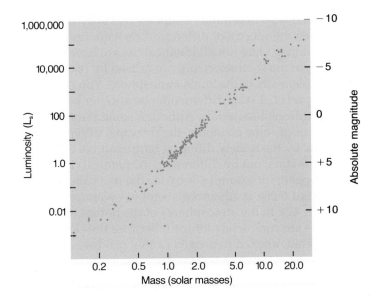

Figure 10.8   The mass-luminosity relation. The plotted points show the masses and luminosities of stars for which both of these quantities are known to an accuracy of 15 to 20 percent. The three points lying below the sequence of points are all white dwarf stars.   *(Adapted from data compiled by D. M. Popper)*

stars differ in mass by a factor of 2, their luminosities would be expected to differ by a factor of about 10.

10.4
TEMPERATURES AND COMPOSITIONS OF STARS

There are several ways to estimate the temperatures of stars. The color of the radiation from a blackbody is a measure of its temperature (see Section 8.2). Because the radiation of stars behaves approximately like the radiation of blackbodies, it is possible to determine stellar temperatures by measuring the colors of stars. Blue stars are hotter than red stars. With the precise instrumentation used by astronomers, it is possible to determine temperatures accurately by measuring how much energy stars emit at different wavelengths (Figure 10.9).

Figure 10.9   A time exposure showing the constellation Orion as it rises over Kitt Peak National Observatory. The colors of the various stars are caused by their different temperatures.   *(National Optical Astronomy Observatories)*

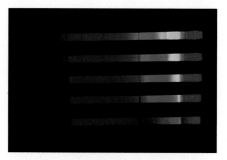

Figure 10.10   Several exposures of the spectrum of Vega, which is the fifth brightest star in the sky and has a temperature of approximately 9500 K. The strongest lines in this spectrum are produced by hydrogen. One of the hydrogen lines is clearly seen near the boundary between blue and green light.   *(National Optical Astronomy Observatories)*

How can spectra be used to estimate the temperature and luminosity of stars?

Another way to estimate temperatures is by analyzing the spectra of stars. When the spectra of different stars were first observed, it was found that they were not all identical. As we have seen (Section 8.3), the dark lines in the solar spectrum are caused by the presence of various chemical elements in the Sun's atmosphere. You might guess, therefore, that the spectra of stars differ from one another because stars are composed of different materials. Although some stars do have unusual abundances of some elements, the principal differences in stellar spectra are caused by the widely differing temperatures in the outer layers of the various stars, not by differences in chemical composition.

Hydrogen, for example, is by far the most abundant element in all stars (except those at advanced stages of evolution, which are discussed in Chapter 12). In the atmospheres of the hottest stars, however, hydrogen atoms are completely ionized. Because the electron and the proton are separated, ionized hydrogen cannot produce absorption lines. In the atmospheres of the coolest stars, hydrogen is neutral and can produce absorption lines; however, practically all of the hydrogen atoms are in the lowest energy state (unexcited) in these stars and thus can absorb only those photons that can lift an electron from that first energy level to a higher level. The photons absorbed in this way produce a series of absorption lines that lie in the ultraviolet part of the spectrum and cannot be studied from the Earth's surface.

In a stellar atmosphere with a temperature of about 10,000 K, an appreciable number of hydrogen atoms are excited to the second energy level. They can then absorb additional photons and rise to still higher levels of excitation. These photons correspond to visible light (Figure 10.10).

Absorption lines stemming from hydrogen, therefore, are strongest in the visible region of the spectrum in stars whose atmospheres have temperatures near 10,000 K. They are less conspicuous in the spectra of both hotter and cooler stars, even though hydrogen is roughly equally abundant in most types of stars. Similarly, every other chemical element, in each of its possible stages of ionization, has a characteristic temperature at which it is most effective in producing absorption lines in any particular part of the spectrum.

Classification of Stellar Spectra

Because the temperature of a star determines which absorption lines are present in its spectrum, arranging spectra according to the patterns of their spectral lines is equivalent to arranging stars in sequence according to temperature. Astronomers have identified seven principal patterns of lines, or **spectral classes.** Arranged in order of decreasing stellar temperature, the seven main spectral classes are designated O B A F G K M—known to generations of college students by the mnemonic "Oh be a fine (guy) (girl), kiss me." Each of these spectral classes is further subdivided into ten subclasses designated by numbers. A B0 star is the hottest type of B star; a B9 star is the coolest type of B star and is only slightly hotter than an A0 star.

TABLE 10.1 Spectral Sequence

Spectral Class	Color	Approximate Temperature (K)	Principal Features	Stellar Examples
O	Violet	>28,000	Relatively few absorption lines in observable spectrum. Lines of ionized helium, doubly ionized nitrogen, triply ionized silicon, and other lines of highly ionized atoms. Hydrogen lines appear only weakly.	10 Lacertae
B	Blue	10,000–28,000	Lines of neutral helium, singly and doubly ionized silicon, singly ionized oxygen and magnesium. Hydrogen lines more pronounced than in O-type stars.	Rigel Spica
A	Blue	7,500–10,000	Strong lines of hydrogen. Also lines of singly ionized magnesium, silicon, iron, titanium, calcium, and others. Lines of some neutral metals show weakly.	Sirius Vega
F	Blue to white	6,000–7,500	Hydrogen lines are weaker than in A-type stars but are still conspicuous. Lines of singly ionized calcium, iron, and chromium, and also lines of neutral iron and chromium are present, as are lines of other neutral metals.	Canopus Procyon
G	White to yellow	5,000–6,000	Lines of ionized calcium are the most conspicuous spectral features. Many lines of ionized and neutral metals are present. Hydrogen lines are weaker even than in F-type stars. Bands of CH, the hydrocarbon radical, are strong.	Sun Capella
K	Orange to red	3,500–5,000	Lines of neutral metals predominate. The CH bands are still present.	Arcturus Aldebaran
M	Red	<3,500	Strong lines of neutral metals and molecular bands of titanium oxide dominate.	Betelgeuse Antares

In the hottest O stars (temperatures over 28,000 K), only lines of ionized helium and highly ionized atoms of other elements are conspicuous. Hydrogen lines are strongest in A stars with atmospheric temperatures of about 10,000 K. Ionized metals provide the most conspicuous lines in stars with temperatures from 6000 to 7500 K (spectral type F). In the coolest M stars (below 3500 K), absorption bands of titanium oxide and other molecules are very strong. The sequence of spectral classes is summarized in Table 10.1 and Figure 10.11. The spectral class assigned to the Sun is G2.

To see how spectral classification works, suppose that you have a spectrum in which the hydrogen lines are about half as strong as those viewed in an early A star. If you look at Figure 10.11, you will see that the star could be either a B star or a G star. If the spectrum also contains helium lines, then it is a B star. If it contains lines of ionized iron and other metals, it must be a G star.

Figure 10.11 Relative intensities of different absorption lines in stars of various spectral types.

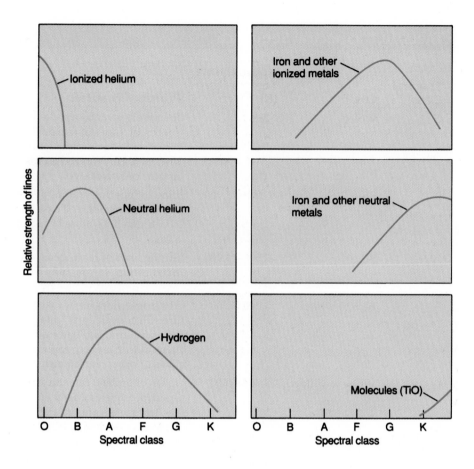

Influence of Pressure on the Spectrum of a Star

Stars come in a wide variety of sizes. At some periods in their evolutionary history, stars can expand to enormous dimensions with correspondingly low densities and atmospheric pressures. Stars of such exaggerated size are called **giants.** Luckily for the astronomer who wishes to distinguish such giants from the run-of-the-mill stars (like our Sun), giants can be identified from their spectra.

At the very low densities expected for the extended tenuous atmospheres of these giant stars, the pressures are also low. Under these conditions, ionized atoms recombine with electrons more slowly than they do in normal stars. These low-density gases, therefore, maintain a higher average degree of ionization than do high-density gases of the same temperature. Ionized lines are therefore slightly weaker in the spectrum of a dwarf star of high atmospheric pressure than they are in the spectrum of a giant star of the same temperature but of low atmospheric pressure. This difference in line strengths enables us to determine whether a star is a dwarf or a giant.

Abundances

Once allowance has been made for temperature and pressure in a star's atmosphere, analyses of the strengths of absorption lines in its spectrum can yield information regarding the relative abundances of the various chemical elements. Such quantitative analyses have shown that the relative abundances of the different chemical elements in most stars are approximately the same as those in the Sun (see Table 9.2).

Hydrogen comprises about 75 percent of the mass of most stars. Hydrogen and helium together make up 96 to 99 percent of the mass; in some stars they amount to more than 99.9 percent. Among the 4 percent or less of "heavy elements," neon, oxygen, nitrogen, carbon, magnesium, argon, silicon, sulfur, iron, and chlorine are the most abundant. Usually, but not always, the elements of lower atomic weight are more abundant than those of higher atomic weight.

What are stars made of?

10.5
DIAMETERS OF STARS

The Sun has an observable angular diameter of about $1/2°$. Thus, we can calculate directly the Sun's true (linear) diameter, which is 1.39 million km, or about 109 times the diameter of the Earth. The Sun is the only star whose angular diameter is easily resolved. Even through the largest telescopes, apart from the distortions introduced by turbulence in the Earth's atmosphere, stars appear to the eye to be points of light. There are several techniques, however, that the astronomer can use to measure the sizes of stars.

Stellar Diameters for Stars Whose Light is Blocked by the Moon

One technique, which gives very precise diameters but can be used for only a few stars, is to observe the dimming of light when the Moon passes in front of a star. The object is to measure the time required for a star's brightness to drop to zero as the edge of the Moon moves across it. Because we know how rapidly the Moon moves in its orbit around the Earth, it is possible to calculate the angular diameter of the star. If the distance to the star is also known, then it is possible to calculate its diameter in kilometers. This method works only for fairly bright stars that happen to lie along the ecliptic where the Moon (or much more rarely a planet) can pass in front of them as seen from the Earth.

How do astronomers measure the diameters of stars?

Stellar Radii from the Orbits of Binary Stars

Another technique works for certain binary stars. Some double stars are lined up in such a way that when viewed from the Earth each star passes in front of the other during every revolution. When one star blocks the

Figure 10.12 Schematic light curve of a hypothetical eclipsing binary star with total eclipses (i.e., one star passes directly in front of and behind the other). The numbers indicate parts of the light curve corresponding to various positions of the smaller star in its orbit. In this example, the smaller star is the hotter star.

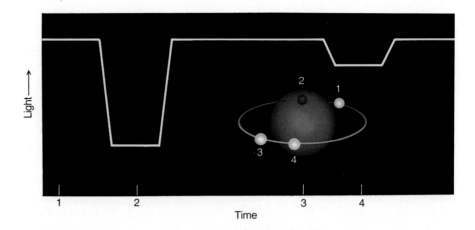

light of the other and prevents it from reaching the Earth, astronomers say that an *eclipse* has occurred. Such double-star systems are called **eclipsing binaries.**

To illustrate how the sizes of the stars are related to the light curve, consider a hypothetical eclipsing binary in which the stars are very different in size, like those illustrated in Figure 10.12. Assume that the orbit is viewed exactly edge-on. When the small star just starts to pass behind the large star (first contact), the brightness begins to drop. The total phase of the eclipse begins at second contact, when the small star has gone entirely behind the large star. At the end of totality (third contact), the small star begins to emerge. When the small star has reached last contact, the eclipse is over.

During the time interval between first and second contacts (or between third and last contacts), the small star has moved a distance equal to its own diameter. During the time interval from first to third contacts (or from second to last contacts), the small star has moved a distance equal to the diameter of the large star. If the spectral lines of both stars are visible in the spectrum of the binary, the speed of the small star with respect to the large star is also known. This speed, multiplied by the time intervals from first to second contacts and from first to third contacts, gives the respective diameters of the small and large stars.

Generally, the orbits are not exactly edge-on, and the eclipses are not central. Further, binary star orbits are ellipses, not circles. However, it is a relatively simple geometry problem, at least in principle, to sort out these effects. From the depths of the minima and the exact instants of the various phases, it is possible to calculate both the inclination of the orbit and the sizes of the stars in relation to their separation. If the eclipses are partial, the analysis is far more difficult, but still can be accomplished.

Stellar Radii From Radiation Laws

The sizes of most stars can be calculated using an indirect method based on observations of brightness and spectrum as follows: The luminosity of a star can be obtained from measurements of its apparent brightness and distance. The temperature of a star can be obtained from its color or spectrum. Because stars are fairly good approximations of blackbodies, we can apply the Stefan-Boltzmann law to them (Section 8.2). Remember that according to the Stefan-Boltzmann law, the energy E emitted by a square meter of material of temperature T is given by

$$E = \sigma T^4$$

To get the luminosity of a star, we must multiply the energy emitted per square meter by the total surface area of a star. Because the surface area of a sphere is equal to $4\pi R^2$, we have a relation for the stellar luminosity L:

$$L = 4\pi R^2 \sigma T^4$$

If we know L and T, we can calculate R.

Stellar Diameters

The results of the measurements of stellar size confirm that most nearby stars are roughly the size of the Sun—typically 1 million km or so in diameter. Faint stars, as might be expected, are generally smaller than more luminous stars. However, there are some dramatic exceptions to this simple generalization.

A few of the very luminous stars, those that are also red in color (indicating relatively low surface temperatures), are truly enormous. These stars are called **giants** or **supergiants.** An example is Betelgeuse, the second brightest star in the constellation of Orion and one of the dozen brightest stars in our sky. The distance of Betelgeuse is about 500 LY—not very close—yet its angular diameter in the sky is about 3×10^{-5} degrees, almost big enough to appear as a disk through a large telescope under the best observing conditions. Translated into linear dimensions, this corresponds to a diameter greater than 10 AU, large enough to fill the entire inner solar system almost as far as Jupiter. Chapter 11 discusses in detail the evolutionary process that leads to the formation of giant and supergiant stars.

10.6
THE HERTZSPRUNG-RUSSELL (HR) DIAGRAM

In this chapter, we have described some of the characteristics of stars and how those characteristics are measured. You may feel somewhat overwhelmed by all of this new information. It will be easier to understand and remember what stars are like if we can now find some patterns that describe the relationships between size, mass, luminosity, and temperature. Fortunately, such patterns do exist.

In 1913, American astronomer Henry Norris Russell plotted the intrinsic luminosities of stars of known distance against their spectral classes. This investigation, and a similar independent study in 1911 by Danish astronomer Ejnar Hertzsprung, led to an extremely important

OBSERVING YOUR WORLD   THE STARS

One of the most beautiful and awe-inspiring sights in all of nature is the dark night sky filled with stars. The stars that are visible above the horizon change with the seasons and with the time of night. The patterns of the stars, however, are fixed, at least over the time scale of a human life. With star maps it is easy to learn your way around the sky.

All stars are assigned to one of 88 different groupings or constellations of stars. Most of these constellations date back to ancient times, but a few—especially in the Southern Hemisphere—were defined during the past few hundred years. Constellations usually have a story associated with them. There is Orion, the Hunter, permanently separated in the sky from the Scorpion, which caused his death. Andromeda is chained to a rock, offered in sacrifice to atone for the boasts of her mother Cassiopeia. There are dogs, a dragon, a ram, a dolphin, scales, crowns, and crosses. To learn to identify a constellation and to know its story is to make a friend for life.

Learn to know the sky in all seasons of the year—its beauty can bring endless hours of pleasure.

Activities

1. Stars rise about 4 minutes earlier every night. Look toward the eastern horizon and note where stars are located at a specific time of night. Wait a week or two to look again, and note the difference in altitude above the horizon. Ancient peoples used the times of rising of specific stars to mark the seasons and the start of a new year.

2. If you look at very bright stars when they are close to the horizon, you may see them appear to change in brightness and possibly in color as well. This phenomenon is called twinkling and is caused by turbulence in the Earth's atmosphere. As small regions in the atmosphere move about, they bend the light from a star, just as a prism does, and cause it to travel in ever-changing directions. The brightness of the star varies with the fraction of its light that is directed precisely toward your eye. The color of the star may also appear to change rapidly if, for example, red light is directed toward your eye and blue light is not, and then, a fraction of a second later, the reverse situation happens.

3. Several very bright stars are identified in Figure 10.13. Select some that are visible at this time of year. What color would you expect each to be? Go outside at night and find these stars. Do they actually have approximately the color that you expected?

---

discovery concerning the relationship between the luminosities and surface temperatures of stars.

Features of the Temperature-Luminosity or HR Diagram

> What is the relationship between luminosity and temperature for most stars?

The relationship between stellar luminosities and temperatures is best illustrated with a plot of stellar temperature (or spectral class) against luminosity. This plot is frequently called the Hertzsprung-Russell diagram, or the **HR diagram,** in honor of the two astronomers who first made such plots. It is one of the most important and widely used diagrams in astronomy, with applications that extend far beyond the purposes for which it was originally developed nearly a century ago.

The two quantities plotted in the HR diagram are readily determined for many stars. The *luminosities* can be found from the known dis-

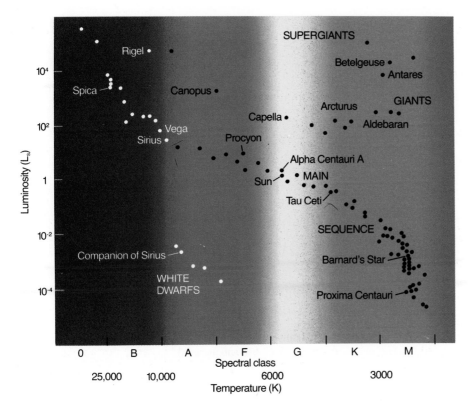

Figure 10.13 The HR diagram for a selected sample of stars. Luminosity is plotted along the vertical axis. Along the horizontal axis, we can plot either temperature or spectral type. Several of the brightest stars are identified by name.

tances and the observed apparent brightnesses. Even if the distance is not well determined, an HR diagram can be plotted for a group of stars all at about the same distance (such as a cluster of stars) by using apparent or relative brightness. The *temperature* of a star is indicated by either its color or its spectral class.

As an example of an HR diagram, luminosities are plotted against temperatures (and spectral classes) for some nearby stars in Figure 10.13. When plotting these diagrams, astronomers always adopt the convention that temperature increases toward the left and luminosity increases toward the top. Figure 10.14, a schematic HR diagram for a large sample of stars, makes the various features more apparent.

The most significant feature of the HR diagram is that the stars are not distributed over it at random, as they would be if the stars exhibited all combinations of luminosity and temperature. Instead, we see that the stars cluster into certain parts of the HR diagram. The majority of stars are aligned along a narrow sequence running from the upper-left (hot, highly luminous) section of the diagram to the lower-right (cool, less luminous) section of the diagram. This band of points is called the **main sequence.** It represents a relationship between *temperature* and *luminosity* that is followed by most stars. The characteristics of main-sequence stars of different spectral types are summarized in Table 10.2.

A substantial number of stars lie above the main sequence on the HR diagram, in the upper-right (cool, high luminosity) section. These

Figure 10.14 Schematic HR diagram for many stars.

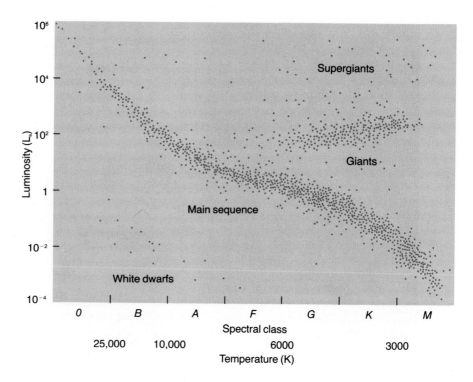

are the *giants*. At the top part of the diagram are stars of even higher luminosity, the *supergiants*. These names are appropriate, because the stars in this part of the HR diagram are enormous. Finally, there are stars in the lower left (hot, low luminosity) corner known as *white dwarfs*. To say that a star lies "on" or "off" the main sequence does not refer to its position in space but only to the location of the point that represents its luminosity and temperature on the HR diagram.

Figure 10.13 does not show the correct relative proportions of various kinds of stars. The stars in Figure 10.13 were selected because their

TABLE 10.2   Characteristics of Main-Sequence Stars

Spectral Type	Mass (Sun = 1)	Luminosity (Sun = 1)	Temperature	Radius (Sun = 1)
O5	40	7×10^5	40,000 K	18
B0	16	2.7×10^4	28,000 K	7
A0	3.3	55	10,000 K	2.5
F0	1.7	5	7,500 K	1.4
G0	1.1	1.4	6,000 K	1.1
K0	0.8	0.35	5,000 K	0.8
M0	0.4	0.05	3,500 K	0.6

distances are known, but this sample omits many intrinsically faint stars that are nearby but have not had their parallaxes measured. To be truly representative of the stellar population, an HR diagram should be plotted for *all* stars within a certain distance. Astronomers estimate that if such a plot were made, we would find that about 90 percent of the stars in our part of space are main-sequence stars and about 10 percent of stars are white dwarfs. Fewer than 1 percent of stars are giants or supergiants.

Stellar Distances from Spectra

Examination of the HR diagram reveals a very important method for determining stellar distances. Suppose, for example, that a star is known to be a spectral class G star on the main sequence. Its intrinsic luminosity could then be read off the HR diagram at once; it would be about 1 L_s. From the estimate of the intrinsic luminosity and the star's apparent brightness, its distance can be calculated.

In general, however, the spectral class alone is not enough to fix, unambiguously, the intrinsic luminosity of a star. The G star described in the last paragraph could have been a main-sequence star of luminosity 1 L_s, a giant star of luminosity 100 L_s, or a supergiant star of even higher luminosity. We recall, however, that pressure differences in the atmospheres of stars of different sizes result in slightly different degrees of ionization for a given temperature. These effects, which result from differences in pressure, produce subtle but observable differences in the spectra of stars of the same temperature. It is thus possible to classify a star by its spectrum, not only according to its temperature (spectral class) but also according to whether it is a main-sequence star, a giant, or a supergiant. With both of these items of information, a star's position on the HR diagram is uniquely determined. Therefore, its luminosity is also known, and its distance can be calculated.

Extremes of Stellar Luminosities, Radii, and Densities

The HR diagram is useful for investigating the extremes in size, luminosity, and density found for stars. The most massive stars are the most luminous, at least for main-sequence stars. These stars have luminosities of about 10,000 L_s to 100,000 L_s. A few stars are known that are 1 million times as luminous as the Sun. These superluminous stars, most of which are at the upper-left section on the HR diagram, are very hot, very blue stars with spectral types O and B. These are the stars that would be the most conspicuous at very great distances in space.

The cool giants and supergiants are located at the upper-right corner of the HR diagram. Giant stars are at least several hundred times as luminous as the Sun, and supergiant stars are some thousands of times as luminous as the Sun. These stars also have much larger diameters than the Sun. Many supergiants are so large that if the solar system could be centered in such a star, the star's surface would lie beyond the orbit of Mars.

It is easy to show that the luminous supergiants must be very large. Consider, as an example, a red, cool supergiant that has a surface temperature of 3000 K and a luminosity of 10,000 L_s. Although this star is 10,000 times more luminous than the Sun, it has only half its surface temperature. The Stefan-Boltzmann law tells us that the surface brightness is proportional to the fourth power of the temperature (Section 8.2). Thus, if the star has one-half the temperature of the Sun, it emits only $(1/2)^4 = 1/16$ as much light per unit area as the Sun. In order to have a luminosity of 10,000 L_s, its total surface area must be greater than the Sun's by 160,000 times. Its radius, therefore, is 400 times the Sun's radius.

In contrast, the very common red, cool stars of low luminosity at the lower end of the main sequence are much smaller and are more compact than the Sun. An example of such a red dwarf is the star Ross 614B, which has a surface temperature of 2700 K and 1/2000 of the Sun's luminosity. Each square meter of this star emits only 1/20 as much light as a square meter of the Sun, but to have only 1/2000 the Sun's luminosity, the star needs to have only about 1/100 the Sun's surface area, or 1/10 its radius. A star with such a low luminosity also has a low mass (Ross 614B has a mass about 1/12 that of the Sun) but still would have a mean density about 80 times that of the Sun. Its density must be higher, in fact, than that of any known solid found on the surface of the Earth.

The faint red main-sequence stars are not the stars of the most extreme densities, however. The white dwarfs, at the lower-left corner of the HR diagram, have densities that are many times greater still.

White Dwarfs

What is a white dwarf?

The first white dwarf stars to be discovered were the companions to much more luminous main-sequence stars. The first such object, found in 1914, forms a binary star with Sirius, the brightest-appearing star in the sky. Since the discovery of the companion of Sirius, hundreds of white dwarfs have been found.

A good example of a typical white dwarf is the nearby star 40 Eridani B. Its temperature is a relatively hot 12,000 K, but its luminosity is only 1/275 L_s. Calculations show that it has a radius of 0.014 times the radius of the Sun and a volume of 2.5×10^{-6} the volume of the Sun. Its mass, however, is 0.43 times that of the Sun, so its density is about 170,000 times the density of the Sun, or more than 200,000 g/cm^3.

A teaspoonful of such material would have a mass of about 1 ton. At such densities, matter cannot exist in its usual state. Although it is still gaseous, its atoms are completely stripped of their electrons. As we shall see in Chapter 12, most stars are believed to become white dwarfs near the end of their evolution. Eventually, after many billions of years, white dwarfs radiate away their internal heat, cooling off to become black dwarfs—cold, dense stars no longer shining. White or black dwarfs, however, are not the only possible final evolutionary states for stars. Some stars, we shall see, become neutron stars, with densities a billion times as great as those of white dwarfs. Still others may collapse to black holes of even greater density. We take up these bizarre objects in Chapter 12.

▶ SUMMARY

10.1 The shift in position of a nearby star relative to very distant background stars is called the **parallax** of that star and is a measure of its distance. Parallax measurements give accurate distances only for stars within a distance of about 60 **light years (LY)** from the Sun. Measurements from space will make it possible to measure parallaxes of stars as distant as 200 LY.

10.2 The total energy emitted by a star is its **luminosity.** The luminosity of stars ranges from about 0.0001 times the luminosity of the Sun to more than 1 million L_s. Stars with low luminosity are much more common than are stars with high luminosity. The relationship that shows how many stars in a given volume of space fall in successive intervals of luminosity is called the **luminosity function.**

10.3 Stellar masses can be determined by analysis of the motions of **double stars** as they orbit around their common **center of mass.** The three types of double stars are **visual binaries, spectroscopic binaries,** and **eclipsing binaries.** Stellar masses range from about 1/12 the mass of the Sun to (rarely) 100 times the mass of the Sun. The most massive stars are, in most cases, also the most luminous, and this correlation is known as the **mass-luminosity relationship.** Objects with masses between 1/65 and 1/12 the mass of the Sun do not become hot enough to fuse hydrogen to helium. Such objects are called **brown dwarfs.**

10.4 The temperatures of stars can be determined from their colors (hotter stars emit more energy at shorter wavelengths) or their **spectral classes.** The seven main spectral classes in order of decreasing temperature are O, B, A, F, G, K, M. Spectra can also be used to determine the composition of stars. Hydrogen and helium combined make up 96 to 99 percent of the mass of stars.

10.5 The size of a star can be determined by measuring the time it takes an object (the Moon, a planet, or a companion star) to pass in front of it and block its light. Measurement of temperature and total luminosity, combined with the Stefan-Boltzmann law, also makes it possible to calculate the radius of a star.

10.6 The **HR diagram** is a plot of stellar luminosity as a function of temperature. Most stars lie on the **main sequence,** which extends diagonally across the HR diagram from high temperature and high luminosity to low temperature and low luminosity. The position of a star along the main sequence is determined by its mass. Main-sequence stars derive their energy from the fusion of hydrogen to helium. About 90 percent of the stars near the Sun lie on the main sequence. About 10 percent of the stars are **white dwarfs.** Fewer than 1 percent of the stars are **giants** or **supergiants.**

▶ REVIEW QUESTIONS

1. Explain how parallax measurements can be used to determine distances to stars. What would be the advantage of making parallax measurements from Pluto rather than from Earth?

2. Create a table relating the following units of astronomical distance: kilometer, Earth radius, solar radius, astronomical unit, and light year.

3. What two factors determine how bright a star appears to be?

4. What are the extreme values of the mass, luminosity, temperature, and diameter of stars?

5. What is the main reason that the spectra of stars are not all identical? Explain.

6. Sketch an HR diagram. Label the axes. Show where cool supergiants, white dwarfs, the Sun, and main sequence stars are found.

▶ THOUGHT QUESTIONS

7. Plot the luminosity functions of the nearest stars (Appendix 10) and of the 20 brightest stars (Appendix 11). Explain how and why these two luminosity functions differ.

8. Star A has lines of ionized helium in its spectrum, and star B has bands of titanium oxide. Which is hotter? Why?

9. The spectrum of the Sun has hundreds of strong lines of un-ionized iron but only a few lines of helium, which are very weak. A star of spectral type B has very strong lines of helium but very weak iron lines. Does this difference in their spectra mean that the Sun contains more iron and less helium than the B star?

10. What are the approximate spectral classes of stars described as follows:
a. Lines of hydrogen in the visible part of the spectrum are very strong; some lines of ionized metals are present.
b. Lines of ionized helium are the strongest in the spectrum.
c. Lines of ionized calcium are the strongest in the spectrum; hydrogen lines are present with only moderate strength; lines of neutral and ionized metals are also present.
d. Lines of titanium oxide (TiO) are the strongest in the spectrum; lines of neutral metals are also present.
Arrange stars a, b, c, and d in order of decreasing temperature.

11. The spectrum of a star shows lines of ionized helium and also molecular bands of titanium oxide. What is strange about this spectrum? Can you suggest an explanation?

12. It is possible to construct the equivalent of an HR diagram for human beings by plotting height against weight. Try doing so for your classmates; obtain additional information for children and babies. Do all combinations of height and weight occur? Can you think of special examples of human beings that deviate from normal relationships?

13. Use Appendix 10 to determine what fraction of nearby stars are single, double, or triple stars.

14. From the data in Figure 10.3, what spectral types would you expect to be most common in the solar neighborhood? Explain your reasoning. Use the data from Appendix 10 to construct a graph that shows the percentage of stars of each of the main spectral classes O, B, A, F, G, K, M.

15. Suppose you wanted to search for main-sequence stars with very low mass with a space telescope. Would you design your telescope to detect light in the ultraviolet or in the infrared part of the spectrum. Why?

16. There are fewer eclipsing binaries than spectroscopic binaries. Explain why. Within 50 LY of the Sun, visual binaries outnumber eclipsing binaries. Why? Which is easier to observe at large distances—a spectroscopic binary or a visual binary?

▶ PROBLEMS

17. Show that the distance to a star is equal to $57.3/p$, where the distance is measured in AU and p is measured in degrees.

18. What is the radius of a star (in terms of the Sun's radius) with the following characteristics:
    a. Twice the Sun's temperature and four times its luminosity?
    b. Three times the Sun's temperature and 81 times its luminosity?

19. Two stars, A and B, appear equally bright when their radiation at all wavelengths is measured. They also have identical diameters. Yet star A has twice the surface temperature of star B. Which is more distant, and by what factor?

20. Verify that the density of a red dwarf with 1/12 the mass of the Sun and 1/10 the radius of the Sun is 80 times the density of the Sun. Calculate the density of the red supergiant described in Section 10.6, which has a mass of 50 times the mass of the Sun and a radius that is 400 times the radius of the Sun. The outer parts of such a star would constitute an excellent laboratory vacuum.

21. Suppose you weigh 150 pounds on the Earth. How much would you weigh on the surface of a white dwarf star, the same size as the Earth but having a mass 300,000 times larger than the mass of the Earth (nearly the mass of the Sun)?

▶ SUGGESTIONS FOR ADDITIONAL READING

Books

Kaler, J. *Stars and Their Spectra.* Cambridge U. Press, 1989. The definitive book on the spectral classification of stars and what their spectra can teach us.

Moore, P. *Astronomers' Stars.* Norton, 1987. Focuses on 15 specific stars that have been important in the development of astronomy.

Cooper, W. & Walker, E. *Getting the Measure of the Stars.* Adam Hilger, 1989. A manual with clear background information and observing activities.

Rowan-Robinson, M. *The Cosmological Distance Ladder.* Freeman, 1985. An astronomer summarizes the variety of methods for measuring distances in the cosmos.

Articles

Davis, J. "Measuring the Stars" in *Sky & Telescope,* Oct. 1991, p. 361. On measuring diameters of stars.

Evans, D., *et al.* "Measuring Diameters of Stars" in *Sky & Telescope,* Aug. 1979, p. 130.

Ferris, T. "A Plumb Line to the Stars" in *Mercury,* May/June 1989, p. 66. On the history of measuring stellar parallax.

Kaler, J. "Origins of the Spectral Sequence" in *Sky & Telescope,* Feb. 1986, p. 129. A fine history.

Hearnshaw, J. "Origins of the Stellar Magnitude Scale" in *Sky & Telescope,* Nov. 1992, p. 494.

Reddy, F. "How Far the Stars" in *Astronomy,* June 1983, p. 6.

Sneden, D. "Reading the Colors of the Stars" in *Astronomy,* Apr. 1989, p. 36.

Hodge, P. "How Far Away Are the Hyades" in *Sky & Telescope,* Feb. 1988, p. 138.

DeVorkin, D. "Henry Norris Russell" in *Scientific American,* May 1989.

Nielsen, A. "E. Hertzsprung—Measurer of Stars" in *Sky & Telescope,* Jan. 1968, p. 4.

Gingerich, O. "The Search for Russell's Original Diagram" in *Sky & Telescope,* July 1982, p. 36.

STARS: BIRTH THROUGH OLD AGE

Stars form in clouds of gas and dust. This image of M16 shows dark clouds within which stars are forming as well as a small cluster of stars that was born about 2 million years ago.

(Canada France Hawaii Telescope)

Astronomers estimate that there are more than 10,000 very luminous young stars in our Galaxy—stars with lifetimes that are measured in only millions of years. If, as seems likely, such highly luminous stars have been present throughout the billions of years that our Galaxy has existed, then as these stars die, they must be replaced by new ones. On average, in fact, one new bright star must be formed somewhere in our Galaxy every 500 to 1000 years if the total number of highly luminous stars is to remain approximately constant. And for every such luminous star that is formed, there are many others of more modest mass and luminosity.

As we shall see in this chapter, star formation is a continuous process that is occurring even now. Stars of all masses, low as well as high, are being formed, and that formation process is taking place in the interiors of clouds of dust and gas, which provide the necessary raw materials.

11.1
THE LIFETIMES OF STARS

It is natural to think of the stars as fixed, permanent, and unchanging. Yet stars are radiating energy at a prodigious rate, and no source of energy can last forever. For example, deep in the interior of the Sun, 600 million tons of hydrogen are converted to helium every second, with the simultaneous conversion of about 4 million of these tons to energy. At this rate, the Sun can continue to shine for more than 10 billion years, so massive is its fuel supply. But what about stars that are consuming their nuclear fuel more rapidly?

Stars that are more massive and more luminous than the Sun exhaust their fuel supply much more rapidly than does the Sun. The most massive stars have only 50 to 100 times the mass of the Sun, yet their luminosities—and correspondingly the rate at which they consume their supply of hydrogen—are a million times greater. Accordingly, these massive stars must exhaust their fuel supply, burn themselves out, and become unobservable in no more than a few million years. The brightest hot star in Orion—Rigel—began to shine about the time that the first human-like creatures walked the Earth.

How do we know that stars do not shine forever?

The HR Diagram and Stellar Evolution

How is the astronomer to study the evolution of stars when the changes we would like to understand require millions or even billions of years to take place? In our brief lifetimes, we have only a single snapshot of the Galaxy. Our problem is similar to that of a hypothetical group of alien scientists who visit the Earth to study the growth and the life cycle of humans but are allowed to stay for only a few seconds. Within those few seconds they might see one or two deaths, but nothing else would change; even the birth of a baby requires longer than a few seconds. So what do they do?

One approach would be to tabulate the characteristics of a large number of humans. The alien scientists would see, for example, humans of many different sizes, from about 40 cm up to about 2 m in length. Some of the variation may reflect differences from one individual to another, but careful examination would probably indicate that there is also an evolutionary process. Small humans become large humans. By noting the numbers of humans in each size range, the aliens could calculate the rate of growth, and the fraction of a human lifetime required to grow to full size, even though they never actually measured any change in an individual during their brief interval of observation.

Astronomers use a similar technique to study stellar evolution. We see a snapshot of the stellar population, consisting of individual stars at all stages of their life cycles. One of the best ways to represent this snapshot is by plotting the properties of stars on an HR diagram, which relates stellar temperatures to their luminosities. Some of the differences between stars result from, as we have seen, their different masses. Others represent changes as a star ages. Even though we do not see dramatic changes in any one star (except for a few violent deaths, as discussed in Chapter 12), we can use the distribution of stars in the HR diagram, together with stellar models, to "trace out" the evolution of a star.

As a star uses up its nuclear fuel, both its luminosity and its temperature change. Thus, the position where it is plotted in the HR diagram changes as the star ages. Astronomers often speak of a star *moving* on the diagram or of its evolution *tracing out a path* in the diagram. Of course, the star does not really move at all in a spatial sense. This is just a short way of saying that its temperature and its luminosity change as it evolves.

11.2
STAR FORMATION

Molecular Clouds: Stellar Nurseries

If we want to find the very youngest stars—stars still in the process of formation—we must look in places where there is plenty of the raw material required to make stars. Stars are made of gas, and so we must look into dense clouds of gas. Observations confirm that most star formation takes place in clouds of gas that are called, because of their size and composition, **giant molecular clouds** (Figure 11.1). The giant molecular clouds are the most massive objects in the Galaxy. A typical cloud has a mass equal to 100 to 1 million times the mass of the Sun, and the diameter of a typical cloud is 50 to 200 LY.

Figure 11.1   The nearest molecular cloud is in Orion and is a region of active star formation. The bright region in this photograph is visible to the unaided eye as the "star" in the "sword" of Orion. Hot stars ionize nearby gas to produce the glow seen in this picture. The dark regions are cool portions of the molecular cloud. *(National Optical Astronomy Observatories)*

ESSAY   Interstellar Matter

By ordinary terrestrial standards the space between the stars is empty. In no laboratory on Earth can so complete a vacuum be produced. Indeed, there is more gas in the Earth's atmosphere in a hypothetical vertical tube with a cross-section of 1 sq m than would be encountered by extending that same tube from the top of the atmosphere all the way to the most distant object yet observed, which is more than 10 billion LY away. Yet it is possible to detect and study tenuous material in interstellar space.

The matter that lies between the stars, which astronomers refer to as the interstellar medium, is made of gas and dust. Hydrogen makes up about three-quarters of the interstellar gas, and helium accounts for all but 1 to 3 percent of the remainder. Most of the gas is cold and nonluminous. Near very hot stars, however, hydrogen is ionized by stellar ultraviolet radiation. When the free electrons recombine with hydrogen nuclei, a series of emission lines is produced. The strongest one lies in the red region of the spectrum. Clouds of hydrogen thus appear red on photographs of gas-rich regions near hot O and B stars (Figure 11A).

Interstellar dust consists of solid particles composed primarily of silicates or carbon. The dust can be detected because it blocks the light of stars that lie behind it (Figure 11B). Dust grains can also reflect starlight (Figure 11C). Interstellar clouds contain a mixture of dust and gas (see Figure 11A). The interstellar medium is dynamic. Clouds form, collide, coalesce, and fragment to form stars.

The density of gas and dust surrounding the Sun and lying between it and the nearest stars is very low. The volume of space filled by this low-density material is, however, so large that the total mass of gas and

Figure 11A   Clouds of luminous gas and opaque dust, such as the nebulosity NGC 3603, are found between the stars.   *(Copyright Anglo-Australian Telescope Board)*

dust between the stars is equal to about 10 percent of the mass of the stars. For the Galaxy as a whole, the interstellar matter has a mass equal to several billion times the mass of the Sun.

The interior of a molecular cloud is cold. Typical temperatures are about 10 K, so most atoms are bound into molecules. Molecular hydrogen (H_2) is the most abundant element, but other molecules, including carbon monoxide (CO), cyanogen (CN), and ammonia (NH_3), are also present. Relatively heavy molecules, such as HC_9N, are found in some

Figure 11C The Trifid Nebula (M20) in the constellation Sagittarius. In the reddish region, the hydrogen is ionized by nearby hot stars. The red light is produced by an emission line of hydrogen. The blue light is light from a hot blue star reflected toward the Earth by interstellar dust. The Trifid Nebula is about 30 LY in diameter and is about 3000 LY distant from the Sun. *(T. Boroson and N.A. Sharp/National Optical Astronomy Observatories)*

Figure 11B Old stars in our Galaxy are yellowish in color and form the brightest part of the Milky Way. Superimposed on the background stars in this picture is a dark cloud (Barnard 86), which is visible only because it blocks out the light from the stars beyond. Also in this picture, and possibly associated with the dark cloud, is a small cluster of young blue stars (NGC 6520). *(Copyright Anglo-Australian Telescope Board, 1980)*

cold clouds, and ethyl alcohol, C_2H_5OH, is plentiful—up to one molecule for every cubic meter in space. The largest of the cold clouds have enough ethyl alcohol to make 10^{28} fifths of 100-proof liquor. Wives or husbands of future interstellar astronauts, however, need not fear that their spouses will become interstellar alcoholics. Even if a spaceship

Figure 11.2 A typical interstellar grain is thought to consist of a core of rocky material (silicates), graphite, or possibly iron surrounded by a mantle of ices.

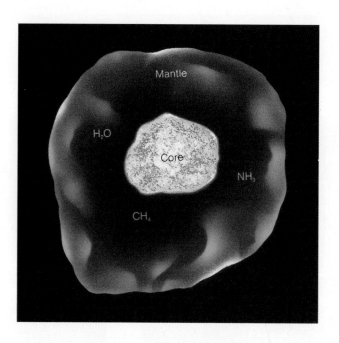

Where are stars born?

were equipped with a giant funnel 1 km across and could scoop through such a cloud at the speed of light, it would take about 1000 years to gather up enough alcohol for one standard martini!

Some molecular clouds also contain cyanoacetylene (HC_3N) and acetaldehyde (CH_3CHO), generally regarded as starting points for the formation of the amino acids necessary for living organisms. The presence of these organic molecules does not imply the existence of life in space. Conversely, as we learn more about the processes by which these complex molecules are produced, we gain an increased understanding of similar processes that must have preceded the beginnings of life on the primitive Earth billions of years ago.

Molecular clouds all contain dust grains, small solid particles that consist largely of *silicates,* including common minerals frequently found in terrestrial igneous rocks and in meteorites. Other grains appear to be nearly pure *carbon* (graphite), and still others are thought to include complex *hydrocarbon* molecules.

The cores of grains are probably formed in shells of cooling gas ejected by red giants and other stars that are nearing the end of their evolution (see Chapter 12). As the gas streams away from these stars, it cools, and solids condense, just as they did in the cooling solar nebula (see Chapter 7). These grain cores may then subsequently be incorporated into an interstellar cloud, where they can grow by accreting other atoms. The most widely accepted model pictures the grains as consisting of rocky cores with icy mantles (Figure 11.2). The most common ices are water (H_2O), methane (CH_4), and ammonia (NH_3)—the same ices that played an important role in the formation of the solar system.

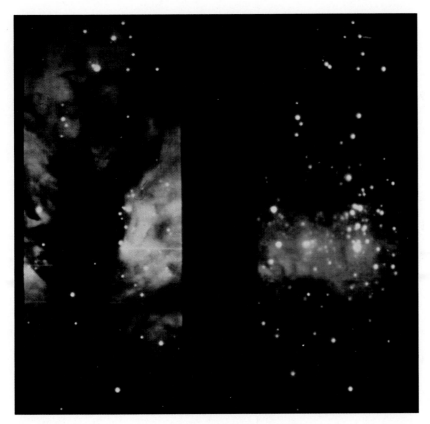

Figure 11.3   Images of the heavily obscured region in the Orion molecular cloud. This region, known as NGC 2024, is located within our Galaxy at a distance of about 1500 LY from Earth. The image on the left was taken in visible light and shows what appears to be an empty sky in the central region of the picture. In fact, this region is not empty at all, as it contains a substantial amount of dust that obscures visible light. The image on the right was made with an infrared detector. Electromagnetic radiation at these long wavelengths can penetrate the obscuring dust easily. Infrared observations, therefore, provide a way to detect the recently formed stars within the cloud. *(I. Gatley and R. Probst/National Optical Astronomy Observatories)*

The interiors of molecular clouds cannot be observed with visible light. The dust in these clouds acts like a thick blanket of interstellar smog, which cannot be penetrated by visible radiation (Figure 11.3). It is only with the new techniques of infrared and millimeter radio astronomy that we have been able to measure the conditions inside these clouds and study directly the very early stages of the births of stars.

Birth of a Star

The earliest stages of star formation are still clouded in mystery. There is a factor of almost 10^{20} difference between the density of a molecular cloud and that of the youngest stars that can be observed. So far we have been unable to observe directly what happens within a cloud as material comes together and collapses gravitationally through this range of densities to form a star. We do not know what causes a single large cloud to fragment and form individual stars, nor do we know why multiple star systems and planets form.

Once the first stars form within a molecular cloud, there is strong observational evidence that these stars can trigger the formation of additional stars, which can lead to the formation of still more stars (Figure

Figure 11.4   Schematic diagram showing how star formation can move progressively through a molecular cloud. The oldest group of stars lies to the left of the diagram and has expanded because of the motions of individual stars. Eventually the stars in the group will disperse and will no longer be recognizable as a cluster. The youngest group of stars lies to the right, next to the molecular cloud. This group of stars is only 1 to 2 million years old. The pressure of the hot, ionized gas surrounding these stars compresses the material in the nearby edge of the molecular cloud, and initiates the gravitational collapse that will lead to the formation of more stars.

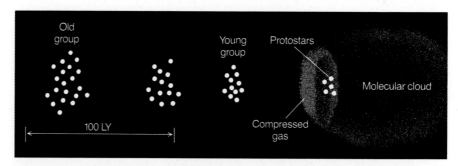

11.4). The basic idea is as follows: When a massive star is formed, it emits copious amounts of ultraviolet radiation, which heats the surrounding gas in the molecular cloud. This heating increases the pressure in the gas and causes it to expand. When massive stars exhaust their supply of fuel, they explode, and the energy of the explosion also heats the gas. The hot gases then explode into the surrounding cold cloud, compressing the material in it until the cold gas is at the same pressure as the expanding hot gas. This compression increases the gas density in the molecular cloud by about a factor of 100. At densities this high, stars can begin to form in the compressed gas.

The Orion Molecular Cloud

The best-studied of the stellar nurseries is the Orion region. A luminous cloud of dust and gas can be seen with binoculars in the middle of the sword in the constellation of Orion. Associated with the luminous material is a much larger molecular cloud, which is invisible in the optical region of the spectrum. Near the center of the optically bright region of the

Figure 11.5   The Trapezium stars. The central "star" of the three forming the sword of Orion is, in fact, a group of four stars known as the Trapezium cluster. The Trapezium stars appear near the center of this picture in the midst of the yellowish nebulosity. These stars are easily visible with binoculars and are the brightest members of a substantial cluster hidden by the dust in the nebula. These stars are at a distance of about 1500 LY from the Sun.
(Copyright Anglo-Australian Telescope Board, 1981)

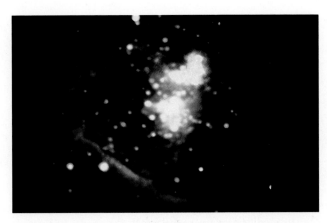

Figure 11.6 An infrared picture of the region of the Trapezium stars. Because infrared radiation penetrates the dust, these observations reveal the cluster of young stars within the molecular cloud. The bright portions of this picture cover about the same area as the yellowish region of Figure 11.5. The red linear feature in the lower left of this infrared picture coincides with the bright region that separates the yellowish and orange regions in Figure 11.5. *(National Optical Astronomy Observatories)*

nebula is the Trapezium cluster of very luminous O-type stars (Figure 11.5). Observations show that these stars are only about 2 million years old. An infrared picture of this same region shows hundreds of stars, invisible on photographs taken at optical wavelengths (Figure 11.6).

The long dimension of the Orion molecular cloud stretches over a distance of about 100 LY and is about 50 times larger than the region that can be seen optically. The total quantity of molecular gas is about 200,000 times the mass of the Sun. The Orion region provides a good example of how star formation can move progressively through a cloud. Star formation began about 12 million years ago at one edge of this molecular cloud and has slowly moved through it, leaving behind groups of newly formed stars. The oldest of these groups is the one farthest from the site of current star formation, and the ages become progressively younger the closer the groups are to the Trapezium cluster. Stars that are even younger are embedded in the molecular cloud and can be observed only in the infrared. The oldest groups of stars can be easily observed because they are no longer shrouded in dust and gas. In these older regions, any material not incorporated into stars has been heated either by stellar radiation or by supernova explosions and blown away into interstellar space (Figure 11.7).

Figure 11.7 The Rosette Nebula. The cluster of blue stars at the center of this picture formed less than 1 million years ago. The gas and dust have been driven away from the bright stars by radiation and intense stellar winds. *(Anglo-Australian Telescope Board)*

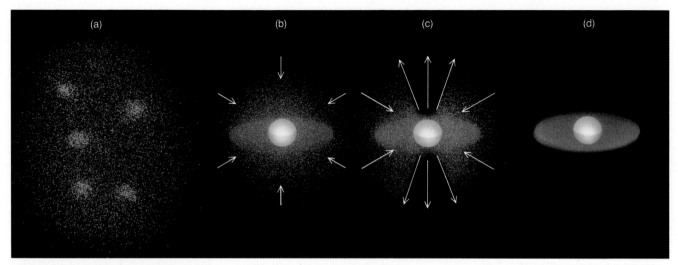

Figure 11.8 The formation of a star. (a) Dense cores form within the molecular cloud. (b) A protostar with a surrounding disk of material forms at the center of a dense core, accumulating additional material from the molecular cloud through gravitational attraction. (c) A stellar wind breaks out along the two poles of the star. (d) Eventually this wind sweeps away the cloud material and halts the accumulation of additional material, and a newly formed star surrounded by a disk becomes observable. *(Based on drawings by F. Shu, F. Adams, and S. Lizano)*

How does the structure of a protostar and the gas cloud that surrounds it change with time?

Gravitational Collapse

The story of stellar evolution is the story of the competition between two forces—*gravity* and *pressure.* As described in Chapter 9, the force of gravity tries to make a star collapse. Internal pressure produced by the motions of the gas atoms tries to force the star to expand. When these two forces are in balance, the star is stable. Major changes in the structure of a star occur when one or the other of these two forces gains the upper hand.

During the period when a condensation of matter in a molecular cloud is contracting to become a true star, we call the object a **protostar.** Because we are unable to observe directly these earliest stages of star formation, we must rely on calculations to tell us how stars are born. In a theoretical study of stellar evolution, we compute a series of *models* for a star, each successive model representing a later point in time. Given one model, we can calculate how the star should change (in the case of the young protostars under discussion, because of gravitational contraction) and hence what the star will be like at a slightly later time. At each step we determine the luminosity and radius of the protostar; from these we determine its surface temperature.

Let us follow the evolution of a stellar condensation within a molecular cloud (Figure 11.8). Because a molecular cloud is more massive than a typical star by a factor of 100,000 or more, many stars must form within each cloud. The first step in the process of creating stars is the formation within the cloud—through a process that we do not yet understand—of cores of material that have a somewhat higher density than the surrounding cloud. These cores then attract additional matter because of the gravitational force that they exert on the cloud material that surrounds them. Eventually, the gravitational force becomes strong enough to overwhelm the pressure exerted by the cold material that forms the dense cores. The material undergoes a rapid collapse, the density of the core increases greatly, and a protostar is formed. Usually

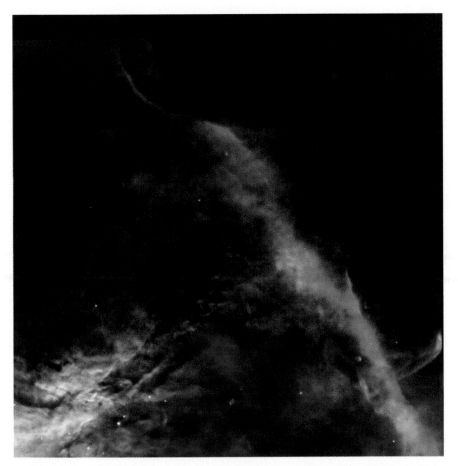

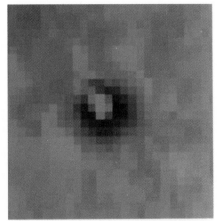

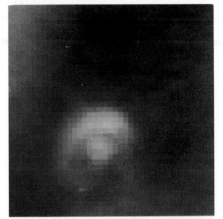

Figure 11.9   An image of a portion of the Orion nebula taken with the Hubble Space Telescope. The image shows a region about 1.3 LY across. Two little doughnut-shaped objects are thought to be protoplanetary disks around newly formed solar-like stars. One of the disks appears near the center of the image and about one-sixth of the way up from the lower edge. The other is just inside the orange rim of the glowing gas about one-quarter of the way up from the lower edge and one-fifth of the way in from the right.   *(C.R. O'Dell/NASA)*

many other dense cores in the same region also form stars nearly simultaneously, thereby producing a cluster of stars.

Calculations predict that a protostar should be surrounded by a disk of material. The protostar and the disk are embedded in an envelope of dust and gas, which is still falling onto the protostar (see Figure 11.8). This envelope of material blocks optical radiation. Therefore, protostars in this phase of evolution are observable only in the infrared.

At some point, for reasons that we do not yet understand, the protostar develops a *stellar wind*, which consists mainly of protons streaming away from the star at velocities of about 200 km/s. This wind eventually sweeps away all but the dense material in the disk, and the protostar becomes visible (see Figure 11.8).

Observations show that most protostars do have disks around them (Figures 11.9 and 11.10), just as theory predicts. The mass contained within a typical disk is in the range 1 to 10 percent of the mass of the

Figure 11.10   HST pictures of three protoplanetary disks in the Orion Nebula. Each picture is about 12 light days across, and each picture element (tiny square) is 50 AU across. The red color shows the central star. The disk with a tail is near a hot star (not shown) that heats the matter in the disk and drives it off into space.   *(C.R. O'Dell/NASA)*

Sun. Observations also show that there are strong stellar winds that flow outward from the polar regions only. These *bipolar flows* are apparently confined to the polar regions by the disk, which surrounds the stellar equator (Figure 11.11).

Formation of Planets (?)

Why does a disk form around a protostar? Rotation is probably the most important factor. Suppose that the core that is going to collapse and form a protostar is rotating slowly. As the collapse proceeds, the material forming the protostar must rotate ever faster. The technical reason for this is that angular momentum is conserved. Figure skating is the most familiar example of conservation of angular momentum. A figure skater can increase her speed of rotation by holding her arms tightly against her body; she can slow her rotation by extending her arms.

In a collapsing gas cloud, rotating particles are subject to an outward force that resists the efforts of gravity to draw those particles toward the center of the newly forming star. The material in the equator of the protostar is rotating most rapidly and resists most strongly the overall gravitational collapse. As it is drawn inward, it rotates even more rapidly and resists further collapse even more strongly. The result is the formation of a rotating disk of material that surrounds the protostar.

This description should sound very much like what happened when the Sun and planets formed (Section 7.5). Do the disks around protostars also form planets? Observations suggest that most protostars with masses similar to the mass of the Sun have disks with masses greater than 1 percent of the mass of the Sun. Therefore, most protostars have more than enough material surrounding them to build a planetary system like the one that surrounds our own Sun. The disks can be detected because they contain dust, which is heated by radiation from the protostar. The dust reradiates this energy in the infrared.

Infrared observations have also been successful in detecting disks around a few nearby main-sequence stars (Figure 11.12). If allowance is made for the difficulty in detecting the faint radiation from the disks, then it is possible that most main-sequence stars will have disks with masses of 10^{-7} times the mass of the Sun, or about 10 percent the mass of the Earth.

Because the disks around protostars start out with typically 100,000 times more mass than that which surrounds main-sequence stars, it is natural to ask what happened to it. Specifically, was the mass missing from the disks around main-sequence stars used to build planets? Unfortunately, we cannot observe planets directly. When the dust surrounding a protostar is distributed in a disk, each individual dust particle is heated by the protostar and radiates its heat in the infrared. We detect the total radiation from all of the individual particles. If those particles then accrete and form a few planets, we detect only the radiation from the planetary surfaces. The overwhelming majority of the particles are hidden in the interior of the planets where we cannot see them. For this reason, the total radiation from a few planets is much less than the

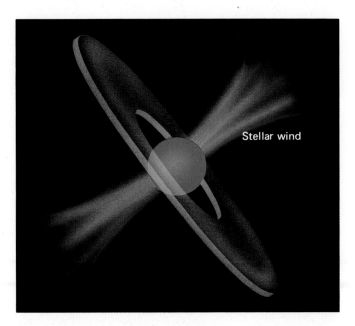

Stellar wind

Figure 11.11 Diagram showing how a disk of material can force a stellar wind to flow outward only from the two poles of a protostar.

radiation emitted from all of the individual particles that combined to form those planets.

There is indirect evidence, however, that planets do indeed form. The disks surrounding a few main-sequence stars have been thoroughly studied, and the evidence is that the disk particles orbit the stars at large distances. That is, the disk is actually shaped like a donut, with very few dust particles in the hole immediately surrounding the central star. One reason for thinking that there is a hole in the disk is that we can actually see the central star. If dust particles extended all the way to the star's surface, they would block the light from the star, making it invisible. The radius of the central hole is typically about 20 AU. The size of our own

What is the evidence that planetary systems are likely to be found around other stars?

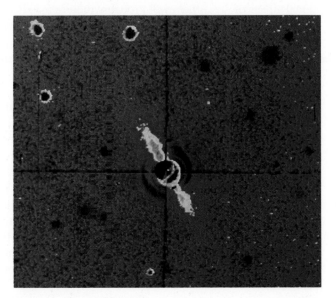

Figure 11.12 Although it is not a planetary system as we understand the term, this optical photograph clearly shows a disk of solid material revolving around the star Beta Pictoris. The light of the star itself has been blocked out in order to show the faint light of the disk more clearly. Evidence of this material was first picked up by IRAS (Infrared Astronomical Satellite), which discovered excess infrared radiation from this star.   *(NASA/JPL)*

ESSAY   The Search for Other Planetary Systems

The discovery that other stars are suns much like our own has naturally led to the question of whether these suns might be accompanied by planets. Are there perhaps millions of Jupiters or Neptunes or even Earths out there in the Galaxy? And if such planets were present, could we detect them?

If detection implies direct imaging, then no planet—even one as large as Jupiter—could be photographed with current telescopes. It is not just that the planets would be faint at stellar distances. The main problem is that the weak reflected light of a planet would be swamped by the relative brilliance of its star. Even the Hubble Space Telescope cannot photograph a planet in the presence of its parent star. Therefore, we must turn to indirect means if we wish to answer our question concerning the existence of other planetary systems.

Contemporary technology has brought the discovery of other planets within our grasp. The trick is to look for the gravitational effect of the planet upon its star. As a planet orbits, it pulls the star first one way and then the other. The star, in effect, executes a miniature orbit of its own, with the same period of revolution as the planet. This motion of the star can be detected using one of two different techniques.

The first technique is to search for the tiny changes in radial velocity of the star as it moves around its orbit. For example, the radial velocity of the Sun changes because of the gravitational effects of Jupiter by about 10 m/s with a period of 12 years. A Jupiter-mass planet at a distance of 1 AU would induce velocity changes of 50 m/s with a period of just one year. Effects of this magnitude in other stars can be detected by very high resolution spectroscopy, and several such surveys are currently underway.

The second technique depends on the very accurate measurement of stellar positions in order to detect the "wobble" of a star induced by one or more orbiting planets. At the current accuracy of these techniques, the wobble caused by a Jupiter-mass planet could be detected for any nearby star (within 20 LY), but if the planet were much smaller than Jupiter, the motion of the star would not be measurable. A specially built telescope in space, however, could easily detect a planet with only 1/10 Jupiter's mass around any of several hundred nearby stars, if such planets are present.

The two techniques are complementary, in that the spectroscopic approach works best for a compact system with the planet close to the star, whereas the positional measurements are most sensitive if the planet is farther out. Both approaches work best for more massive planets orbiting less massive stars. If there are planets as large as Jupiter accompanying low-mass stars in the solar neighborhood, we should be able to find them. However, if the largest planets are no bigger than Uranus or Neptune, and if they are limited to approximately solar-type stars, finding them will probably require a dedicated telescope in space. Neither technique can find planets as small as the Earth.

solar system is about 40 AU and thus is rather similar. The most likely explanation for the fact that the disk does not reach all the way to the surface of the star is that the particles that originally filled the central hole accreted to form planets.

Observations suggest that it takes only about 10 million years to clear away the dust from the central regions of the disks that surround protostars with masses similar to the Sun. If this clearing occurs because the dust is being incorporated into planets, then the time to form planets is very short relative to the total stellar lifetimes. Remember that the Sun is about 4.5 billion years old and will survive for another 5 billion years.

To date, the results of these searches are mixed. Many decades of positional measurements with small refracting telescopes yielded claims in the 1960s of planetary detection, but all of these claims have since been found to be erroneous. More recent applications of this technique with higher-precision instruments have turned up nothing. The spectroscopic approach, however, has had better luck. In 1988, astronomers discovered an object with a mass that may be as small as 11 times the mass of Jupiter in an 84-day orbit around the nearby star HD 114762. Although this object is probably a brown dwarf, its mass is certainly getting close to that of a true planet. Meanwhile, a Canadian monitoring project carried out at Mauna Kea in Hawaii has found some evidence of changing radial velocity for about 25 percent of the 20 stars being studied, and several of these are tentatively associated with "planets" in the Jupiter-Saturn size range. As this project continues, it should soon be possible to determine with confidence whether the observed effects are really due to orbiting planets.

The one definite detection of extrasolar system planets is the result of radio observations of a pulsar. This discovery is described in the next chapter.

The search for planets around other stars seems to be a project whose time has come. If there are numerous large planets out there, it is within the capability of modern astronomy to find them. If suc-cessful, the search will allow astronomers for the first time to consider the formation of our own planetary system within a broader cosmic context. If the search is carried out with negative results, however, the implication will be almost equally interesting. An absence of planets accompanying single, solar-type stars would challenge our current understanding of the process of star and planet formation and would send astronomers "back to the drawing boards" to develop better theories. A negative result would also raise the profound question of the uniqueness of our own solar system and hence of life in the universe.

The search for evidence of extraterrestrial life is far more challenging, but in October 1992, NASA initiated a major program to do just that. The idea is to look for radio signals from an advanced civilization. The rationale is that with the Arecibo radio telescope we could detect a signal beamed directly at us from a similar radio telescope that was as much as 14,000 LY distant. The best strategy for detecting a signal, however, is to hope that some other civilization is already trying to contact us—one close enough to have detected our own radio and television broadcasts, which are already making their way through interstellar space. NASA will use the Arecibo telescope to make a targeted high sensitivity search of 1000 nearby stars and to carry out a low sensitivity survey of the entire sky. The experiment is scheduled to take about 10 years.

11.3
FROM PROTOSTAR TO RED GIANT

Evolution to the Main Sequence

We have now described the evolution of the material that surrounds the protostar. What happens to the protostar itself? In the beginning, this young stellar object derives its energy not from nuclear reactions but from its gravitational contraction—by the sort of process proposed for the Sun by Helmholtz and Kelvin in the last century (Section 9.3).

Initially, the protostar remains fairly cool, its radius is very large, and its density is very low. It is transparent to infrared radiation, and the heat generated by gravitational contraction can be radiated away freely into space. Because heat builds up slowly inside the protostar, the gas pressure remains low and the outer layers fall almost unhindered toward the center; thus, the protostar undergoes very rapid collapse, which stops only when the protostar becomes dense and opaque enough to trap the heat released by gravitational contraction.

When the central temperature exceeds about 10 million degrees, it is high enough to fuse hydrogen into helium, and we say that the star has *reached the main sequence.* It is now approximately in equilibrium, and its rate of evolution slows dramatically. Only the slow depletion of hydrogen as it is transformed into helium in the core gradually changes its properties.

Evolution in the HR Diagram

We have now described in words the early evolutionary history of the Sun and other similar stars. Theoretical calculations, however, give us *quantitative* information about how the luminosity, radius, and temperature of protostars change with time. We can thus find where any star (or its embryo) should be represented on the HR diagram. By plotting the temperature and the luminosity of the collapsing protostar as it changes with time, we can trace the track that the star follows on the HR diagram. The evolutionary tracks of newly forming stars are shown in Figure 11.13.

In general, the pre-main-sequence evolution of a star slows down as the star moves along its evolutionary track toward the main sequence. The numbers labeling the points on each evolutionary track in Figure 11.13 are the times, in years, required for the protostars to reach those stages of contraction. The time for the whole evolutionary process is highly mass-dependent. Protostars of mass much higher than the Sun's reach the main sequence in a few thousand to 1 million years; the Sun required several millions of years; tens of millions of years are required for less massive stars to evolve to the main sequence. For all stars, however, we should distinguish three evolutionary time scales:

> How do the luminosity and temperature of a star change as it evolves from protostar to main sequence star to red giant?

1. The initial gravitational collapse from interstellar matter is relatively quick. A condensation that is approximately 1000 AU in diameter collapses to form a protostar in a few *thousand* years.

2. As the protostar collapses, the internal pressure increases until it almost balances the weight of the star. When this approximate balance is achieved, the contraction of the protostar slows down. This phase of slow contraction lasts typically for *millions* of years and ends when the interior of the star becomes hot enough to ignite hydrogen fusion. For the lowest mass stars, this phase of evolution can take as long as 100 million years.

3. When hydrogen fusion begins, the star is on the main sequence. Subsequent evolution is very slow, for a star changes only as nuclear

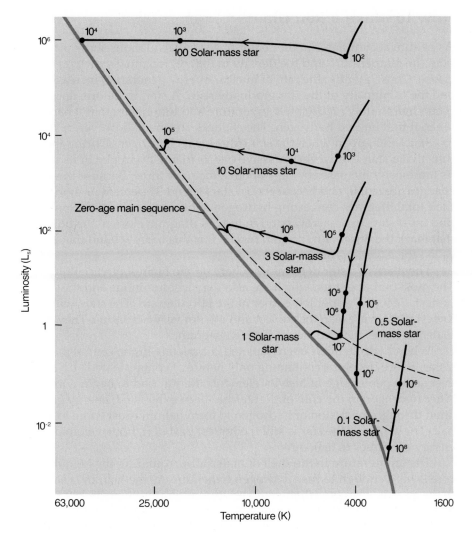

Figure 11.13 Theoretical evolutionary tracks of contracting stars, or protostars, on the HR diagram. When protostars first form, they are very cool but radiate a substantial amount of energy as they contract. Protostars, therefore, first appear on the right-hand side of the HR diagram, with more massive stars having higher luminosities. According to calculations, protostars lying above the dashed line are still surrounded by infalling matter and would be hidden by it. Protostars first become visible at optical wavelengths when their temperatures and luminosities place them below the dashed line. The numbers along each track indicate the number of years required to reach that stage of evolution. For example, it takes 10^7 years for a 1 solar mass star to reach the main sequence, but only 10^5 years for a 10 solar mass star to reach the main sequence. *(Based on calculations by R. Larson)*

reactions alter its chemical composition. For a star of one solar mass, this gradual process requires *billions* of years.

Once a star has reached the main sequence, it derives its energy almost entirely from the thermonuclear conversion of hydrogen to helium. It remains on the main sequence for most of its "life." Because only 0.7 percent of the hydrogen used is converted to energy, the star does not change its mass appreciably. In its central regions, where nuclear reactions occur, the chemical composition gradually changes as hydrogen is depleted and helium accumulates. This change of composition forces the star to change its structure, including its luminosity and size. Eventually, the point that represents it on the HR diagram evolves away from the main sequence.

All evolutionary stages are relatively faster in stars of high mass and slower in those of low mass.

How to Make a Red Giant

As helium accumulates in the center of a star, calculations show that both the temperature and the density in that region must slowly increase. Consequently, the rate of nuclear-energy generation increases, and the luminosity of the star gradually rises. A star, therefore, does not retain indefinitely *exactly* the temperature and luminosity that it had when it first ignited hydrogen. The changes are small, however, relative to what will happen when the star exhausts its hydrogen. During the time that a star, even a very massive one, is fusing hydrogen in its core, its luminosity increases by only about a factor of three. Because these changes are small, and because most stars spend 90 percent or more of their total lifetimes converting hydrogen to helium, we would expect to find most stars in a narrow band in the HR diagram close to the points that mark the onset of hydrogen fusion. It is this narrow band that is called the *main sequence.*

The mass of a star determines where on the main sequence it lies. The most massive stars, which are also the most luminous and the hottest, lie in the upper left corner of the HR diagram. The stars with the lowest masses are also the coolest and are not very luminous. They populate the lower right corner of the HR diagram.

When hydrogen has been depleted completely in the central part of a star, a core develops containing only helium, "contaminated" by whatever small percentage of heavier elements the star had to begin with. After hydrogen in the core of the star has been exhausted, energy generation through the fusion of hydrogen to form helium must come to a halt. The center of the star is still the hottest part of it, however, and heat energy continues to leak out.

The temperature in the shell of material surrounding the helium core is high enough to fuse hydrogen to helium. As the helium is added to the core, it shrinks in diameter, and its temperature increases. The luminosity of the star also increases.

These changes result in a substantial and rather rapid readjustment of the star's entire structure, so that the star leaves the vicinity of the main sequence altogether. About 10 percent of a star's hydrogen must be converted to helium before the star evolves away from the main sequence. The more luminous and massive a star, the sooner this happens. Because the total rate of energy production in a star must be equal to its luminosity, the core hydrogen is used up first in the very luminous stars. The most massive stars spend only a few million years on the main sequence. A star of one solar mass remains there for about 10^{10} years. A small red star of about 0.4 times the mass of the Sun has a main-sequence life of some 2×10^{11} years, a value much greater than the age of the universe. Therefore, such low-mass stars would not have had time to complete their main-sequence phase and go on to the next stage of evolution. The main-sequence lifetimes for stars of various masses are listed in Table 11.1.

As the central helium core contracts, the energy that it releases is absorbed in the surrounding material, thereby heating this portion of the star and forcing it to expand. The surface layers of the star are pushed

TABLE 11.1   Lifetimes of Main-Sequence Stars

Spectral Type	Mass (Mass of Sun = 1)	Lifetime on Main Sequence
O5	40	1 million years
B0	16	10 million years
A0	3.3	500 million years
F0	1.7	2,700 million years
G0	1.1	9,000 million years
K0	0.8	14,000 million years
M0	0.4	200,000 million years

outward as well and reach enormous proportions. The expansion of the surface layers causes them to cool, and the star becomes red. Meanwhile, some of the gravitational energy released from the contracting core heats up the hydrogen surrounding it to ever higher temperatures. In these hot regions, the conversion of hydrogen to helium accelerates, causing most stars to *increase* in total luminosity. Stars with high luminosity and cool surface temperatures are what we call red giants. After leaving the main sequence, the stars move to the upper-right section of the HR diagram, into the region of the red giants.

Figure 11.14 shows the tracks of evolution for stars of several representative masses and with chemical composition similar to that of the

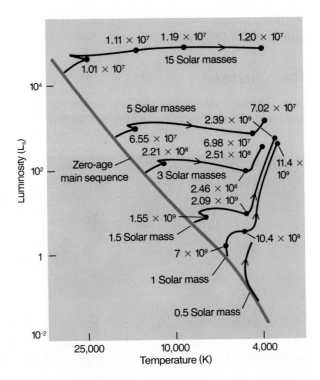

Figure 11.14   Predicted evolution of stars from the main sequence to red giants. See text for explanation. *(Based on calculations by I. Iben)*

Sun. These tracks, calculated from theoretical models of stars, show how luminosity and temperature change as stars evolve away from the main sequence to become red giants. The numbers along the tracks indicate the times, in years, required for the stars to reach those points in their evolution after leaving the main sequence.

11.4
CHECKING OUT THE THEORY

The description of stellar evolution given above is based entirely on calculations of stellar models. No star completes its main-sequence lifetime or its evolution to a red giant quickly enough for us to observe these structural changes as they happen. Fortunately, nature has provided us a way to test the calculations. Instead of observing the evolution of a single star, we can look at a group or cluster of stars.

As we have already seen, observations of stars in molecular clouds show that it is typical for many stars to form in a specific region nearly simultaneously. This group of stars is likely to remain close together, kept from separating by the force of gravity. These stars will have nearly the same age and, because they formed within a single cloud, they will have identical compositions. The stars in a cluster do, however, have different masses.

Suppose we now observe a cluster some time after it formed. If our theory is correct, stars with higher masses evolve more quickly. We can therefore hope to find clusters in which massive stars have already completed their main-sequence phase of evolution and have become giants, while lower-mass stars in the same cluster are still on the main sequence or even undergoing pre-main-sequence gravitational contraction.

Star Clusters

Two distinct types of star clusters permit us to check theoretical calculations of evolution for a wide range of stellar ages. **Globular clusters,** so

Figure 11.15 The globular cluster M5. This cluster contains about 10^6 stars and is about 25,000 LY from the Sun. *(Canada France Hawaii Telescope)*

Figure 11.16 The Pleiades open star cluster. This cluster contains hundreds of stars and is located about 400 LY from the Sun. The brightest six (or for some people seven) stars can be seen with the unaided eye. *(National Optical Astronomy Observatories)*

called because of their appearance (Figure 11.15), contain the oldest stars that we have found thus far. About 100 globular clusters are known in our Galaxy. All are very far from the Sun. One of the most famous globular clusters is called M13, in the constellation of Hercules. Through a good pair of binoculars this cluster resembles a tiny cotton ball. A small telescope reveals its brightest stars. A large telescope shows it as a beautiful, globe-shaped system of stars.

The diameters of typical globular star clusters are 50 LY to more than 100 LY. Globular clusters usually contain hundreds of thousands of member stars. If the Earth revolved not about the Sun but about a star in the densest part of a globular cluster, the nearest neighboring stars, just light *months* away, would appear as points of light brighter than any stars in our sky. Many thousands of stars would be scattered uniformly over the sky, and even on the darkest of nights the brightness of the sky would be comparable to faint moonlight.

Open clusters are often associated with interstellar matter and contain stars that are much younger than the stars in globular clusters. Several open clusters are visible to the unaided eye. Most famous among them is the Pleiades (Figure 11.16), which appears as a group of six stars (some people can see more than six) arranged like a tiny dipper in the constellation of Taurus. A good pair of binoculars shows dozens of stars in the cluster, and a telescope reveals hundreds. (The Pleiades is not the Little Dipper; the latter is part of the constellation of Ursa Minor, which also contains the North Star.) The Hyades is another famous open cluster in Taurus. To the naked eye, the cluster appears as a V-shaped group of faint stars, marking the face of the bull. Telescopes show that the Hyades cluster actually contains more than 200 stars.

Typical open clusters contain several dozen to several hundred member stars, although a few contain more than 1000. Compared to globular clusters, open clusters are small, usually having diameters of

TABLE 11.2 Characteristics of Star Clusters

	Globular Clusters	Open Clusters
Number known in Galaxy	125	>1000
Location in Galaxy	Halo and nuclear bulge	Disk (and spiral arms)
Diameter (LY)	50 to 300	<50
Mass (solar masses)	10^4 to 10^5	10^2 to 10^3
Number of stars	10^4 to 10^5	50 to 10^3
Color of brightest stars	Red	Red or blue
Integrated luminosity of cluster (L_s)	10^4 to 10^6	10^2 to 10^6
Examples	Hercules Cluster (M13)	Hyades, Pleiades

less than 50 LY. The most important characteristics of open and globular clusters are summarized in Table 11.2.

Evolution of Young Star Clusters

What should the HR diagram be like for a cluster whose stars have recently condensed from an interstellar cloud? After a few million years, the most massive stars should have completed their contraction phase and should be on the main sequence, while the less massive ones should be off to the right, still on their way to the main sequence. Figure 11.17

Figure 11.17 The HR diagram of hypothetical cluster M2001 at 3 million years of age.

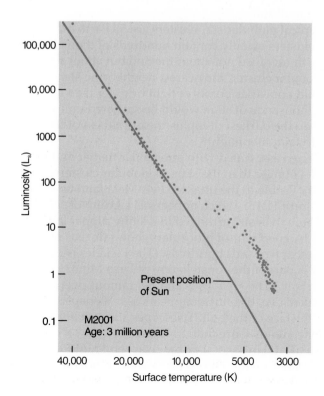

Figure 11.18   The cluster NGC 2264, which is at a distance of 2700 LY from the Sun. This region of newly formed stars is a complex mixture of red hydrogen gas, ionized by hot embedded stars, and dark obscuring dust lanes.   *(Copyright Anglo-Australian Telescope Board, 1981)*

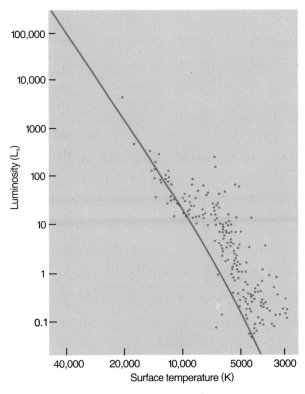

Figure 11.19   The HR diagram of NGC 2264.   *(Data by M. Walker)*

shows the HR diagram calculated by R. Kippenhahn and his associates at Munich for a mythical 3 million year old cluster we will call M2001.

There are real star clusters that fit this description, too. The first to be studied (about 1950) was NGC 2264, a cluster still associated with nebulosity (Figure 11.18). Figure 11.19 shows its HR diagram. Among the several other star clusters in such an early stage is one in the middle of the Orion Nebula (see Figures 11.5 and 11.6).

Older Star Clusters

After a short time—less than 1 million years after reaching the main sequence—the most massive stars use up the hydrogen in their cores and evolve off the main sequence to become red giants. As more time passes, stars of successively lower mass leave the main sequence, making it seem to burn down, like a candle. Meanwhile, even after 30 million years, the least massive stars will not have completed their grav-

How can observations of star clusters be used to test theories of stellar evolution?

itational contraction to the main sequence, and thus these stars still lie above it in the diagram.

Figure 11.20 shows the HR diagram of the real cluster NGC 3293 (Figure 11.21), which we judge to be a little less than 10 million years old. Note the bright red star in the cluster. This massive star is the first one that has exhausted the hydrogen in its core, completed its main sequence evolution, and become a red giant star. The Hyades has a similar HR diagram but with four red giant stars and a main sequence that is truncated at a lower mass, suggesting an age of a few hundred million years.

Note the gap that appears in the HR diagram for NGC 3293 between the stars near the main sequence and the red giants. In the snapshot of stellar evolution represented by the HR diagram, a gap does not necessarily represent a locus of temperatures and luminosities that stars avoid. In this particular case, it just represents a domain of temperature and luminosity through which a star moves very quickly as it evolves. We see a gap because at this particular moment we have not caught a star in the process of scurrying across this part of the diagram. There are theoretical reasons to expect stars of relatively high mass to become un-

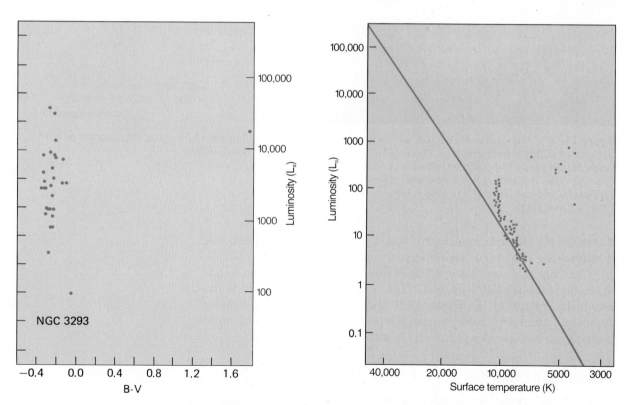

Figure 11.20 *Left,* the HR diagram of NGC 3293. Note the one red giant at the far right of the diagram. Only the brightest stars were observed, so only a portion of the main sequence is shown. *Right,* a more complete HR diagram for M41, a cluster that is only slightly older than NGC 3293. Note that M41 has several red giants.

Figure 11.21 The open cluster of stars NGC 3293. All of the stars in this cluster formed at about the same time. The most massive stars, however, exhaust their nuclear fuel more rapidly and hence evolve more quickly than the stars of low mass. As stars evolve, they become redder. The bright orange star in NGC 3293 is the member of the cluster that has evolved most rapidly. *(Copyright Anglo-Australian Telescope Board, 1977)*

stable as they leave the main sequence and then to readjust themselves to the more stable red giant configuration very rapidly. The absence of stars in the gap thus suggests that stars of high mass evolve quickly from the main sequence to the red giant domain.

At 4 billion years, stars only a few times as luminous as the Sun begin to leave the main sequence. The total main-sequence lifetime for the Sun is expected to be about 10^{10} years. Still older than this are the globular clusters, whose ages are calculated to be between 13 and 15 billion years. A picture of the globular cluster 47 Tucanae is shown in Figure 11.22, and the HR diagram for this cluster is illustrated in Figure 11.23. A few open clusters (for example, NGC 188 and M67) are older than the Sun, but they do not approach globular clusters in age.

Figure 11.22 The globular cluster 47 Tucanae (NGC 104) is one of the nearest globular clusters and is at a distance of 16,000 LY from the Sun. This group of old stars has a diameter of about 200 LY. *(National Optical Astronomy Observatories)*

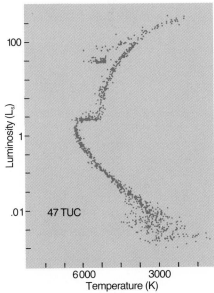

Figure 11.23 HR diagram of the globular 47 Tucanae. *(Data by J. Hesser and collaborators)*

Note that globular clusters (and, presumably, other very old systems) have main sequences that terminate at a luminosity of about three times the luminosity of the Sun. Star formation in these systems evidently ceased billions of years ago. Open clusters, however, are often located in regions of interstellar matter, where star formation can still take place. Indeed, we find open clusters of all ages from less than 1 million to several billion years.

Nucleosynthesis of Heavy Elements

When stars become red giants, a new source of energy becomes available. When the temperature in the central core of a red giant reaches 10^8 K, fusion of three helium atoms to form a single carbon nucleus can take place. Except in the most massive stars, once this process begins, the entire core is reignited in a flash.

Calculations indicate that when this *helium flash* occurs, the star's surface temperature increases and its luminosity decreases. The point representing the star on the HR diagram takes on a new position to the left of and somewhat below its place as a red giant. At this time, the core of the star is stable, fusing helium to form carbon. Surrounding this helium-fusing core is a shell in which hydrogen is fusing to form helium.

In the stellar core, a newly formed carbon nucleus is often joined by another helium nucleus to produce a nucleus of oxygen. Over time, the helium in the core is converted into carbon and oxygen. During later stages of evolution of stars with masses similar to that of the Sun, this carbon-oxygen core will be transformed into a white dwarf (Chapter 12).

After some time, the helium in the core of the star will be exhausted, and energy production by fusion of helium will be shut off. The situation is analogous to that of a main-sequence star when its central hydrogen is used up. Energy is still being generated outside the core, however. In the region immediately surrounding the carbon-oxygen core, the star is hot enough to fuse helium. Farther out in the star is another shell where hydrogen is fusing to form helium. The star now moves back to the red giant domain on the HR diagram. In stars with masses similar to that of the Sun, the formation of a carbon-oxygen core marks the end of the generation of nuclear energy at the center of the star.

In massive stars, the weight of the outer layers of the star is sufficient to compress the carbon-oxygen core until it becomes hot enough to ignite the carbon. Nuclear reactions can then form neon, still more oxygen, and finally silicon. After each of the possible sources of nuclear fuel is exhausted, the core contracts until it reaches a temperature high enough to lead to the fusion of still heavier nuclei. Each cycle lasts for a shorter period of time than the one that preceded it, and each releases energy. When the temperature in the core finally becomes hot enough to fuse silicon to form iron, energy generation must cease. Up to this point, each fusion reaction has *released* energy. Iron nuclei are, however, so tightly bound that they *require* energy to fuse with any other atomic nucleus.

Astronomers now believe that virtually all chemical elements of weight up to that of iron are built up by this **nucleosynthesis** in the cen-

Where are the elements other than hydrogen and helium created?

ters of the more massive red giant stars. Most elements heavier than iron originate in nuclear reactions that occur in the outbursts of supernovae, which are described in the next chapter. All of the carbon, nitrogen, oxygen, iron—indeed all of the elements other than hydrogen and helium—found on Earth were manufactured deep inside stars. But ultimately, all stars exhaust their supply of energy. At this point some truly remarkable things can happen—as we shall see in the next chapter.

▶ SUMMARY

11.1 It is possible to estimate the lifetime of a star by determining its total mass and then calculating how rapidly its hydrogen must be converted to helium in order to account for its luminosity. The most massive stars live only a few million years. A star like the Sun will spend about 10 billion years on the main sequence.

11.2 Giant molecular clouds have masses as large as 10^6 times the mass of the Sun and have typical diameters of 50 to 200 LY. They are the most massive objects in the Galaxy. At the cold temperatures (10 K) that characterize the interiors of these clouds, such molecules as H_2 and CO are abundant; organic molecules are also present. Most star formation occurs within giant molecular clouds. The nearest stellar nursery is the molecular cloud seen in the constellation of Orion, where star formation began about 12 million years ago and is moving progressively through the cloud. The formation of a star inside a molecular cloud begins with a dense core of material, which accretes matter and collapses because of gravity. The accumulation of material halts when the **protostar** develops a strong stellar wind. It is likely that nearly all protostars are surrounded by a disk containing an amount of mass that may be as large as 10 percent the mass of the Sun—more than enough to form a planetary system.

11.3 The evolution of a star can be described in terms of its changes in temperature and luminosity. These changes can best be followed by plotting them on an HR diagram. The higher the mass of a star, the shorter the time it spends in each stage of evolution. A star spends most of its life on the main sequence. The fusion of hydrogen to form helium changes the interior composition of a star, which in turn results in changes in temperature, luminosity, and radius. As stars age, they evolve away from the main sequence to become red giants.

11.4 Calculations that show what happens as stars age can be checked by measuring the properties of stars in clusters. The members of a given cluster all formed at about the same time and all have the same composition, so that comparison of theory and observations is fairly straightforward. There are two types of star clusters. **Globular clusters** contain hundreds of thousands of old stars. **Open clusters** typically contain hundreds of young to middle-aged stars. Calculated HR diagrams match those observed for these two types of clusters. Because the most massive stars evolve the most rapidly, highly luminous blue stars are on the main sequence in the youngest clusters; stars with the lowest masses lie to the right of the main sequence in the youngest clusters and are still contracting toward it. With passing time, stars of progressively lower mass evolve away from the main sequence. In globular clusters, which typically have ages of 13 to 15 billion years, there are no luminous blue stars at all. Elements heavier than hydrogen and helium have been produced by **nucleosynthesis** in the interiors of stars.

▶ REVIEW QUESTIONS

1. How do we know that the lifetime of an individual star cannot be infinitely long?

2. What is the main factor that determines where a star falls along the main sequence?

3. Describe the characteristics of a typical giant molecular cloud.

4. Is it likely that there are planetary systems around other stars? Describe the evidence.

5. Describe what happens when a star forms. Begin with a dense core of material in a molecular cloud and trace the evolution up to the point at which the newly formed star reaches the main sequence.

6. Describe how pressure and gravity influence the structure of a star. What happens when a star exhausts hydrogen in its core and the generation of energy by nuclear fusion stops?

7. Explain how an HR diagram can be used to determine the age of a cluster of stars.

▶ THOUGHT QUESTIONS

8. Why is star formation more likely to occur in cold molecular clouds than in regions where the temperature of the interstellar medium is several hundred thousand degrees?

9. The HR diagram for the nearby stars that are not in clusters shows very luminous main-sequence stars and also various kinds of red giants and supergiants. Explain these features, and interpret the HR diagram for field stars.

10. In the HR diagrams for some young clusters, stars of both very low and very high luminosity are off to the right of the main sequence, whereas those of intermediate luminosity are on the main sequence. Can you offer an explanation? Sketch an HR diagram for such a cluster.

11. If a star with the mass of the Sun is a member of the cluster NGC 2264, would it be on the main sequence yet? Why?

12. Explain how you could decide whether red giants seen in a star cluster probably had evolved away from the main sequence or were still evolving toward the main sequence.

13. Open clusters often contain some dust and gas, while globular clusters do not. Is this what you would expect given the difference in age of the two types of clusters? Explain.

14. Suppose a star cluster were at such a large distance that it appeared as an unresolved spot of light on telescopic photographs. What would you expect the color of the spot to be if it were the image of the cluster immediately after it formed? How would the color differ after 10^{10} years. Why?

15. Problem 12 in Chapter 10 asked you to construct a kind of HR diagram for humans by plotting height against weight. Use that same diagram to show how people change as they age. In the terms used for stars, show how a person "moves" in the height-weight diagram as he or she evolves from infancy to old age.

▌ PROBLEMS

16. If a giant molecular cloud is 100 LY across and has a mass equivalent to 10,000 Suns, what is its average internal density? Suppose that half of the mass of the cloud is made up of dust grains, each with a mass of 10^{-13} kg; how many such grains are there per cubic centimeter in the cloud?

17. Suppose a star spends 10×10^9 years on the main sequence and uses up 10 percent of its hydrogen. Then it quickly becomes a red giant with a luminosity 100 times as great as the luminosity it had on the main sequence and remains a red giant until it exhausts the rest of its hydrogen. How long a time would it be a red giant? Ignore helium fusion and other nuclear reactions and assume that the star brightens from main sequence to red giant almost instantaneously.

18. From the data in Table 11.2, estimate the average mass of the stars in a typical globular cluster. Is the average mass of stars in an open cluster likely to be larger or smaller than the average mass in a globular cluster?

19. The star Rigel is nearly 100,000 times more luminous than the Sun. Its mass is uncertain but is probably no more than 50 times the mass of the Sun. Estimate the main-sequence lifetime of Rigel. (Assume that the main-sequence lifetime of the Sun is 10 billion years.)

▌ SUGGESTIONS FOR ADDITIONAL READING

Books

Cohen, M. *In Darkness Born: The Story of Star Formation.* Cambridge U. Press, 1988. A fine introduction to the birth of stars, by an astronomer.

Hartmann, W., *et al. Cycles of Fire.* Workman, 1987. Lavishly illustrated introduction to stellar evolution.

Kaler, J. *Stars.* Scientific American Library/W. H. Freeman, 1992. An excellent book on stellar evolution by an astronomer who has written extensively on this subject.

Wynn-Williams, G. *The Fullness of Space: Nebulae, Stardust, and the Interstellar Medium.* Cambridge U. Press, 1992. A thorough exposition by an astronomer.

Verschuur, G. *Interstellar Matters.* Springer Verlag, 1989. Essays on dust and the interstellar medium.

Articles

On Starbirth and the Interstellar Medium

Lada, C. "Deciphering the Mysteries of Stellar Origins" in *Sky & Telescope,* May 1993, p. 18.

Hartley, K. "A New Window on Star Birth" in *Astronomy,* Mar. 1989, p. 32.

Gingerich, O. "Robert Trumpler and the Dustiness of Space" in *Sky & Telescope,* Sept. 1985, p. 213.

Stahler, S. & Comins, N. "The Difficult Births of Sunlike Stars" in *Astronomy,* Sept. 1988, p. 22.

Verschuur, G. "Interstellar Molecules" in *Sky & Telescope,* Apr. 1992, p. 379.

Higgins, D. & Eicher, D. "The Secret World of Dark Nebulae" in *Astronomy,* Sept. 1987, p. 46. A guide for amateur astronomers.

Kanipe, J. "Inside Orion's Stellar Nursery" in *Astronomy,* Aug. 1989, p. 40.

Mumford, G. "The Legacy of E. E. Barnard" in *Sky & Telescope,* July 1987, p. 30.

Shore, L. & S. "The Chaotic Material Between the Stars" in *Astronomy,* June 1988, p. 6.

On the Lives of the Stars

Kaler, J. "Journeys on the H–R Diagram" in *Sky & Telescope,* May 1988, p. 483. A good summary of stellar evolution.

Croswell, K. "Galactic Archeology Tracing the History of Oxygen and Iron" in *Astronomy,* July 1992, p. 28.

Darling, D. "Breezes, Bangs, and Blowouts: Stellar Evolution Through Mass Loss" in *Astronomy,* Sept. 1985, p. 78; Nov. 1985, p. 94.

Fortier, E. "Touring the Stellar Cycle" in *Astronomy,* Mar. 1987, p. 49. Observing objects that show the stages of stellar evolution.

Bennett, G. "The Cosmic Origins of the Elements" in *Astronomy,* Aug. 1988, p. 18.

Kaler, J. "The Coolest Stars" in *Astronomy,* May 1990, p. 20. On red dwarfs and brown dwarfs.

Wyckoff, S. "Red Giants: The Inside Scoop" in *Mercury,* Jan./Feb. 1979, p. 7.

On the Search for Planets Elsewhere

Bruning, D. "Desperately Seeking Jupiters" in *Astronomy,* July 1992, p. 37.

Black D. "Worlds Around Other Stars" in *Scientific American,* Jan. 1991, p. 76.

Croswell, K. "Does Barnard's Star Have Planets?" in *Astronomy,* Mar. 1988, p. 6.

Field, G. "Are There More Than Nine Planets in the Universe?" in *Mercury,* Mar./Apr. 1982, p. 42.

STARS: OLD AGE AND DEATH

A photograph of SN 1987A and the region of the Large Magellanic Cloud where it occurred. The supernova is the bright star just below the center of the picture. The large cloud of gas in the upper left is called the Tarantula Nebula, and is a region of active star formation. *(National Optical Astronomy Observatories)*

Before dawn on February 24, 1987, Ian Shelton, a Canadian astronomer working at an observatory in Chile, pulled a photographic plate from the photo developer. Two nights earlier he had begun a survey of the Large Magellanic Cloud, a small galaxy that is the Milky Way's nearest neighbor in space. Shelton planned to study variable stars in the Cloud, but the photographic plates taken the two previous nights did not have sharp images. It was difficult to guide long exposures with the telescope that Shelton was using, which was more than 50 years old. The plate from the third night was a success. The images were sharp. Shelton examined the Tarantula nebula, a region of bright glowing gas where active star formation is occurring. Nearby, where there should have been only faint stars, he saw a large spot on his plate. Concerned at first that his photograph was flawed, Shelton went outside to look at the Large Magellanic Cloud—and saw that a very bright new object had actually appeared in the sky.

What Shelton had discovered was an exploding star, a **supernova.** *Now known as SN 1987A, because it was the first supernova discovered in 1987, this brilliant newcomer to the southern sky gave astronomers for the first time an opportunity to study in detail the death of a star.*

Sooner or later all stars must die, because eventually every star exhausts its store of nuclear energy. Without an internal energy source, a dying star can only collapse until it attains an enormous density. We know of three possible high-density end states for stars: white dwarfs, neutron stars, *and* black holes. *Which of these three a star becomes depends on only one thing—its mass at the time it collapses.*

As a star contracts toward one of these end points, its structural changes are sometimes accompanied by loss of mass. This loss of mass may occur gently, in the form of a stellar wind, although it is often explosive and capable of destroying most or all of the star. Any major loss of mass will influence the star's evolution, and the final state of the object after collapse may be very different from what we would have expected from the initial mass of the star on the main sequence. We must therefore be careful to distinguish between the initial, main-sequence mass of a star, and the mass of the final collapsed object.

12.1
WHITE DWARFS

Most stars complete their evolution by becoming white dwarfs. White dwarfs (Section 10.6) are the end point of stars that begin their main-sequence evolution with masses up to about eight times the mass of the Sun. When such a star exhausts its nuclear fuel, it shrinks under the pressure of its overlying layers until its internal pressure becomes great enough to support its own weight. Equilibrium is established only when the star has reached an enormous density, typically about 1 million times the density of water.

Structure of White Dwarfs

White dwarf stars are far more dense than any substance with which we are familiar on Earth. Their peculiar structure is a result of certain rules that govern the behavior of electrons. According to these rules, which have been verified by studying the behavior of electrons in the laboratory, no two electrons can be in the same place at the same time doing the same thing. We specify the *place* of an electron by its precise position in space, and specify what it is doing by its momentum and by the way it is spinning.

Imagine the electrons in the interior of a star. The temperature inside a star is so high that the atoms are stripped of virtually all their electrons. If the temperature is high enough and the density is low enough, as it is in normal stars, the electrons are moving rapidly. It is highly unlikely that any two of them will be in the same place moving in exactly the same way at the same time. But what happens when a star exhausts its store of nuclear energy?

How do white dwarfs form?

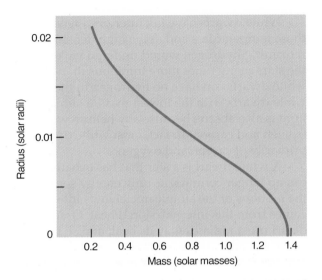

Figure 12.1   Theoretical relation between the masses and the radii of white dwarf stars.

When a star has exhausted its store of energy, the star begins to shrink. In the absence of a source of heat, the stellar interior starts to cool and its pressure is no longer high enough to resist the crushing mass of the outer layers of the star. The collapse halts only when the electrons are crowded so closely together that further collapse would require two or more electrons to violate the rule against occupying the same place and having the same momentum. A gas in such a state is said to be **degenerate.** The electrons in a degenerate gas resist being crushed closer together. Thus they exert a tremendous new kind of pressure that halts the gravitational collapse.

Therefore, white dwarfs are stars whose electrons have become degenerate. Given the mass of a star, we can calculate how small it can be before the degenerate electrons stop its collapse. With this information, we can determine the final radius of the white dwarf. A white dwarf with a mass similar to that of the Sun has a diameter similar to that of the Earth (several thousand kilometers). The theoretically calculated relation between the masses and the radii of white dwarfs is shown in Figure 12.1. Note that the larger the mass of the star, the smaller is its radius.

According to these calculations, which were first carried out by astrophysicist S. Chandrasekhar (Figure 12.2), a white dwarf with a mass of about 1.4 times the mass of the Sun would have a radius of zero! What this result means is that even the pressure exerted by degenerate electrons is not enough to halt the collapse of stars that are more massive than 1.4 solar masses when they reach this stage of evolution. When a star runs out of nuclear fuel, it can become a white dwarf only if its mass is less than 1.4 times the mass of the Sun. This maximum mass of 1.4 solar masses is called the **Chandrasekhar limit.** Stars with masses exceeding this limit must collapse even further. These are the stars that become supernovae.

Figure 12.2   Subrahmanyan Chandrasekhar was born in India in 1910, and was educated at Madras and Cambridge Universities. He has spent most of his career at the University of Chicago, where he has made fundamental contributions in almost every area of astrophysics, including the physical theory of white dwarfs. Chandrasekhar received the Nobel Prize in 1983.

White dwarfs have hot interiors—tens of millions of Kelvins. At those temperatures and at the high densities of white dwarf stars, any remaining hydrogen would undergo violent fusion into helium, making the stars many times more luminous than observed. Consequently, white dwarfs can have no hydrogen in their interiors. For all but the white dwarfs with the lowest mass, a similar argument shows that helium is also absent, because any helium would have been converted to oxygen and carbon. In fact, most white dwarfs are probably composed primarily of carbon and oxygen.

A white dwarf is a star that has exhausted its nuclear fuel. Because it can no longer contract, its only energy source is the heat represented by the motions of the atomic nuclei in its interior. The light of a white dwarf comes from this internal, stored heat. Gradually, however, the white dwarf must radiate away its heat, and so it must slowly fade. After many billions of years, that heat will be gone, the nuclei will cease their motion, and the white dwarf will no longer shine—a final cold body with the mass of a sun and the size of a planet.

12.2
MASS LOSS

What kinds of stars eventually end their lives as white dwarfs? The critical determining factor is **mass.** White dwarfs can be no more massive than the Chandrasekhar limit of 1.4 times the mass of the Sun. Yet, somewhat surprisingly, observations indicate that stars that have masses larger than 1.4 times the mass of the Sun *at the time they are on the main sequence* also complete their evolution by becoming white dwarfs. White dwarfs have been found in young, open clusters—clusters so young that only stars with masses greater than five times the mass of the Sun have had time to exhaust their supply of nuclear energy and complete their evolution to the white dwarf stage. In the Pleiades, for example, stars with masses four to five times the mass of the Sun are still on the main sequence. Yet this cluster also has at least one white dwarf, and the star that has become this dwarf must have had a main-sequence mass that exceeded five solar masses.

If stars that initially have masses as large as five times the mass of the Sun (or, according to theoretical calculations, as large as eight solar masses) are to become white dwarfs, then somehow they must get rid of enough matter so that their total mass at the time nuclear energy generation ceases is less than 1.4 solar masses. Let us look at one of the ways in which stars can lose mass.

Planetary Nebulae

Stars with masses no greater than a few times the mass of the Sun lose mass by forming **planetary nebulae.** Planetary nebulae are shells of gas ejected from, and expanding around, certain extremely hot stars (Figure 12.3). The name is derived from the fact that a few planetary nebulae

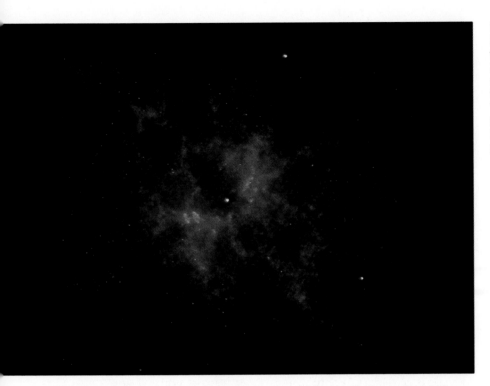

Figure 12.3 The planetary nebula NGC 2440. The star at the center of the nebula has a temperature of 200,000 K and is one of the hottest stars known. This image, which was taken with the Hubble Space Telescope, shows filaments and streamers in the nebula that cannot be seen in pictures taken with groundbased telescopes. *(NASA)*

bear a superficial resemblance, when viewed through a small telescope, to a planet. Actually they are thousands of times larger than the entire solar system and have nothing whatsoever to do with planets.

The most famous planetary nebula is the Ring Nebula in Lyra (Figure 12.4). It is typical of many planetary nebulae in that—although it is actually a hollow shell of material emitting light—it appears as a ring. The explanation is that near the center we are looking through the thin

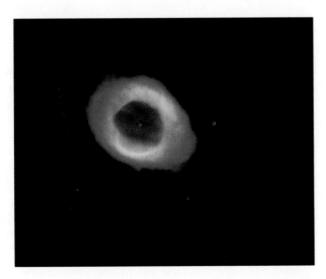

Figure 12.4 The Ring Nebula. The image is coded in such a way that red corresponds to emission from singly ionized nitrogen, green corresponds to emission from doubly ionized oxygen, and blue corresponds to emission from ionized helium.
(Courtesy of Bruce Balick, University of Washington)

What is a planetary nebula?

dimensions of the front and rear parts of the shell, while along its periphery our line of sight encounters a long path through the glowing material. Similarly, a soap bubble often appears to be a thin ring. Altogether, about a thousand planetary nebulae have been catalogued in our own Galaxy. Doubtless there are many distant planetary nebulae that have escaped detection, so there must be some tens of thousands. From spectroscopy we calculate that the shells must have masses of 10 to 20 percent that of the Sun. The shells typically expand about their parent stars at speeds of 20 to 30 km/s.

A typical planetary nebula has a diameter of about 1 LY. If we assume that its gas shell has always expanded at the speed with which it is now enlarging about its parent star, we can determine that its shell has been ejected within the past 50,000 years. After about 100,000 years, the shell is so enlarged that it is too thin and tenuous to be seen. When we take into account the relatively short period of time over which a planetary nebula exists, we find that planetary nebulae are very common and that an appreciable fraction of all stars must sometime evolve through the planetary nebula phase.

The central stars of planetary nebulae have surface temperatures as high as 100,000 K and are among the hottest stars known. Despite their high temperatures, the central stars of planetary nebulae do not have exceedingly high luminosities—some emit little more total energy than does the Sun. They must, therefore, be stars of small size. Thus, a planetary nebula may be the last ejection of matter by a star before it collapses and becomes a white dwarf.

Almost certainly, planetary nebulae originate from red giants. We have already seen (see Section 11.3) that red giants have small, dense cores. As a core contracts, it reaches a density so high that the electrons become degenerate; therefore, it is really like a small white dwarf at the center of a red giant star. At some point in the giant star's evolution, the outer envelope detaches and is ejected as a planetary nebula, leaving behind its core as a white dwarf.

Consequences of Mass Loss

As we have seen, many stars manage to qualify for white-dwarfhood by ejecting their outer layers into space. Very significantly, however, the material they shed is not the same as that from which they were formed. The nuclear reactions by which they shine alter the chemical composition of their constituent gases.

We saw in Chapter 11 that stars convert hydrogen into helium. Moreover, at least some stars in some stages of their evolution are building up helium into carbon and heavier elements. Thus, within stars, lighter elements are gradually being converted into heavier elements. The technical term for this process is *nucleosynthesis*. During the red giant phase of evolution, material from the central regions is dredged up and mixed with the outer layers of the star. When a star nearing the end of its evolution ejects these outer layers into the interstellar medium, that matter is richer in heavy elements than was the material from which the star formed. In other words, a gradual *enrichment* of the heavy-element abundance in interstellar matter is taking place. The heavy-element abundance in stars that are forming now is thus higher than in those that formed in the past. The fact that the oldest known stars (those in globular clusters) are the stars with the lowest known abundance of heavy elements provides evidence to support this scenario.

As we shall see in Chapter 16, we think that the first generation of stars formed in the Galaxy began their evolution with a composition of nearly pure hydrogen and helium and that all of the other elements were synthesized in the hot centers of stars at advanced stages of their evolution. Stars such as the Sun, in whose outer layers heavy elements are observed spectroscopically, thus have to be of the second (or higher) generation; that is, they have been formed from matter that was once part of other stars. It is a grand concept that the planets (and we ourselves!) are composed of atoms that were synthesized in earlier generations of stars—that we are literally made of "stardust."

12.3
EVOLUTION OF MASSIVE STARS

Stars of masses up to about eight times the mass of the Sun probably end their lives as white dwarfs. But stars can have masses as large as 100 times the mass of the Sun. What is their fate?

The initial stages of evolution of a massive star are quite similar to what happens to stars with masses comparable to that of the Sun. On the main sequence, massive stars are, of course, fusing hydrogen to form helium, converting mass to energy in the process. After the hydrogen in the core of the star is exhausted, the core contracts because there is no longer enough pressure produced by the generation of heat to resist the force of gravity. The contraction releases gravitational energy and so heats both the core and its environs. Hydrogen fusion continues in the shell surrounding the core, while helium fusion in the core begins to produce carbon and then oxygen.

Up to this point, the evolution of massive stars is very similar to the evolution of stars like the Sun, but there is one important difference. Massive stars evolve much more rapidly. A star similar to the Sun may spend 10 billion years in its main-sequence phase of evolution fusing hydrogen in its core. A star with a mass that is 60 to 100 times greater than the mass of the Sun can exhaust its supply of hydrogen in a few million years. Massive stars have much more available fuel, but they use it up at such a prodigious rate that their lifetimes are short.

It is only after helium in the core is exhausted that the evolution of a massive star takes a very different course from that of solar-type stars. In a massive star, the weight of the outer layers is sufficient to force the core of carbon and oxygen to contract, heating both the core and its surroundings. Helium fusion begins in a shell surrounding the carbon-

How does the evolution of massive stars differ from that of stars with masses similar to that of the Sun?

Figure 12.5 Just before its final gravitational collapse, a massive star resembles an onion. The iron core is surrounded by layers of silicon, sulfur, oxygen, neon, carbon mixed with some oxygen, helium, and finally hydrogen.

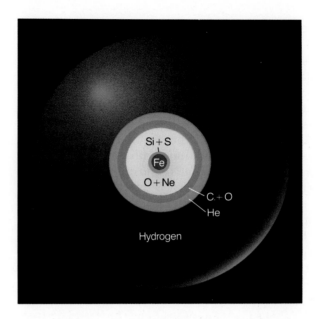

oxygen core, and outside this shell there is a second shell in which hydrogen is fusing to form helium. The carbon-oxygen core continues to shrink until it becomes hot enough to ignite carbon, which can then form neon, still more oxygen, and finally silicon. After each of the possible sources of nuclear fuel is exhausted, the core contracts until it reaches a temperature high enough to lead to the fusion of still heavier nuclei. When the temperature in the core finally becomes hot enough to fuse silicon to form iron, energy generation ceases. Up to this point, each fusion reaction has *produced* energy. Iron nuclei are, however, so tightly bound, that energy must be *absorbed* to fuse iron with any other atomic nucleus.

At this stage of its evolution, a massive star resembles an onion with an iron core and, at progressively larger distances from the center, shells of decreasing temperature in which nuclear reactions involving silicon, oxygen, neon, carbon, helium, and hydrogen are successively taking place (Figure 12.5).

What happens next to a star with an iron core? The computations become very complicated, but we can trace the events in a schematic way. In effect, a massive star builds a white dwarf in its center where no nuclear reactions are taking place. For stars that begin their evolution with masses of at least 10 to 12 times the mass of the Sun, this white dwarf is made of iron. For stars with initial masses in the range of 8 to 10 times the mass of the Sun, the white dwarf that forms the core is made of oxygen, neon, and magnesium because the star never gets hot enough to form elements as heavy as iron. Whatever its composition, the white dwarf embedded in the center of the star is supported against further gravitational collapse by degenerate electrons.

Although no energy is being generated within the white dwarf core of the star, fusion still occurs in shells surrounding the core. As the core

TABLE 12.1   Fate of Stars of Different Masses

Initial Mass (Mass of Sun = 1)	Final Evolutionary State
$M < 0.01$	Planet
$0.01 < M < 0.08$	Brown dwarf
$0.08 < M < 0.25$	Helium white dwarf
$0.25 < M < 8$	Carbon-oxygen white dwarf
$8 < M < 12$	Oxygen-neon-magnesium white dwarf
$12 < M < 40$	Supernova; neutron star
$40 < M$	Supernova; black hole

accretes the ashes of the fusion reactions going on in the surrounding shell, the mass of the core grows. Ultimately, it is pushed over the Chandrasekhar limit of 1.4 times the mass of the Sun. That is, it becomes so massive that the force exerted by degenerate electrons is no longer great enough to resist gravity. The electrons merge with protons inside the nuclei of iron and other atoms to produce neutrons. The removal of electrons removes the main source of support for the core, and it collapses.

This collapse occurs very rapidly. In less than 1 second, the core, which originally was approximately the same diameter as the Earth, collapses to a diameter that is less than 100 km. The speed with which material falls inward reaches one-fourth the speed of light. The collapse halts only when the density of the core equals the density of an atomic nucleus. In effect, the matter in the core has merged to form a single gigantic nucleus. This nuclear material strongly resists further compression, abruptly halting the collapse. The shock of the abrupt jolt generates waves throughout the outer layers of the star and causes the star to blow off those outer layers in a violent supernova explosion.

Table 12.1 summarizes the discussion so far about what happens to stars of different initial masses. The mass limits that correspond to various outcomes may change somewhat as models improve. It is also not certain whether stars with masses in the range of four to eight times the mass of the Sun form white dwarfs. The nuclear reactions in these stars may go one step beyond helium fusion to reactions involving carbon, which may occur explosively and produce a supernova. For stars with initial masses between eight and twelve times the mass of the Sun, theory suggests that nuclear fusion may produce a core of neon, oxygen, and magnesium rather than one of carbon and oxygen. Mass loss, probably during a planetary nebula phase, subsequently exposes this core, which we observe as a white dwarf. If the uncertainties are kept in mind, Table 12.1 provides a useful guide to what theorists think is the ultimate result of stellar evolution.

ESSAY   Evolution of the Sun

With the information we have been assembling on stellar evolution, we can look again at the evolution of our own Sun and its influence on the planets. Figure 12A summarizes our current ideas on the evolution of a star of approximately one solar mass. In its early stages, the star contracts and moves to the left of the HR diagram (toward higher temperatures), reaching the main sequence near the present location of the Sun. The time required for the Sun to contract to the main sequence was probably a few tens of millions of years.

Since it reached the main sequence, the Sun has increased somewhat in luminosity, probably by about 30 to 50 percent. Interestingly, the surface temperature of the Earth has not varied much over this same interval of time. Apparently changes in our planet's atmosphere have compensated approximately for the increasing luminosity of the Sun. During this main-sequence interval of 4 to 5 billion years, the Sun has depleted much of the hydrogen at its very center, but a pure helium core has not yet had time to form. It is not certain how much more time the Sun has before starting to evolve to the red giant stage, but a good guess is that it has lived out about half of its main-sequence life. We can probably look forward to at least another 5 billion years before the Sun's structure undergoes large changes.

All available evidence leads us to expect that some time in the future the Sun will leave the main-sequence stage and evolve into a red giant. Thus, about 5 billion years from now, the Sun will expand to a radius of tens of millions of kilometers—and it will reach nearly to the orbit of Mars. The Earth, well inside the Sun and exposed to temperatures of thousands of degrees, will gradually vaporize. The gases in the greatly distended outer layers of the Sun will be very tenuous, but they should still offer enough resistance to the partially vaporized Earth to slow its orbital motion. Calculations suggest that the Earth will spiral inward toward the very hot interior of the Sun, reaching its end about 10,000 years after being swallowed by the Sun.

Figure 12A   The evolutionary track on the HR diagram for a star with a mass equal to 1.2 times the mass of the Sun. The Sun will pass through the stages of evolution shown in the diagram. In about 5 billion years, the Sun, which is now in its main-sequence phase of evolution, will become a red giant. After the helium flash occurs in its core, the Sun will become hotter and less luminous and will occupy a position on the HR diagram near that of the RR Lyrae variable stars (Section 13.2). After that, the Sun will again become a red giant and will lose mass. The mass loss will expose the Sun's hot inner core, which will appear at the center of a planetary nebula. The Sun will then become a white dwarf and will slowly cool until all of its remaining store of energy is radiated away.

12.4
SUPERNOVAE

Historical Supernovae

Supernovae were discovered long before astronomers realized that these spectacular cataclysms marked the deaths of stars. Five supernovae in our own Galaxy have been observed in the last 1000 years, all before the invention of the telescope. The Chinese reported the temporary appearance of "guest stars" in the years A.D. 1006, 1054, and 1181. The two remaining galactic supernovae occurred in 1572 (Figure 12.6) and 1604.

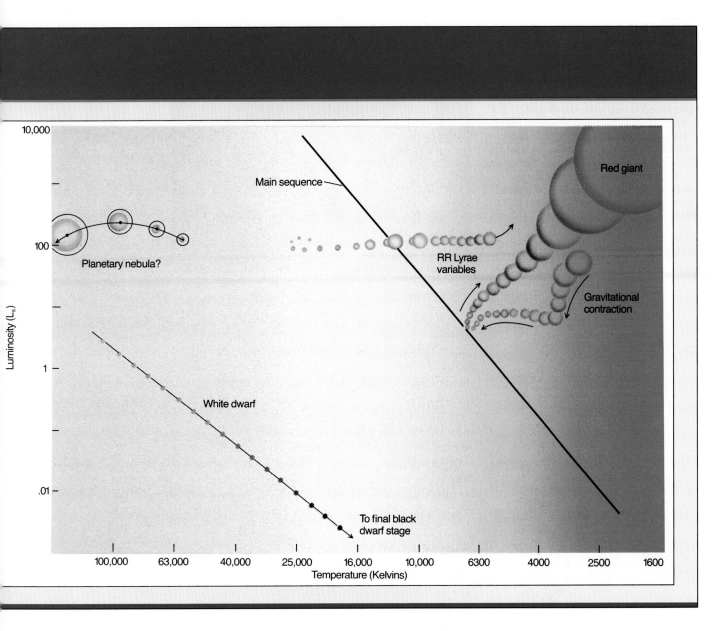

The latter was observed in considerable detail by Johannes Kepler. From historical records, from studies of the remnants of supernova explosions in our own Galaxy, and from analyses of supernovae in other galaxies, we estimate that, on average, one supernova explosion occurs somewhere in the Milky Way Galaxy every 25 to 100 years.

   At their maximum brightness, supernovae are about 10 billion times the luminosity of the Sun. For a brief time, a supernova may outshine the entire galaxy in which it appears. After maximum brightness, the star fades in light until it disappears from telescopic visibility within a few months or years after its outburst. Supernovae eject material at the time of their outbursts. Ejection velocities may reach 10,000 km/s.

Figure 12.6   A map of the radio emission from the expanding shell of gas ejected by the supernova explosion observed in 1572. The different colors correspond to different intensities of radio emission, produced by extremely energetic electrons gyrating in a magnetic field. This supernova is believed to have resulted from the explosion of a white dwarf that accreted enough matter so that its mass exceeded 1.4 solar masses. Apparently nothing remains of the star that exploded.   *(National Radio Astronomy Observatory/AUI)*

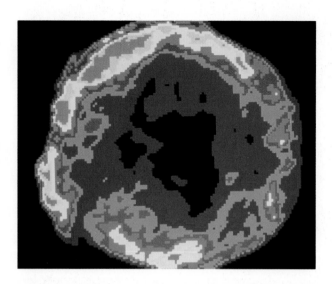

Available evidence suggests that there are two distinct types of supernovae. One type, which has just been described, marks the death of a massive star. The second type occurs in a binary system that initially contains a white dwarf and a nearby companion. The intense gravitational force exerted by the white dwarf attracts matter from the companion star. The mass of the white dwarf, which initially must have been less than 1.4 times the mass of the Sun, begins to build up, and eventually it may exceed the Chandrasekhar limit. At this point, the white dwarf must begin to collapse. As it does, it heats up, new nuclear reactions begin, and the energy released is so great that it completely disrupts the star. Gases are blown out into space at velocities of several thousand kilometers per second. No central star remains behind (see Figure 12.6). The explosion completely destroys the white dwarf.

Supernova 1987A

The 1987 explosion of a massive star in the Large Magellanic Cloud was the first bright supernova since the invention of the telescope and at long last provided astronomers with an opportunity to test their calculations of how stars die (Figure 12.7). Also, for the first time, we know what the star was like before it exploded.

By combining theory and observation, astronomers have reconstructed the life story of the star that became SN 1987A. Formed about 10 million years ago, the star originally had about 20 times the mass of the Sun. For 90 percent of its life it lived quietly on the main sequence, converting hydrogen to helium. At this time its luminosity was about 60,000 L_s and it was spectral type O. The temperature in its core was about 40 million K, and its central density was about 5 g/cm^3, or about five times the density of water.

By the time hydrogen was exhausted at the center of the star, a helium core of about six times the mass of the Sun had developed, and hy-

drogen fusion was proceeding in a shell surrounding this core. The core contracted and grew hotter, until it reached a temperature of 170 million K and a density of 900 g/cm^3, at which time helium began to fuse to form carbon and oxygen. The surface of the star expanded to a radius of about 10^8 km, or about the distance from the Earth to the Sun. The star's luminosity nearly doubled to 100,000 L_s and the star became a red supergiant. While it was a red supergiant, the star lost some mass in the form of a stellar wind (Figure 12.8).

Figure 12.8 This image, which was taken with the Hubble Space Telescope, shows (in yellow) a ring of stellar material ejected by the star that ultimately became SN 1987A. This material was ejected long before the supernova explosion. Within the next 100 years, the expanding debris from the supernova, which appears in red in this image, will plow into the ring and tear it apart. The blue stars at the left and the right of the ring are not associated with the supernova. *(NASA)*

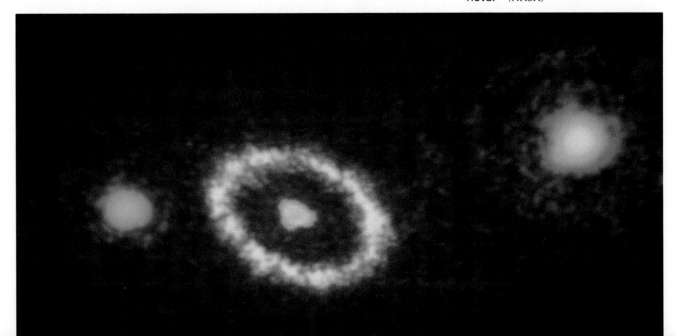

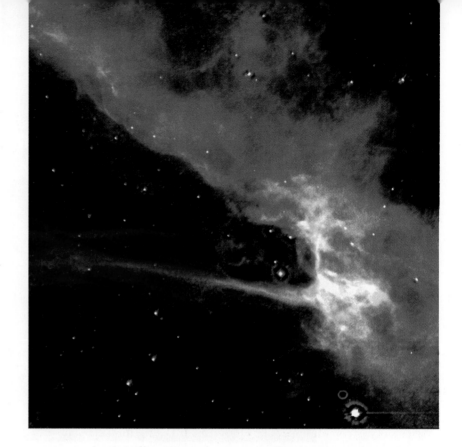

Figure 12.9 An image of a small portion of the supernova remnant known as the Cygnus Loop. This image was taken with the Hubble Space Telescope. The glowing gases mark the edge of a bubble-like blast wave produced by a stellar explosion, which occurred about 15,000 years ago. The bluish streak of light stretching from left to right may mark the passage of a knot of gas ejected by the supernova at a velocity of about 1500 km/s. The matter ejected by supernovae enriches space with heavy elements.
(J.J. Hester/NASA)

What is a supernova?

Helium fusion lasted for only about 1 million years, forming a core of carbon and oxygen about four times the mass of the Sun. When helium was exhausted at the center of the star, the core contracted again, the surface of the star also decreased in radius, and the star became a blue supergiant with a luminosity still about equal to 100,000 L_s. It was then that the explosion occurred.

When the contracting core of the blue supergiant reached a temperature of 700 million K and a density of 150,000 g/cm^3, nuclear reactions began to convert carbon to neon, sodium, and magnesium. This phase lasted only about 1000 years. The core, having exhausted carbon as a fuel, again contracted and heated—this time to a temperature of 1.5 billion K and a density of 10^7 g/cm^3—at which point the production of oxygen and magnesium, and then the conversion of oxygen to silicon and sulfur, lasted for several years. At temperatures of 3.5 billion K and densities of 10^8 g/cm^3, the silicon began to melt into a sea of helium, neutrons, and protons, which then combined with some of the remaining silicon and sulfur nuclei to form iron.

When the silicon is converted to iron, and the mass of the core exceeds 1.4 times the mass of the Sun, the collapse begins. In the case of SN 1987A, the collapse lasted only a few tenths of a second, and the velocity of the collapse in the outer portion of the iron core reached 70,000 km/s. The outer shells of neon, helium, and hydrogen, however, do not know about the collapse. Information travels through the star at the speed of sound and cannot reach the surface in the few tenths of a second that is required for the collapse to occur. The surface layers hang suspended, much like a cartoon character that dashes off the edge of a cliff and

TABLE 12.2 Evolutionary Phases That Preceded SN 1987A

Phase	Central Temperature (K)	Central Density (g/cm³)	Duration of Phase
Hydrogen fusion	40×10^6	5	9×10^6 years
Helium fusion	170×10^6	900	10^6 years
Carbon fusion	700×10^6	150,000	10^3 years
Neon fusion	1.5×10^9	10^7	Several years
Oxygen fusion	2.1×10^9	10^7	Several years
Silicon fusion	3.5×10^9	10^8	Days
Core collapse	200×10^9	2×10^{14}	Tenths of a second

hangs momentarily in space before he realizes that he no longer is held up by anything.

The collapse of the core continues until the densities rise to several times that of an atomic nucleus. The resistance to further collapse becomes very great, and the core rebounds. Infalling material runs into the brick wall of the rebounding core and is blown outward. The material that is ejected in the ensuing explosion is rich in heavy elements, and studies confirm that the composition is what would be predicted from the models of the composition of the stellar interior immediately prior to the explosion (Figure 12.9).

Table 12.2 summarizes the steps that led inexorably to the explosion that was SN 1987A. The data are taken from detailed calculations by S. Woosley (University of California at Santa Cruz) and his collaborators.

Neutrinos from SN 1987A

Brilliant as SN 1987A was in the southern skies, less than 0.1 percent of the energy of the explosion appeared as optical radiation. One of the predictions of models of supernovae is that a large number of neutrinos should be ejected from the core of the star at the time of the collapse. When the collapse occurs, the electrons merge with protons to form neutrons, and this reaction also releases neutrinos, which escape from the star at the speed of light. The energy carried away by these neutrinos is truly astounding. In the first second, the total luminosity of the neutrinos is 10^{46} watts, which exceeds the luminosity of all the stars in all the galaxies in the part of the universe that we can observe. And the supernova generated this energy in a volume less than 50 km in diameter! Supernovae are by far the most violent events in the universe.

One of the most exciting results from observations of SN 1987A is that astronomers detected the neutrinos at the right time and in the expected quantity, thereby obtaining strong confirmation of the theoretical calculations. The neutrinos were detected by two instruments, which might be called "neutrino telescopes," about 3 hours before the brightening of the star was first detected in the optical region of the spectrum. Both "telescopes," one in Japan and the other under Lake Erie, consist of several thousand tons of purified water surrounded by several hundred detectors that are sensitive to light. Incoming neutrinos interact with the water to produce positrons and electrons, which move rapidly through the water and emit deep blue light.

The Japanese system detected 11 neutrino events over an interval of 13 seconds, and the instrument beneath Lake Erie measured 8 events at the same time. Because the neutrino telescopes are located in the Northern Hemisphere and the supernova occurred in the Southern Hemisphere, the neutrinos detected had already passed through the Earth and were on their way back out into space when they were captured!

Only a few neutrinos were detected because the probability that they will interact with matter is very, very low. It is estimated that the supernova actually released 10^{58} neutrinos. About 500 trillion of these neutrinos passed through every square meter on the Earth, and about a million people experienced a neutrino interaction within their bodies. Of course, this interaction had absolutely no biological effect and went completely unnoticed.

Because the neutrinos come directly from the heart of the supernova, their energies provide a measure of how hot the star was at the time of the explosion. The central temperature was about 200 billion K.

What is left now that the explosion is over? According to theory, there should be an object with a density approximately equal to that of an atomic nucleus composed essentially entirely of neutrons at the center of the gaseous cloud that has been blown out into space. Despite careful searches, astronomers have not yet (October 1993) found evidence that the predicted neutron star is actually there. A neutron star has been found at the heart of the remnant of the much older Crab supernova, and studies of the Crab offer the best evidence that our ideas about supernovae producing neutron stars are actually correct.

The Crab Nebula

The best-studied example of the remnant of a supernova explosion is the Crab Nebula in Taurus, a chaotic, expanding mass of gas (Figure 12.10). The Doppler shifts of light from the center of the Crab Nebula show the gases there to be moving toward us at speeds up to 1450 km/s. If we assume that the nebula has always expanded at this same rate, we can derive its age by calculating how long it would take for the nebula to reach its present size. It turns out that both the location and computed time of formation of the Crab Nebula are in good agreement with the occurrence of the supernova of 1054. The Crab Nebula, therefore, must be the material ejected during that stellar explosion.

Observations of the center of the Crab Nebula have demonstrated conclusively that supernova explosions leave behind a neutron star. The first neutron star to be discovered, however, did not lie in the Crab, and initially it presented astronomers with a mystery.

Figure 12.10 The Crab Nebula, in the constellation Taurus. This nebula is the remnant of a supernova explosion, which was seen on Earth in the year 1054 A.D. It is located some 6300 LY away and is approximately 6 LY in diameter, still expanding outward. In this true color picture, the red filaments are tendrils of hot gas that emits strongly at the wavelength corresponding to the red Balmer line of hydrogen. The pulsar is clearly visible just below the center.
(W. Schoening and N. Sharp/National Optical Astronomy Observatories)

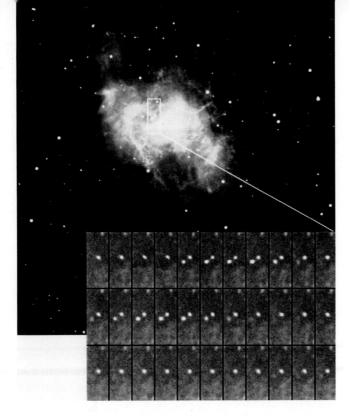

Figure 12.11 A series of photographs of the central part of the Crab Nebula taken by Nigel Sharp at Kitt Peak National Observatory. Note the star that seems to blink on and off; it is the pulsar, which has a period of 0.033 seconds. *(National Optical Astronomy Observatories)*

The Discovery of Neutron Stars

In 1967, Jocelyn Bell, a graduate research student at Cambridge University, was studying distant radio sources with one of the Cambridge radio telescopes. In the course of Bell's investigation she made a remarkable discovery—one that won her advisor, Antony Hewish, the Nobel Prize in physics, because his analysis of the object (and other similar ones) revealed the first evidence for neutron stars.

What Bell had found, in the constellation of Vulpecula, was a source of rapid, sharp, intense, and extremely regular pulses of radio radiation, the pulses arriving exactly every 1.33728 s. For a time there was speculation that they might be signals from an intelligent civilization. Radio astronomers half-jokingly dubbed the source "LGM," for "little green men," and withheld announcement pending more careful study. Soon, however, three additional similar sources were discovered in widely separated directions in the sky. When it became apparent that this type of source was fairly common, astronomers concluded that it was highly unlikely that they were signals from other civilizations. By now hundreds of such sources have been discovered. They are called **pulsars,** for pulsating radio sources.

The pulse periods of different pulsars range from a little longer than 1/1000 s to nearly 10 s. One pulsar is in the middle of the Crab Nebula, and it has a pulse period of 0.033 s. The period is observed to be very slowly increasing, showing that pulsars evolve, pulsing gradually more slowly as they age. The source of the pulses is what appears to be a star at the center of the nebula. In addition to pulses of radio energy, there are pulses of optical and x-ray radiation from the Crab as well (Figure 12.11).

The Crab pulsar emits considerably more energy than does the Sun. Yet the energy that we detect from the Crab pulsar is not constant, but arrives in pulses or bursts that occur 30 times each second. What type of object can emit such large bursts of energy?

One clue comes from measurements of the masses of pulsars. Some pulsars that emit x rays are also members of binary star systems for which enough information is available to calculate their masses using Kepler's laws. These x-ray pulsars have masses in the range of 1.4 to 1.8 times that of the Sun. This range is consistent with the range of masses theorists predict neutron stars should have.

Pulsars are, in fact, *spinning* neutron stars—just what would be expected from the theory of evolution of very massive stars. Neutron stars are the most dense objects in the universe. The force of gravity at the surface of a neutron star is 10^{11} times the force of gravity at the surface of the Earth. The interior of a neutron star is composed of about 95 percent neutrons, with a small number of protons and electrons mixed in. In effect, a neutron star is a giant atomic nucleus with a mass about 10^{57} times the mass of a proton. There are approximately 100 million neutron stars in our Galaxy.

A neutron star itself is so small—typically about 10 km in radius—that it cannot be observed. What we "see," instead, are pulses of x rays, light, and radio waves, all produced by the magnetic field embedded in the rapidly spinning neutron star. The rapid spin is another example of the conservation of angular momentum. Even if the star begins with a very slow rotation when it is on the main sequence, it speeds up in the process of collapse until it spins in just a fraction of a second.

Any magnetic field that existed in the original star is highly compressed if the core collapses to a neutron star. At the surface of the neutron star, protons and electrons escape but are caught up in this spinning field and accelerated nearly to the speed of light, where they emit energy over a broad range of the electromagnetic spectrum. The pulsar radiation is, however, confined to a narrow rotating beam like a lighthouse beacon. As the rotation carries first one and then the other magnetic pole of the star into our view, we see a rapidly varying emission—the pulses that make the object a pulsar. Table 12.3 compares the characteristics of a typical white dwarf and a typical neutron star.

What are pulsars and how are they related to supernovae?

TABLE 12.3   Properties of Typical White Dwarfs and Neutron Stars

	White Dwarf	Neutron Star
Mass	1.0	1.5
(Mass of Sun = 1)	(Always <1.4)	(Always <3)
Radius	5000 km	10 km
Density	$5 \times 10^5 \text{g/cm}^3$	10^{14}g/cm^3

ESSAY   Planets Around a Pulsar

Recently, astronomers have reported that one pulsar is accompanied by at least two planets. The pulsar itself spins hundreds of times each second, and we can measure the rate of spin by measuring the time interval between successive pulses. If there are planets present, then the pulsar will also revolve in a small orbit around the center of mass of the system defined by the positions and masses of the pulsar and planets. As the pulsar moves in its orbit, the intervals between successive pulses will vary, being slightly longer when the pulsar is moving away from us and shorter when the pulsar is moving toward us. By analyzing the intervals between pulses, astronomers have concluded that the pulsar PSR1257 + 12 is being orbited by two objects with masses of at least 2.8 and 3.4 times the mass of the Earth. The orbital periods are 98 and 67 days, and the sizes of the orbits are similar to the size of Mercury's orbit. This system may also contain a third planet with an orbital period of about one year.

These planets would certainly not be hospitable abodes for life. The pulsar emits most of its energy in the form of an intense stellar wind, which would blast the planet with high energy particles moving at nearly the speed of light. Planets this close to the pulsar could certainly not have survived the supernova explosion that preceded the formation of the pulsar. Therefore, they must have been formed since the explosion, either from the debris blown off in the explosion or, more likely, from material blasted away from a nearby companion star, which has now been completely vaporized by radiation from the pulsar. Theoretical calculations show that such planets could form on a time scale of 10^5 to 10^6 years.

The significance of this result is that it suggests that the formation of planets may be relatively easy. If planets can form so quickly in a disk of material around a pulsar, it may be equally easy to form planets in the disks that surround protostars. If that is true, then potentially habitable planets may be very common. The next question astronomers want to answer is whether most pulsars that rotate hundreds of times per second have planets or whether the particular pulsar that has been studied is unique. Detailed studies of additional pulsars are now being made, but because the orbital periods are months to years long and multiple cycles must be observed to make sure the results are correct, it will be several years before we know whether most pulsars, or only a few, have planets around them.

Evolution of Pulsars

The energy radiated by a pulsar robs the star of rotational energy. Thus, theory predicts that rotating neutron stars should gradually slow down and that the periods of pulsars should slowly increase as time passes. Indeed, as we have already mentioned, the Crab pulsar is actually observed to be increasing the interval between its pulses. According to present ideas, the Crab pulsar is rather young and has a short period (we know it is only about 900 years old) while the other, older pulsars have already slowed to longer periods. Typical pulsars thousands of years old have lost too much energy to emit appreciably at visible and x-ray wavelengths, and are observed only as radio pulsars; their periods are 1 second or more.

12.5
BLACK HOLES

We have seen that most stars end their lives as white dwarfs. Yet some stars, evidently through processes connected with the supernova phenomenon, end up as neutron stars. White dwarfs must have masses less than 1.4 times that of the Sun, and calculations have shown that neutron stars probably cannot exceed three solar masses. A star that contains a core whose mass exceeds three times the mass of the Sun at the time that it runs out of nuclear energy and begins to collapse probably becomes a **black hole.**

Definition of a Black Hole

To understand what a black hole is, we will do a thought experiment—that is, an experiment that is just in our heads, because it is impossible to do the experiment in a laboratory. We already know that a rocket must be launched from the surface of the Earth at a very high velocity if it is to escape the pull of the Earth's gravity. In fact, any object—rocket, bullet, ball—that is thrown into the air with a velocity that is less than 11 km/s will fall back to the Earth's surface. Only those objects launched with a velocity greater than 11 km/s can escape into space.

For the Sun, the escape velocity is even higher—618 km/s. Now imagine that we begin to compress the Sun and force it to shrink in diameter. When the Sun reaches the diameter of a neutron star (less than 20 km), the velocity required to escape the gravitational pull of the shrunken Sun is about half the speed of light. Suppose we continue to compress the Sun to a smaller and smaller diameter. Ultimately, the escape velocity will exceed the speed of light. What happens then? If no light can escape, the object will be invisible.

In modern terminology, we call such an object a black hole, a name suggested by the American scientist John Wheeler in 1969. The idea that such objects might exist is, however, not a new one. Cambridge professor and amateur astronomer John Michell wrote a paper in 1783 about the possibility that stars might exist for which the escape velocity exceeds that of light. In 1796, the French mathematician Pierre Simon, Marquis de Laplace, wrote about similar objects, which he termed "dark bodies."

Einstein's theory of general relativity is required to calculate what actually happens when the gravitational force becomes so large. The calculations show that when the star becomes so dense that the escape velocity equals the speed of light, everything, including light itself, is trapped inside. Nothing at all can escape through that surface where the escape velocity equals the speed of light. That surface is called the **event horizon,** and its radius is the Schwarzschild radius, named for Karl Schwarzschild, who first described the situation a few years after Einstein introduced the theory of general relativity. This surface is the boundary of the black hole. All that is inside the event horizon is hidden

What is the evidence that black holes exist?

forever from us. As the star shrinks through the event horizon, it virtually disappears from the universe.

Size of a Black Hole

The size of the event horizon is proportional to the mass of the *collapsed object*. (The radius of the event horizon is equal to $2\ GM/c^2$, where G is the gravitational constant, M is the mass of the black hole, and c is the speed of light.) For a black hole of one solar mass, the event horizon is about 3 km in radius; thus the entire black hole is about one-third the size of a neutron star of the same mass. For a star of three solar masses, the radius of the event horizon is about 9 km, the same as the radius of a three-solar-mass neutron star. If a collapsing star is much more massive than three solar masses, it reaches its event horizon before it can shrink to neutron star dimensions. This is another way of saying that a neutron star cannot exist for masses above approximately three solar masses.

The event horizons of larger and smaller black holes—if they exist— have greater and lesser radii, respectively. For example, if a globular cluster of 100,000 stars could collapse to a black hole, it would be 300,000 km in radius, a little less than half the radius of the Sun. If the entire Galaxy could collapse to a black hole, it would be about 10^{12} km in radius—about 8000 AU. Conversely, for the Earth to become a black hole, it would have to be compressed to a radius of only 1 cm—about the size of a golf ball.

But do black holes exist? Theoretically, they are certainly possible, and stellar evolution calculations suggest that a black hole is inevitable for a star that reaches its final collapse with a mass greater than about three solar masses. As far as we know, no form of matter can support its own weight if its mass is greater than three solar masses. However, the number of black holes that will form is uncertain. Because massive stars lose a great deal of their mass as they evolve, we really do not know what fraction of stars end their evolution with more than three solar masses of material left. The most obvious way to answer this question is to seek *observational evidence* for black holes.

The observational search for black holes is not simple. How do we find an object we cannot see? The answer is that we can detect it by its gravitational effects on other stars, and this is most easily accomplished in a binary star system.

Candidates for Black Holes

To find a black hole we must first find a star whose motion (found from the Doppler shift of its spectral lines) shows it to be a member of a binary star system with a companion of mass too high to be a white dwarf or a neutron star. Second, that companion star must not be visible, for a black hole, of course, gives off no light. But being invisible is not enough, for a relatively faint star might be unseen next to the light of a

Figure 12.12 Accretion disk. Mass lost from a giant star through a stellar wind streams toward a black hole and swirls around it before it finally falls in. In the inner portions of the accretion disk, the matter is revolving so fast that internal friction heats it to very high temperatures and x rays are emitted.

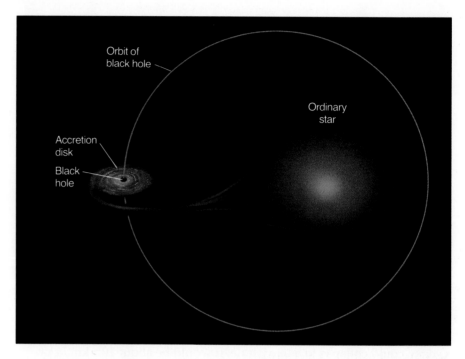

brilliant companion. Therefore, we must also have evidence that the unseen star is a collapsed object—one of extremely small size.

Modern space astronomy provides a means to determine if a candidate object is collapsed. If matter falls toward or into a small object of high gravity (either a neutron star or a black hole), the matter is accelerated to high speed. For example, near the event horizon of a black hole, matter is moving at velocities that approach the speed of light. Internal friction can heat it to very high temperatures—up to 100 million K or more. Such hot matter emits radiation in the form of x rays. Orbiting x-ray telescopes have revealed such intense sources of x radiation. What we are looking for then are x-ray sources associated with binary stars with invisible companions that have masses greater than is permitted for neutron stars. We cannot prove that such a system contains a black hole, but at present we have no other theory for what the invisible massive companion can be.

We can easily understand the origin of the infalling gas that produces the x-ray emission. Stars in close binary systems can exchange mass, especially as one of the members evolves into a red giant and overflows into the gravitational field of the smaller companion. Suppose that a massive star in such a double-star system has already exploded as a supernova, leaving behind either a neutron star or a black hole, and that the second star has become so large that its outer layers pass through the point of no return between the stars and fall toward the collapsed companion. The mutual revolution of the giant star and its collapsed companion causes the material from the former to flow not directly onto the companion but, because of conservation of angular momentum, to spiral around it, collecting in a disk of matter called the **accretion disk**. In the inner part of the accretion disk, the matter is revolving so fast that internal friction heats it up to temperatures at which it emits x rays.

Another way to form an accretion disk in a binary star system is from material ejected from the companion of the black hole or neutron star as a stellar wind. Some of that ejected gas will flow close enough to the collapsed companion to be captured by it into the disk (Figure 12.12). Such a case, we think, is the binary system containing the first x-ray source discovered in Cygnus—Cygnus X-1. The visible star is spectral type B. Measurements of the Doppler shifts of the B star's spectral lines show that it has an unseen companion with a mass nearly ten times that of the Sun. That companion must be a black hole if it is a small, collapsed object. The x rays from it strongly suggest that it is, for we have no other explanation for the source of those x rays than gas heated by infall toward a tiny massive object.

There are a few other binary systems that meet all the conditions to be prime candidates for black holes. One of these is in the Large Magellanic Cloud.

Myths About Black Holes

Much of the modern folklore about black holes is misleading. One idea is that black holes are monsters that go about sucking things up with their gravity. Actually, the gravitational attraction of a black hole at a large distance from it is the same as that around any other star (or object) of the same mass. Even if another star, or a spaceship, were to pass one or two solar radii from a black hole, Newton's laws would give an excellent description of what would happen to it. It is only *very* near the surface of a black hole that gravitation is so strong that Newton's laws break down; for a black hole of the mass of the Sun, light would have to come within 1.5 km of its surface to be trapped. A solar mass black hole, remember, is only 3 km in radius—a very tiny target. Even collisions between ordinary stars, hundreds of thousands of times bigger in diameter, are so rare as to be essentially nonexistent. A star would be far, far safer to us as an interloping black hole than it would have been in its former stellar dimensions.

▶ SUMMARY

12.1 Stars that begin their lives with masses less than about eight times the mass of the Sun end their evolution as **white dwarfs.** A typical white dwarf has a mass comparable to that of the Sun and a diameter comparable to that of the Earth. The pressure exerted by **degenerate electrons** keeps white dwarfs from contracting to still smaller diameters. The maximum possible mass for a white dwarf is 1.4 times the mass of the Sun. This maximum mass is referred to as the **Chandrasekhar limit.**

12.2 During the course of their evolution stars may shed their outer layers and lose a significant fraction of their initial mass. Stars with masses up to eight times the mass of the Sun can lose enough mass to become white dwarfs. White dwarfs have masses less than the **Chandrasekhar limit,** which is 1.4 times the mass of the Sun. **Planetary nebulae** are shells of gas ejected by stars. The mass that is lost is enriched with heavy elements produced by **nucleosynthesis** in the stellar interior.

12.3 In a massive star, hydrogen fusion in the core is followed by several other fusion reactions involving heavier elements. Just before it exhausts all sources of energy, a massive star has an iron core surrounded by shells of (in order of increasing distance from the center and decreasing temperature) silicon, oxygen, neon, carbon, helium, and hydrogen.

12.4 When the mass of the iron core of a star exceeds the Chandrasekhar limit, the core collapses until its density exceeds that of an atomic nucleus. The core rebounds and transfers energy outward, blowing off the outer layers of the star in a **supernova** explosion. Studies of SN 1987A, including the detection of neutrinos, have confirmed theoretical calculations of what happens during the explosion. At least some supernovae leave behind a rotating **neutron star,** which is called a **pulsar.** Pulsars emit pulses of radiation at regular intervals. Their periods are approximately in the range of 0.001 to 10 seconds and are related to the period of rotation of the neutron star. Pulsars lose energy as they age, their rotation slows, and their periods increase.

12.5 Theory suggests that stars more massive than three times the mass of the Sun at the time they exhaust their nuclear fuel and collapse will become **black holes.** Observations of binary stars with **accretion disks** that emit x rays suggest that a few black holes do exist.

▶ REVIEW QUESTIONS

1. What is a white dwarf? Describe its structure.

2. From the time it is in the main-sequence phase of its evolution until it becomes a white dwarf, describe the evolution of a star with a mass of 1.2 times the mass of the Sun.

3. Describe the evolution of a massive star up to the point at which it becomes a supernova. How does the evolution of a massive star differ from that of the Sun? Why?

4. What is a supernova?

5. What is the evidence that a supernova can produce a neutron star?

6. What is the evidence that black holes exist?

▶ THOUGHT QUESTIONS

7. You observe an expanding shell of gas through a telescope. What measurements would you make to determine whether you have discovered a planetary nebula or the remnant of a supernova explosion?

8. Arrange the following stars in chronological order:

a. A star with no nuclear reactions occurring in its core, which is made primarily of carbon and oxygen
b. A star of uniform composition from center to surface; the star contains hydrogen but has no nuclear reactions occurring in its core
c. A star that is fusing hydrogen to form helium in its core
d. A star that is fusing helium to carbon in its core and hydrogen to helium in a shell around its core
e. A star that has no nuclear reactions occurring in its core but is fusing hydrogen to form helium in a shell around its core

9. Would you expect to find any white dwarfs in the Orion nebula?

10. Suppose no stars had ever formed that were more massive than about two times the mass of the Sun. Would life as we know it have been able to develop?

11. Would you be more likely to observe the explosion of a massive star in a globular cluster or in an open cluster? Why?

12. If the Sun could suddenly collapse and become a black hole, how would the period of the Earth's revolution about it differ from what it is now?

13. Indicate what the final stage of evolution—white dwarf, neutron star, or black hole—will be for each of the following kinds of stars. Look through this book for the necessary data.

a. Spectral-type O main-sequence star
b. B main-sequence star
c. A main-sequence star
d. G main-sequence star
e. M main-sequence star

14. Which is likely to be more common in our Galaxy—neutron stars or black holes?

15. Show the evolutionary track in the HR diagram for the star that exploded as SN 1987A. Begin when the star was on the main sequence and end just before the explosion.

▶ **PROBLEMS**

16. The gas shell of a particular planetary nebula is expanding at the rate of 20 km/s. Its diameter is 1 LY. Find its age. For this calculation, assume that there are 3×10^7 s/yr, and 10^{13} km/LY.

17. Prepare a chart or diagram that exhibits the relative sizes of a typical red giant, the Sun, a typical white dwarf, and a neutron star of mass equal to the Sun's. You may have to be clever to devise such a diagram.

18. Suppose the central star of a planetary nebula is 16 times as luminous as the Sun and 20 times as hot (about 110,000 K). Find its radius, in terms of the Sun's. Compare this radius with the radius of a typical white dwarf.

19. The radius R of the event horizon of a black hole is given by the formula $R = 2GM/c^2$ where M is the mass of the black hole, G is Newton's gravitational constant, and c is the speed of light. Calculate the radius of a black hole with the mass of the Earth; the mass of the Sun; the mass of a globular cluster; and the mass of the Milky Way Galaxy.

20. Suppose that the Earth were collapsed to the size of a golf ball, becoming a small black hole.
a. What would be the revolution period of the Moon around it, at a distance of 400,000 km?
b. What would be the revolution period of a spacecraft orbiting at a distance of 6000 km?
c. What would be the revolution period of a miniature spacecraft orbiting at a distance of 10 cm?
d. For the mini-spaceship in (c), calculate its orbital speed and compare it with the speed of light.

▶ SUGGESTIONS FOR ADDITIONAL READING

Books

Greenstein, G. *Frozen Star*. Freundlich Books, 1983. A superbly written introduction to the death of stars and the strange objects they leave behind.

Marschall, L. *The Supernova Story*. Plenum, 1988. The introduction of choice to supernovae and Supernova 1987A.

Kaufmann, W. *Black Holes and Warped Spacetime*. W. H. Freeman, 1979. Wonderful nontechnical introduction to the world of black holes.

Cooke, D. *The Life and Death of Stars*. Crown, 1985. A journalist summarizes our understanding of stellar evolution.

Wheeler, J. *A Journey Into Gravity and Spacetime*. W. H. Freeman/Scientific American Library, 1990. Not easy, but a most rewarding exploration of relativity and black holes by the scientist who coined the term "black holes."

Murdin, P. *End in Fire: The Supernova in the Large Magellanic Cloud*. Cambridge U. Press, 1990. Introduction to exploding stars by an Australian astronomer.

Articles

On white dwarfs and planetary nebulae and the fate of the Sun

Kaler, J. "Extraordinary Spectral Types" in *Sky & Telescope*, Feb. 1988, p. 149.

Kaler, J. "The Smallest Stars in the Universe: White Dwarfs and Neutron Stars" in *Astronomy*, Nov. 1991, p. 51.

Whitmire, D. & Reynolds, R. "The Fiery Fate of the Solar System" in *Astronomy*, Apr. 1990, p. 20. Future evolution of the Sun into a red giant.

Kaler, J. "Giants in the Sky: The Fate of the Sun" in *Mercury*, Mar./Apr. 1993, p. 34.

Kaler, J. "Planetary Nebulae and Stellar Evolution" in *Mercury*, July/Aug. 1981, p. 114.

Kaler, J. "Bubbles from Dying Stars" in *Sky & Telescope*, Feb. 1982, p. 129.

Kawaler, S. & Winget, D. "White Dwarfs: Fossil Stars" in *Sky & Telescope*, Aug. 1987, p. 132.

Trimble, V. "White Dwarfs: The Once and Future Suns" in *Sky & Telescope,* Oct. 1986, p. 348.

Soker, N. "Planetary Nebulae" in *Scientific American,* May 1992.

On supernovae and pulsars

Graham-Smith, F. "Pulsars Today" in *Sky & Telescope,* Sept. 1990, p. 240.

Naeye, R. "Supernova 1987A Reconsidered" in *Sky & Telescope,* Feb. 1993, p. 39.

Bruning, D. "Lost and Found: Pulsar Planets" in *Astronomy,* June 1992, p. 36.

Thorpe, A. "Giving Birth to Supernovae" in *Astronomy,* Dec. 1992, p. 46. Which stars will become supernovae in the future.

Kirschner, R. "Supernova: The Death of a Star" in *National Geographic,* May 1988, p. 618.

Bell-Burnell, J. "Little Green Men, White Dwarfs, Or What?" in *Sky & Telescope,* Mar. 1978, p. 218. The story of the discovery of pulsars.

Hewish, P. "Pulsars After 20 Years" in *Mercury,* Jan./Feb. 1989, p. 12. By one of the discoverers of pulsars.

deVaucouleurs, G. "The Supernova of 1885 in Messier 31" in *Sky & Telescope,* Aug. 1985, p. 115.

Lattimer, J. & Burrows, A. "Neutrinos From Supernova 1987A" in *Sky & Telescope,* Oct. 1988, p. 348.

Talcott, R. "Insight into Star Death" in *Astronomy,* Feb. 1988, p. 6. Review of what we learned from Supernova 1987A.

Verschuur, G. "On the Trail of Exotic Pulsars" in *Astronomy,* Dec. 1988, p. 22.

Tucker, W. "Exploding Stars, Superbubbles, and HEAO Observations" in *Mercury,* Sept./Oct. 1984, p. 130.

On black holes

Overbye, D. "God's Turnstile: The Work of John Wheeler and Stephen Hawking" in *Mercury,* July/Aug. 1991, p. 98.

McClintock, J. "Do Black Holes Exist?" in *Sky & Telescope,* Jan. 1988, p. 28.

Croswell, K. "The Best Black Hole in the Galaxy" in *Astronomy,* Mar. 1992, p. 30. On the binary system in Monoceros.

Parker, B. "In and Around Black Holes" in *Astronomy,* Oct. 1986, p. 6.

THE MILKY WAY GALAXY

This full-sky view of the Milky Way was taken from the summit of Mount Graham, which is a 3200 m high mountain in southeastern Arizona. Note the dark rift through the center of the Milky Way. Around the horizon, in a counterclockwise direction beginning from the bottom, are the lights of Wilcox, Tucson, and Phoenix, and the silhouettes of test towers and fir trees. The brightest point-like object in the lower left of the picture is the planet Jupiter.   *(Roger Angel, Steward Observatory/University of Arizona)*

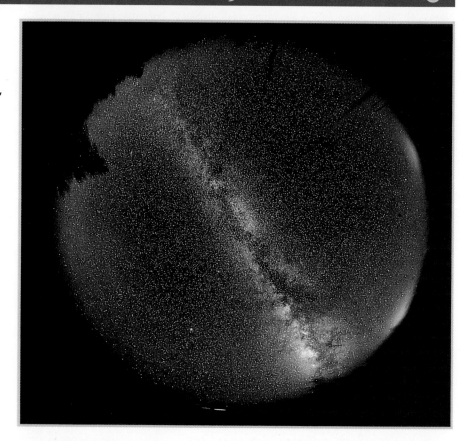

One of the most striking features in a truly dark sky is a band of faint white light that stretches from one horizon to the other. Because of its appearance, this band of light is called the Milky Way. In the Northern Hemisphere the band is brightest in the region of the constellation Cygnus and is best viewed in the summer. In the Southern Hemisphere, the Milky Way is even brighter—so bright, in fact, that Indians in South America gave names to various portions of it just as northern astronomers gave constellation names to conspicuous groupings of stars. Unfortunately, the Milky Way is not bright enough to be seen from urban areas because of their large numbers of artificial lights. Therefore, many city-dwellers have not seen it.

In 1610, Galileo made the first telescopic observations of the Milky Way Galaxy and discovered that it is composed of a multitude of individual stars. We now know that the Sun is located within a disk-shaped system of stars. The

Milky Way is the light from nearby stars that lie more or less in this disk. We call this great stellar system, which includes all of the individual stars that we can see except with the largest telescopes, the Milky Way Galaxy or, more simply, the Galaxy.

13.1
THE ARCHITECTURE OF THE GALAXY

In 1785, William Herschel showed that the stellar system to which the Sun belongs has the shape of a wheel or disk. Herschel used a telescope to count the numbers of stars that he saw in various directions. He found that the numbers of stars were about the same in any direction around the Milky Way, but that there were many fewer stars in directions perpendicular to the Milky Way. This result seemed to show that the Sun was near the center of a flattened system of stars (Figure 13.1). We now know that Herschel was right about the shape of the Galaxy but wrong about the location of the Sun. Interstellar dust absorbs the light from stars and restricts optical observations in the plane of the Milky Way to stars within about 6000 LY. Herschel was able to observe only a tiny fraction of the system of stars that surrounds us.

Overview of the Galaxy

With modern instrumentation, astronomers can make observations at radio and infrared wavelengths, and electromagnetic radiation at these long wavelengths penetrates the dust easily. Based on these new observations, astronomers now picture the Galaxy as a thin circular **disk** of luminous matter distributed across a region about 100,000 LY in diameter, with a thickness of about 1000 LY. A schematic diagram of the

Figure 13.1 A copy of a diagram by William Herschel, showing a cross-section of the Milky Way system based on counting stars in various directions looking outward from the solar system.

Figure 13.2   Schematic representation of the Milky Way Galaxy. The face-on view shows the locations of the spiral arms. The Sun is located on the inside edge of the short Orion spur. The globular clusters are located in the halo of the Galaxy.

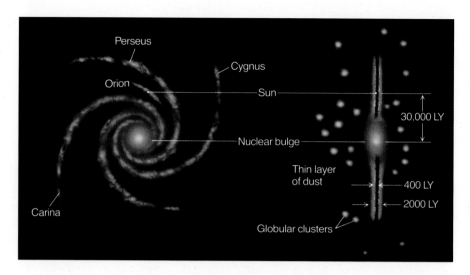

What does our own Milky Way Galaxy look like?

Galaxy is shown in Figure 13.2. The Sun is in the disk and orbits the center of the Galaxy at a distance of nearly 28,000 LY. Young stars are also located in the disk of the Galaxy and are concentrated in a series of **spiral arms**. Dust and gas, the raw materials from which stars are made, are mostly found in the galactic disk as well. The globular clusters are distributed in a much larger spherical **halo** around the galactic center. The halo may also contain a substantial amount of nonluminous or dark matter that extends to a distance of at least 150,000 LY from the galactic center.

Within 3000 LY of the galactic center, the stars are no longer confined to the disk but rather form a spheroidal **nuclear bulge** of old stars. In ordinary visible radiation it is possible to observe stars in the bulge

Figure 13.3   Although the nucleus of our Galaxy is completely hidden by dust at visible wavelengths, there are regions where the obscuration is low and we can see into the cloud of old, population II stars that are concentrated toward the center and form the nuclear bulge. This is a picture of such a region, which is called Baade's window. The density of stars in this photograph peaks at a distance of about 28,000 LY from the Sun. The brightest star in the picture is Gamma Sagittarii, a deep yellow star that appears in the foreground at a distance of about 100 LY and can be seen with the unaided eye. *(Anglo-Australian Telescope Board)*

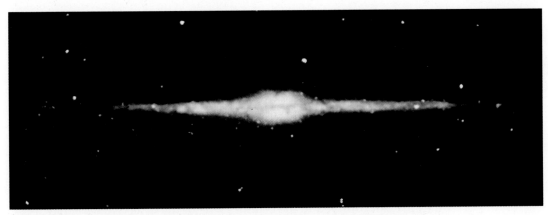

Figure 13.4   This image presents a view of the Milky Way Galaxy obtained by an experiment aboard the Cosmic Background Explorer Satellite (COBE). This image was taken in the near-infrared part of the spectrum and permits us to see, for the first time, the bulge of old stars that surrounds the center of our Galaxy. Redder colors correspond to regions where there is more dust, and this dust forms a thin disk of material. In optical regions of the spectrum, this dust absorbs so much radiation that we cannot see the bulge of old stars.   *(NASA)*

only in directions where the obscuration by interstellar dust is unusually low (Figure 13.3). The first picture that actually succeeded in showing the bulge as a whole was taken at infrared wavelengths from a satellite orbiting the Earth (Figure 13.4).

In its overall characteristics, the Milky Way Galaxy resembles the Andromeda Galaxy (M31), which is at a distance of about 2.3 million LY (Figure 13.5). However, there are some differences between these two galaxies. Andromeda has more globular clusters than does the Milky

Figure 13.5   The spiral galaxy in Andromeda (M31) looks very much like our own Galaxy. Note the bulge of older yellowish stars in the center, the bluer and younger stars in the outer regions, and the dust in the disk that blocks some of the light from the bulge.   *(California Institute of Technology/Palomar Observatory)*

Way, and the rate at which stars are forming is slower in Andromeda. Recent studies also suggest that the nuclear bulge of the Milky Way Galaxy may not be spherical, as Andromeda's is, but may be shaped like a bar. Some other barred spirals are illustrated in Figure 14.5.

This picture of the Galaxy is one of the major achievements of 20th century astronomy. Until early in this century, astronomers accepted Herschel's conclusion that the Galaxy was centered approximately at the Sun and extended only a few thousand light years from it. The shift from the "heliocentric" to the "galactocentric" view of the Milky Way Galaxy, as well as the first measurement of its true size, came about largely through the efforts of Harlow Shapley, who investigated the distribution of globular clusters. Because of their brilliance, and the fact that they are not confined to the central plane of the Galaxy where they would otherwise be largely obscured by interstellar dust, globular clusters can be observed (with telescopes) to very large distances.

13.2
GLOBULAR CLUSTERS AND THE SIZE OF THE GALAXY

In Section 10.1 we discussed the use of stellar parallaxes to determine the distances to stars within about 60 LY of the Sun. Entirely new techniques are required for measuring distances of remote globular clusters.

Distances Within the Galaxy

There is a special class of stars that is particularly useful for calculating distances. These stars are the **pulsating variables.** A pulsating variable star actually changes its diameter with time—periodically expanding and contracting as your chest does when you breathe. This expansion and contraction can be measured by using the Doppler effect to see spectral lines shift toward the blue as the surface of the star moves toward the Earth and then to the red as the surface of the star shrinks back. As it pulsates, the star changes color, indicating that its temperature is also varying. By far the most obvious effect of pulsation is the changing luminosity of the star, which results in periodic variations in the observed brightness.

The two types of pulsating stars that are most useful for measuring distances are the **cepheid variables** and the **RR Lyrae variables.** We will first discuss the properties of these two classes of stars and then show how to use them to derive distances.

Cepheid and RR Lyrae Variables

Cepheids are large, yellow, pulsating stars named for the prototype and first known star of the group, Delta Cephei. The variability of Delta Cephei was discovered in 1784 by the young English astronomer John Goodricke just two years before his death at the age of 21.

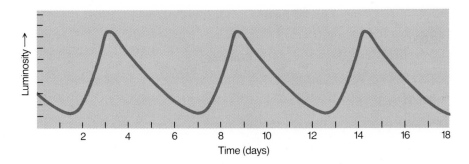

Figure 13.6 Light curve (plot of brightness as a function of time) of Delta Cephei, the prototype after which the class of cepheid variables is named.

A graph that shows how the brightness of a variable star changes with time is called a **light curve** of that star (Figure 13.6). The *maximum* is the point of the light curve where the star is brightest; the *minimum* is the point on the light curve where the star is faintest. If the light variations repeat themselves periodically, the interval between successive maxima is called the **period** of the star. The **amplitude** of the variation is the difference in brightness between the maximum and the minimum. The amplitudes of cepheid variable stars range from less than 10 percent up to a factor of about 10.

The brightness of Delta Cephei varies by about a factor of two in a period of 5.4 days. The star brightens rather rapidly to maximum light and then dims more slowly to minimum light (see Figure 13.6).

More than 700 cepheid variables are known in our Galaxy. Most cepheids have periods in the range 3 to 50 days and luminosities about 10,000 times greater than that of the Sun (10,000 L_s). Polaris, the North star, is a small-amplitude cepheid variable that varies in luminosity by 10 percent in a period just under 4 days.

The cepheid variables are massive young stars that are useful for measuring distances to open clusters and to other groups of recently formed stars. The RR Lyrae variables provide a method for measuring distances to older groups of stars. The RR Lyrae variables also pulsate, but they have periods of only about 12 hours, and their luminosity is only about 50 L_s. Most globular clusters contain at least a few RR Lyrae stars.

How do we use the relationship between period and luminosity for cepheids and RR Lyrae variables to measure distances?

The Period-Luminosity Relation for Cepheid Variables

The importance of cepheid variables lies in the fact that a relation exists between their periods of pulsation (or light variation) and their average luminosity. Simply by measuring the period of a cepheid, we can estimate its true luminosity. Because its average apparent brightness can be measured, we can calculate how far away it is.

The relation between period and luminosity was discovered in 1912 by Henrietta Leavitt, an astronomer of the Harvard College Observatory (Figure 13.7). Some hundreds of cepheid variables had been discovered in the Large and Small **Magellanic Clouds** (Figure 14.9 and Figure 14.10), two great stellar systems that are actually neighboring galaxies (although they were not known to be galaxies in 1912).

Figure 13.7 Henrietta Swan Leavitt (1868–1921) joined the staff of the Harvard College Observatory in 1892. While studying variable stars in the Magellanic Clouds, she discovered the period-luminosity relation for cepheids, which later made it possible for Edwin Hubble to demonstrate that our Galaxy is but one among billions in the universe. *(Harvard College Observatory)*

Figure 13.8 The period-luminosity relation for cepheid variables. Also shown are the period and luminosity of RR Lyrae stars.

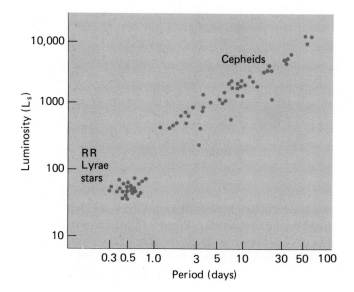

Leavitt found that the brighter-appearing cepheids always have the longer periods of light variation. Because the stars themselves in each Magellanic Cloud must be at approximately the same distance from the Earth, the relationship between period and luminosity was readily apparent. In principle, the same conclusion could have been reached by measuring individual parallaxes, and hence luminosities, for cepheids in our own Galaxy. Unfortunately, however, no cepheids are close enough for accurate parallaxes to be measured.

Figure 13.8 shows the **period-luminosity** relationship for cepheid variables. The location of the RR Lyrae stars in this diagram is also shown. All of the RR Lyrae variables have nearly identical luminosities and periods.

Harlow Shapley (Figure 13.9) was one of the astronomers who recognized the importance of pulsating variable stars as distance indicators. From the known luminosities of RR Lyrae variables, Shapley was able to determine the distances to the globular clusters. He found that the clusters form a roughly spherical system. The center of that system is not at the Sun, however, but at a point in the direction of Sagittarius at a distance of some 25,000 to 30,000 LY (Figure 13.10). Shapley made the bold—and correct—assumption that the system of globular clusters is centered upon the center of the Galaxy. Today this assumption has been verified by many pieces of evidence, including the observed distributions of globular clusters in other spiral galaxies.

Figure 13.9 Harlow Shapley (1885–1972) began his career as a newspaper reporter. In his twenties, he enrolled in the University of Missouri, where he searched through the catalog for a suitable major and found Astronomy. After earning his bachelor's degree, he went to Princeton for his Ph.D. Subsequently, at Mount Wilson, his study of globular clusters revealed the true extent of our Galaxy. *(Yerkes Observatory)*

The Galactic Halo

Detailed observations show that the globular clusters are distributed in a sphere centered on the center of the Galaxy. A sparse "haze" of individual stars—not members of clusters but far outnumbering the cluster stars—also exists in the region outlined by the globular clusters. This

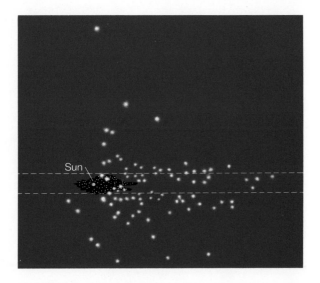

Figure 13.10 A copy of a diagram by Shapley, showing the distribution of globular clusters in a plane perpendicular to the Milky Way and containing the Sun and the center of the Galaxy. Herschel's diagram (Figure 13.1) is shown centered on the Sun, approximately to scale.

haze of stars and clusters forms the galactic halo, a region whose volume exceeds that of the main disk of the Galaxy by many times. Individual RR Lyrae stars have been found in significant numbers as far away as 30,000 to 50,000 LY on either side of the galactic plane, which shows that the halo must have an overall diameter of at least 100,000 LY.

13.3
INTERSTELLAR MATTER AND SPIRAL STRUCTURE

In addition to stars, the disk of the Galaxy contains interstellar gas and dust. Neutral atomic hydrogen in the Galaxy is confined to an extremely flat layer, which is only about 400 LY thick. In the plane of the Galaxy, this hydrogen extends well beyond the Sun to a distance of about 80,000 LY from the center of the Galaxy.

Dust is also confined to the disk, with the highest concentrations occurring in the spiral arms. The thickness of the dust layer is also about 400 LY. There is very little emission from dust lying outside the Sun's orbit around the center of the Galaxy.

Because we are surrounded by interstellar dust, which absorbs starlight, it is difficult for us to study the structure of our own Galaxy by means of optical observations. Fortunately, radio waves penetrate the dust easily. Thus observations with radio telescopes have played a key role in studies of galactic structure.

What do studies of the distribution of interstellar matter tell us about the shape of the Milky Way?

Atomic Hydrogen

Hydrogen accounts for about 75 percent of the interstellar gas, and hydrogen and helium together compose from 96 to 99 percent of it by mass. Most of the gas is cold and nonluminous. Gas clouds at temperatures from a few Kelvins to about 30 K consist primarily of hydrogen

Figure 13.11 NGC 3576 and NGC 3603. Clouds of luminous gas surround compact clusters of very hot stars. These hot stars ionize the hydrogen in the gas clouds. When electrons then recombine with protons and move back down to the lowest energy orbit, emission lines are produced. The strongest line in the visible spectrum is red, and is responsible for the color of the photograph. *(Copyright Anglo-Australian Telescope Board)*

molecules (H_2). Clouds at temperatures from a few tens to hundreds of Kelvins contain primarily atomic hydrogen. Near very hot stars, hydrogen is ionized by the ultraviolet radiation from those stars. The light emitted from regions of ionized gas consists largely of emission lines, including a very strong line of hydrogen in the red region of the spectrum (Figure 13.11). This line corresponds to the transition from the third energy level to the second energy level in hydrogen, and is a member of the Balmer series (see Section 8.4).

Emission of energy from interstellar material was the first radio radiation of astronomical origin to be detected. Some of the most important radio observations are of the spectral line of hydrogen at the radio wavelength of 21 cm. In addition to the energy levels discussed in Section 8.4, the lowest level of hydrogen is divided into two closely spaced levels. A transition between these levels produces a very small change in energy, corresponding to a photon of 21-cm wavelength (Figure 13.12).

Studies of the 21-cm line have played a key role in determining the nature of the spiral structure in our own Galaxy. Although the interpretation of the measurements is somewhat controversial, it appears likely that the Galaxy has four major spiral arms, with some smaller spurs (see Figure 13.2). The Sun appears to be near the inner edge of a short arm or spur called the *Orion arm,* which is about 15,000 LY long and contains such conspicuous features as the Cygnus Rift (the great dark nebula in the summer Milky Way) and the Orion Nebula. More distant, and therefore less conspicuous, are the *Sagittarius-Carina* and *Perseus arms,* which

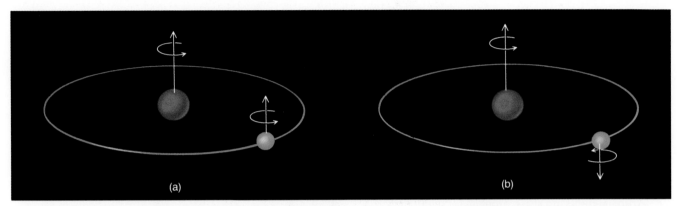

(a)

(b)

Figure 13.12 Formation of the 21-cm line. When the electron in a hydrogen atom is in the orbit closest to the nucleus, the proton and the electron may be spinning either in the same direction (left) or in opposite directions (right). If the spins of the two particles are in opposite directions, the atom as a whole has a very slightly lower energy than if the two spins are aligned. When the electron flips over, the atom either gains or loses a tiny bit of energy, absorbing or emitting electromagnetic energy with a wavelength of 21 cm.

are located, respectively, about 6000 LY inside and outside the Sun's position relative to the galactic center. Both of these arms, and the Cygnus arm, are about 80,000 LY long. The fourth arm is unnamed and is difficult to detect because emission from it is confused with strong emission from the central regions of the Galaxy, which lie between it and the Earth.

Formation and Permanence of Spiral Structure

At the Sun's distance from its center, the Galaxy does not rotate like a solid wheel. Individual stars obey Kepler's third law. Remember that Pluto takes longer than the Earth to complete one full circuit around the Sun. In just the same way, stars in larger orbits take longer to complete their orbits around the center of the Galaxy than do stars in smaller orbits.

Because stars in larger orbits trail behind those in smaller orbits, the existence of spiral patterns is not surprising. No matter what the original distribution of gas and dust might be, the rotation pattern of the Galaxy will form the material into spirals. Figure 13.13 shows the development of spiral arms from two irregular blobs of interstellar matter. A spiral pattern forms because the portions of the blobs closest to the galactic center move fastest, while those farther away trail behind.

Figure 13.13 Hypothetical formation of two spiral arms from irregular clouds of interstellar material, based simply on the rotation of the Galaxy.

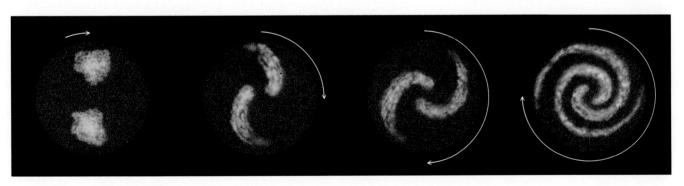

OBSERVING YOUR WORLD   LIGHT POLLUTION

Bright lights have robbed most city dwellers of the pleasure of looking at the stars. Many people have never seen the Milky Way. "Light pollution" has brightened the night sky and made it difficult or impossible to see the constellations, watch for meteor showers, or identify the occasional passing comet. Artificial lights are also a problem for many major observatories, making it impossible to detect faint stars and galaxies against the bright sky background. Lights, you may say, are the price we must pay for living in large cities. Lights are essential for safety and security.

Fortunately, there is a solution that provides high-quality lighting while minimizing the amount of light pollution. The key is first of all to direct the light to where it is most useful. That means that outdoor lighting should be shielded so that no light is emitted directly upward, where it does no good, but down toward streets and sidewalks and sports fields, where it is needed. Engineers have designed light fixtures that shield street lights in such a way that no light is projected upward.

The second step is to choose the proper source of light. Figure 13A shows the spectra of several types of light sources. Note that the low-pressure sodium lamp emits light at only a few discrete wavelengths. The rest of the spectrum is free of contaminating light, and this is the kind of light that astronomers prefer. Fortunately, low-pressure sodium lights are also extremely energy-efficient, are the most cost-effective form of street lighting, and are the best light for visual acuity, because they emit most of their energy in the yellow part of the spectrum, where the eye is most sensitive.

Activities

1. Check the lights in your own city. Are they shielded? What type of lights are they? You should

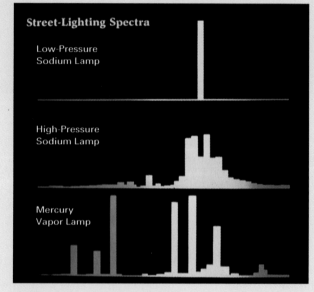

Figure 13A Comparison of the spectra of three forms of street lighting. Mercury vapor lights are the most common, but much of the energy is emitted in the red and blue regions of the spectrum, at wavelengths to which the eye is not very sensitive. In contrast, low-pressure sodium lamps emit all of their energy in a very narrow wavelength region where the eye is very sensitive. This type of lighting leaves most of the spectrum free of the radiation that would contaminate observations of faint stars and galaxies.

be able to tell from their color. Low-pressure sodium lights emit a pure yellow; high-pressure sodium are yellow with a pinkish cast; mercury vapor lights are white. Use the information in Figure 13A to explain why the colors of the three types of lights differ. If possible, obtain an inexpensive diffraction grating from your school or a local nature or science store and look at the spectrum of the lights and confirm your identification.

2. Look at the city lights the next time you are in a plane at night. Do you see the lights themselves? If so, then some of the light is being emitted directly upward, where it does no good. Or do you see only the light reflected from the pavement on the streets? In this case, the lighting is properly shielded.

3. If you live in or near a large city, check the sky. What part of it is contaminated with artificial light? In what directions is it brightest? Now drive away from the city and notice how the appearance of the sky changes. How far do you have to go before the sky appears to be truly dark? Calculations have been made to estimate how far away from a typical city one must be before the glow from the city's lights is only 10 percent brighter than a dark sky at a point halfway between the horizon and the zenith in the direction of the city. For a city of 30,000 people, one must be about 16 miles distant; for a city of

200,000 people, the distance is about 30 miles; for a city of 1,000,000 people, the distance is about 60 miles; and for a city of 5.5 million, the distance is about 125 miles. How do these numbers compare with your estimates?

4. Estimate how many stars have been lost to view because of light pollution where you live. One good constellation to use is the Little Dipper. What is the faintest star you can see in and around the Dipper (see Figure 13B)? To estimate the number of stars lost, assume that in a truly dark sky you can see to magnitude 6.0. There are approximately 4900 stars over the whole sky that are brighter than magnitude 6.0; 1700 stars are brighter than magnitude 5.0; 540 stars are brighter than magnitude 4.0; 150 stars are brighter than magnitude 3.0; only 45 are brighter than magnitude 2.0; and 16 are brighter than magnitude 1.0.

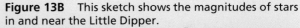

Figure 13B   This sketch shows the magnitudes of stars in and near the Little Dipper.

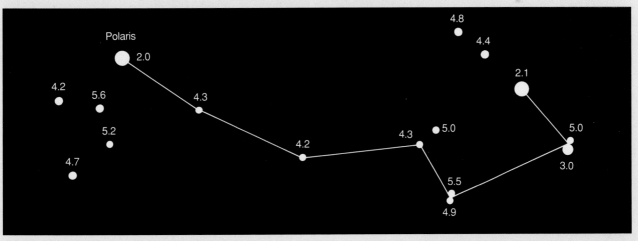

Figure 13.14 The spiral galaxy NGC 2997. The two spiral arms, which appear to originate in the yellow nucleus, are peppered with bright red blobs of ionized hydrogen that are similar to regions of star formation in our own Milky Way Galaxy. The hot blue stars that generate most of the light in the arms of the galaxy form within these hydrogen clouds. *(Copyright Anglo-Australian Telescope Board, 1980)*

Calculations have been made to show how stars and gas clouds would move if they were on circular paths around the galactic center, and if they were influenced both by the gravitational fields produced by the Galaxy as a whole and by the matter forming the spiral arms themselves. The calculations show that objects should slow down slightly in the regions of the spiral arms and linger there longer than elsewhere in their orbits, thus building up a wave of higher density where the spiral arms are. The regions of higher density rotate more slowly than does the actual material in the Galaxy, so that the stars, gas, and dust pass slowly through the spiral arms.

As a good analogy for what happens, suppose you are driving on a freeway with three lanes. Suppose that there are cars that are moving unusually slowly in all three lanes ahead of you. Traffic behind these three cars will be forced to slow down, and the density of cars will increase. Some individual cars may manage to get past the three slowly moving cars, but others will take their place in the traffic jam. Viewed from high above the freeway, the point of maximum density of cars would appear to move along the freeway at the same speed as that of the three slowly moving cars. The place of maximum density would thus be moving more slowly than the cars either well in front of the traffic jam or behind it. In just the same way, stars and interstellar clouds slow down when they pass through the spiral arms, which are the places of maximum density.

As gas and dust clouds approach the inner boundaries of an arm and encounter the higher density of more slowly moving matter, they collide with it. It is here, where the shock of the collision occurs, that theory predicts star formation is most likely to take place. We know from our own Galaxy that the youngest stars are found in the spiral arms. In some other galaxies, where we can view the spiral arms face-on, we see young stars, along with the most dense dust clouds, near the inner boundaries of spiral arms, just as theory predicts.

Molecular Clouds

The coldest hydrogen gas in the Galaxy has a temperature of less than about 30 K and is found mostly in the form of hydrogen molecules (H_2). The most massive molecular clouds are found in the spiral arms. In many cases, individual clouds have gathered into large complexes containing a dozen or more discrete clumps. Because the large molecular clouds and molecular cloud complexes are the sites where star formation occurs, most young stars are also to be found in spiral arms. That is why the spiral arms shine so brilliantly in photographs of spiral galaxies (Figure 13.14).

13.4
STELLAR POPULATIONS IN THE GALAXY

Striking correlations are found between the velocities of stars and their other characteristics, including age, chemical composition, velocities, and location in the Galaxy. Some classes of stars, which are referred to as a group as **population I** stars, are found only in regions of interstellar matter. Such stars are restricted to the disk and are especially concentrated in the spiral arms. Examples include bright supergiants, main-sequence stars of high luminosity (spectral classes O and B), and young, open-star clusters. Molecular clouds are found in similar places.

The distributions of some other classes of objects, which are collectively referred to as **population II,** show no correlation with the location of spiral arms. These objects are found throughout the disk of the Galaxy, with the greatest concentration toward the nuclear bulge. They also extend into the galactic halo. Examples are planetary nebulae, since most planetary nebulae are formed by old stars with masses similar to that of the Sun, and RR Lyrae variables. The globular clusters, which are also classified as population II, are found almost entirely in the halo and central bulge of the Galaxy.

Figure 13.15 The orbits of stars in the Galaxy. The Sun and other population I stars orbit in the plane of the Galaxy. Population II stars have orbits that move high above and below the plane.

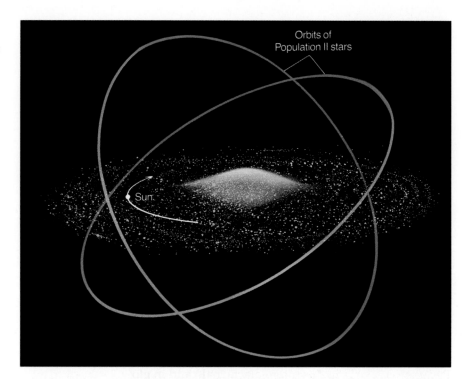

The majority of the stars near the Sun are population I stars, and they move nearly parallel to the Sun's path around the galactic nucleus. Their speeds with respect to the Sun are generally less than 50 km/s. The interstellar material near the Sun shares this motion also. Therefore, population I objects are often referred to as *low velocity* objects.

Conversely, population II objects typically have speeds relative to the Sun in excess of 80 km/s. They move along orbits of rather high eccentricity that cross the Sun's orbit in the plane of the Galaxy at large angles (Figure 13.15). Stars in the galactic halo and globular clusters also appear to move at high speed with respect to the Sun's orbit. For this reason, population II objects are referred to as *high velocity* objects. A globular cluster must pass through the plane of the Galaxy twice during each revolution, but the large distances between stars, both within the cluster and within the Galaxy itself, make stellar collisions during the cluster's passage through the galactic disk exceedingly improbable.

Discovery of Two Types of Stellar Populations

The terms "population I" and "population II" were first applied to different classes of stars by Walter Baade, of Mount Wilson Observatory. During World War II, many astronomers devoted most of their time to war research. As a German national, Baade was not allowed to work for the military, so he was able to make a great deal of use of the Mount Wilson telescopes. Aided by reduced sky brightness resulting from the wartime blackout of Los Angeles, Baade was able to distinguish the two

How do population I stars differ from population II stars?

populations of stars in the nearby Andromeda Galaxy. He was impressed by the similarity of the stars in the nuclear bulge of the Andromeda Galaxy to those in the globular clusters and the halo of our own Galaxy, and by the differences between all of these stars and those found in the spiral arms. On this basis, he referred to all the stars in the halo, nuclear bulge, and globular clusters as population II. The bright blue stars in the spiral arms were called population I.

Today we can interpret the phenomenon of different stellar populations in light of stellar evolution. Thus, population I contains stars of many different ages, including some that were recently formed or are still forming from gas and dust in the galactic disk. Population II, however, consists entirely of old stars, formed early in the history of the Galaxy.

Chemical Composition

Measurements show that there are differences in the chemical compositions of population I and population II stars. Nearly all stars appear to be composed mostly of hydrogen and helium, but the abundance of the heavier elements is not the same in all stars. In the Sun and in other population I stars, the heavy elements (elements heavier than hydrogen and helium) account for about 1 to 4 percent of the total stellar mass. Population II stars in the outer galactic halo and in globular clusters have much lower abundances of the heavy elements—often less than one-tenth or even one-hundredth that of the Sun.

The abundance of heavy elements in stars varies systematically with the time that stars were born. Old population II stars, which were formed 10 billion or more years ago, have a lower abundance of heavy elements than do the Sun and other population I stars. This result indicates that the heavy-element content of the Galaxy has increased over time. As discussed in earlier chapters, we expect heavy elements to be created in stars. These heavy elements are returned to the interstellar medium when stars die, only to be recycled by becoming part of a new generation of stars.

Since Baade's pioneering work, we have learned that the notion that all stars can be characterized as either old, with low abundances of elements heavier than hydrogen and helium, or young and rich in heavy elements is an oversimplification. The stars in the nuclear bulge of the Galaxy are all fairly old. Their mean age is in the range 11–14 billion years and none is younger than about 5 billion years, yet the abundance of heavy elements in these stars is about twice that of the Sun. Astronomers think that star formation in the nuclear bulge occurred very rapidly shortly after the Milky Way Galaxy formed, so even the 11–14 billion year old stars were enriched with heavy elements expelled in supernova explosions of the first generations of massive stars.

A completely opposite situation occurs in the Small Magellanic Cloud, a small satellite galaxy that lies close to the Milky Way Galaxy. Even the youngest stars in this galaxy are deficient in heavy elements, presumably because star formation has occurred so slowly that there have been so far relatively few supernova explosions. Therefore, the abundance of a star depends not only on when it forms but also on how many generations of nearby stars have already completed their evolution prior to the time that the star began its gravitational collapse from interstellar matter.

THE MASS OF THE GALAXY

The Galactic Year

Like a gigantic planetary system, the entire Galaxy is spinning, albeit rather slowly. Like the solar system, it does not turn as a rigid solid mass. Rather, each star or dust cloud, obeying the rules of Newtonian gravitation, follows its own orbit around the center.

The motion of the Sun in the Galaxy is deduced from the apparent motions of objects surrounding us that do not move in similar, nearly circular orbits with the same speed as the Sun. The globular clusters are one such group of objects. These clusters are moving, but the fact that they are found in a spheroidal distribution, rather than being confined to the flat plane of the Galaxy, is evidence that the system of globular clusters as a whole is not rotating as rapidly as the disk of the Galaxy. Their orbits about the galactic center can be likened to the orbits of comets around the Sun, whereas the Sun and other objects in the disk have more nearly circular orbits, like the planets in the solar system (see Figure 13.15).

By analyzing the radial velocities of the globular clusters in various directions, we can determine the motion of the Sun with respect to them. In one direction, we seem to be approaching the globular clusters; in the opposite direction, we seem to be receding from the globular clusters. This observation immediately tells us the direction in which the Sun is revolving about the center of the Galaxy. The motion of the Sun in the Galaxy can also be deduced from radial velocities of nearby external galaxies.

When the data from various sources are combined, they indicate that the Sun and most other stars in the solar neighborhood are moving in the direction of the constellation Cygnus, with a speed of about 220 km/s. This direction lies in the Milky Way Galaxy and is about 90° from the direction of the galactic center, which shows that the Sun's orbit is nearly circular and lies in the main plane of the Galaxy. As viewed from the north side of the galactic plane, the orbital motion of the Sun is clockwise (the opposite of the direction of planetary orbits around the Sun).

The period of the Sun's revolution about the nucleus, the *galactic year*, can be found by dividing the circumference of the Sun's orbit by its speed. The period turns out to be roughly 200 million of our terrestrial years. We can observe, therefore, only a "snapshot" of the Galaxy in rotation; we do not actually see stars traverse appreciable portions of their orbits. Since the formation of the solar system, only about 20 galactic years have passed.

The Mass of the Galaxy

We can make an estimate of the mass of the inner part of the Galaxy (the part that lies inside the Sun's orbit) with an application of Kepler's third law (as modified by Newton). Assume the Sun's orbit to be circular and the Galaxy to be roughly spherical so we can treat it as though the mass inside the orbit of the Sun were concentrated at a point at the galactic center. If the Sun is 28,000 LY from the center, its orbit has a radius of 1.8×10^9 AU (there are 6.3×10^4 AU in one LY). Because its period is 2×10^8 yr, we have

$$\text{Mass}_{\text{Galaxy}} = (1.8 \times 10^9)^3/(2 \times 10^8)^2 = 10^{11} \text{ solar masses}$$

More sophisticated calculations based on complicated models give a similar result.

It must be emphasized that this is only the mass contained in the volume inside the Sun's orbit. It is a good estimate for the total mass of the Galaxy if and only if no more than a small fraction of its mass is to be found beyond the radius that marks the Sun's distance from the galactic center. For many years, astronomers thought this assumption was reasonable, because the number of bright stars and the amount of luminous matter drop dramatically at distances more than about 30,000 LY from the galactic center.

A Galaxy of Mostly Invisible Matter

In science, reasonable assumptions can turn out to be wrong. Observations now show that although there is relatively little luminous matter lying beyond 30,000 LY, there must be a lot of nonluminous, invisible, **dark matter** at large distances from the galactic center.

We can understand how astronomers reached this conclusion by remembering that according to Kepler's third law, objects orbiting at large distances from a massive object move more slowly in their orbits than do objects closer to that central mass. We have already seen an example of this idea in the case of the solar system. The outer planets move more slowly in their orbits than do the ones close to the Sun.

There are a few objects, including globular clusters, that lie well outside the luminous boundary of the Milky Way. If all of the mass of our Galaxy were concentrated within the luminous region, then these very distant objects should travel around their galactic orbits at lower speeds than, for example, the Sun does. In just the same way, the velocity of Pluto in its orbit is slower than the velocity of the Earth.

In fact, globular clusters and other objects at large distances from the luminous boundary of the Milky Way Galaxy are *not* moving more slowly than the Sun. As we look outward to the few observable objects between 30,000 and 150,000 LY from the center of the Galaxy, we find that their orbital velocities remain nearly constant at values between

How do astronomers measure the mass of the Galaxy?

What is the evidence that most of the mass in the Galaxy is invisible?

Figure 13.16 Relationship between velocity of rotation of the Galaxy and distance from its center. The red line shows observed velocities obtained from measurements of carbon monoxide (CO) gas. The blue curve shows what the rotation curve would look like if all of the matter in the Galaxy were located inside a radius of 30,000 LY.

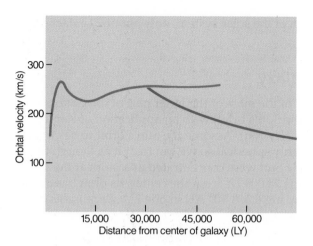

about 250 and 300 km/s (Figure 13.16). The only way that this can happen is if there is a huge amount of additional mass beyond the *visible* boundary of the Milky Way Galaxy that can hold these objects in their high speed orbits. Studies of the motions of very remote globular clusters show that the total mass of the Galaxy out to a radius of 150,000 LY is about 10^{12} solar masses, 10 times greater than the amount of matter inside the solar orbit. Theoretical arguments suggest that this **dark matter** is distributed in a spherical halo around the Galaxy. But what is it? This matter is, except for its gravitational force, entirely invisible and undetectable!

Because this matter is invisible, it cannot be in the form of ordinary stars. It cannot be gas in any form. If it were neutral hydrogen, its 21-cm radiation would have been detected. If it were ionized hydrogen, it would emit visible radiation. If it were molecular hydrogen, absorption bands would have been observed in ultraviolet spectra of objects lying beyond the Galaxy. The halo cannot consist of interstellar dust, because dust in the required quantities would block the light from distant galaxies. The dark matter cannot be black holes of stellar mass or neutron stars, or accretion of interstellar matter onto such objects would produce more x rays than are observed. Also, formation of black holes and neutron stars is preceded by supernova explosions, which scatter heavy elements into space to be incorporated into subsequent generations of stars. If the halo mass consisted of stellar size black holes and neutron stars, then the young stars that are shining today would have to contain much larger abundances of heavy elements than they actually do.

The possibilities that remain to account for the dark matter in the Galaxy are low-mass objects such as brown dwarfs, white dwarfs that formed from an early generation of stars and that have now cooled and ceased to shine, black holes that have masses a million times the mass of the Sun, or exotic subatomic particles of a type not yet detected on Earth. Stop a moment to think how startling this conclusion is. About 90 percent of the mass in our Galaxy is invisible, and we do not even know

Figure 13.17   This wide-angle picture covers over 50° of the sky in the direction of the center of the Milky Way. The galactic center is totally obscured at optical wavelengths.   *(Copyright Anglo-Australian Telescope Board, 1980)*

what it is made of! As we shall see, dark matter is a major constituent of other galaxies as well (Chapter 15).

13.6
THE NUCLEUS OF THE GALAXY

The center of the Galaxy lies in the direction of the constellation Sagittarius. We cannot see the nucleus in visible light or in the ultraviolet, because those wavelengths are absorbed by the intervening interstellar dust (Figure 13.17). In the optical region of the spectrum, light from the central region of the Galaxy is dimmed by a factor of a trillion (10^{12}). High-energy x rays and gamma rays, however, force their way through the interstellar medium and can be recorded by instruments orbiting above the Earth's atmosphere. Also, infrared and radio radiation, whose wavelengths are long compared with the sizes of the interstellar grains, flows around the dust particles and reaches us from the center of the

Figure 13.18 Infrared radiation can penetrate the dust that lies between us and the center of the Milky Way Galaxy. This photograph, at a wavelength of 2200 nm, shows the cluster of cool stars at the center of the Galaxy. *(National Optical Astronomy Observatories)*

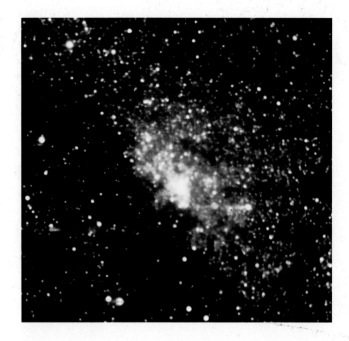

Galaxy (Figure 13.18). The very bright radio source in that region, known as Sagittarius A, was the first cosmic radio source discovered.

Astronomers have used observations at all of these wavelengths to try to determine what lies at the center of the Milky Way. The question of most interest to astronomers is whether or not there is a black hole at the center of the Galaxy. A great deal of effort has been made trying to answer this question, and the jury is still out.

The Central Few Light Years

What do these observational techniques tell us about the structure of the nucleus of the Milky Way Galaxy? The center of the Galaxy is surrounded by a clumpy and rather irregular disk of molecular clouds. These clouds form a donut-shaped ring that has an inner diameter of about 10 LY. The outer boundary of the ring extends to a distance of at least 25 LY from the galactic center. The individual molecular clouds are typically 0.5–1.5 LY in size (Figure 13.19). The ring appears to be rotating about the galactic center, but the motions of the individual clumps are turbulent, and the clouds must sometimes collide with each other.

It is possible to observe the spectrum of the gas in the ring and to use the Doppler effect to measure its orbital velocity. Just as the mass of the Sun provides the gravitational force that holds the planets in their orbits, so too there must be enough mass inside the ring to force it to orbit the center of the Galaxy. From the observed rotational velocity of the dust ring, we estimate that the total mass inside it must be 2 to 5 million times the mass of the Sun.

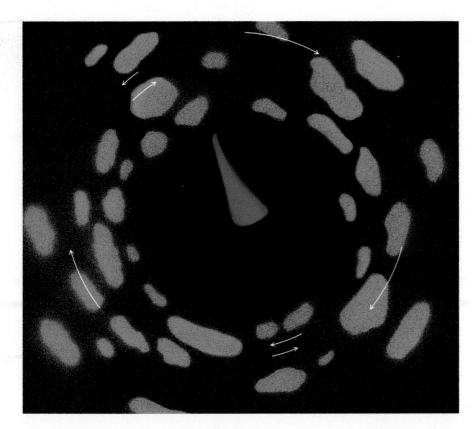

Figure 13.19 Schematic drawing of the central 20 LY of the Galaxy. Streamers of ionized gas are falling into the center of the Galaxy. Surrounding the central regions is a ring of dust, and beyond this lie individual clouds of gas.

This large mass is not composed of either gas or dust. Measurements show there is very little dust or cold gas inside the ring. There are some streamers of hot, ionized gas (Figure 13.20), but the total mass of the ionized gas in the central few LY of the Galaxy is only about 100 times the mass of the Sun. The most widely accepted idea is that this hot gas is falling from the ring toward the center of the Galaxy. As clouds in the ring collide, the velocity of some of the material in the clouds is slowed to the point that it can no longer remain in orbit. This material then spirals inward toward the galactic center.

The region inside the ring cannot remain empty of gas and dust for long. Over a period of only 100,000 years, material falling inward will blur the sharp edge that now separates the ring from its nearly empty interior. Over about that same time interval, collisions between the clouds will tend to eradicate the individual clumps and produce a smooth ring of material. In the last 100,000 years, some remarkable event must have severely disturbed the ring of gas and created the cavity inside it. One possibility is an explosion of some kind at the center of the Galaxy. According to this hypothesis, the explosion blew material outward and cleared the cavity of dust and gas.

If gas and dust do not account for most of the mass—a few million times the mass of the Sun—in the center of the Galaxy, is this mass in the

Figure 13.20 A radio image of a region approximately 10 LY across at the center of our Galaxy shows radio emission from hot gas. The most likely interpretation is that this gas is falling into the nucleus. The emission is strongest (red color) where the gas is most dense. The mass of this gas is about 100 times the mass of the Sun. The bright red region in the middle is the compact radio source at the center of the Galaxy. *(National Radio Astronomy Observatory/AUI)*

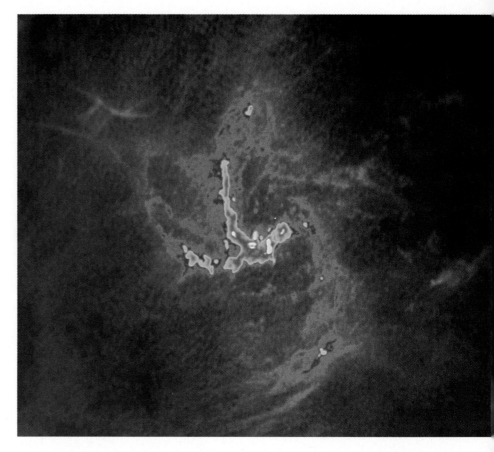

form of stars? There are stars inside the ring (Figure 13.21), and they are relatively close together. The average distance between stars is perhaps 1/300 the average distance between stars in the vicinity of the Sun. But even the mass of the stars is not enough to add up to the 2 to 5 million solar masses that we are looking for. Is this mass hidden inside a black hole?

The Central Energy Source—A Black Hole?

A black hole itself emits no electromagnetic radiation. We cannot hope to *see* a black hole in the center of the Galaxy even if one is there. We must use circumstantial evidence, as they say in mystery stories, to decide whether or not a black hole is the only plausible explanation for the kinds of things we *do* see.

The most direct way to prove that there is a black hole in the center of the Galaxy is to show that there is a large amount of mass within a very tiny volume. We can measure the velocities of stars and the small amounts of gas that lie very close to the galactic center—inside the ring—and then determine what gravitational force is exerted by the mass concentrated at the center. These measurements indicate that the

What is the evidence that a massive black hole lies at the center of the Galaxy?

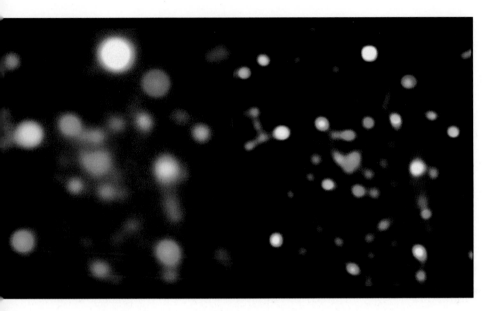

Figure 13.21 Infrared image of the stars in a region 2 LY across surrounding the galactic center. The picture shows the original image on the left, and a computer-enhanced version on the right, of the center of the Milky Way. The observations were made at 1.65 μm, 2.2 μm, and 3.45 μm, which are represented in the color image by blue, green, and red, respectively. The different colors of the stars are indicative of age, composition, and absorption by obscuring gas and dust clouds. *(Darren DePoy and Nigel Sharp/National Optical Astronomy Observatories)*

amount of mass located within one to two light years of the center of the Galaxy is equal to 3 to 4 million times the mass of the Sun.

If this mass is actually contained within a black hole, then the size of the black hole must be much smaller than one to two light years. A black hole with a mass of about 3 million solar masses would have a radius similar to that of our own Sun. Unfortunately, we cannot see fine enough detail with existing telescopes to measure the velocities of gas and stars that are only a few tens of astronomical units from the center of the Galaxy. Therefore, we do not yet know if the mass at the center of the Galaxy is as highly concentrated as we would expect for a black hole.

There is indirect evidence, however, that the mass may indeed be very highly concentrated. There is an intense radio source, which is called Sagittarius A*, at or very near what we believe is the exact center of the Galaxy. Measurements with the VLA radio telescope show that the diameter of this radio source is no larger than the diameter of Jupiter's orbit (10 AU), and it may be even somewhat smaller.

If there were a black hole at the center of the Galaxy, it could account for the intense radio emission from Sagittarius A*. Matter—gas, dust, and even perhaps stars—is attracted by the gravitational force of a black hole. Material spirals in toward the black hole and forms an *accretion disk* around it. As the material spirals ever closer to the black hole, it accelerates and heats through compression to millions of degrees. This hot matter then would be the source of the radio emission from the galactic center.

Suppose there is a black hole of several million solar masses in the center of the Galaxy. Where did the mass come from? At the present time, matter from such sources as colliding gas clouds in the ring is falling into the galactic center at the rate of about one solar mass per 1000 years. If matter had been falling in at the same rate for about 5 billion years, and we have no idea whether or not this is a reasonable assumption, then it would have easily been possible to accumulate the matter needed to form a black hole with a mass of several million solar masses.

The density of stars near the galactic center is such that we would expect a star to pass near the black hole and be disrupted by it every few thousand years. As this material falls into the black hole there should be a brilliant outburst. Perhaps such an outburst drove gas and dust out of the galactic center a few tens of thousands of years ago, leaving behind the cavity and the clumpy ring of molecular clouds that we now see. Between outbursts, according to this idea, the black hole should be fairly quiescent, as it is now.

Thus, scientists find a great deal of circumstantial evidence that there is a black hole in the center of the Galaxy. But a black hole may not be the only possibility. For example, gamma rays and radio emission can also be produced in an accretion disk around a neutron star. Over the next few years, astronomers will turn their new observational tools—gamma ray, x-ray, infrared, and radio telescopes—to the galactic center to try to prove or disprove the existence of a black hole at the heart of our own Galaxy. As we shall see in later chapters, there is evidence that black holes exist in other galaxies and in quasars as well.

13.7
THE FORMATION OF THE GALAXY

The flattened shape of our own Galaxy suggests that it formed through a process that was similar to the way in which the Sun and solar system formed (see Section 7.5). According to the traditional model, the Galaxy formed from a single rotating cloud. Because the oldest stars—stars in the halo and in globular clusters—are distributed in a sphere centered on the nucleus of the Galaxy, we can assume that the protogalactic cloud was spherical in shape. The oldest stars in the halo have ages of about 15 billion years. Therefore we estimate that the collapse of the gas cloud occurred about that long ago. As the cloud collapsed, it formed a thin disk. The stars that were formed before the cloud collapsed did not participate in the collapse but continue to orbit in the halo (Figure 13.22).

Gravitational forces within the disk caused the gas in the disk to fragment into clouds or clumps with masses like those of star clusters. These individual clouds then fragmented further to form individual stars. Because the oldest stars in the disk are nearly as old as the stars in the halo, the collapse was rapid (astronomically speaking) and required perhaps no more than a few hundred million years.

Some low mass members of the earliest generations of stars are still shining. The abundance of elements heavier than helium in these very old stars is 1 percent or less of the abundance of heavy elements in the Sun. More massive members of these earliest generations of stars have already exhausted their store of nuclear energy and have ceased to

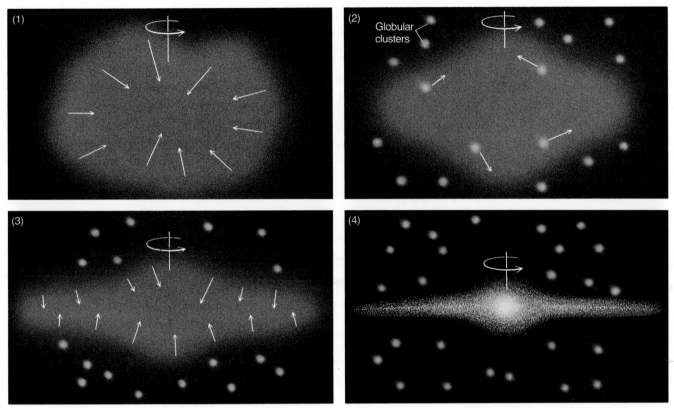

Figure 13.22 The Galaxy probably formed from an isolated, rotating cloud of gas that collapsed because of gravity. Halo stars and globular clusters either formed prior to the collapse or formed elsewhere and were attracted to the Galaxy early in its history by gravity. Stars in the disk formed late, and the gas from which they were made was contaminated with heavy elements produced in early generations of stars.

shine. These stars produced heavy elements in their interiors that were then recycled into the interstellar medium and subsequently incorporated into new generations of stars, which accordingly have a higher content of heavy elements. Most of the stars in the disk are several billion years younger than the halo stars and have high abundances of elements heavier than hydrogen and helium.

Astronomers have recently begun to find hints that the evolution of the Galaxy has not been quite as peaceful as this traditional model suggests. Globular clusters differ in age by as much as 3 billion years; they were not all formed during the few hundred million years before the interstellar matter collapsed to form the galactic disk, as the traditional model predicts. Studies also show that some globular clusters orbit the center of the Galaxy in a direction opposite to that of the majority of the globulars and of the galactic disk. These "backward" globular clusters appear to be 2–3 billion years younger than globular clusters that rotate in the normal direction, and they tend to lie farther from the center of the Galaxy.

Based on this evidence, some astronomers have suggested that the Galaxy formed in two stages. The first stage proceeded as shown in Figure 13.22. About 2 billion years after the formation of the majority of globular clusters, which rotate in the forward direction, the Galaxy acquired additional stars and globular clusters from one or more satellite galaxies, which ventured too close to the Milky Way Galaxy and were captured by it. (Remember that some of the satellites in our own solar system have retrograde orbits and are thought to be captured asteroids or comets.)

It appears that the Milky Way is on the brink of another collision. The Large and Small Magellanic Clouds, our two nearest satellite galaxies, are spiralling ever closer to the Milky Way Galaxy and will merge with it some time in the future. If such mergers can happen now, there seems to be no reason why they could not have happened in the past.

▶ SUMMARY

13.1 The Sun is located in the outskirts of the **Milky Way Galaxy.** The **Galaxy** consists of a **disk,** which contains dust, gas, and young stars, a **nuclear bulge,** which contains old stars, and a spherical **halo,** which contains very old stars, including the members of globular clusters and RR Lyrae variables.

13.2 Distances of clusters can be determined by use of the **period-luminosity** relationship for two types of pulsating variable stars: **cepheids** and **RR Lyrae** stars. Measurements of cepheids in the **Magellanic Clouds** were used to derive the period-luminosity relationship for these stars. Analysis by Shapley of the distribution of globular clusters gave the first indication that the Sun is not located at the center of the Galaxy.

13.3 Interstellar matter is found in the disk of the Galaxy. Hydrogen and helium account for 96 to 99 percent of interstellar matter. Radio observations at 21 cm show that atomic hydrogen is confined to a flat disk, which has a thickness of only 400 LY. Atomic hydrogen extends beyond the Sun to a distance of at least 80,000 LY from the galactic center. Dust is found in the same locations as atomic hydrogen inside the Sun's orbit; there is very little dust outside the Sun's orbit. Radio observations of atomic hydrogen show that the Galaxy has four main **spiral arms.** Molecular clouds and star formation tend to be concentrated in the spiral arms.

13.4 The properties of stars are closely related to their positions in the Galaxy. Old stars are referred to as **population II** stars and are found in the halo, in globular clusters, and in the nuclear bulge. Young **population I** stars are found in the disk and are especially concentrated in the spiral arms. Relative to the Sun, population I stars are low-velocity objects and population II stars are high-velocity objects. The Sun is a member of population I. Population I stars, which formed after previous generations of stars had produced heavy elements and ejected them into the interstellar medium, have higher metal abundances than do population II stars.

13.5 The mass of the Galaxy can be determined by measuring the orbital velocities of stars or interstellar matter. The Sun revolves completely around the center of the Galaxy in about 200 million years. The total mass of the Galaxy is about 10^{12} solar masses, and about 90 percent of this mass consists of **dark matter** that emits no electromagnetic radiation and can be detected only because of the gravitational force it exerts on visible stars and interstellar matter. This dark matter is mostly located in the halo; candidates for what the dark matter might be include brown dwarfs, white dwarfs, massive black holes, or some type of subatomic particle not yet detected on Earth.

13.6 The nucleus of the Galaxy contains a compact radio source. Measurements of the velocities of stars and gas show that the central 1 to 2 LY of the Galaxy contains a mass that is 2 to 5 million times the mass of the Sun. A massive black hole can explain these observations, but it is not yet certain that a black hole is the only possibility.

13.7 Models suggest that the stars in the halo and globular clusters formed first, while the Galaxy was spherical in shape. Some globular clusters may have been captured from satellite galaxies about 2 billion years after the formation of the majority of globulars. The gas, somewhat enriched in heavy elements by the first generation of stars, collapsed from a spherical distribution to a disk-shaped distribution. Stars are still forming today from the gas and dust that remain in the disk. Star formation occurs most rapidly in the spiral arms, where the density of interstellar matter is highest.

▶ REVIEW QUESTIONS

1. Explain why we see the Milky Way Galaxy as a faint band of light stretching across the sky.

2. Where in the Galaxy would you expect to find dust? molecular hydrogen? ionized hydrogen?

3. Describe several characteristics that distinguish population I stars from population II stars.

4. Describe the orbits of population I stars and population II stars around the center of the Galaxy.

5. Why is it difficult to determine whether or not there is a black hole in the center of the Galaxy?

6. Explain why the heavy element abundances of stars correlate with their positions in the Galaxy.

▶ THOUGHT QUESTIONS

7. Suppose the Milky Way were a band of light extending only halfway around the sky (that is, in a semicircle). What then would you conclude about the Sun's location in the Galaxy? Give your reasoning.

8. The globular clusters probably have highly eccentric orbits and either oscillate through the plane of the Galaxy or revolve about its nucleus. Suppose the latter is the case; where would the clusters spend most of their time? (Think of Kepler's laws.) At any given time, would you expect most globular clusters to be moving at high or low speeds with respect to the center of the Galaxy? Why?

9. Consider the following five types of objects: (1) open cluster, (2) giant molecular cloud, (3) globular cluster, (4) a group of O and B stars, (5) planetary nebula.

a. Which one or ones are found only in spiral arms?

b. Which one or ones are found only in the parts of the Galaxy that are not in the spiral arms?

c. Which are thought to be very young?

d. Which are thought to be very old?

e. Which have stars that are of the highest temperatures?

10. Where in the Galaxy do you suppose undiscovered globular clusters may exist?

11. Why does star formation occur primarily in the disk of the Galaxy?

12. Where in the Galaxy would you expect to find supernovae of Type II, which are the explosions of massive stars? Where would you expect to find supernovae that involve explosions of white dwarfs?

13. According to Chapter 12, astronomers believe that there is one supernova explosion somewhere in our Galaxy every 25 to 100 years, yet we have not seen a galactic supernova in nearly 400 years. Is it likely that we would be able to observe *every* galactic supernova explosion? Why or why not?

14. Based on the information in the text, would you classify cepheid variables as population I stars or population II stars? Which class of variable stars do you think has a higher abundance of heavy elements—RR Lyrae stars or cepheids?

15. Suppose that stars evolve without losing mass. Once matter is incorporated into a star it remains there forever. How would the appearance of the Galaxy be different from what it is now? Would there be population I and population II stars? What other differences would there be?

▶ **PROBLEMS**

16. Suppose that the mean mass of a star in the Galaxy were only one-third solar mass. Use the value for the mass of the Galaxy found in the text to find how many stars the system contains.

17. The Sun orbits the center of the Galaxy at a speed of 220 km/s, 30,000 LY from the center.
a. Calculate the circumference of the Sun's orbit, assuming it to be approximately circular.
b. Calculate the Sun's period, the "galactic year."
c. Use Newton's formulation of Kepler's third law to calculate the mass of the Galaxy inside the orbit of the Sun.
d. It is estimated that the total mass of the Galaxy is five times that within the Sun's orbit. If this mass were to collapse inside the orbit of the Sun, what would be the length of the galactic year?
e. Based on your answer to (d), what would be the Sun's new orbital speed?

18. Construct a rotation curve for the solar system like that shown in Figure 13.16. Use the orbital velocities of the planets (see Appendix 7). How does this curve differ from the rotation curve for the Galaxy? What does this curve tell you about where most of the mass in the solar system is concentrated?

19. Suppose that all of the mass in our Galaxy lies inside the orbit of the Sun, which for convenience in calculation we will assume to be circular with a radius of 30,000 LY; assume that the Sun's orbital velocity is 220 km/s. Use Kepler's third law to find the orbital speed of a cloud of atomic hydrogen at a distance of 60,000 LY. In other words, verify the data in Figure 13.16. What would be the orbital speed of a globular cluster at a distance of 150,000 LY?

20. Calculate how much mass would fall into the black hole at the center of the Galaxy in 1 billion years at the current rate of infall of one solar mass per 1000 years.

▌ SUGGESTIONS FOR ADDITIONAL READING

Books

On the Galaxy

Davis, J. *Journey to the Center of Our Galaxy.* 1991, Contemporary Books. A journalist summarizes our current picture of the Milky Way Galaxy.

Verschuur, G. *The Invisible Universe Revealed.* Springer Verlag, 1987. Has chapters on galactic structure and the galactic center as revealed through radio astronomy.

Goldsmith, D. & Cohen, N. *Mysteries of the Milky Way.* Contemporary Books, 1991. A series of vignettes about Milky Way research issues.

Whitney, C. *The Discovery of the Galaxy.* A. Knopf, 1971. A history of how we found out about the properties of our Galaxy.

On Dark Matter

Krauss, L. *The Fifth Essence: Dark Matter in the Universe.* Basic Books, 1989.

Parker, B. *Invisible Matter and the Fate of the Universe.* Plenum Press, 1989.

Trefil, J. *The Dark Side of the Universe.* Scribners, 1988.

Tucker, W. & K. *The Dark Matter.* Morrow, 1988.

Articles

On the Galaxy

Verschuur, G. "Journey into the Galaxy" in *Astronomy*, Jan. 1993, p. 32.

Chaisson, E. "Journey to the Center of the Galaxy" in *Astronomy*, Aug. 1980, p. 6.

Dame, T. "The Molecular Milky Way" in *Sky & Telescope*, July 1988, p. 22.

van den Bergh, S. & Hesser, J. "How the Milky Way Formed" in *Scientific American*, Jan. 1993, p. 72.

Palmer, S. "Unveiling the Hidden Milky Way" in *Astronomy*, Nov. 1989, p. 32.

Verschuur, G. "Is the Milky Way an Interacting Galaxy?" in *Astronomy*, Jan. 1988, p. 26.

Gingerich, O. "The Discovery of the Milky Way's Spiral Arms" in *Sky & Telescope*, July 1984, p. 10.

Jones, B. "William Herschel: Pioneer to the Stars" in *Astronomy*, Nov. 1988, p. 40.

Hoskin, M. "William Herschel and the Making of Modern Astronomy" in *Scientific American*, Feb. 1986.

Verschuur, G. "The Magnetic Milky Way" in *Astronomy*, June 1990, p. 32.

Townes, C. & Genzel, R. "What Is Happening at the Center of the Galaxy?" in *Scientific American*, Mar. 1990.

Kaufmann, W. "Our Galaxy" in *Mercury*, May/June 1989, p. 79; July/Aug. 1989, p. 117.

On Dark Matter

Bartusiak, M. "Wanted: Dark Matter" in *Discover*, Dec. 1988, p. 62.

Trimble, V. "The Search for Dark Matter" in *Astronomy*, Mar. 1988, p. 18.

Tucker, W. & K. "Dark Matter in Our Galaxy" in *Mercury*, Jan./Feb. 1989, p. 2; Mar./Apr. 1989, p. 51.

GALAXIES AND THE STRUCTURE OF THE UNIVERSE

The nearby spiral galaxy M83. This galaxy is about 10 million LY away from Earth and has a diameter of 30,000 LY. *(Anglo-Australian Telescope Board)*

The final three chapters of this book consider the entire universe. Looking outward from the Milky Way Galaxy, of which the Earth is a part, we will probe to ever greater distances until we reach the limits of what we can observe—at the same time looking backward in time to the period before stars or galaxies formed.

The study of the structure, origin, and evolution of the universe is called **cosmology**. This is the same word we introduced in Chapter 1 to describe theories of the structure and motions of the solar system. It might seem strange that the term "cosmology" would be applied to such different subjects, but the evolution of the term reflects our changing understanding of the cosmos. To

Copernicus and Galileo, the solar system was the cosmos. Today we have a much grander concept of the universe, and the term cosmology has grown to accommodate this concept. Therefore, this chapter focuses on galaxies and their distribution in space. Subsequent chapters will describe the formation and evolution of galaxies and of the universe as a whole.

14.1
DISCOVERY OF GALAXIES

Faint star clusters, glowing gas clouds, dust clouds reflecting starlight, and galaxies all appear as faint, unresolved patches of light when viewed visually with telescopes of moderate size. Because the true natures of these various objects were not known to early observers, all of them were called "nebulae." *Nebula* (plural *nebulae*) literally means "cloud."

The "analogy [of the nebulae] with the system of stars in which we find ourselves . . . is in perfect agreement with the concept that these elliptical objects are just [island] universes—in other words, Milky Ways . . ." So wrote Immanuel Kant (1724–1804) concerning the faint patches of light that telescopes revealed in large numbers. Despite Kant's speculation that these patches of light are giant systems of stars like our own Milky Way Galaxy, their true nature remained a subject of controversy until 1924.

Galactic or Extragalactic?

One of the earliest catalogues of nebulous-appearing objects was prepared in 1781 by French astronomer Charles Messier (1730–1817). Messier was a comet hunter, and he made a list of 103 fuzzy-looking objects that might be mistaken for comets. Messier's list contains some of the most conspicuous star clusters, nebulae, and galaxies in the sky, and these objects are often referred to by their numbers in his catalogue—for example, M31, the great galaxy in Andromeda.

By 1908, nearly 15,000 nebulae had been catalogued and described. Some had been correctly identified as star clusters and others were identified as gaseous nebulae (such as the Orion Nebula). The nature of most of them, however, still remained unexplained. Were they luminous clouds of gas in our own Galaxy? Or were they remote unresolved systems of thousands of millions of stars—galaxies in their own right?

The Resolution of the Controversy

The controversy was finally resolved by the discovery of variable stars in some of the nearer nebulae in 1923 and 1924. Edwin Hubble (Figure 14.1), working with the 100-in. (2.5-m) telescope at the Mt. Wilson Observatory, analyzed the light curves of variables that he had discovered in M31, M33, and NGC 6822 and found that they were cepheids. Although cepheid variables are supergiant stars, those studied by

Is the Milky Way Galaxy the only galaxy in the universe?

Figure 14.1 Edwin Hubble (1889–1953) left a career in law to study astronomy. In 1919 he joined the staff of the Mount Wilson Observatory, where he discovered that the nebulae are galaxies like our own Milky Way. He was also the discoverer of the expansion of the universe, as described by the Hubble law. *(California Institute of Technology)*

Hubble were very faint—near apparent magnitude 18. Those stars, therefore, and the systems in which they were found, must be very far away. He determined that the "nebulae" must lie far beyond the boundary of the Milky Way and therefore must be galaxies.

14.2
TYPES OF GALAXIES

The majority of optically bright galaxies fall into two general classes: **spirals** and **ellipticals**. A minority are classified as **irregular**.

Spiral Galaxies

Our own Galaxy and the Andromeda Galaxy (Figure 14.2), which is believed to be much like ours, are typical, large, spiral galaxies. Like our Galaxy (Chapter 13), a spiral consists of a nucleus, a disk, a halo, and spiral arms. Interstellar material is usually spread throughout the disks of spiral galaxies. Bright emission nebulae are present, and absorption of light by dust is also often apparent, especially in those systems turned almost edge-on to our line of sight (Figure 14.3). The spiral arms contain the young stars, which include luminous supergiants. These bright stars and the emission nebulae make the arms of spirals stand out like the arms of a fourth-of-July pinwheel (see the photograph at the beginning

What are the three types of galaxies?

Figure 14.2 The nearby spiral galaxy in Andromeda, which is similar in size and structure to the Milky Way Galaxy. Its distance is about 2.1 million LY and its diameter is about 100,000 LY. This galaxy contains over 3 billion stars. M32, a companion galaxy, also appears in the picture. *(National Optical Astronomy Observatories)*

of this chapter). Open star clusters can be seen in the arms of nearer spirals, and globular clusters are often visible in their halos. In M31, for example, more than 200 globular clusters have been identified. Spiral galaxies contain both young and old stars.

Perhaps one-third or more of spiral galaxies display conspicuous bars running through their nuclei. The spiral arms of such a system usu-

Figure 14.3 The Sombrero Galaxy, NGC 4594. This galaxy, which is more than 24 million LY away, is surrounded by a prominent dust ring almost 50,000 LY across. The central spheroidal region has the orange appearance characteristic of an older population of stars. *(N. Sharp/National Optical Astronomy Observatories)*

ally begin from the ends of the bar, rather than winding out directly from the nucleus. Such galaxies are called barred spirals (Figure 14.4). Recent observations suggest that the Milky Way Galaxy also has a bar.

In both normal and barred spirals we observe a gradual transition of shapes. At one extreme, the nuclear bulge is large and luminous, the arms are faint and tightly coiled, and bright emission nebulae and supergiant stars are inconspicuous. At the other extreme are spirals in which the nuclear bulges are small—almost lacking—and the arms are loosely wound, or even wide open. In these latter galaxies, population I objects such as luminous stars, star clusters, and emission nebulae are common. Our Galaxy and M31 are intermediate between these two extremes. Photographs of spiral galaxies, illustrating this transition of types, are shown in Figure 14.5.

Spiral galaxies range in diameter from about 20,000 LY to more than 100,000 LY, and the atomic hydrogen in the disks often extends to far greater diameters. The masses of spiral galaxies range from 10^9 to 10^{12} times the mass of the Sun. The total luminosities of most spirals fall in the range of 10^8 to 10^{11} L_s. Our Galaxy and M31 are relatively large and massive, as compared with an average spiral galaxy.

Figure 14.4 NGC 4650, a barred spiral in the constellation Centaurus. *(National Optical Astronomy Observatories)*

Figure 14.5 (a) Types of spiral galaxies. (b) Types of barred spiral galaxies. *(CIT/Palomar Observatory)*

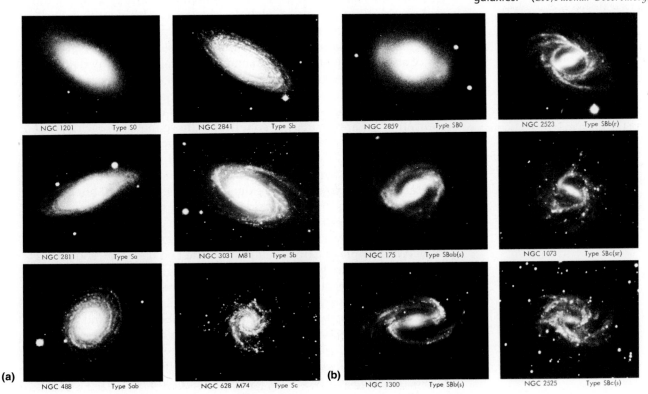

(a)

NGC 1201 Type S0	NGC 2841 Type Sb
NGC 2811 Type Sa	NGC 3031 M81 Type Sb
NGC 488 Type Sab	NGC 628 M74 Type Sc

(b)

NGC 2859 Type SB0	NGC 2523 Type SBb(r)
NGC 175 Type SBab(s)	NGC 1073 Type SBc(sr)
NGC 1300 Type SBb(s)	NGC 2525 Type SBc(s)

Figure 14.6 Types of elliptical galaxies. *(Palomar Observatory, California Institute of Technology)*

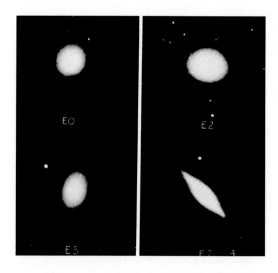

Elliptical Galaxies

Elliptical galaxies are spherical or ellipsoidal systems that consist almost entirely of old stars. They contain no traces of spiral arms. Their light is dominated by red stars (population II), and in this respect, ellipticals resemble the nuclear bulge and halo components of spiral galaxies. In the larger nearby ellipticals, many globular clusters can be identified. Dust and emission nebulae are not conspicuous in elliptical galaxies, although many ellipticals contain a small amount of interstellar matter. Narrow lanes of absorbing dust are seen in many ellipticals, and x-ray data indicate that 1 to 2 percent of the total mass of ellipticals may be in the form of gas at a temperature that exceeds 1 million degrees.

Elliptical galaxies show various degrees of flattening, ranging from systems that are approximately spherical to those that approach the flatness of spirals (Figure 14.6). The rare giant ellipticals (for example, M87, shown in Figure 14.7) reach luminosities of 10^{11} L_s. The mass of the stars in giant ellipticals is typically at least 10^{12} times the mass of the Sun. The diameters of these large galaxies extend over at least several hundred thousand light years and are considerably larger than the largest spirals.

Elliptical galaxies range all the way from the giants, which we have just described, to dwarfs, which we think are the most common kind of galaxy. An example of a dwarf elliptical is the Leo II system, shown in Figure 14.8. There are so few bright stars in this galaxy that even its central regions are transparent. However, the total number of stars (most of which are too faint to show in Figure 14.8) is probably at least several million. The luminosity of this typical dwarf is about 10^6 L_s and is about equal to the luminosity of the brightest known individual stars or to the brightest globular clusters.

Intermediate between the giant and dwarf elliptical galaxies are systems such as M32 and NGC 205, two companions of M31. They can be seen in the photograph of M31 (see Figure 13.5); NGC 205 is the system that is farther from M31.

Figure 14.7   The giant elliptical galaxy called M87, which has an active nuclear region that is a prominent source of radio and x-ray emission. *(Copyright Anglo-Australian Telescope Board)*

Irregular Galaxies

As many as 25 percent of all galaxies fall into the class of irregular galaxies, a term that includes a wide variety of sizes and shapes. Irregular galaxies contain stars of both population I and population II.

The two best-known irregular galaxies are the Large and Small Magellanic Clouds (Figures 14.9 and 14.10), our nearest extragalactic neighbors. Although they are invisible from the United States or Europe, these two systems are prominent when viewed from the Southern Hemisphere, where they look like wispy clouds detached from the

Figure 14.8   Leo II, a dwarf elliptical galaxy.   *(R. Schild, Center for Astrophysics)*

Figure 14.9   The Large Magellanic Cloud, a satellite of our own Galaxy, is visible to the naked eye from the Southern Hemisphere. The large red nebula (the Tarantula Nebula) is a site of active star formation and contains many young supergiant stars. *(National Optical Astronomy Observatories)*

Milky Way. They are only about one-tenth as distant as the Andromeda spiral. The Large Cloud contains the 30 Doradus complex, which is one of the largest and most luminous groupings of supergiant stars and associated gas known in any galaxy. Supernova 1987A (Section 12.4) occurred near 30 Doradus.

The Small Magellanic Cloud is greatly elongated and considerably less massive than the Large Magellanic Cloud. The length of this narrow wisp of material is six times longer than its width, and it is pointed directly toward our Galaxy, stretching from about 150,000 LY out to 250,000 LY. Apparently the Small Cloud is the victim of a near-collision with the Large Cloud that took place some 200 million years ago. It is now being pulled apart by the gravitation of the Milky Way.

Figure 14.10   The Small Magellanic Cloud. This dwarf irregular galaxy is a satellite of the Milky Way Galaxy. *(Cerro Tololo Inter-American Observatory/ National Optical Astronomy Observatories)*

Masses of Galaxies

We determine the masses of galaxies, like those of other astronomical bodies, by measuring their gravitational influences on other objects or on the stars within them. Internal motions in galaxies provide the most reliable methods for measuring their masses. The procedure for the measurement of spiral galaxies is to observe the rotation of a galaxy from the Doppler shifts of radio lines of neutral hydrogen (H) or carbon monoxide (CO), and then to compute its mass with the help of Kepler's third law. For example, such observations of M31, the Andromeda Galaxy, show it to have a mass (within the main visible part of the galaxy, out to a distance of 100,000 LY from its center) of about 4×10^{11} solar masses, which is about the same as the mass of our own Galaxy. (The total mass of M31 is higher than 4×10^{11} solar masses because we have not included the material that lies more than 100,000 LY from its center. Like our own Galaxy, Andromeda appears to have a large amount of dark matter that lies beyond its luminous boundary.)

Elliptical galaxies are not highly flattened and are not in rapid rotation. Nevertheless, the velocities of the stars in such a galaxy depend on its gravitational attraction for them, and thus on its mass. The spectrum of a galaxy is a composite of the spectra of its many stars, whose different motions produce different Doppler shifts. The lines in the composite spectrum, therefore, are broadened, and the amount by which they are broadened indicates the range of speeds with which the stars are moving with respect to the center of mass of the galaxy. The range of speeds depends, in turn, on the force of gravity that holds the stars within the galaxies. With information about the speeds, it is then possible to calculate the mass of the galaxy.

How do we measure the masses of galaxies?

Mass-to-Light Ratio

A shorthand method of characterizing a galaxy is by giving the ratio of its mass, in units of the solar mass, to its light output, in units of the solar luminosity. For the Sun, of course, this ratio would be unity. Galaxies, however, are not composed entirely of stars that are like the Sun. The overwhelming majority of stars are less luminous than the Sun, and usually these stars contribute most to the mass of a system, without accounting for very much light. Thus, the **mass-to-light ratio** is generally greater than one. Galaxies in which star formation is still occurring tend to have mass-to-light ratios between 1 and 10 while in galaxies consisting mostly of an older stellar population, the ratio is between 10 and 20.

The comments in the preceding paragraph refer only to the inner, more conspicuous parts of galaxies. In Chapter 13 we discussed the evidence for invisible matter in the outer disk and halo of our own Galaxy, extending much farther from the galactic center than do the more conspicuous bright stars. Recent measurements of the rotations of outer parts of nearby galaxies, such as the Andromeda spiral, suggest that they too have extended distributions of dark matter around the visible disk of stars and dust. This largely invisible matter adds to the mass of the galaxy while contributing nothing to its luminosity, resulting in high mass-to-light ratios. If dark, invisible matter is present in a galaxy, its mass-to-light ratio can be as high as 100. The mass-to-light ratios measured for various types of galaxies are given in Table 14.1.

The outer parts of at least some galaxies, therefore, contain matter that has not yet been identified by its light but only by means of its gravitational influence. The anomaly is sometimes called the missing mass problem. In fact, the mass is there, it is the light that is missing.

What is this dark matter? Extensive searches have been made for both hot and cold gas, and neither is present in sufficient quantity to account for the dark matter. The most likely explanation seems to be that the dark matter in galaxies is composed of very massive collapsed ob-

TABLE 14.1 Characteristics of Galaxies of Different Types

	Spirals	Ellipticals	Irregulars
Mass (solar masses)	10^9 to 10^{12}	10^6 to 10^{13}	10^8 to 10^{11}
Diameter (thousands of LY)	20 to 200	5 to 500	5 to 30
Luminosity (solar units)	10^8 to 10^{11}	10^6 to 10^{11}	10^7 to 2×10^9
Population content of stars	Old and young	Old	Old and young
Composite spectral type	A to K	G to K	A to F
Interstellar matter	Both gas and dust	Almost no dust; little gas	Much gas; some are deficient in dust; others contain large quantities of dust
Mass-to-light ratio	2 to 20	100	1

jects (black holes), stellar-like objects that are not massive enough to burn hydrogen (brown dwarfs), or massive neutrinos or exotic subnuclear particles (see Chapter 16).

Whatever the composition of the dark matter, these measurements of other galaxies support the conclusion reached already from studies of the rotation of our own Galaxy—namely that probably 90 percent or more of all the material in the universe cannot be observed directly in any part of the electromagnetic spectrum.

The light that we see from galaxies does not trace the bulk of material that is present in space, and thus may present us with a very misleading picture of the large-scale structure of the universe. An understanding of the properties and distribution of this invisible matter is crucial, however. Through the gravitational force that it exerts, dark matter probably plays a dominant role in the formation of galaxies. As we shall see in Chapter 16, it may also determine the ultimate fate of the universe.

Galaxy Summary

The main features of the different kinds of galaxies are summarized in Table 14.1. Many of the figures are approximate only, and individual galaxies may differ greatly from the values given.

14.3
THE EXTRAGALACTIC DISTANCE SCALE

One of the most important, difficult, and controversial problems in modern observational astronomy is that of measuring the distances to galaxies. Galaxies are much too far away to display parallaxes or proper motions. Until very recently, the only way to determine the distance to a galaxy was to identify objects within it whose intrinsic luminosities were already known. Hubble, for example, used the cepheids in M31. Measurements of the apparent luminosities of these objects then made it possible to calculate the distance. Remember that the apparent brightness of an object decreases inversely as the square of its distance (Section 8.1).

In the past few years, astronomers have devised three new techniques for measuring distances to galaxies. These new techniques all give the same answer for the distances to galaxies with an accuracy of about 10 percent, and so it appears at long last that we really do have a fairly good estimate for the size of the visible universe.

Distances from Standard Candles

To determine distances to galaxies, we must resort to a multistep process. Roughly, the traditional approach is as follows. First, we derive distances to individual nearby stars in our own Galaxy by measuring parallaxes. With knowledge of the intrinsic luminosities of these nearby

ESSAY   The Scale of the Universe

In Chapter 10 we discussed a model for the universe in which all dimensions were reduced by a factor of 1 trillion (10^{12}). In this model, the Sun is the size of a mustard seed, and the nearest star is 10 km distant. In order to provide a more comprehensible model for our own Galaxy, let us now shrink our model by another thousandfold, thus reducing all dimensions by a factor of 10^{15}.

In this new model universe, the diameter of the Sun is only a little more than the wavelength of light, and the typical spacing between stars in the solar neighborhood is a few meters. Most of the stars visible to the unaided eye would fit into a volume about 5 km on a side—still much larger than any building. To contain our entire Galaxy, we would need a space about the volume of the Moon. Clearly, our model universe needs to shrink still more in order for us to visualize it.

Suppose we take 10^{18} as a reduction factor. The Sun is now reduced to the size of a single atom, the naked-eye stars occupy an ordinary room, and the Galaxy fits into a space about 1 km across—it would drop nicely into Meteor Crater in Arizona. The nearest other galaxies similar to our own are a few kilometers away—about like the spacing of farming villages in England or Germany, where the settlements are within walking distances of the fields. In our model, you could walk to the Andromeda Galaxy in half a day, and a day hike would take you across the Local Group of galaxies. But to reach the nearest large cluster of galaxies—the Virgo Cluster—you would have to find some better form of transportation, because it is several hundred kilometers distant.

We could continue to shrink our model, but the illustrations are sufficient to show the scale of the universe of galaxies. Perhaps more impressive than the distances themselves is the fact that the light we see from these galaxies left them long ago. As we survey far beyond the Milky Way, we look back farther and farther, probing the depths of time as well as those of space.

How can observations of objects of known luminosity (standard candles) be used to measure the distances to galaxies?

stars, we can then determine distances to clusters, which contain stars similar to those with known luminosities. Once we measure the distance to a cluster, we know the luminosity of every star within the cluster. Fortunately, clusters contain some stars, including cepheid variables, that are much more luminous than any of the nearby stars for which we can obtain parallaxes by direct measurement. We then search for similar highly luminous stars in other galaxies. Because we can measure the apparent brightness of these stars, and because we already know their true luminosity from studies of stars in clusters in our own Galaxy, we can use the inverse-square law for the propagation of light to determine the distances to the galaxies to which they belong. Any object whose intrinsic luminosity is known is referred to as a **standard candle**.

Several different types of stars have been used as standard candles. The most accurate distances for nearby galaxies are based on measurements of cepheid variable stars. Cepheids have the advantage of being relatively luminous (maximum luminosity about $2 \times 10^4 \, L_s$). Because they vary in brightness, cepheids are easily identified from multiple images of a galaxy. The luminosity of a cepheid can be determined from its period through the period-luminosity relation. Comparison of their known absolute magnitudes and observed apparent magnitudes enables

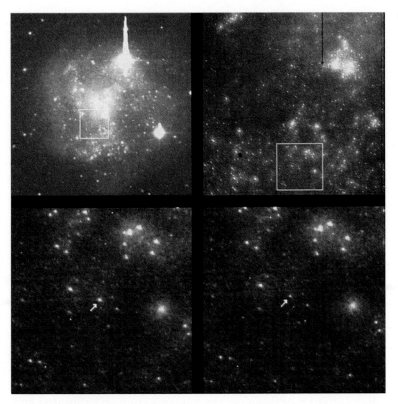

Figure 14.11 A series of images showing a cepheid in the spiral galaxy IC 4182, which is at a distance of 16 million LY. (*Top left*) A groundbased image obtained with the 200-inch Hale Telescope on Palomar. The inset box shows the field of view of the Hubble Space Telescope. (*Top right*) An HST view of a small region of IC 4182. The inset box marks the position of the enlarged region in the two lower frames. (*Bottom left*) The arrow points to a cepheid variable star. (*Bottom right*) The same region of the galaxy imaged by HST five days earlier. The arrow points to the same cepheid, which is clearly fainter. (*NASA*)

us to find their distances and, therefore, the distances to the galaxies in which they occur, with the inverse-square law of light.

The amount of work involved in finding cepheids and measuring their periods can be enormous. Hubble, for example, obtained 350 photographs of M31 over a period of 18 years and identified only 40 cepheids. Even though cepheids are fairly luminous stars, they can be detected in only about 30 of the nearest galaxies with the world's largest ground-based telescopes. One of the main goals of the Hubble Space Telescope is to measure cepheids in more distant galaxies (out to distances of at least 50 million LY) to improve the accuracy of the extragalactic distance scale (Figure 14.11).

Other standard candles that are even more luminous than cepheids have been used to measure distances that are more than 50 million LY away. Supergiant stars, globular clusters, and supernovae have all been used as standard candles.

Particularly useful for measuring distances are supernovae that involve the explosion of a white dwarf (Section 12.4). Supernovae of this type all reach the same luminosity (about $10^{10} L_s$) at maximum light, and can be detected out to distances of 600 million LY. The explosion occurs when a white dwarf star, which already has a mass near the Chandrasekhar limit, accretes additional matter from a companion star. When the mass of the star becomes too great to be supported by the pressure of degenerate electrons, the star collapses. A thermonuclear explosion then occurs and disrupts the entire star, emitting copious amounts of energy. The major problem with using this type of supernova as a standard candle is that none has so far been observed in nearby galaxies whose distances are already known from some other technique. Therefore, although all such supernovae seem to have about the same intrinsic luminosity, we do not yet know very accurately what that luminosity actually is.

At very large distances, we can use the total light emitted by an entire galaxy as a standard candle. Galaxies span an enormous range in intrinsic luminosity. Furthermore, most galaxies do not have distinguishing characteristics that enable us to estimate their luminosities. A dwarf elliptical galaxy as seen through a telescope looks about the same as a giant elliptical galaxy. We can only tell which galaxies are highly luminous and which ones are not if we see a collection of them of various brightnesses, side by side in a cluster. Therefore, we can use the apparent brightness of the brightest members (say, the average of the brightest five) in a large cluster to estimate the distance to the cluster.

New Techniques

What are the three most useful new techniques for measuring distances to galaxies?

Three new techniques have recently been devised for measuring distances to galaxies. All of these techniques assume that the distances measured to nearby galaxies through the use of the cepheids are correct. Characteristics of these nearby galaxies are then calibrated and used to measure distances to galaxies that lie beyond the limits where measurements of cepheids are possible.

The first technique makes use of the observed fact that the luminosity of a spiral galaxy, like the Milky Way, is related to its rotational velocity. This method is called the Tully-Fisher technique, named for the two astronomers who devised it in the late 1970s. As we have seen in the case of the Milky Way, we can estimate the mass of a galaxy by measuring the rotational velocity of stars or nebulae in its outer regions (Section 13.5). If the ratio of mass to luminosity is the same in all spirals, then a measurement of the mass can be used to derive the intrinsic luminosity.

It is somewhat surprising that this technique works, because much of the mass associated with galaxies is dark matter, which does not contribute at all to the luminosity. There is also no obvious reason why the mass-to-light ratio should be the same for all spiral galaxies. Nevertheless, observations show that measuring the rotational velocity of a galaxy is enough to determine its intrinsic luminosity and therefore its distance.

The second of the new techniques makes use of planetary nebulae. If we look at the location of the central stars of planetary nebulae in an HR diagram (see Figure 12A), it does not seem as if the stars would be bright enough to be seen in other galaxies. However, the central stars are very bright in the ultraviolet part of the spectrum. The most luminous stars have total luminosities, including ultraviolet radiation, of 10^4 to 10^5

L_s. The ultraviolet light emitted by the central star is absorbed in the surrounding nebula, and much of it is re-emitted in the form of emission lines in the optically visible part of the spectrum, where it can be easily observed.

Planetary nebulae are easy to find in other galaxies. All one has to do is photograph the galaxy in the light of one of the emission lines. Ordinary stars will not emit strongly at this wavelength, and the planetary nebulae will therefore stand out from the background light of other stars.

Observations show that the number of planetary nebulae in a galaxy depends on their intrinsic luminosity. Suppose that you were to count all of the nebulae in a galaxy whose distance you already know from a study of the cepheids in it. If you then plot the number of planetary nebulae that fall within each interval of luminosity, you will find that there will be relatively few very bright nebulae. You will also discover that the number of planetary nebulae increases with decreasing brightness. What you have done is determine a luminosity function for planetary nebulae (Section 10.2). Studies of planetary nebulae in many galaxies show that this luminosity function is the same for *all* galaxies.

All we have to do to determine a distance to a galaxy whose distance is not already known is count the number of planetary nebulae as a function of apparent brightness. Because the luminosity function is the same in every galaxy, we already know how many planetary nebulae there are as a function of intrinsic luminosity. Therefore, we can immediately convert the observed brightness of these nebulae to intrinsic luminosity. The distance can then be calculated from the inverse square law for the propagation of light. This particular approach to finding distances was developed by George Jacoby at Kitt Peak National Observatory.

The third new technique involves measuring fluctuations in the apparent surface brightness of elliptical galaxies. This technique has been explored by John Tonry at MIT. Elliptical galaxies contain mostly very old stars and very little gas or dust. A perfectly sharp picture of an elliptical galaxy would look much like that of a globular cluster, with many individual stars appearing as discrete points of light. Even with the blurring of the image caused by the Earth's atmosphere, the image of an elliptical galaxy does not appear to be perfectly smooth. It is rather mottled or bumpy because of the lumpy distribution of the light emitted by the individual stars that belong to this galaxy. Furthermore, the amount of the bumpiness depends on the distance to the galaxy. For a nearby galaxy, we can see individual stars or clusters of stars, and the image has many bumps of varying brightness. For a very distant galaxy, the stars cannot be resolved at all, and the image will be smooth.

In order to estimate the distance to an elliptical galaxy, it is necessary only to measure the degree of bumpiness in the distribution of light. This technique will not work for spiral galaxies because they contain large amounts of dust, which also causes fluctuations in surface brightness.

TABLE 14.2 Methods for Estimating the Distance to Galaxies

Method	Reliability	Galaxy Type for Which Method is Useful	Approximate Distance Range over which Method is Useful (10^6 LY)
Cepheids	Very	Spirals; Irregulars	0–50
Brightest stars	Moderate	Spirals; Irregulars	0–150
Planetary nebulae	Very	All	0–70
Supernovae	Moderate	All	0–600
Rotation Tully-Fisher	Very	Spirals; Irregulars	0–75
Surface brightness fluctuations	Very	Ellipticals	0–300
Total light of galaxies	Low	Spirals; Irregulars	0–300
Brightest galaxy in cluster	Very	Ellipticals	70–13,000
Redshifts	Very	All	300–13,000

Accuracy of Extragalactic Distances

Table 14.2 lists the type of galaxy for which each of the techniques described here is useful, the range of distances over which the technique can be applied, and the reliability of the distance estimates derived with each technique.

With so many ways of finding distances to galaxies, one might think that the distance scale is well settled. Unfortunately, it is not. The measurements are difficult to make. Experts differ in their judgment of the proper interpretation of the observations, the calibration of the standard candles, and what standards are the most reliable. That is why the three new techniques described here represent such a major advance. The three techniques are all independent of one another and all give the same distances within about 10 percent. While not every astronomer has accepted the new results as correct, the tests of these new methods that have been made to date are very encouraging. It appears that astronomers are very close to having solved the problem of determining just how far away the galaxies are. The work begun by Hubble with his observations of M31 in the 1920s may now be nearing a conclusion.

14.4
THE EXPANDING UNIVERSE

Early in the 20th century, astronomers began to collect evidence that all galaxies (except for a few of the nearest ones) are moving away from us. This expansion of everything away from the Milky Way Galaxy has many implications for the structure and history of the universe, as we will further discuss in Chapters 15 and 16. In this chapter, we describe how the observational evidence was obtained and how it is used to help determine the distances to remote galaxies and clusters of galaxies.

The Hubble Law

V.M. Slipher, a young astronomer working at Lowell Observatory in the years around World War I, discovered that the spectral lines of galaxies are shifted to the red by amounts that are greater than those generally measured for objects within our own Milky Way. A spectrum of a galaxy is a composite of the spectra of the individual stars in the galaxy, each with its own motion. The galaxy as a whole, however, has an *average* radial velocity relative to the Sun. By measuring these average radial velocities, Slipher found that most of the 40 galaxies he measured were moving *away* from us at speeds up to 2000 km/s. The profound implications of this work became apparent during the 1920s, when Edwin Hubble at Mount Wilson began using cepheid variables to measure the distances to some of these same galaxies.

Hubble carried out the key observations in collaboration with a remarkable man: Milton Humason (Figure 14.12). Humason began his astronomical career by driving a mule train up the trail on Mount Wilson to the observatory. In those early days, supplies had to be brought up that way; even astronomers hiked up to the mountaintop for their turns at the telescope. Humason became interested in the work of the astronomers and took a job as janitor at the observatory. After a time he became a night assistant, helping the astronomers run the telescope and collect data. Eventually, he made such a mark that he became a full astronomer at the observatory.

By the late 1920s, Humason was collaborating with Hubble by photographing the spectra of faint galaxies whose relative distances could be estimated fairly accurately, extending the earlier spectroscopy by Slipher. In 1931, Hubble and Humason jointly published their classic paper, which compared distances and velocities of galaxies. The most distant of these galaxies was found to be moving away from us at speeds of up to nearly 20,000 km/s. Their result (Figure 14.13), now known as the

Figure 14.12 Milton Humason (1891–1972) began his astronomical career driving a mule team up the old dirt trail to Mt. Wilson during the construction days of the observatory. His collaboration with Hubble in establishing the expansion of the universe won him many honors, including an honorary doctorate degree. *(California Institute of Technology)*

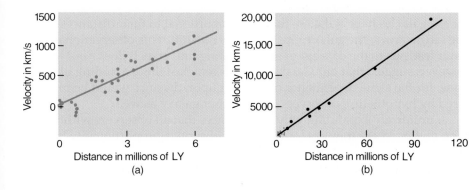

Figure 14.13 (a) Hubble's original velocity-distance relation, published in 1929. (b) Hubble and Humason's velocity-distance relation, published in 1931 in *The Astrophysical Journal*. The red dots at the lower left of (b) are the points in (a). Comparison of the two graphs shows how rapidly the determination of distances and redshifts of galaxies progressed in the two years between these two publications.

Hubble law, established that the velocities of recession of galaxies are proportional to their distances from us. Written as an algebraic equation, the Hubble law is

$$v = H \cdot d$$

where d is the distance, v is the recession speed, and H is a number called the **Hubble constant**. Because all of the recessions produce shifts of the galaxies' spectral lines toward longer (redder) wavelengths, the Hubble law is also sometimes called the Law of the Redshifts.

This remarkable result—that redshift or recession speed is proportional to distance—implies an entirely new way to determine the distance to a galaxy—from its velocity. If we measure the redshift in the spectrum of a distant galaxy, this information, together with the Hubble law, is sufficient to determine distance, independent of concerns about the constancy of standard candles or the possibilities of unknown sources of extinction between the galaxies.

To use the Hubble law as a measure of distance, we must calibrate it—that is, we must determine the Hubble constant (H), the constant of proportionality between distance and redshift or speed. This cannot be done using nearby galaxies because their orbital velocities, which result from motion within a group or cluster of galaxies, are comparable to the recession because of the expansion of the universe. Rather, we must use galaxies with recessional speeds of thousands of kilometers per second. As we have seen, it is not simple to find standard candles or other means to measure the distances of objects out to 100 million LY or more from the Earth.

The values determined for the Hubble constant depend on the assumptions made by different astronomers concerning the extragalactic distance scale. Modern estimates of this constant range from about 15 km/s to 30 km/s of speed for every 1 million light years' distance. The new techniques for measuring distances described in the previous section indicate that H is about 25 km/s per 1 million LY, a value that is likely to be correct within a factor of 25 percent either way. Some additional consequences of this value for the Hubble constant will be explored in Chapter 16.

Meaning of the Hubble Law

At this point, we must be clear about what the Hubble law tells us. All distant galaxies have redshifts that are proportional to their distances. The farther away the galaxy, the higher its redshift. This observational result leads us to the concept of an expanding universe. It does not, however, suggest that our Galaxy is at the center of this expanding universe, for the same dependence of redshift upon distance would be measured by hypothetical observers on any other galaxy as well.

A uniformly expanding universe requires that all observers within it, no matter where they are located, must observe a proportionality between the redshifts and the distances of remote galaxies. Imagine a ruler made of flexible rubber, with the usual lines marked off at each centime-

What do astronomers mean when they say that the universe is expanding?

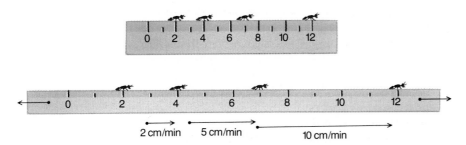

Figure 14.14   Stretching a ruler. See text for explanation.

ter. Now suppose someone with strong arms grabs each end of the ruler and slowly stretches it, so that it doubles in length in 1 minute (Figure 14.14). Consider an intelligent ant sitting on the mark at 2 cm—intentionally not at either end or in the middle. This ant measures how fast other ants, sitting at the 4-, 7-, and 12-cm marks, move away from her as the ruler stretches. The ant at 4 cm, originally 2 cm away, has doubled its distance; it has moved 2 cm/min. Similarly, the ones at 7 cm and 12 cm, originally 5 and 10 cm distant, have had to move away at 5 and 10 cm/min, respectively. All ants move at speeds proportional to their distance. Now repeat the analysis, but put the intelligent ant on some other mark, say on 7 cm or 12 cm, and you'll find that in all cases, as long as the ruler stretches uniformly, this ant finds that every other ant moves away at a speed proportional to its distance.

For a three-dimensional analogy, look at the raisin bread in Figure 14.15. The baker has put too much yeast in the dough, and when he sets the bread out to rise, it doubles in size during the next hour and all the raisins move farther apart. Some representative distances from one of the raisins (chosen arbitrarily, but not at the center) to several others are shown in the figure. Because each distance doubles during the hour, each raisin must move away from every other raisin at a speed proportional to its distance. The same is true, of course, no matter which raisin you start with. But the analogy must not be carried too far; in the bread

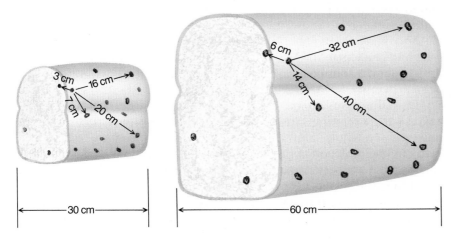

Figure 14.15   Expanding raisin bread. See text for explanation.

it is the expanding dough that carries the raisins apart, but in the universe no pervading medium is presumed to separate the galaxies.

From the foregoing, it should be clear that if the universe is uniformly expanding, all observers everywhere, including us, must see all other objects moving away from them at speeds that are greater in proportion to their distances. As Hubble and Humason showed, that is precisely what observers on Earth do see.

Note that the expansion of the universe does not imply that the galaxies and clusters of galaxies themselves are expanding. The raisins in our raisin bread analogy do not grow in size as the loaf expands. Similarly, mutual gravitation holds galaxies and clusters together, and they simply separate as the universe expands, just as do the raisins in the bread. Galaxies in clusters do, of course, have individual motions of their own superimposed on the general expansion. Galaxies in pairs, for example, revolve around each other, and those in clusters move about within the clusters. In fact, a few galaxies in nearby groups and clusters move fast enough within those systems that they are actually approaching us even though the clusters of which they are a part are moving away.

For studies of nearby galaxies, the general expansion of the universe is of little significance. But at large distances, the velocities with which galaxies rush away from us can become very great—as much as several tenths of the speed of light. Remember, also, that as we look to such great distances, billions of light years away, we are also looking backward in time. We see distant galaxies not as they are today, but as they were when the light left them billions of years ago. The most distant observed galaxies emitted the light we now detect long before the Sun or the Earth even existed.

14.5
THE DISTRIBUTION OF GALAXIES IN SPACE

Celestial objects rarely travel through space alone. The Earth is but one of nine planets orbiting the Sun. The Sun itself seems somewhat unusual in that, as far as we know, it is a single star. Most stars are at least double, and many are members of either open or globular clusters. All of the stars and clusters in the Milky Way Galaxy are gravitationally bound, orbit around a common center, and will complete their evolution in close proximity.

Does this cosmic togetherness persist on still larger scales? Are most galaxies to be found in clusters of galaxies? What is the structure of the universe as a whole? How are galaxies distributed in space? Are there as many in one direction of the sky as in any other? And if we look at galaxies that are very distant, do we find that the density of galaxies is about the same as it is near the Milky Way Galaxy?

In order to determine how galaxies are distributed through space, it is necessary to measure their distances. This measurement requires a

Figure 14.16 Schematic diagram of the Local Group of galaxies, approximately to scale. The three largest galaxies in the Local Group are all spirals.

spectrum of the galaxy so that its velocity can be determined. The distance is then calculated by using the Hubble law. It is only in the past decade that astronomers have had instruments sensitive enough to measure the velocities of thousands of galaxies, and so build up a picture of the structure of the universe.

The studies of the locations of galaxies in space show that galaxies tend to form groups with sizes of several million light years (a small group) up to 50 to 100 million LY (a supercluster). There is evidence that superclusters themselves may form much larger structures, perhaps up to a billion light years in length. Between the superclusters, there are great voids where few if any galaxies can be found.

The Local Group

The region of the universe for which we have the most detailed information is, as you would expect, our own local neighborhood. It turns out that our own Galaxy is a member of a small group of galaxies, which is called the **Local Group**. It is spread over about 3 million light years and contains at least 30 members. There are three spiral galaxies (our own, the Andromeda Galaxy, and M33, which is much smaller and less massive than the other two), at least 12 dwarf irregulars, two intermediate ellipticals, and 14 known dwarf ellipticals. Appendix 13 gives the properties of the galaxies that are generally accepted to be members of the Local Group. Figure 14.16 is a plot of the Local Group.

The average of the motions of all the galaxies in the Local Group indicates that its total mass is about 5×10^{12} solar masses. Although this mass estimate is uncertain, it implies that extensive amounts of dark matter must be present in the Local Group.

How big are clusters of galaxies?

Figure 14.17 The central region of the Virgo cluster of galaxies. Virgo, the nearest large cluster (60 million LY) with its hundreds of bright galaxies, is the dominant feature of the Local Supercluster of galaxies. *(Copyright ROE/AAT Board)*

The Neighboring Clusters

The nearest moderately rich cluster is the famous Virgo Cluster, a system with thousands of members (Figure 14.17). This cluster contains a concentration of mostly elliptical galaxies that includes M87, which is both a radio source and x-ray source (see Figure 14.7). Within several degrees of M87 are many spirals as well as ellipticals and, associated with the brightest galaxies, very many dwarfs, like the dwarf ellipticals in the Local Group. The distance to the concentration of galaxies around M87 is about 50 million LY. The diameter of the Virgo Cluster is about 5 million LY.

There are clusters that are much larger than the Virgo Cluster. One nearby example is the Coma Cluster, which has a diameter of at least 10 million LY and thousands of observable galaxies (Figure 14.18). This cluster is centered on two giant ellipticals, whose luminosities are about $4 \times 10^{11} L_s$. The Coma Cluster contains many more faint galaxies than bright ones. The total number of galaxies in the cluster might be as large as tens of thousands. The mass of this cluster is about 4×10^{15} solar masses.

Rich clusters like the Coma Cluster usually show a great deal of spherical symmetry and a high concentration of galaxies at the center of

Figure 14.18 The central part of the Coma cluster of galaxies. The two dominant galaxies in this image are giant elliptical galaxies, and they have the yellowish color characteristic of old stars. (The bright blue object is a star in our own galaxy.) Thousands of galaxies belong to this cluster. *(National Optical Astronomy Observatories)*

the cluster. These rich clusters contain few, if any, spiral galaxies in the cluster core (Figure 14.19), but rather have a membership dominated by ellipticals. Clusters of galaxies, particularly rich clusters like Coma, are usually sources of x rays. The x rays from a cluster are thermal radiation from gas at a temperature of 10^7 to 10^8 K. The gas is located between the galaxies.

There is probably a significant relation between the presence of hot gas and the absence of spiral galaxies in rich clusters. x-ray emission

Figure 14.19 Cluster of galaxies in Fornax. Most of the galaxies in this cluster are elliptical galaxies, which are composed of rather old, yellowish stars with little or no gas and no evidence of star formation. This photograph illustrates an important characteristic of ellipticals—namely, that they are "gregarious" and tend to be found in regions of high concentrations of other galaxies. Spirals usually are isolated or lie on the outskirts of clusters. There is one spiral in the corner of the picture. *(Anglo-Australian Telescope Board)*

Figure 14.20 A radio image of NGC 1265. The two jets extend about 60,000 LY through space. The galaxy itself is moving through the Perseus cluster of galaxies at a speed of about 2000 km/s. Particles ejected from the galaxy into the two jets are swept backward by the pressure of the intergalactic gas within the cluster to form the "U" shape. *(NRAO/AUI)*

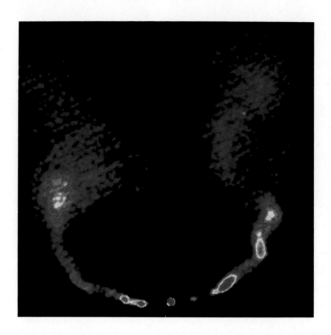

lines of heavy elements such as iron have been observed and suggest that the heavy-element abundance in the hot gas is similar to that in the Sun. Current theories predict that the primordial matter from which the clusters formed was nearly all hydrogen and helium. Therefore, at least some of the x-ray emitting gas must have undergone nucleosynthesis in stellar interiors. This processed matter was then ejected into interstellar space within the cluster galaxies by such mechanisms as supernova outbursts. Finally, this material was swept from the galaxies by collisions between them and by their moving through the intracluster gas, or possibly by internal processes such as stellar winds and supernova explosions (Figure 14.20). Such sweeping of interstellar matter from galaxies stops star formation in them, and the spiral arms gradually disappear. The swept gas is hot because the galaxies collide with each other or pass through intracluster gas at speeds up to thousands of kilometers per second.

The Local Supercluster

After astronomers discovered clusters of galaxies, they naturally wondered whether there were still larger structures in the universe. Are there clusters of clusters of galaxies? The answer is yes, and these very large-scale structures, which contain one or more clusters of galaxies, are called **superclusters**. Between the superclusters, there are great voids where few, if any, galaxies can be found.

The best studied of the superclusters is the one that includes the Milky Way Galaxy. The most prominent grouping of galaxies within the Local Supercluster is the Virgo Cluster. The Milky Way Galaxy lies in the outskirts of the Local Supercluster. The diameter of the Local

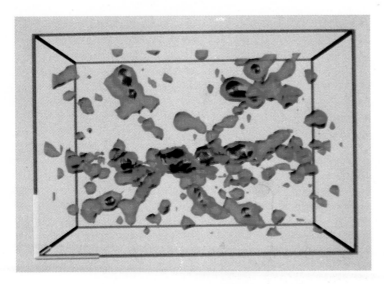

Figure 14.21   The distribution of galaxies in the Local Supercluster, a volume of space approximately 150 million LY across. Galaxies are found in clumps and small groups, although much of space contains no galaxies at all.   *(Courtesy Brent Tully, University of Hawaii)*

Supercluster is at least 60 million LY, and its mass is probably about 10^{15} solar masses.

The vast majority of bright galaxies in the Local Supercluster lie in a small number of clumps or groups, and most of these groupings are narrowly confined to a plane. In fact, 60 percent of all of the galaxies contained within the Local Supercluster are to be found within a disk-like region whose diameter is about six times its thickness. Both the Local Group and the Virgo Cluster itself lie in the plane of this disk.

Perhaps the most surprising fact revealed by study of the Local Supercluster is that space is mostly empty (Figure 14.21). Most of the galaxies are concentrated into individual clusters, and these clusters occupy only about 5 percent of the total volume of space contained within the boundaries of the Local Supercluster. A major problem for any theory of the formation of galaxies and of large-scale structure in the universe is to explain why galaxies are so closely clumped and why most of the universe is devoid of luminous matter.

Voids

The Local Supercluster is not unique in being mostly empty. Surveys in other directions in the sky yield the same result. Galaxies are found preferentially in large filamentary superclusters that extend over distances of at least a few hundred million LY. The largest structure seen so far in the universe is a sheet of galaxies that is at least 500 million LY long, 200 million LY high, and about 15 million LY thick. Referred to as the "Great Wall," this sheet of galaxies is about 250 million LY from our own Galaxy. The mass of the Great Wall is estimated to be 2×10^{16} solar masses, a factor of 10 greater than the mass of the Local Supercluster.

These filamentary structures are separated by voids, that is, by large holes where few galaxies can be found. You can imagine the voids as gi-

What observations show that most of space is empty?

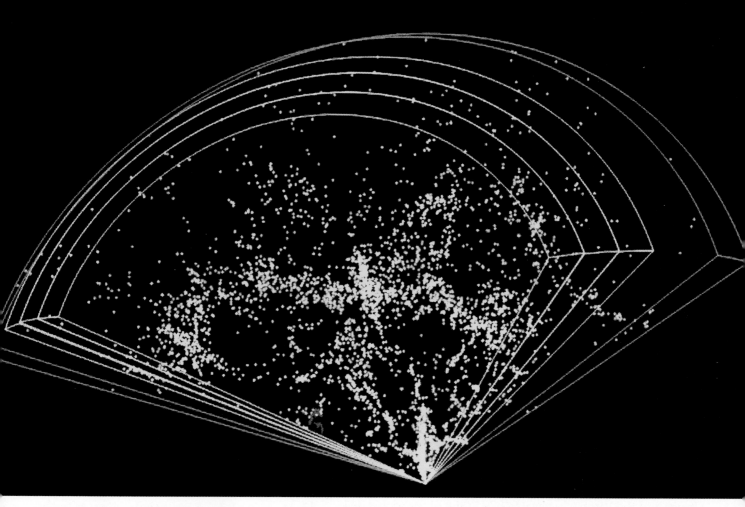

Figure 14.22 A three-dimensional "slice of the universe" showing the distribution of galaxies in space as seen from our location, which is the point at the bottom of the figure where all of the slices come together. Note the concentration of galaxies in narrow bands or lanes with large voids between them. An analogous distribution would be obtained if we took a slice through a collection of bubbles of various sizes. The Great Wall is the band of galaxies stretching from left to right across the middle of the picture. *(Courtesy Margaret Geller/Harvard-Smithsonian Center for Astrophysics)*

ant bubbles, with the clusters of galaxies concentrated along their boundaries.

One way to visualize these bubbles is by plotting the distribution of galaxies in a "slice" taken through space, as in the example shown in Figure 14.22. In such a picture, it is clear that most of the galaxies lie along the boundaries of the voids, which are roughly circular. The image is similar to what you would obtain by taking a slice through a sponge or a piece of Swiss cheese with large holes.

The discovery that most of space consists of voids has come as a surprise to astronomers. Most would probably have predicted that the regions between giant clusters of galaxies are filled with many small groups of galaxies or even with isolated individual galaxies. Careful searches within these voids have so far confirmed that they are not—few galaxies of any kind are found there. Apparently, 90 percent of the galaxies occupy less than 10 percent of the volume of space.

Knowledge of exactly *how* galaxies are distributed may provide clues as to *why* they are distributed that way. We know that the universe, only a few hundred thousand years after it was formed, was extremely smooth (Chapter 16). The challenge for the theoretician is to find a way to turn that featureless universe into the complex one that we see today, with dense regions containing many galaxies and nearly empty regions containing almost none.

For example, based on data like that plotted in Figure 14.22, some astronomers argued that galaxies were distributed on the walls of bubble-like structures and that the matter originally in the interiors of these bubbles had been somehow cleaned out and pushed toward the walls. But how? One possibility is that there were giant explosions during the first few billion years of the universe that swept the gas into bubble-like shells, and that galaxies subsequently formed in these shells. Unfortunately, despite considerable ingenuity, astronomers have yet to devise a way to produce energetic enough explosions to account for the largest voids.

Although we do not have a good explanation for why the voids exist, this discussion should serve to show how measurements of the distribution of galaxies in space can help generate ideas about what kinds of events must have occurred when superclusters, clusters, and galaxies were just beginning to form. The next chapter will describe in more detail what we know about the universe when it was young.

▶ SUMMARY

14.1 Faint star clusters, clouds of glowing gas, dust clouds reflecting starlight, and galaxies all appear as faint patches of light in telescopes of the quality available at the beginning of the twentieth century. It was established that there are other galaxies similar to the Milky Way in size and content only when the discovery of cepheid variables in the Andromeda Galaxy was announced in 1924. Observations of galaxies are an essential part of the subject of **cosmology**, a term used to refer to the study of the structure, origin, and evolution of the universe.

14.2 The majority of bright galaxies are either **spirals** or **ellipticals**. Spiral galaxies contain both old and young stars, as well as interstellar matter, and have typical masses in the range 10^9 to 10^{12} solar masses. Our own Galaxy is a large spiral. Ellipticals are spheroidal or elliptical systems that consist almost entirely of old stars, with very little interstellar matter. Elliptical galaxies range in size from giant ellipticals, which are more massive than any spiral, to dwarf ellipticals, which have masses of only about 10^6 solar masses. A small percentage of galaxies are classified as **irregular**. The Milky Way's nearest neighbors in space are the Large and Small Magellanic Clouds, which are both irregular galaxies. The masses of spiral galaxies are determined from measurements of their rates of rotation. The masses of elliptical galaxies are estimated from analyses of the random motions of the stars within them. Galaxies are characterized by their **mass-to-light** ratios. The luminous parts of galaxies with active star formation have mass-to-light ratios typically between 1 and 10; the luminous parts of elliptical galaxies, which contain only old stars, have mass-to-light ratios of typically between 10 and 20. The mass-to-light ratios of whole galaxies, including their halos, are as high as 100, indicating that a great deal of dark matter is present.

14.3 Astronomers determine the distances to galaxies by measuring the apparent magnitudes of objects whose intrinsic luminosities are known, and then using the inverse square law for light. Such objects are known as **standard candles**.

Some useful standard candles are cepheids, supernovae, and globular clusters. New techniques for determining distances use the rotational velocities of spiral galaxies, which are correlated with their absolute magnitudes; the luminosities of planetary nebulae; and variations in the surface brightness of elliptical galaxies.

14.4 The universe is expanding. Observations show that the lines in the spectra of distant galaxies are **redshifted**, and their velocities of recession are proportional to their distances from the Milky Way Galaxy. The relationship between velocity and distance is known as the **Hubble law**. The rate of recession, which is known as the **Hubble constant**, is approximately 25 km/s per 1 million LY. We are not at the center of this expansion; an observer in any other galaxy would see the same expansion that we do.

14.5 Galaxies tend to group together to form clusters. The **Local Group** contains at least 30 members. Large clusters (such as Virgo and Coma) contain thousands of galaxies. Clusters of galaxies often group together with other clusters to form large-scale structures called **superclusters**, which can extend over distances of several hundred million LY. Clusters and superclusters fill only a small fraction of space. Most of space consists of voids between superclusters, with nearly all galaxies confined to less than 10 percent of the total volume.

▶ REVIEW QUESTIONS

1. What is a galaxy? Describe the main distinguishing features of spiral, elliptical, and irregular galaxies.

2. Describe some of the techniques for determining the distances to galaxies.

3. What is the Hubble law and what does it mean?

4. What is meant by "the expansion of the universe"?

5. Describe the hierarchy of clusterings of galaxies, from the Local Group to superclusters.

▶ THOUGHT QUESTIONS

6. Why can we not determine distances to galaxies by measuring their parallaxes in the same way as we measure the parallaxes of stars?

7. Starting with the determination for the size of the Earth, outline all of the steps that must be taken to obtain the distance to a remote cluster of galaxies.

8. Suppose that a supernova explosion occurred in a galaxy at a distance of 10^8 LY. If we are only now detecting it, how long ago did the supernova actually occur? According to the Hubble law, what is the radial velocity for this galaxy?

9. Use the data in Appendix 13 to determine which is more common in the Local Group—large, luminous galaxies or faint, small galaxies. Which is more common—spirals or ellipticals?

10. Based on the data in Appendix 13, would you describe the Milky Way Galaxy as a typical galaxy? Why or why not?

11. Why can the redshifts in the spectra of galaxies not be explained by the absorption of their light by intergalactic dust?

▶ PROBLEMS

12. A cluster of galaxies is observed to have a radial velocity of 60,000 km/s. Find the distance from the Sun to the cluster.

13. The Andromeda Galaxy and the Milky Way Galaxy are observed to be approaching each other with a radial velocity of about 300 km/s. If there is no change in this motion, how long will it be before the two galaxies collide?

14. Plot the velocity-distance relation for the "raisins in the bread" analogy from the numbers given in Figure 14.15.

15. Repeat Problem 14, but use some other raisin for a reference and measure the distances with a ruler. Is your new plot the same as the last one?

16. If the Large Magellanic Cloud is a satellite galaxy of the Milky Way and its distance (semimajor axis) is 200,000 LY, find its period of revolution about the Galaxy. For this calculation, assume that the mass of the Milky Way is 10^{12} solar masses.

17. Consider the possibility that the Milky Way and Andromeda are in circular orbit about each other. From their separations and masses as given in this chapter, calculate their period of revolution. Compare this period with the result of Question 13. How do both of these times compare with the age of the universe (10 to 15 billion years)?

18. Suppose on one survey you count galaxies to a certain limiting faintness. On a second survey, you count galaxies to a limit that is four times fainter.
   a. To how much greater distance does your second survey probe?
   b. How much greater is the volume of space you are reaching in your second survey?
   c. If galaxies are distributed homogeneously, how many times as many galaxies would you expect to count on your second survey?

19. Suppose we imagine superclusters to be flat (or stringy), to have an average thickness of 50 million LY in the line of sight, and to be separated from each other by voids 200 million LY across. How many superclusters would lie overlapping in projection in a typical line of sight out to a distance of 10 billion LY (about as far as we could hope to see galaxies)?

20. Calculate the mass-to-light ratios for the stars listed in Table 10.2. Can stars alone explain a mass-to-light ratio of 100, which is measured for elliptical galaxies?

▶ SUGGESTIONS FOR ADDITIONAL READING

Books

Hodge, P. *Galaxies.* Harvard U. Press, 1986. A fine introduction by a noted galactic astronomer; now a bit dated.

Ferris, T. *The Red Limit.* Morrow, 1983. A superb history of how we established the large-scale properties of the cosmos.

Lemonick, M. *The Light at the Edge of the Universe.* Random House, 1993. A journalist takes readers on a tour through extragalactic astronomy and cosmology.

Wright, A. & H. *At the Edge of the Universe.* Horwood/J. Wiley, 1989. Excellent introduction to extragalactic astronomy by an astronomer.

Cornell, J., ed. *Bubbles, Voids, and Bumps in Time.* Cambridge U. Press, 1989. Articles by experts on galaxies and cosmology.

Shipman, H. *Black Holes, Quasars, and the Universe,* 2nd ed. Houghton Mifflin, 1982. Chapter 14 has a good discussion of the pyramid of distance methods astronomers use.

Rowan-Robinson, M. *The Cosmological Distance Ladder.* W. H. Freeman, 1985. Somewhat technical, but very clear exposition of how we measure the scale of the universe.

Smith, R. *The Expanding Universe: Astronomy's Great Debate.* Cambridge U. Press, 1982. A history of the Shapley-Curtis Debate, with extensive background.

Articles

On Galaxies and Their Characteristics

Trefil, J. "Galaxies" in *Smithsonian,* Jan. 1989, p. 36. Nice long review article.

Dressler, A. "Galaxies Far Away and Long Ago" in *Sky & Telescope,* Apr. 1993, p. 22. Observations with the Hubble Space Telescope.

Lake, G. "Understanding the Hubble Sequence" in *Sky & Telescope,* May 1992, p. 515.

Hodge, P. "The Andromeda Galaxy" in *Mercury,* July/Aug. 1993, p. 99.

Comins, N. & Marschall, L. "How Do Spiral Galaxies Spiral" in *Astronomy,* Dec. 1987, p. 6.

Davies, J., *et al.* "Are Spiral Galaxies Heavy Smokers?" in *Sky & Telescope,* July 1990, p. 37.

Parker, B. "Celestial Pinwheels: The Spiral Galaxies" in *Astronomy,* May 1985, p. 14.

Silk, J. "Formation of the Galaxies" in *Sky & Telescope,* Dec. 1986, p. 582.

On the History of Establishing Galaxies and Their Distances

Corwin, M. & Wachowiak, D. "Discovering the Expanding Universe" in *Astronomy,* Feb. 1985, p. 18.

Jones, B. "The Legacy of Edwin Hubble" in *Astronomy,* Dec. 1989, p. 38.

Osterbrock, D., *et al.* "Young Edwin Hubble" in *Mercury,* Jan./Feb. 1990, p. 2.

Smith, R. "The Great Debate Revisited" in *Sky & Telescope,* Jan. 1983, p. 28.

Osterbrock, D., *et al.* "Edwin Hubble and the Expanding Universe" in *Scientific American,* July 1993.

Parker, B. "The Discovery of the Expanding Universe" in *Sky & Telescope,* Sept. 1986, p. 227.

On the Cosmic Distance Scale

Hodge, P. "The Extragalactic Distance Scale: Agreement at Last?" in *Sky & Telescope,* Oct. 1993, p. 16.

DeVaucouleurs, G. "The Distance Scale of the Universe" in *Sky & Telescope,* Dec. 1983, p. 511.

Odenwald, S. & Tresch-Fienberg, R. "Galaxy Redshifts Reconsidered" in *Sky & Telescope,* Feb. 1993, p. 31.

McCarthy, P. "Measuring Distances to Remote Galaxies and Quasars" in *Mercury,* Jan./Feb. 1988, p. 19.

Smith, D. "Supernovae: Mileposts of the Universe" in *Sky & Telescope,* Jan. 1985, p. 18.

On Groups of Galaxies and Cosmic Structure

Lake, G. "The Cosmology of the Local Group" in *Sky & Telescope,* Dec. 1992, p. 613.

Geller, M. "Mapping the Universe: Slices and Bubbles" in *Mercury,* May/June 1990, p. 66.

Dyer, A. A New Map of the Universe" in *Astronomy,* Apr. 1993, p. 38; Kanipe, J. "A Cross Section of the Universe" in Astronomy, Nov. 1989, p. 44. On the work of Huchra and Geller.

Burstein, D. & Manly, P. "Cosmic Tug of War: The Local Structure of the Universe" in *Astronomy,* July 1993, p. 40.

Gregory, S. "The Structure of the Visible Universe" in *Astronomy,* Apr. 1988, p. 42.

Hodge, P. "The Local Group: Our Galactic Neighborhood" in *Mercury,* Jan./Feb. 1987, p. 2.

Struble, M. & Rood, H. "Diversity Among Galaxy Clusters" in *Sky & Telescope,* Jan. 1988, p. 16.

Marschall, L. "Superclusters: Giants of the Cosmos" in *Astronomy,* Apr. 1984, p. 6.

Vogel, S. "Star Attractor" in *Discover,* Nov. 1989, p. 20. On the Great Attractor.

15

THE EVOLUTION OF GALAXIES AND QUASARS

Four of the five galaxies in the group known as Stephan's quintet. This true-color picture was constructed by combining images in red, green, and blue light. The blue color in the arms of the spiral galaxy at the top of the picture indicates that young stars are present. *(W. Schoening and N. Sharp/NOAO)*

O nly in the last decade have observational and theoretical studies of the evolution of galaxies begun to make real progress in developing a picture of how galaxies change over the lifetime of the universe. Progress has been slow for several reasons. For example, galaxies are made up of stars, and it was only after the evolution of individual stars was well understood that astronomers could sensibly begin to explore how whole systems of stars change with time. The ideas presented in this text about how stars age and die are the result of research carried out during the past 40 years, and our understanding of star formation has developed rapidly just in the past decade. Twenty years ago, any attempt to describe the evolution of galaxies would have been pointless—we simply did not know enough about the life histories of stars.

Studying galaxies is difficult because they are very, very faint. Even with the biggest telescopes in the world, we cannot see individual stars or even determine the shapes of the most remote galaxies. We do have one advantage, however, in studying galaxy evolution. The universe itself is a kind of time travel machine, which permits us to observe galaxies as they were when the universe was young. When we look at distant galaxies, we see them as they were when the light that we now measure left them. If we observe a galaxy that is 1 billion LY distant, we are seeing it as it was when the light left it 1 billion years ago. By observing more and more distant objects, we look ever further backward toward a time when the galaxies and the universe were young.

In Chapter 14 we discussed "ordinary" galaxies. However, the most luminous objects that we know are not ordinary galaxies but something much more rare and exotic. Quasars are brilliant beacons that mark the greatest depths into space that we can penetrate. We will begin this chapter by describing what we know about quasars and their incredible sources of energy. We will then turn to an exploration of what our observations of quasars and distant galaxies tell us about the evolution of the universe.

15.1
QUASARS

Discovery

The Sun is very faint at radio wavelengths. Since the Sun is a typical star, we would not expect to see strong radio emission from other stars. Astronomers were, therefore, surprised when in 1960 two radio sources were identified with what appeared to be stars. By 1963, the number of such "radio stars" had increased to four (Figure 15.1). These radio stars were especially perplexing because their optical spectra showed emission lines that at first could not be identified with known chemical elements.

The breakthrough came in 1963 when Maarten Schmidt, at Caltech's Palomar Observatory, recognized the emission lines in one of the objects to be the Balmer lines of hydrogen (Chapter 8) shifted far to the red from their normal wavelengths. If, as most astronomers believe, the redshift is caused by the Doppler effect, the object is racing away from us at about 15 percent the speed of light! With this hint, the emission lines in the other "radio stars" were re-examined to see if they too might be well-known lines with large redshifts. Such proved to be the case, but the other objects were found to be receding from us at even greater speeds. Obviously, they are not neighboring stars; their stellar appearance is due to the fact that they are very distant. They are called, therefore, quasi-stellar radio sources, or simply **quasars.** Later, similar objects were found that were not sources of strong radio emission, and these objects are usually called QSOS, or quasi-stellar objects. In fact, it has turned out that only about 1 percent of all quasars are radio sources. Some astronomers think that radio-emitting quasars are a temporary phase in the evolution of quasars.

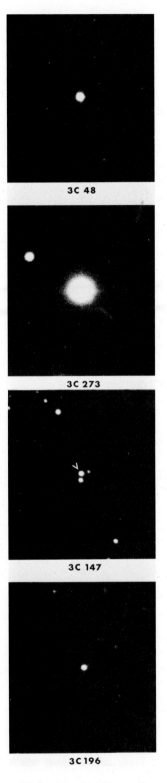

Figure 15.1 Quasi-stellar radio sources photographed with the 5-m telescope. *(Palomar Observatory, California Institute of Technology)*

What are the observational
characteristics of a quasar?

Thousands of quasars have now been discovered. All of their spectra show large redshifts. We define the redshift, which we call z, of a quasar to be the difference between the observed wavelength of a spectral line and the wavelength that line would have if it were produced in a gas in a stationary laboratory test tube ($\Delta\lambda$) divided by the laboratory (rest) wavelength (see Chapter 8). The largest redshift measured to date (April 1993) corresponds to a relative shift in wavelength of $\Delta\lambda/\lambda = 4.9$. The Lyman α line of hydrogen, which has a laboratory wavelength of 121.5 nm in the ultraviolet portion of the spectrum, is shifted all the way through the visible range to 700 nm! This redshift corresponds to a velocity of more than 94 percent the speed of light. (At such high velocities, the simple formula for the Doppler effect given in Chapter 8 does not apply, and we must use the theory of relativity to convert Doppler shift to velocity.)

What are the quasars? They are at the distances of galaxies, but they are certainly not normal galaxies. Galaxies contain stars, so the spectra of normal galaxies have absorption lines, just as do the spectra of stars. Quasar spectra are dominated by emission lines. Accounting for the energy emitted by quasars presents another problem. If quasars obey the Hubble law and are at the distances that correspond to their redshifts, then they are more luminous than the brightest galaxies. The following sections describe the observational clues that astronomers have assembled in order to solve this mystery.

Luminosities of Quasars

We can determine the distance to a galaxy if we know its redshift. Let us assume for the moment that quasars also obey the Hubble law, and that we can estimate their distances accurately by measuring their velocities. If this assumption is true, then quasars are *extremely* luminous compared with ordinary galaxies. In visible light, most are far more energetic than the brightest elliptical galaxies. Quasars also emit energy at x-ray and gamma-ray wavelengths, and many are radio sources as well. Some quasars have total luminosities as large as 10^{14} L_s, or 100 to 1000 times the brightness of the brighter elliptical galaxies.

Finding a mechanism to produce this much energy would be difficult under any circumstance. But quasars present an additional problem. Quasars vary in luminosity on time scales of months, weeks, or even in some cases, days. This variation is irregular, evidently at random, and can amount to a few tens of percent. Since quasars are highly luminous, a change in brightness by, for example, a factor of two means an extremely large amount of energy is released rather suddenly—equivalent to 10^{14} L_s or to the total conversion of about 10 Earth-masses per minute from mass into energy. Moreover, because the fluctuations occur in such short times, the part of a quasar responsible for the light (and radio) variations must be smaller than the distance light travels in a month or so.

To see why this must be so, consider a cluster of stars 10 LY in diameter (Figure 15.2) at a very large distance from Earth. Suppose that every

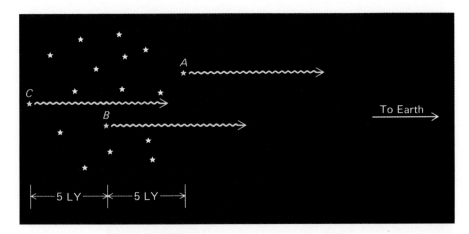

Figure 15.2   A diagram showing why light variations from a large region in space appear to last for an extended period as viewed from Earth. Suppose that all of the stars in this cluster brighten simultaneously and instantaneously. In this example, star A will appear to the observer to brighten five years before star B, which, in turn, appears to brighten five years earlier than star C.

star in this cluster brightens simultaneously and remains bright. We would first see the light from stars on the near side; five years later we would see light from stars at the center. It would be ten years before we detected light from stars on the far side. Even though this cluster brightened instantaneously, from Earth it would appear that ten years had elapsed before maximum brightness was reached. In other words, if an object brightens suddenly, it will seem to us to brighten over a period of time that is equal to the time it takes light to travel across the object from its far side.

In general, the time scale for significant changes in brightness sets an upper limit on the size of the region that brightened. Because quasars vary on time scales of months, the region where the energy is generated can be no larger than a few light months. Some quasars vary on even shorter time scales, and for them the energy must be generated in a region that is even smaller. The challenge then, is to devise a power source that can generate more energy than an entire galaxy in a volume of space that, in some cases, is no larger than our solar system.

Origin of the Redshifts of Quasars

The great difficulty in devising a physical model to account for this flood of energy led some astronomers to suggest that the redshifts of quasars are not the result of the Doppler effect. The spectral lines in galaxies are shifted to the red because the universe is expanding. These astronomers argued that rather than being a consequence of the expansion of the universe, the redshift of the spectral lines in quasars was produced by some physical mechanism that Earth-bound scientists had not previously observed. If this hypothesis were correct, then the measured redshift could not be used to estimate the distances of quasars. Because there were at the time no alternative methods for estimating their distances, the quasars could be assumed to be close enough to us so that their energy output was similar to that of a normal galaxy.

As support for this point of view, some astronomers, most notably Halton Arp, who is now at the European Southern Observatory, have

Figure 15.3 Quasar 3C 275.1, the first to be found at the center of a cluster of galaxies, appears as the brightest object near the center of this image. The quasar nucleus is surrounded by a gas cloud that is elliptical in shape. Its redshift indicates that this quasar is 7 billion LY away and that the light we now observe left the quasar more than 2 billion years before our solar system formed. *(National Optical Astronomy Observatories)*

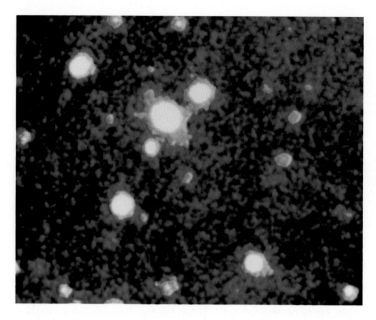

sought evidence for physical associations between high-redshift quasars and low-redshift normal galaxies. If two objects are physically associated, they must be at the same distance. If they also have very different redshifts, then we would be forced to conclude that redshifts are not always a reliable indicator of distance.

Indeed there are many cases in which quasars with large redshifts appear on the sky close to galaxies with small redshifts. It is always possible, however, that these are chance superpositions of two objects that are really at very different distances. There remain too few examples of an apparent association between a quasar and a galaxy with discordant redshifts to convince most astronomers that the redshifts of quasars are not the result of the expansion of the universe.

In fact, several astronomers have turned this argument around and have searched for clusters of galaxies in the vicinity of quasars. If redshifts can be measured and distances can be derived for these normal types of objects, and if the redshifts turn out to be the same as that of the nearby quasar, then we would have compelling evidence that the quasar also obeys the Hubble law. This task is not easy observationally because normal galaxies are fainter than quasars and are therefore more difficult to detect. Nevertheless, studies to date show that quasars are often surrounded on the sky by small clusters of galaxies, and the cluster galaxies exhibit the same redshift as the quasar (Figure 15.3). It is highly improbable that the apparent velocities of quasar and galaxy would coincide unless the two objects were physically associated and at the same distance.

There have been other observations that support the hypothesis that quasars are located at the distances indicated by their redshifts. One key result is the discovery that many relatively nearby quasars ($\Delta\lambda\backslash\lambda = 0.5$) are not true point sources but rather are embedded in a faint, fuzzy-looking patch of light. The color of this fuzz is like that of spiral galaxies. In a few cases, spectra have been obtained that indicate that the light of the fuzz is derived from stars, demonstrating that quasars are located in galaxies.

Violent Activity in Galaxies

When quasars were first discovered, it was thought that they were much more luminous than galaxies. They are indeed more luminous than normal galaxies, but we have now found bona fide galaxies—albeit peculiar ones—that fill in the luminosity gap. These peculiar galaxies share many of the properties of the quasars, although to a less spectacular degree. Members of this class of galaxy are referred to as **active galaxies.** Since the production of abnormal amounts of energy occurs in their centers, they are said to have **active galactic nuclei.** In effect, active galaxies have mini-quasars embedded in their nuclei. Studies of these relatively nearby active galaxies have given astronomers clues about what powers the quasars.

What are the similarities between quasars and active galaxies?

Seyfert galaxies, which are spirals with starlike nuclei (Figure 15.4), are one type of active galaxy. Like quasars, Seyferts have strong, broad emission lines, which indicate that there are gas clouds near their nuclei and that these clouds are moving at high velocity. The width of the lines indicates that the gas is moving at speeds up to thousands of kilometers per second.

Figure 15.4   The Seyfert galaxy NGC 1566, which is at a distance of about 50 million LY, appears on this photograph to be a normal spiral. However, it has a very luminous nucleus, which has many of the characteristics of a quasar, although it is much less energetic. The active region at the center of NGC 1566 has recently been found to vary on a time scale of less than a month, which indicates that it is extremely compact. Spectra show that hot gas near the tiny nucleus is moving at an abnormally high velocity, suggesting that it may be in orbit around a massive black hole.   *(Copyright Anglo-Australian Telescope Board, 1987)*

Figure 15.5 An optical picture of the Seyfert galaxy NGC 1068 shows the bright nucleus at the center of a spiral galaxy. The inset is a picture taken with the Hubble Space Telescope (HST), which shows clouds of ionized gas at the very center of this galaxy. *(NASA)*

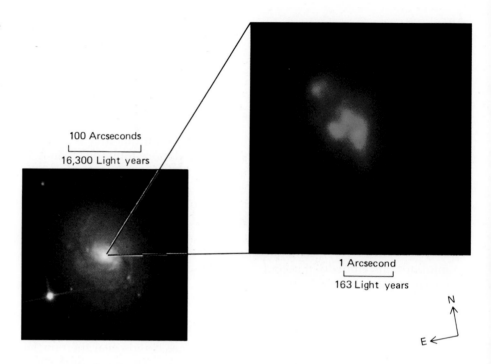

100 Arcseconds
16,300 Light years

1 Arcsecond
163 Light years

N

E

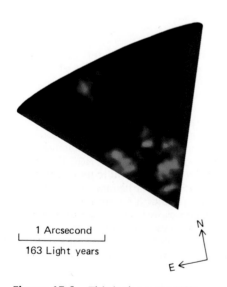

1 Arcsecond
163 Light years

N

E

Figure 15.6 This is the same HST image of NGC 1068 as shown on the right in Figure 15.5, but now the image has been computer-processed to show more detail. The cone was artificially added to the image to illustrate how radiation is beamed from the hidden nucleus, illuminating only in a particular direction. *(NASA)*

One of the earliest images taken with the Hubble Space Telescope (HST) shows the nuclear region of the Seyfert galaxy NGC 1068 (Figure 15.5). The galaxy itself is a barred spiral galaxy, but it contains an extraordinarily bright nucleus. The HST image, freed from the blurring of the Earth's atmosphere, shows the clouds of hot gas in the very center of the galaxy. These clouds are ionized by radiation emitted from the nucleus in a direction perpendicular to the plane of the galaxy (Figure 15.6).

Some Seyferts show brightness variations over a period of a few months, and so, as we concluded for quasars, the region from which the radiation comes can be no more than a few light months across.

The Seyfert and other peculiar galaxies tend to be more luminous than normal galaxies but less luminous than quasars. Their bright but pointlike nuclei indicate that enormous amounts of energy are being emitted from a small region at their centers. The crucial point about Seyferts and other active galaxies is that a significant fraction of their power output comes from a source other than individual stars.

Elliptical Galaxies

It has been known since 1948 that many giant elliptical galaxies that appear comparatively normal in the optical region of the spectrum are powerful emitters of radio energy. Some, M87 (Figure 14.7) being one example, emit thousands of times as much radio energy as is typical of bright galaxies. In some radio galaxies, the bulk of the radio emission comes from small regions within them, while in some others there are bright sources in the nucleus of the galaxy surrounded by larger extended regions of radio emission. In about three-quarters of the radio

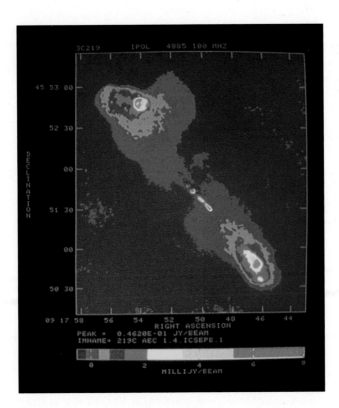

Figure 15.7 This radio image shows the extended lobes of radio emission and a short jet emanating from the core of the elliptical radio galaxy 3C 219. *(NRAO/AUI)*

galaxies, the radio source is double, with most of the radiation coming from extended regions on opposite sides of the galaxy (Figure 15.7). Typically, the two emitting regions are far larger than the galaxy itself and are centered a few hundred thousand LY away from it. Radio observations often reveal two well-delineated jets of radio radiation pointing away from the galaxy toward the large, extended sources. These jets can be more than 1 million light years long (Figure 15.8).

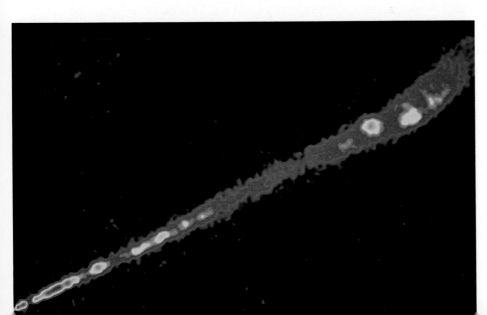

Figure 15.8 The radio jet associated with NGC 6251. This unusually long jet is over 300,000 LY in length. The flow direction is from lower left to upper right, and the jet expands as it moves away from the nucleus of the galaxy. *(National Radio Astronomy Observatory/AUI)*

TABLE 15.1 Numbers of Galaxies of Different Types	
Type	**Number per 10 Billion LY^3**
Luminous spirals	10^7
Seyfert galaxies	10^5
Radio galaxies	10^3
Quasars without radio emission	10^2
Quasars with radio emission	1

Presumably, ionized gases are shot out along the jets into the radio "clouds" by an extremely intense source of energy in the nucleus of the galaxy. The ionized gas eventually collides with neutral gas and slows to a stop, defining the sometimes rather sharp outer edges of the emitting regions. Similar jets are seen in more than 50 percent of all quasars.

Most astronomers now view quasars as simply the most extreme examples of galaxies with active galactic nuclei. As we have seen, even our own Galaxy has a compact source of energy at its center, and broad emission lines indicate that high velocity gas is present there as well. Although the energy emitted by this compact source is tiny relative to that generated by a quasar, the Galaxy may represent the low energy extreme of a continuum of levels of activity in compact nuclei in galaxies. In the next section, we will discuss what causes this activity.

Table 15.1 shows how many of several different types of galaxies we would expect to find if we were to identify all of the objects within 10^{10} (ten billion) cubic LY. Note that radio-emitting quasars are very rare indeed relative to the number of luminous spirals or Seyfert galaxies.

15.2
BLACK HOLES—THE POWER BEHIND THE QUASARS?

Quasars and all of the other types of galaxies that are unusually active emitters of optical, x-ray, and radio radiation have in common a compact source of enormous energy, buried within the nucleus of each. Many models have been offered to account for this energy source, including stellar collisions in dense galactic cores, supermassive stars, extraordinarily powerful supernovae, and others.

The most widely accepted model at the present time is that quasars, and presumably other types of active galaxies, derive their energy output from an enormous black hole at the center of what would otherwise be a normal galaxy. The black hole must be very large—perhaps 1 billion solar masses. Given such a massive black hole, relatively modest amounts of additional material—only about ten solar masses per year—falling into the black hole would be adequate to produce as much energy as a thousand normal galaxies and could account for the total energy of a quasar.

A black hole itself can, of course, radiate no energy. The energy comes from material very close to the black hole. The black hole attracts matter—stars, dust, and gas—which is orbiting around in the dense nuclear regions of the galaxy. This material then spirals in toward the black hole and forms an accretion disk of material around it. As the material spirals ever closer to the black hole, it accelerates and heats through compression to millions of degrees. This hot matter can radiate prodigious amounts of energy as it falls into the black hole.

Observational Evidence for Black Holes in the Centers of Galaxies

One of the strongest pieces of observational evidence that massive black holes do indeed exist in the centers of galaxies has been obtained with the Hubble Space Telescope. Figure 15.9 shows that the stars at the center of the giant elliptical M87 become densely concentrated toward the center, forming a bright, sharp core. The central density of stars in M87 is at least 300 times greater than that expected for a normal giant elliptical galaxy, and over 1000 times more dense than the distribution of stars in the neighborhood of our own Sun. Detailed analysis of the distribution of light in M87 shows that a black hole with a mass of 2.6 billion times the mass of the Sun could have caused this concentration of stars.

This black hole may have formed from the merger of small black holes created by the explosion of massive stars when M87 was a young galaxy. Once formed, the black hole would grow by feeding on gas and stars that passed too close to it. Stars orbiting near the nucleus of the galaxy would be pulled into orbits around the black hole, thereby pro-

What is the source of the prodigious amount of energy radiated by quasars?

Figure 15.9   An image of M87 obtained with the Hubble Space Telescope. The strong concentration of light at the center of the galaxy probably indicates that a 2.6 billion solar mass black hole is located in the nucleus of the galaxy. Both the nucleus and the jet ejected from it are sources of radio emission.   *(NASA)*

ducing the concentration of light we now see. Some of these stars may ultimately fall into the black hole, thereby increasing its mass even more.

Several other galaxies, including our neighbor the Andromeda Galaxy, also appear to have black holes in their centers.

Comparison of the Black Hole Model with Observations

A number of the phenomena that we observe can be explained naturally if quasars contain black holes. First and foremost, it is possible for a black hole to produce the amount of energy that is observed to be emit-

Figure 15.10 Schematic drawings of two accretion disks around large black holes. (a) A thin accretion disk. (b) A "fat" disk, of the type needed to account for channeling outflow of hot material into narrow jets oriented perpendicular to the disk.

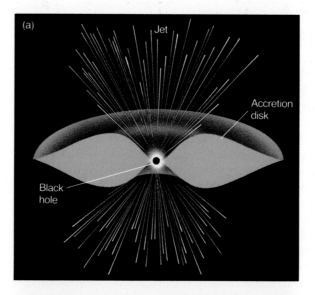

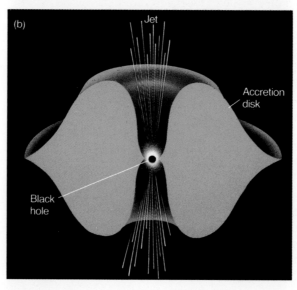

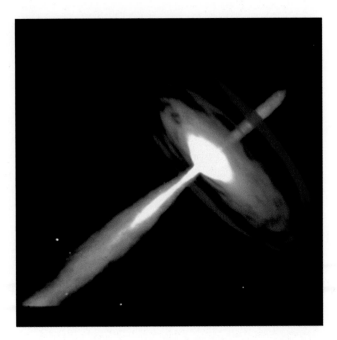

Figure 15.11 Radio jets associated with galaxies and quasars are powered by material falling into a massive, spinning black hole. The high pressures and temperatures generated in the accretion disk surrounding the black hole cause some of the infalling gas to be ejected along the direction of the black hole's spinning axis, creating the galactic jet. *(Artist's concept by Dana Berry, STScI)*

ted by quasars and active galactic nuclei. Detailed calculations show that about 10 percent of the mass of matter falling into a black hole is converted to energy. Remember that during the entire course of the evolution of a star like the Sun, only a tiny fraction of its rest mass will be converted to energy by nuclear fusion. Infall into a black hole is a very efficient way to produce energy.

Because the black hole is also fairly compact in terms of its circumference, the emission produced by infalling matter comes from a small volume of space. As we recall, this condition is required to explain the fact that quasars vary on a time scale of weeks to months.

As we have seen, quasars and other active galaxies emit jets that extend far beyond the limits of the parent galaxy. Observations have traced these jets to within 3 to 30 LY of the parent quasar or galactic nucleus. The black hole and its accretion disk are much smaller than 1 LY, but it is presumed that the jets originate from the vicinity of the black hole.

Why are energetic particles ejected into jets rather than in all directions? It may be that the accretion disk around the black hole is dense enough to prevent radiation from escaping in all but the two directions perpendicular to the disk (Figure 15.10). The basic idea behind the formation of jets is that matter in the accretion disk will move inward toward the black hole. Some of this matter will not fall into the black hole but will feed the jets. That is, some infalling matter will be accelerated by the intense radiation pressure in the vicinity of the black hole and will be blown out into space along the rotation axis of the black hole in a direction perpendicular to the plane of the accretion disk (Figure 15.11). The detailed mechanism for converting the energy associated with infall into

Figure 15.12 The active galaxy NGC 4261, which is located in the Virgo cluster, is shown at the left. The elliptical galaxy, which is the white circular region in the center, is observed at optical wavelengths, while the jets are observed at radio wavelengths. An HST image of the central portion of the galaxy is shown on the right. It contains a ring of dust and gas about 400 LY in diameter surrounding what may be a supermassive black hole. Note that the jets emerge from the galaxy in a direction perpendicular to the plane of the ring, consistent with the models shown in Figure 15.10.

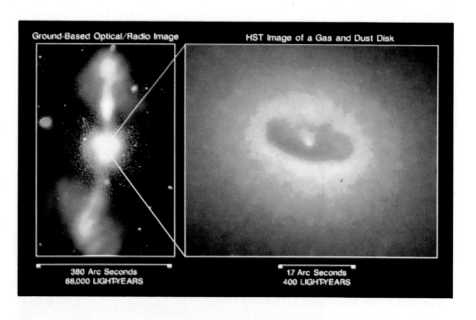

Figure 15.13 A graph showing the number of quasars as a function of the age of the universe. An age of zero corresponds to the beginning of the universe. An age of one corresponds to the present time. Note that quasars were most abundant when the universe was about 20 percent of its current age.

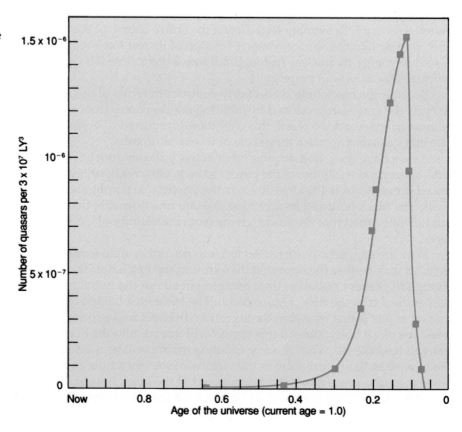

Figure 15.14 An image obtained with the Hubble Space Telescope of the core of a Seyfert galaxy (Markarian 315). The core of this galaxy actually has two nuclei separated by about 6000 LY. The brighter of the two nuclei probably contains a black hole. The fainter nucleus is the core of a galaxy that is merging with the Seyfert galaxy. This image provides support for the idea that gas from a merging galaxy can fuel the black holes that are believed to lie within the centers of quasars and within active galactic nuclei. *(NASA)*

an outward flowing jet remains a matter of controversy for theorists. Figure 15.12 shows a galaxy which contains both jets and an accretion disk in its nucleus.

If matter in the accretion disk is continually being depleted by falling into the black hole or being blown out from the galaxy in the form of jets, then a quasar can continue to radiate only as long as there is gas available to replenish the accretion disk. Where does this matter come from? One possibility is that very dense star clusters form near the centers of galaxies. These stars might then supply the fuel, either through gas that is lost during the normal course of stellar evolution by means of stellar winds and supernova explosions or because the tidal forces exerted by the black hole are strong enough to tear the stars apart.

Observations show that very bright quasars were much more common a few billion years ago than they are now (Figure 15.13). There are many more quasars at great distances, where we are seeing the universe as it was several billion years back in time, than there are nearby. Many astronomers believe that quasars are newly formed galaxies passing through some violent stage of formation. After this stage of formation, the quasar becomes quiescent. One possible explanation is that, as quasars age, they simply run out of fuel—that as time passes, all of the gas, dust, and stars available to fuel the black hole are consumed by it. Indeed, many of the relatively nearby, still-active quasars appear to be embedded in galaxies that have recently been involved in collisions with other galaxies. Gas and dust from this second galaxy have apparently been swept up by a dormant black hole and so have provided the new source of fuel required to rekindle it (Figure 15.14).

Figure 15.15 Elliptical galaxy NGC 1199 and companions. NGC 1199 is the large E3 elliptical galaxy at bottom left. It is accompanied by a face-on barred spiral galaxy (NGC 1189) seen toward the upper left and an S0 galaxy (NGC 1190) of spindly appearance near the right edge. We can see the yellow-orange color characteristic of older population II stars in the elliptical galaxy and in the central bar and bulge of the barred spiral. Younger population I stars produce the bluish color of the disk of the barred spiral. This picture shows how colors can be used to determine the ages of stars found in distant galaxies. *(N.A. Sharp/NOAO)*

15.3
EVOLUTION OF GALAXIES—THE OBSERVATIONS

The study of distant galaxies has been a major focus of recent astronomical research. The goal is to observe directly what galaxies were like shortly after, and conceivably even while, they formed. Three types of observations have provided most of the clues to how galaxies evolve— spectroscopy, measurements of the colors of galaxies, and determination of their shapes.

Spectra, Colors, and Shapes

A spectrum of a galaxy provides a great deal of information. Measurement of the radial velocity can be used to estimate the distance. Studies of the rotation of galaxies can be used to estimate their masses. Detailed analysis of the spectral lines can determine what types of stars inhabit a galaxy, what their composition is, and whether a galaxy contains large amounts of interstellar matter.

Unfortunately, many galaxies are so faint that it is impossible to collect enough photons, even with the world's biggest telescopes, to produce a measurable spectrum. Astronomers use colors to estimate what kinds of stars inhabit the faintest galaxies (Figure 15.15). To understand why this works, remember that hot luminous blue stars have lifetimes of only a few million years. If we see a very blue galaxy, we know that it must have many hot luminous blue stars, and that star formation must have occurred within the past few million years. If we see a reddish galaxy, however, it must contain mostly old stars, formed billions of years before the light that we now see was emitted by the galaxy.

Another important clue to the nature of a galaxy is its shape. As we have seen, spiral galaxies contain young stars and large amounts of interstellar matter. Elliptical galaxies have mostly old stars and very little interstellar matter. For whatever reason, elliptical galaxies turned most of their interstellar matter into stars many billions of years ago, while star formation has continued until the present day in spiral galaxies. Consistent with this overall picture, spiral galaxies are bluer than elliptical galaxies. Unfortunately, galaxies at very large distances also appear small in the sky. Very often we cannot tell whether a distant galaxy is a spiral or an elliptical. One of the main goals of the Hubble Space Telescope is to obtain images of distant galaxies, unblurred by the Earth's atmosphere, in order to determine whether the same types of galaxies that we see nearby in the present day universe existed billions of years ago.

Quasars are intrinsically more luminous than ordinary galaxies, but for very distant quasars we can detect only the radiation from the nucleus, not from the surrounding galaxy. Since the source of radiation in the nucleus is not well understood, it is difficult to interpret the observations of quasars in terms of how galaxies as a whole change with time. Despite these problems, the observations of distant quasars provide important clues to guide theorists in their attempts to develop models of the universe as it was when galaxies were young.

The Ages of Galaxies

One starting point for all theories of galaxy formation is the fact that most galaxies are very old indeed, and there are several observations that lead to this conclusion. For example, there are stars in globular clusters in our own Galaxy that are 13 to 15 billion years old. Therefore, the Milky Way must be at least this old.

The universe itself is not significantly older than the globular cluster stars. The age of the universe is derived from the observed rate of expansion. As we have seen (see Section 14.4), galaxies are moving farther and farther apart. If we project this expansion back in time, we find that all of the galaxies were very close together sometime between 10 and 15 billion years ago. The major uncertainty in the age of the universe is caused by uncertainties in estimates of how far away galaxies are from us at the present time. These ideas are discussed in detail in Chapter 16. It appears that the globular cluster stars in the Milky Way Galaxy must have formed during the first 2 billion years after the expansion of the universe began.

The most distant elliptical galaxies for which we have some information on composition emitted the light that we observe when the universe was only about half its present age. Yet some of these galaxies

How can we determine what types of stars are found in very distant galaxies?

How old are the galaxies?

have about the same luminosity and colors, and hence about the same stellar content, as do galaxies that are only a few million LY distant, which are therefore about twice as old. The similarity of ellipticals that span half the age of the universe suggests that star formation in this type of galaxy has either been absent or nearly so for the last several billion years. Star formation probably began about 1 billion years after the universe began, and new stars continued to form for at most a few billion years.

We can probe still farther back in time, and still closer to the beginning of the universe, by observing quasars, which are much brighter than normal galaxies and can be seen at larger distances. As we have seen, quasars were most common when the universe was only about one-third of its present age. Remember that quasars are found in the centers of galaxies. The mere fact that quasars existed in such large numbers when the universe was only one-third as old as it is now is another clue that galaxies were formed very early.

Another clue comes from the study of the gas in quasars. Astronomers find that the composition of the gas in distant quasars is very much like the composition of the gas in our own Galaxy. Specifically, the gas in quasars contains not only hydrogen and helium, but also heavier elements such as carbon, nitrogen, and oxygen. We think that these heavy elements were not present when the universe began but were manufactured in the first generation of stars that evolved within newly formed galaxies. Because quasars contain large amounts of heavy elements, at least one generation of stars had already completed its evolution even before the light that we now see was emitted. Given the distance to quasars, this means that some galaxies must have formed when the universe was less than 20 percent as old as it is now.

One question that we do not know the answer to is whether *all* galaxies formed at the same time during the first 1 or 2 billion years after the universe began, or whether some galaxies formed billions of years later.

How Galaxies Change with Time

The galaxies that we see today appear to differ from the types of galaxies that were most common several billion years ago. Studies show that rich clusters at distances of about 5 billion LY contain many more blue galaxies than do nearby rich clusters. Recent observations with the Hubble Space Telescope indicate that these blue galaxies are mainly spiral galaxies (Figure 15.16). Because a blue galaxy must contain young stars, this difference in color indicates that more spiral galaxies were actively forming stars 5 billion years ago than now. The rate of star formation has, on average, declined dramatically during the past 5 billion years or so.

Two processes have been suggested for producing a high rate of star formation. First, there is evidence that collisions of galaxies can compress the gas within them and stimulate the formation of stars (see essay, "Colliding Galaxies"). Another possibility is that rich clusters of

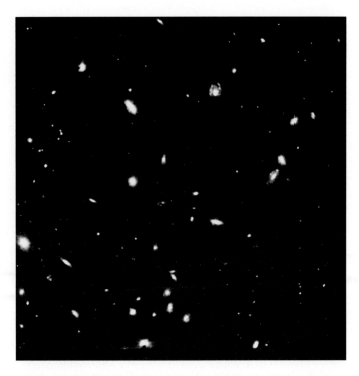

Figure 15.16   A Hubble Space Telescope image of a remote cluster of galaxies. We are seeing this cluster as it was when the universe was only two-thirds as old as it is now (redshift = 0.4). This cluster contains many more spirals than do clusters in the present era. During the past few billion years, many spiral galaxies have either disappeared, probably through mergers with other galaxies, or grown much more faint as star formation has faded away.   *(A. Dressler/NASA)*

galaxies may also contain gas between the galaxies that has a high temperature and high pressure. A galaxy, as it moves on its orbit through the cluster, may suddenly run into some of this high temperature gas. In the ensuing collision, the cold molecular clouds in the galaxy may be compressed, again accelerating the rate of star formation.

Why have these blue spirals vanished from clusters of galaxies over the past 5 billion years? One possibility is that mergers of galaxies in the rich clusters have reduced the likelihood of collisions, and therefore reduced the likelihood of enhanced rates of star formation. Indeed, counts of galaxies suggest that there were many more galaxies several billion years ago than there are now. Perhaps the number has decreased through mergers.

Star Formation in Galaxies

As we have seen, the evolution of spiral and elliptical galaxies differs in a fundamental way. In spirals, star formation is a continuous process that is still occurring today, although on average, it is occurring at a somewhat lower rate than it was several billion years ago. In elliptical galaxies, even the youngest stars are older than the Sun. Since there is very little dust or gas in ellipticals, star formation cannot take place in the present era.

Where did the gas and dust go? Much of it must have been consumed very rapidly in the formation of the first generations of stars. But

What is the evidence that galaxies, like stars, change as they age?

Figure 15.17 The Virgo Cluster of galaxies imaged in x rays (blue) and hydrogen radio emission (red). The x-ray data were obtained at the Einstein Observatory. Note that the x-ray and hydrogen emissions arise from different parts of the cluster. It seems likely that galaxies near the center of the cluster have been stripped of their hydrogen gas. *(National Radio Astronomy Observatory/AUI)*

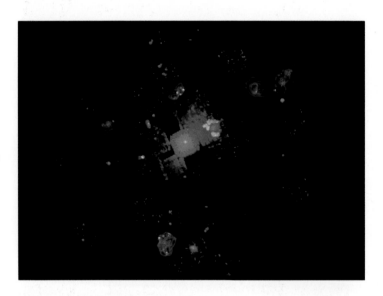

star formation alone would not be efficient enough to consume all of the gas and dust originally present in elliptical galaxies. In any case, as stars evolve, they lose mass either via stellar winds, by forming planetary nebulae, or by exploding. In the process, they inject dust and gas into the space between the stars, replenishing the interstellar material.

It must be that gas and dust are somehow efficiently removed from elliptical galaxies. One possibility is that the gas is swept out. Ellipticals occur in clusters of galaxies, not in isolated and otherwise empty regions of space. In these clusters, gas is present between the galaxies, as we know from x-ray observations. As an elliptical galaxy orbits about within a cluster, it moves rapidly (typical velocities are 1000 km/s) through the gas that lies within the cluster but outside the galaxies. This intergalactic gas bombards whatever small amount of gas may lie within the boundaries of the elliptical and drives the gas from the galaxy (Figure 15.17). The pressure of the intergalactic gas is too small to affect the motions of the stars in the galaxy.

Observations show that about 80 to 90 percent of the galaxies in the high-density environments in the centers of clusters of galaxies are ellipticals and disk-shaped galaxies that have very little gas, no spiral arms, and no recent star formation. In the present era, relatively few spirals are found near the center of rich clusters. Conversely, isolated galaxies found in regions outside of clusters or groups of galaxies, where the density of material is low, are mostly spirals.

Spiral galaxies that have retained their gas and dust until the present era have managed to do so because they lie isolated in regions of space where the distances to other galaxies are fairly large and the density of intergalactic gas is too low to sweep them clean. The Milky Way and the Andromeda galaxies are examples.

The Masses of Galaxies

The mass of the Milky Way Galaxy, excluding the invisible matter that makes up the large dark halo, is about 10^{11} times the mass of the Sun. Most other spiral galaxies have masses within about a factor of 100 of this same value. The largest ellipticals are somewhat more massive than the largest spirals, but any theory of galaxy formation must explain why galaxies that are very much more massive do not occur.

Another important characteristic of galaxies is that small galaxies, which are also the fainter galaxies, are more common than large galaxies. In this respect, galaxies are much like stars; remember that there are many more faint stars with low masses than there are bright stars with large masses.

15.4
EVOLUTION OF GALAXIES—THE THEORIES

The universe is expanding. At the beginning, as we shall see in the next chapter, the universe was very smooth. Matter was distributed uniformly. Now, however, the universe is certainly no longer smooth. Matter has clumped into stars. Stars themselves are not found everywhere in space but clump together to form galaxies. Even the galaxies are not uniformly distributed but have congregated to form clusters and superclusters with great voids in between. The challenge for the theorist is to understand how an initially smooth (or nearly smooth) distribution of matter in the universe gives rise to the complex structure that we now see.

Top-Down or Bottom-Up

There are many ideas about how structure might have formed. Here we will look at the two possibilities that have been explored in the most detail. Top-down theories assume that large structures, that is, supercluster-sized concentrations of matter, formed first and then fragmented to form galaxies. Bottom-up theories hypothesize that small structures formed first and then merged to build larger ones. If this bottom-up picture is correct, galaxies formed first and then gradually assembled to build clusters and then superclusters of galaxies.

Both top-down and bottom-up theories assume that the universe was not initially absolutely smooth, but rather contained small fluctuations in density. As the universe expanded, the regions of higher density accumulated additional mass because they exerted a slightly larger than average gravitational force on surrounding material.

As in the case of star formation, the fate of these regions of higher density depended on the balance between pressure and gravity. If the gravitational force exceeded the pressure force, the individual region would ultimately stop expanding in diameter with the expansion of the

Which type of structure formed first—large-scale concentrations of matter containing the mass of a supercluster of galaxies or small concentrations with masses similar to that of a globular cluster of stars?

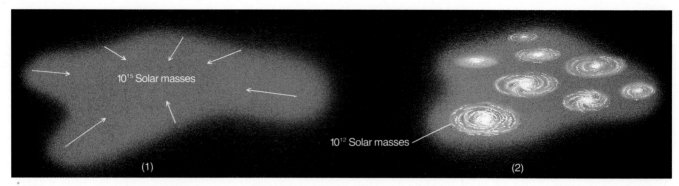

10^{15} Solar masses

(1)

10^{12} Solar masses

(2)

Figure 15.18 Schematic diagram showing how galaxies might have formed if large-scale supercluster structures formed first and then fragmented to form galaxies.

universe. It would then begin to collapse to form a cluster of stars, a galaxy, a cluster of galaxies, or even a supercluster of galaxies.

Unfortunately, the calculations of what might happen are very difficult, and the most likely size of the first high-density regions to collapse is not clear. There are two possibilities. The typical initial condensations may have been very large and contained total masses equal to 10^{15} times the mass of the Sun, which is the mass contained within a supercluster. Alternatively, they may have been rather small and contained only 10^6 times the mass of the Sun, which is about the mass of a large globular cluster. Intermediate size condensations are not likely to have been formed initially.

Superclusters First?

Top-down theories calculate the consequences if only the very large-scale density fluctuations were able to collapse. Initially, gas clouds with masses of about 10^{15} times the mass of the Sun began to collapse. The collapse was irregular, and the structure formed was a pancake-shaped blob. Within the pancake, many individual regions of high density formed, and these too began to collapse. Calculations show that stable structures could be formed only for masses less than 10^{12} times the mass of the Sun and with diameters less than 300,000 LY—just the size of galaxies (Figure 15.18). Larger fragments were extremely diffuse and were destroyed in collisions with other fragments before stars could form within them.

This top-down model has the advantage that it can explain why galaxies are no larger than they are observed to be, but it has one fatal flaw. It takes a long time for the original pancake to collapse and fragment into still smaller structures the size of galaxies. This model predicts that galaxies should still be forming in the present era. Although astronomers have found a few nearby galaxies that may be young, it is clear that most galaxies formed billions of years ago. We have also found quasars at such large redshifts that they must have formed when the universe was less than 10 percent of its current age. The top-down model cannot account for such rapid formation of quasars, which are galaxies with massive black holes in the centers.

Bottom-Up?

The alternative approach is to assume that small-scale structures formed first. This idea is the one that is most widely accepted as of this writing, but it will very likely be modified as both theory and observation improve.

The basic assumption of the bottom-up models is that the first regions of higher density to begin to collapse had masses of about 10^6 times the mass of the Sun, or about the size of a large globular cluster. Their collapse began when the universe was no more than 1 or 2 percent of its present age. As time passed, regions containing ever larger mass, possibly as large as the mass of a giant elliptical, began to collapse as well. Galaxies were formed either from the collapse of single galaxy-sized clouds or through the merger of several smaller structures. Clusters of galaxies formed as individual galaxies congregated, drawn together by their mutual gravitational attraction (Figure 15.19). First, a few galaxies came together to form a group, much like our own Local

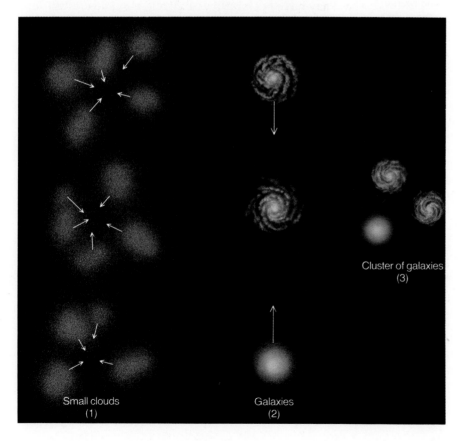

Figure 15.19 Schematic diagram showing how galaxies might have formed if small clouds formed first and then congregated to form galaxies and, subsequently, clusters of galaxies.

ESSAY Colliding Galaxies

Over the past 20 years, astronomers have found mounting evidence that galaxies frequently collide with one another and that these collisions play an important role in determining how galaxies evolve.

What happens when galaxies collide? As far as the stars are concerned, not very much. Since the stars are very far apart, a direct collision of two stars is highly unlikely. The orbits of a few stars may be altered by the collision. In the case of spiral galaxies, long tails of stars may be drawn out away from the central regions of the galaxies (Figure 15A).

If the collision is slow, the two colliding galaxies may form a binary system, with the two nuclei surrounded by a common envelope of stars. Eventually the nuclei may coalesce to form a single galaxy. The large elliptical galaxies found in the centers of clusters of galaxies are probably formed by the mergers of several smaller galaxies. Calculations show that slow collisions and mergers can transform spiral galaxies into elliptical galaxies.

The interstellar matter in galaxies is much more affected by galaxy interactions than are the stars. Interstellar gas clouds are large and are likely to experience direct impacts with other clouds. These violent collisions compress the gas in the clouds, and the increased density can lead to vigorous star formation (Figure 15B). In some interacting galaxies, the star formation is so intense that all of the available gas will be exhausted in only a few million years, so that the burst of star formation is clearly only a temporary phenomenon. Bursts of star formation are very rare in isolated galaxies.

Additional evidence that galaxy collisions stimulate star formation comes from the Hubble Space Telescope. Figure 15C shows an HST image of the peculiar galaxy NGC 1275. There are about 50 globular clusters in this picture. The surprising thing about these clusters is that they contain young, hot, massive stars. Remember that all of the stars in globular clusters in the Milky Way Galaxy are billions of years old. Astronomers think that NGC 1275 may actually be two galaxies—a giant elliptical galaxy and a smaller spiral galaxy—colliding with each other. The globular clusters may have formed recently as a result of this collision.

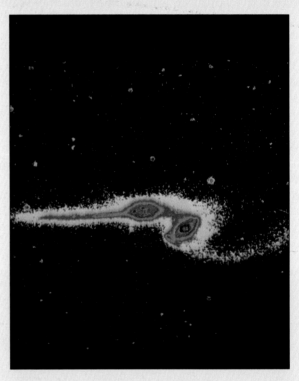

Figure 15A NGC 4676AB (the Mice), a classic system of colliding galaxies that have produced narrow tails as a consequence of the interaction. In this computer-processed image, the different colors correspond to different intensities. *(Courtesy of W. Keel and R. Kennicutt/NOAO)*

Galaxy collisions also play a role in activating those quasars that are closest to the Milky Way. About one-third of all nearby quasars, in which we can see the surrounding host galaxy, have a companion galaxy. The interaction with the companion appears to be responsible for providing a supply of gas that can be accreted by the black hole in the center of the host galaxy. Just how the gas is provided to the black hole is not clear. Gas may be transferred from the companion galaxy to the host galaxy. Alternatively, the orbits of gas clouds in the host galaxy may be disrupted, and the clouds may fall toward the black hole in the center.

Figure 15B   Optical (*left*) and infrared (*right*) images of M51, the Whirlpool Galaxy. The outlying arm, which reaches to the companion galaxy, is much brighter in the optical than in the infrared. This difference in color suggests that most of the stars in this arm are hot, young stars, and that the formation of these stars was stimulated by the interaction of the two galaxies.   *(NOAO)*

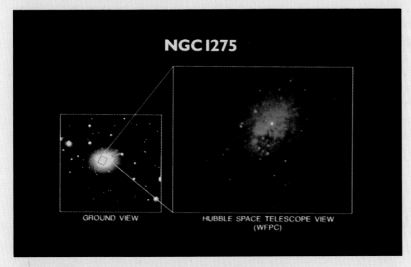

Figure 15C   An image from HST of the Galaxy NGC 1275. About 50 bluish globular clusters can be seen in this picture.   *(NASA)*

Group. The groups then began to combine to form clusters and eventually superclusters, like the Local Supercluster.

This model explains in a natural way why there are more small galaxies than large galaxies. The collapsing clouds are initially gaseous and collide and merge to form galaxies. The more collisions and mergers that occur, the larger the galaxy that finally emerges. Since it is more likely that a given cloud will experience only a few collisions and mergers rather than many, very large galaxies should be rare.

Giant elliptical galaxies are round and are found in the regions of highest density. It is likely that they were formed through the collision and merger of many fragments. Any collision of two systems of gas and stars will tend to stir up the orbits of the individual stars within each system, and will also tend to strip away from the outer regions any matter that is not strongly held to the system by gravity. The result of many collisions is the building of round systems that do not have extended disks.

According to this theory, spiral galaxies are formed in relatively isolated regions. A single cloud of gas collapses undisturbed to a disk, in which stars are then formed. A spiral might, through collisions with smaller systems, acquire some of the stars that populate its halo and its nuclear bulge. These stars are distributed in a spherical fashion, as are the stars in ellipticals, which are also built through mergers. As the isolated cloud collapses to form a spiral, it may leave behind some fragments that become dwarf galaxies. Many spirals, including the Milky Way Galaxy, are surrounded by a swarm of small galaxies.

The bottom-up model predicts that there should be very few young galaxies in the present era, and indeed young nearby galaxies appear to be rare. This model also predicts that clusters and superclusters should still be in the process of forming. Observations of the motions of galaxies in clusters suggest that they are still in the process of accumulating their member galaxies.

Neither the top-down nor the bottom-up model can explain all of the observations of galaxies in detail. The bottom-up model is more consistent with the fact that galaxies seem to have formed very early in the history of the universe, while larger structures like superclusters of galaxies are still forming. However, this is a field of research that is in its infancy, and these simple models will surely be modified over the next few years.

15.5
A UNIVERSE OF (MOSTLY) DARK MATTER

In this text so far we have focused nearly all of our attention on matter that radiates electromagnetic energy—x rays, optical radiation, radio waves, etc. But not all types of matter emit photons. Indeed, we have already discussed briefly some observations that suggest that galaxies contain large amounts of *dark matter* (Section 14.2).

The idea that much of the universe is filled with dark matter may seem like an unusual and bizarre concept, but there is one historical example of dark matter that is very familiar to astronomers. In the middle of the nineteenth century, measurements showed that the planet Uranus did not follow exactly the orbit predicted on the basis of adding up the gravitational forces of all of the known objects in the solar system. The deviations in Uranus' orbit were attributed to the gravitational effects of an (at the time) invisible planet. Calculations showed where that planet had to be, and Neptune was discovered very near the predicted location.

In just the same way, astronomers today are trying to determine the location and amount of dark matter in galaxies by measuring the gravitational effects it produces on objects we can see. It now appears that dark matter makes up at least 90 percent—and perhaps as much as 99 percent—of all the matter in the universe. The following sections describe the evidence for the dark matter and offer some speculations about what it might be made of.

Local Neighborhood

The first place to look for dark matter is in our own solar system. Astronomers have examined the orbits of the known planets and of spacecraft as they journeyed to the outer planets and beyond. No deviations have been found from the orbits predicted on the basis of the objects already discovered in our solar system. Thus there is no evidence for large amounts of nearby dark matter.

Astronomers have also looked for evidence of dark matter in the region of the Milky Way Galaxy within a few hundred light years of the Sun. In the vicinity of the Sun, most of the stars are restricted to a thin disk. It is possible to calculate how much mass must be in the disk to keep the stars from wandering away to large distances above or below the disk. The total matter in the disk is not more than twice the amount of luminous matter, so again there is little evidence for unseen matter close to the Sun.

What is the evidence that most of the matter in the universe does not emit electromagnetic radiation at any wavelength?

Dark Matter Around Galaxies

Although the local neighborhood may not contain much dark matter, we have seen evidence that 90 percent of the mass in the Milky Way Galaxy may be in the form of dark matter. The stars in the outer region are revolving very rapidly around the center of the Galaxy—so rapidly that the mass contained in all the stars and all the interstellar matter in the Galaxy cannot exert enough gravitational force to keep those distant stars in their orbits (Chapter 13). The same result is found for other spiral galaxies as well. Although less convincing, there is some evidence for halos of dark matter around elliptical galaxies.

Mathematical analyses of the rotation of spiral galaxies suggest that the dark matter is found in large halos surrounding the luminous parts

Figure 15.20 Drawing of a spiral galaxy surrounded by a halo of dark matter. The dark matter halo is much larger in diameter than is the disk of luminous matter.

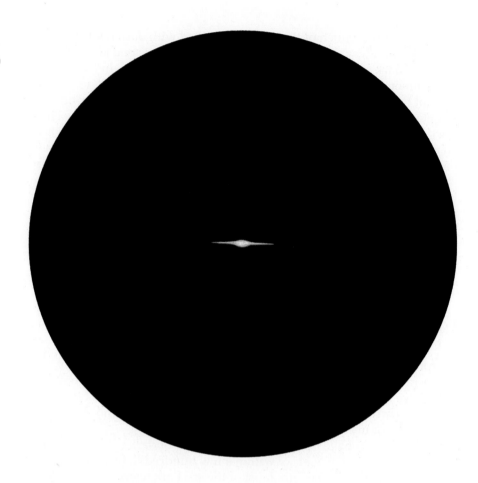

of the galaxy. The radius of the halo may be as large as 300,000 LY (Figure 15.20).

Dark Matter in Clusters of Galaxies

Galaxies in clusters orbit around the center of mass of the cluster. It is not possible to follow a galaxy around its entire orbit. For example, it takes 10 billion years or more for the Andromeda and Milky Way galaxies to complete a single orbit around each other. It is possible, however, to measure the velocities with which galaxies in a cluster are moving and then estimate what the total mass in the cluster must be to keep the individual galaxies from flying off into space. The amount of dark matter contained within giant clusters of galaxies appears to be at least equal to the fraction of dark matter in individual galaxies and could be as much as ten times larger.

Dark Matter in Superclusters

The universe is expanding, and all galaxies participate in that expansion. The expansion is not perfectly uniform, however. Some galaxies are

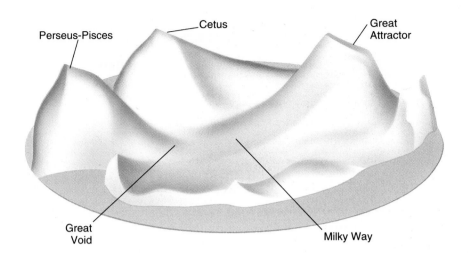

Perseus-Pisces

Cetus

Great
Attractor

Great
Void

Milky Way

Figure 15.21   A drawing showing the concentrations of matter within about 150 million LY of the Milky Way Galaxy, which is located at the center of the drawing. The Galaxy is located in a region of relatively low density. The mass concentration that corresponds to the Great Attractor is shown at the right-hand side of the drawing. All of the galaxies, including those in the Perseus-Pisces region, are flowing toward the Great Attractor at a velocity of about 425 km/s—a giant river in the sky containing thousands of galaxies.

moving away from us at a slightly faster than average rate. Others are moving away at a slower than average rate. Suppose, for example, that a galaxy lies outside of, but relatively close to, a rich cluster of galaxies. The gravitational force of the cluster will tug on that neighboring galaxy and slow down the rate at which it moves away from the cluster as a result of the expansion.

As a specific example, consider the Milky Way Galaxy and the other members of the Local Group. They lie on the outskirts of the Virgo supercluster. The mass concentrated at the center of the Virgo cluster exerts a gravitational force on the Local Group. As a result, the Local Group is moving away from the center of the Virgo cluster at a velocity that is a few hundred kilometers per second slower than one would predict from the Hubble law.

Astronomers have now measured accurate distances and velocities for a few thousand galaxies within about 150 million LY of the Milky Way Galaxy (Figure 15.21). They have found that the galaxies tend to be flowing toward a large concentration of mass, which has been dubbed the "Great Attractor." From the rate of the flow it is possible to calculate how much matter must be located in the Great Attractor. The mass of the Great Attractor is estimated to be 3×10^{16} solar masses, equivalent to tens of thousands of galaxies. Current studies, which may be modified as more observations are obtained, indicate that the amount of dark matter in the Great Attractor may be as much as 100 times the amount of luminous matter. Presumably, new observational studies will find other similar concentrations of matter, but astronomers have just begun to look for them.

Mass-to-Light Ratio

Section 14.2 described the use of the mass-to-light ratio to characterize the matter in galaxies or clusters of galaxies. For systems containing mostly old stars, the mass-to-light ratio is typically between 10 and 20.

TABLE 15.2 Mass-to-Light Ratio	
Type of Object	**Mass-to-Light Ratio**
Sun	1
Matter in vicinity of the Sun	2
Total mass in Milky Way	10
Small groups of galaxies	50–150
Rich clusters of galaxies	250–300

Mass-to-light ratios of 100 or more are a signal that a substantial amount of dark matter is present. Table 15.2 summarizes the results of measurements of mass-to-light ratios for various classes of objects. Very large mass-to-light ratios are found for all systems of galaxy size and larger, and this result indicates that dark matter is present in all of these types of objects. There is persuasive evidence that dark matter makes up most of the total mass of the universe. If the observations of the motions of galaxies toward the Great Attractor prove to be correct, then 99 percent of the mass in the universe is dark.

What is the dark matter? If there is only five to ten times as much dark matter as luminous matter, then the dark matter could consist of normal particles—protons and neutrons. These protons and neutrons are not assembled into stars, or we would see them. Neither can the dark matter be in the form of dust and gas, or we could detect it. The protons and neutrons could be in the form of black holes, brown dwarfs, or white dwarfs—objects too faint to be observed directly at large distances.

Recently a new technique has been devised to search for these faint objects, which have been dubbed MACHOs (massive compact halo objects) in the halo of our own Galaxy. If an invisible MACHO passes directly between a distant star and the Earth, the distant star will appear to brighten over a time interval of several days and then return to its normal brightness. The MACHO acts as a kind of magnifying glass, temporarily increasing the apparent brightness of the star as it passes in front of it. This effect was one of the predictions of Einstein's theory of general relativity.

Two research teams, both making observations of millions of stars in the Large Magellanic Cloud, have reported brightenings of the type expected if MACHOs are present in the halo of the Milky Way. More observations will be required to confirm these early results, to establish just how massive the MACHOs are, and how much they contribute to the total mass of the Galaxy.

If it turns out that there is 100 times more dark matter than bright matter, as the observations of the flow toward the Great Attractor suggest, then for reasons described in the next chapter, the dark matter cannot be made entirely of MACHOs or of any other objects composed of neutrons and protons. Alternative possibilities for the composition of dark matter will be discussed in the next chapter.

▶ SUMMARY

15.1 **Quasars** were first discovered because of their strong radio emission. Optical spectroscopy of the star-like objects has shown that quasars have redshifts ranging from 15 to (so far) nearly 95 percent of the speed of light. Some quasars are members of small groups or clusters. Others have fuzz around them that has the spectrum of a normal galaxy. In such cases, the quasars obey the Hubble law, which relates velocity or redshift to distance. On the basis of this evidence, most astronomers believe that the redshifts of the quasars result from the expansion of the universe.

15.2 Most astronomers now view quasars as the most extreme example of a class of peculiar or **active galaxies** that generate large amounts of energy in a small **active galactic nucleus**. Both active galactic nuclei and quasars are thought to derive their energy from material falling toward, and forming a hot accretion disk around, a massive (up to 10^9 solar masses) black hole. Quasars were much more common billions of years ago than they are now, and astronomers speculate that quasars mark some violent stage in the formation of galaxies. Quasar activity can apparently be retriggered by collisions between galaxies, which provide a new source of fuel to feed the black hole.

15.3 Observations provide important constraints on models of the formation of galaxies. Galaxies were formed when the universe was no more than 1 or 2 billion years old. Star formation in spirals was much more active 5 billion years ago than it is today. There were probably more galaxies several billion years ago than there are today, with the number being reduced by collisions and mergers. Low-mass galaxies are much more common than high-mass galaxies. Elliptical galaxies tend to be found in the centers of dense clusters of galaxies. Spiral galaxies tend to be relatively isolated from other galaxies.

15.4 The challenge for theories of galaxy formation is to show how an initially smooth distribution of matter can develop the structure—galaxies and galaxy clusters—that we see today. Calculations show that the first condensations of matter are likely to have contained either the mass of a supercluster of galaxies or of a globular cluster. Observations seem to favor the initial condensation of globular cluster-size masses, which then congregate to form galaxies and clusters of galaxies.

15.5 The visible matter in the universe does not exert a large enough gravitational force to hold stars in their orbits within galaxies or to hold galaxies in their orbits around other galaxies. There is at least 5 to 10 times, and perhaps as much as 100 times, more dark matter than luminous matter. Astronomers do not know yet whether the dark matter is made of ordinary matter—protons and neutrons, for example—or made of some totally new type of particle not yet detected on Earth.

▶ REVIEW QUESTIONS

1. What is a quasar?

2. Do quasars obey the Hubble law, which relates redshift and distance? Describe the evidence.

3. How can a black hole, which itself can emit no energy, account for the tremendous energy output of quasars?

4. What is the evidence that galaxies formed 1 or 2 billion years after the universe began?

5. Describe two possible ways in which galaxies might form. Which possibility seems more likely? Why?

6. What is the evidence that a large fraction of the matter in the universe is invisible?

▶ THOUGHT QUESTIONS

7. Quasars are much rarer in the immediate vicinity of the Milky Way Galaxy than they are at large redshifts. What does this tell you about the evolution of quasars?

8. Suppose you observe a star-like object in the sky. How would you determine whether it is actually a star or a quasar?

9. Describe how you might use the color of a galaxy to determine something about what kinds of stars it contains.

10. Suppose a galaxy forms stars over a time interval of a few million years and then all star formation stops. What would be the most massive stars on the main sequence after 500 million years? after 10 billion years? How would the color of the galaxy change over this time interval? (Refer to Table 11.1.)

11. Suppose that the Milky Way Galaxy were truly isolated and that there were no other galaxies within 100 million LY. Suppose that galaxies are observed in large numbers at distances greater than 100 million LY. Why would it be more difficult to determine accurate distances to those galaxies than if there were also galaxies relatively close by?

12. Given the ideas presented here about how galaxies form, would you expect to find a giant elliptical galaxy in the Local Group? Why or why not? Is there a giant elliptical in the Local Group?

13. Can an elliptical galaxy evolve into a spiral?

14. Why do we know less about the formation of galaxies than we know about the formation of stars?

15. Suppose that you were to develop a theory to account for the evolution of the population of New York City. Would your theory most closely resemble a bottom-up or a top-down theory as we have applied those terms to galaxy evolution?

16. Suppose that you observe a quasar surrounded by a cluster of galaxies. If the redshifts of the galaxies match the redshift of the cluster, astronomers argue that the galaxies and the quasar are physically associated and are at the same distance. If the redshifts do not match, most astronomers conclude that there is no physical association. Do you think these conclusions are justified? Why or why not?

▶ PROBLEMS

17. Rapid variability in quasars indicates that the region in which the energy is generated must be small. Show why this is true. Specifically, suppose that the region in which the energy is generated is a transparent sphere 1 LY in diameter. Suppose that in 1 second this region brightens by a factor of ten and remains bright. How long would it take, as it appears to you on Earth, for the quasar to reach maximum brightness?

18. Theory suggests that elliptical galaxies may form by accumulating small condensations of matter, but we do not suggest that stars grow by accumulating matter from other stars. Why is it more likely that one galaxy will collide with another than that two stars will collide? (Hint: Compare the separations between stars in the solar neighborhood with their diameters. Then compare the diameter of galaxies in the Local Group with their separations.)

▶ SUGGESTIONS FOR ADDITIONAL READING

Books

Preston, R. *First Light.* Atlantic Monthly Books, 1987. An intriguing book on doing astronomy with the 200-inch telescope at Palomar, with a number of sections on quasar research.

Verschuur, G. *The Invisible Universe Revealed: The Story of Radio Astronomy.* Springer Verlag, 1987. Part II is a good introduction to quasars and active galaxies.

(See also several of the books cited for Chapter 14 on galaxies and Chapter 13 on dark matter.)

Articles

Preston, R. "Beacons in Time: Maarten Schmidt and the Discovery of Quasars" in *Mercury,* Jan./Feb. 1988, p. 2.

Croswell, K. "Have Astronomers Solved the Quasar Enigma?" in *Astronomy,* Feb. 1993, p. 29.

Finkbeiner, A. "Active Galactic Nuclei: Sorting Out the Mess" in *Sky & Telescope,* Aug. 1992, p. 138.

Wilkes, B. "The Emerging Picture of Quasars" in *Astronomy,* Dec. 1991, p. 35.

Barnes, J., *et al.* "Colliding Galaxies" in *Scientific American,* Aug. 1991.

Keel, W. "Crashing Galaxies, Cosmic Fireworks" in *Sky & Telescope,* Jan. 1989, p. 18.

Hartley, K. "Elliptical Galaxies Forged by Collision" in *Astronomy,* May 1989, p. 42.

Gregory, S. "Active Galaxies and Quasars: A Unified View" in *Mercury,* July/Aug. 1988, p. 111.

Rees, M. "Black Holes in Galactic Centers" in *Scientific American,* Nov. 1990.

Schendel, J. "Looking Inside Quasars" in *Astronomy,* Nov. 1982, p. 6.

Schorn, R. "The Extragalactic Zoo" in *Sky & Telescope,* Jan. 1988, p. 23; Apr. 1988, p. 376; July 1988, p. 36. Three articles on various active galaxies and how they got their names.

Weedman, D. "Quasars: A Progress Report" in *Mercury,* Jan./Feb. 1988, p. 12.

THE BIG BANG

The open dome of the Keck 10-m telescope on Mauna Kea in Hawaii. Note the person standing at the lower corner of the open slit. Much of the observing time on this telescope will be devoted to observational cosmology.   *(Roger Ressmeyer for CARA)*

Cosmology, *which is the study of the organization and evolution of the universe, is one of the most fascinating fields of modern research. Through a combination of theory and observation, and physics and astronomy, scientists now believe that they can trace the evolution of the universe back to within tiny fractions of a second of the instant when it began.*

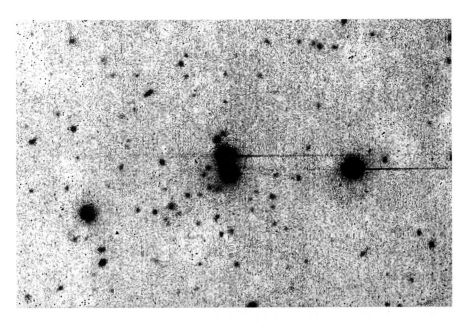

Figure 16.1 Studies of distant galaxies and quasars provide clues to the early evolution of the universe. This image shows a quasar (bright central object) in a cluster of fainter galaxies at a distance of about 6 billion LY. *(Howard Yee and Richard Green/Canada-France-Hawaii Telescope Corporation)*

16.1

THE EXPANDING UNIVERSE

The universe is *expanding*. If we were to make a movie of galaxies as they move farther apart, and then run the movie backward, we would find that all of the matter in the observable universe was once concentrated in an infinitesimally small volume. We call this time the *beginning of the universe.* From the rate of expansion, we calculate that this beginning occurred between 10 and 15 billion years ago.

As telescopes have grown larger and astronomical detectors have become more sensitive, it has become possible to observe more and more remote galaxies with greater and greater speeds of recession (Figure 16.1). One of the most distant galaxies detected so far is shown in Figure 16.2. A quasar with a velocity of 94 percent of the speed of light has recently been discovered.

There is no way to *prove* that the redshifts of these distant objects are not due to some cause other than the expansion of the universe. If there

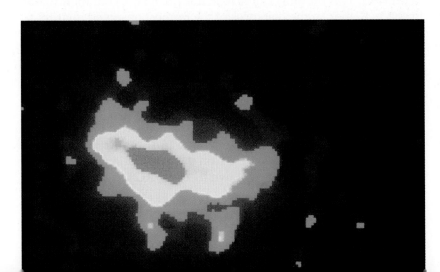

Figure 16.2 A photograph of the most distant galaxy known. The different colors represent different levels of brightness. The galaxy is obviously irregular in shape, but it is so faint that we know very little about its type, stellar content, or composition. Light left this galaxy when the universe was only about 20 percent of its present age. *(Courtesy of Ken Chambers and National Optical Astronomy Observatories)*

What is the evidence that the
universe is expanding?

is another cause, however, then some physical principles unknown to modern scientists are at work. It is characteristic of the scientific method to adopt the more straightforward explanation—that the redshifts are a result of the expansion of the universe—especially because no reasonable alternative has been devised.

In fact, the universe *must* be either expanding or contracting. The universe cannot be static. According to Newton's law of gravity, all objects exert a mutual attraction on one another. If the universe is of finite size, it is obvious that even if galaxies were initially stationary, they would inevitably begin to move closer together, unable to resist the pull of gravity. Newton was not able to prove whether or not an *infinite* universe could be static.

In 1917, Einstein modified Newton's law of gravity according to his theory of *general relativity* and then applied this new theory to the universe as a whole. Einstein proved that static universes could not exist, even if they were infinite. Since astronomers at that time had not discovered the expansion of the universe, Einstein altered his equations with the introduction of a new term, the **cosmological constant.** The cosmological constant represents a repulsion that can balance gravitational attraction over large distances and permit a static universe.

Einstein included the cosmological constant in order to force his model of the universe to conform to the widely held philosophical notion that the universe is neither expanding nor contracting. When Einstein learned that subsequent observations demonstrated that the universe is expanding, he is reported to have said that the introduction of the cosmological constant was "the biggest blunder of my life."

The Size of the Observable Universe

In all directions we see galaxies and clusters of galaxies. At greater and greater distances they appear ever fainter, and beyond a few billion light years we detect only the greatest giants among the galaxies—the most luminous members of great clusters. With the assumption that the Hubble constant H is equal to 25 km/s per 1 million LY, the current distance record for a galaxy is about 10 billion LY. Quasars can be seen to somewhat greater distances, but at these large distances, even they seem to thin out. This thinning out is real; it is not an effect of their remoteness in space. Rather, at the distances of the farthest observed quasars, we are looking back to a time when the universe was only about 10 to 20 percent of its present age. At a still earlier time, quasars evidently did not exist, or at least were exceedingly rare. Perhaps galaxies had not formed yet or were only recently formed and had not yet had time to produce massive black holes.

The actual distance to which we can see depends on the Hubble constant and on how much the expansion is slowed over time by the mutual gravitational attraction of galaxies. Table 16.1 summarizes how far we can see for $H = 25$ km/s per 10^6 LY on the assumption that the expansion of the universe is not slowing down (Section 16.2).

TABLE 16.1	Velocity–Distance Relationship for $H = 25$ km/s per 10^6 LY in a Nearly Empty Universe	
Velocity (speed of light = 1)	Distance (billions of light years)	Least Luminous Observable Objects
0.001	0.01	Ordinary galaxies
0.40	4.3	Clusters of galaxies
0.60	6.5	Radio galaxies
0.78	8.4	Radio galaxies
0.88	9.8	Radio galaxies/quasars
0.93	10.6	Quasars

16.2
COSMOLOGICAL MODELS

What does cosmological theory say about the past history of the universe? And what is its future?

Common to most cosmological theories is the **cosmological principle**—the assumption that within large volumes of space the universe is homogeneous and isotropic—and we will make that assumption here. Although all of the arguments will be made in terms of Newton's law of gravity, the results are also correct according to Einstein's theory of *general relativity,* which is the best description of gravitation yet found.

The models that we will consider all presume that the universe began at a particular time in the past and that the universe is evolving today. These models make solid predictions that can be tested about such observable quantities as the age of the universe and the abundance of helium relative to hydrogen. As we shall see, these predictions appear to be consistent with what is actually observed.

The Age of the Universe

As the universe expands, galaxies and clusters of galaxies separate from each other. Thus, if we extrapolated backward in time, we would find them coming together. If we look far enough back into the distant past, we would find a point in time when all matter was crowded to an extreme density. We call this point in time the beginning of the universe, or at least the beginning of that part of the universe we can know about. The universe suddenly began its expansion with a phenomenon called the **big bang.**

How can the rate of expansion of the universe be used to estimate its age?

The total amount of matter in the universe creates gravitation, whereby all objects pull on all other objects (including light). This mutual attraction must slow the expansion. How much the expansion slows depends on how much matter there is in the universe. At the extreme, if the total density of mass (and energy) has always been low enough that gravitation is ineffective (an essentially "empty" universe), the deceleration would be zero. In that case, the universe would have been expanding at the present rate ever since the big bang. If, however, there is a substantial amount of matter in the universe, then the expansion must be slower now than it was just after the universe began.

If we assume that the universe is nearly empty and that the rate of expansion has been constant since the universe began, we can then estimate how long it took for the galaxies to reach their present separations from one another. Alternatively, if we assume that the universe contains significant amounts of mass, the expansion would have been faster in the past. In such models, galaxies reach their present separations in a shorter time than if the expansion occurs at a constant rate equal to the current velocity. Consequently, we can obtain an estimate of the *maximum age* of the universe by calculating how long it would take for distant galaxies, always moving away from us at their present rates, to have reached those distances.

Call the maximum possible age T_0. The Hubble law can be written as the simple equation

$$v = H \cdot d$$

where v is the radial velocity of a galaxy at distance d. But velocity is just distance divided by time; that is, $v = d/T_0$. Hence, combining these two expressions, we have

$$d/T_0 = H \cdot d$$

or

$$T_0 = 1/H.$$

We see, then, that the maximum age of the universe is just the *reciprocal of the Hubble constant*. For a Hubble constant of 25 km/s per 1 million LY, this time is 13 billion years (Problem 17). For another way to derive this same result, see Problem 18.

Maximum and Minimum Ages

The age of the universe that is calculated in this way depends, of course, on the present rate of expansion—the value of the Hubble constant. Radial velocities of galaxies can be determined from the Doppler shifts of their spectral lines. Thus the accuracy of H depends on the accuracy of our measurements of the distances to galaxies. If the distance estimates were in error by a factor of 2, the Hubble constant would be wrong by a factor of 2. The estimate of the age of the universe would, as a result, also be wrong by a factor of 2 (Figure 16.3).

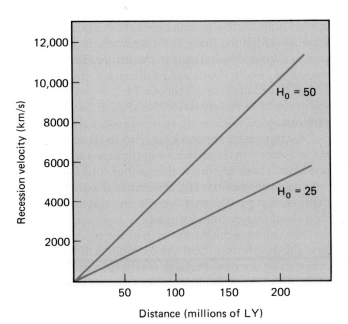

Figure 16.3 Suppose that we measure a velocity of 5000 km/s for a distant galaxy. If that galaxy is actually at a distance of 200 million LY, then the Hubble constant is 25 km/s per million LY. But suppose that we estimate the distance incorrectly by a factor of 2 and think that the galaxy is at a distance of 100 million LY. Then we will think the Hubble constant is 50 km/s per million LY and that the age of the universe is only one-half its true value.

These calculated ages are *maximum* values. The true age of the universe must be less than 1/H because the expansion has to be decelerating. As we will see, the age of a universe that is decelerating just enough to halt its expansion eventually is two-thirds of its maximum age. Therefore, the range of acceptable ages for a Hubble constant of 25 km/s per 1 million LY is between 9 and 13 billion years. Given the considerable uncertainty in the possible age, in many places in this book we have referred to the age of the universe as 10 to 15 billion years. This range probably contains the correct value but allows for our uncertainties in both the extragalactic distance scale and the degree of deceleration.

The best alternative method of estimating the age of the universe is from calculations of the evolution of the oldest stars we know, namely, the members of globular clusters. Most estimates yield ages in the range 13 to 15 billion years. If these estimates are correct, then the age of the oldest stars is just barely consistent with the age of the universe estimated from the Hubble constant adopted for this book, provided that the expansion rate has always been about the same as it is now.

The fact that the age of the universe estimated in these two different ways is about the same may seem comforting, but this apparent agreement may be deceiving. It is possible to make the universe appear tidy and well understood by choosing from the many values estimated by various researchers those measurements and calculations that yield the same ages for stars and for the universe as a whole. Although theorists do obtain ages of 13 to 15 billion years for most globular clusters, there are some clusters for which the age appears to be at least 16 billion years. At the same time, many astronomers argue that the Hubble constant should be 30 km/s per 1 million LY, not 25 km/s as we have assumed here. Furthermore, many theorists argue that the best models are those in which the mass in the universe is just enough to halt the expansion at a time infinitely far in the future (Section 16.4). With this much mass and $H = 30$ km/s per 1 million LY, the age of the universe becomes $2/3 \times 10$ billion years, or about 7 billion years. In this case, the oldest stars would be nearly twice as old as the universe—and that, of course, cannot be!

Astronomers are working hard to determine whether there is a real inconsistency in the different approaches to estimating the age of the universe. There are many things that could go wrong. The observational determinations of the Hubble constant are still somewhat uncertain. The theories that predict that there is enough mass in the universe to slow the rate of expansion ultimately to zero may not be right. The models that we use to estimate the ages of the oldest stars may be inaccurate. And the assumption that the cosmological constant is equal to zero may prove to be in error. Perhaps it will turn out that Einstein did not make such a blunder after all!

One of the major challenges of modern astrophysics is to reconcile the various estimates of the age of the universe. Good agreement would support standard models of cosmology. Proof that the disagreement is real could lead to a revolution in our understanding of the universe comparable to what occurred when Hubble discovered the redshift-distance relation.

The Scale of the Universe

By assumption (the cosmological principle), the universe is homogeneous if we look at a large enough region of space. Consequently, the expansion rate must be uniform (the same in all directions), so that the universe undergoes a uniform change in size or *scale* with time. It is customary to represent that scale by R, which is a function of time. The actual value of the scale is arbitrary. We could think of it as being the distance between any two representative objects or points in space, since R changes in the same way everywhere. R plays the same role as does the bar on a terrestrial map that shows how many inches correspond to a certain number of kilometers. This bar provides a scale for relating inches and kilometers. In just the same way, R serves as a scale that tells us how much the universe has expanded (or contracted) at any time. For a static universe, R would be constant. In an expanding universe, R increases with time. One way of describing the expansion and evolution of the universe is to describe the variation of R since the universe began.

The equations of general relativity provide a mathematical description of how R changes as the universe evolves. But even without general relativity we can understand how the scale of the universe varies with time. In a uniformly expanding (nearly zero density) universe, R grows in direct proportion to elapsed time. In a higher density universe, gravitational forces retard the expansion, and the rate of change of R decreases as the expansion is slowed.

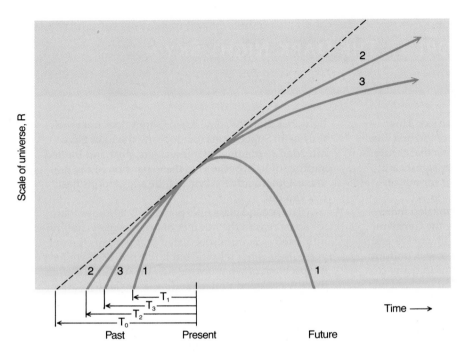

Figure 16.4   A plot of *R*, the scale of the universe, against time for various cosmological models. Curve 1 represents a closed universe, curve 2 represents an open universe, and curve 3 represents a universe with a critical density. The dashed line is for an empty universe, one in which the expansion is not slowed by gravity.

The behavior of *R*—the rate of change of distance in the expanding universe—is shown for several possible models in Figure 16.4. Time increases to the right and the scale length, *R*, increases upward in the figure. At the present time (marked along the time axis), *R* is increasing at the rate indicated by the Hubble constant. The straight dashed line corresponds to the nearly empty universe with no deceleration. In this case, *R* has been increasing uniformly with time since the big bang occurred about 13 billion years ago. This initial time, the reciprocal of the Hubble constant, is labeled T_0 in the figure. The other curves represent varying amounts of deceleration, although they all correspond to the observed rate of expansion at present. Decelerating universes must, of course, have formed more recently than universes with no deceleration.

The degree of deceleration depends on two very uncertain observational parameters: the Hubble constant (or distance scale) and the mean density of matter in space. By "mean density" we mean the mass of matter (including the equivalent mass of energy) that would be contained in each unit of volume (say, $1/cm^3$) if all of the stars, galaxies, and other objects were taken apart, atom by atom, and if all of those particles, along with the light and other energy, were distributed throughout all space with absolute uniformity.

If the mean density of the universe is high enough and the deceleration is large, *R* is given by curve 1. In this case, the universe stops expanding some time in the future and begins contracting. Eventually, the scale drops to zero, with what Princeton physicist John A. Wheeler calls the "big crunch." In this case, the universe is said to be **closed,** for it cannot expand forever.

What are the possible futures for the universe?

OBSERVING YOUR WORLD   THE DARK NIGHT SKY

Go outside at night and look at the sky. How would you describe it? No doubt one of the first characteristics you would mention is the fact that *the sky is dark at night*. This turns out to be a very significant observation—one that has implications for our understanding of cosmology.

The fact that the sky is dark has puzzled many scientists over the centuries, including the German physician and amateur astronomer Wilhelm Olbers. In 1826, Olbers argued that if the universe is uniform and infinite, both standard assumptions in many cosmologies, and populated by eternal and unchanging stars, then we should see an infinite number of stars. In fact, in every direction we look, our line of sight, if we extend it far enough, should run into a star. As a good analogy, imagine that you are standing in a forest. No matter what direction you look in, provided that the forest is large enough, you will see a tree.

So, too, with stars in an infinite universe. Indeed, it is possible to show mathematically that in an infinite and eternal universe filled with stars like the Sun, every point in the sky should appear to us to be as bright as the surface of the Sun. It would seem as if we were living inside a fiery, hot furnace that poured down on the Earth 180,000 times as much energy as the Sun does. The fact that the sky is dark is often referred to as *Olbers' paradox*.

Why then is the sky not blazing with heat and light? Scientists have offered two classes of explanations. The first class hypothesizes that the universe is infinite, that the stars do emit the expected amount of light, but that, for some reason, light does not reach the Earth. Suggestions for stopping the light have included reddening by intergalactic dust and the redshifting of light because of the expansion of the universe. Quantitatively, none of the suggestions has worked out.

The second class of explanations assumes that for some reason the number of bright stars is not infinite. Early scientists speculated that perhaps the universe itself is not infinite, but if it is not, then we have the problem of what lies outside the universe. Before there was a good understanding of stellar evolution, some scientists suggested that perhaps many stars were dark and radiated no energy.

It turns out that the number of observable bright stars is indeed not infinite, but for a reason that became apparent only in this century. *The sky is dark at night because the universe is young.* It takes time for light to traverse the distance from a star or a system of stars—that is, from a galaxy—to us. Accordingly, we can see a galaxy only if it is near enough so that the light emitted by it has had time to reach us during the finite lifetime of the universe. Because the universe is not infinitely old, we do not see an infinite number of stars or galaxies.

Olbers' paradox offers an excellent example of how a very simple observation may have profound implications. The challenge for the scientist is to look at the world with fresh eyes and see that the things we take for granted may offer important clues about the universe in which we live.

It is tempting to speculate that another big bang might follow the "crunch," giving rise to a new expansion phase, and ensuing contraction, perhaps oscillating indefinitely between successive big bangs in the past and future. Such speculation is sometimes referred to as the *oscillating theory* of the universe, but it is not really a theory. We know of no mechanism that can produce another big bang. General relativity (and other theories) predict, instead, that the universe will collapse into a universal black hole at the time of the crunch. (Of course, we have no complete theory for the first big bang either!) In any case, the oscillating theory is a speculation on a possible variation of the closed model of the universe.

TABLE 16.2	Evolving Relativistic Cosmological Models for which $H = 25$ km/s per 1 million LY	
Kind of Universe	**Age (T) (units 10^9 years)**	**Mean Density (ρ) (g/cm³)**
Closed	$T < 9$	$\rho > 10^{-29}$
Open	$9 < T < 13$	$\rho < 10^{-29}$
Flat	$T = 9$	$\rho = 10^{-29}$

Alternatively, if the mean density is too low (curve 2 in Figure 16.4), gravitation is never important enough to stop the expansion. In this case the universe expands forever. Such a universe is said to be **open.**

At a *critical value* of the density (curve 3), the universe can just barely expand forever. At some time infinitely far in the future, the rate of expansion will become very close to zero, but the universe will never reverse direction and begin to contract. This **critical density** marks the boundary between the families of open and closed universes. The age of the critical-density universe is exactly $2/3\ T_0$. Open universes have ages between $2/3\ T_0$ and T_0, and closed universes have ages less than $2/3\ T_0$.

The various possibilities for the evolution of the universe are analogous to those for a rocket launched from the Earth. If the rocket has just the critical velocity of escape from the Earth (about 11 km/s), it barely escapes the Earth; as the rocket reaches a distance infinitely far from Earth, its velocity approaches zero on a parabolic orbit. This case corresponds to the critical-density universe. At lower and higher launch velocities, the rocket falls back to Earth or escapes with energy to spare—analogous to closed and open universes, respectively.

Summary of Models of the Universe

The possibilities described in the previous paragraphs are summarized in Table 16.2. To be specific, a particular value of the Hubble constant, H, has been assumed: 25 km/s per 1 million LY. This table also lists the value of the mean density of the universe that corresponds to the open, closed, and critical-density models.

Note particularly the value of the critical density. This value, the density at which the universe is just barely closed, is about 10^{-29} g/cm³. If the Hubble constant is larger than 25 km/s per 1 million LY, the critical density is also larger, perhaps up to 2×10^{-29} g/cm³. With a small Hubble constant it could be as small as 0.5×10^{-29} g/cm³. But some value in this range—near 10^{-29} g/cm³—is the density we should look for if we expect the universe to eventually stop its expansion.

The Future of the Universe

There are several observational tests by which we hope to be able to determine whether we live in an open or closed universe. One is the determination of the mean density of matter in space. We can estimate the mean density from the number of galaxies and clusters we observe out to a given distance and from a knowledge of the masses of these objects. There is considerable uncertainty about the masses of clusters of galaxies, and we do not know how much matter (if any) may exist in intergalactic space. Nevertheless, such estimates indicate a mean density less than 10^{-30} g/cm^3, and probably near 2×10^{-31} g/cm^3. This is below the critical density by a factor of 5 to 10 (or possibly more) and suggests an open universe, although the estimates are too uncertain to be sure of the conclusion. Only the observations of the motion of the Local Group galaxies toward the Great Attractor (Section 15.5) give any indication that the total amount of matter approaches the critical density, and those observations will remain controversial until they are confirmed by independent evidence.

As we shall see in the following section, the production of deuterium in the early universe is very sensitive to the density of the universe within the first few minutes of the expansion. The proportion of deuterium in interstellar space is thought to be a measure of how much deuterium formed in the big bang, for in stellar interiors deuterium is rather quickly converted to helium. It is very difficult to detect deuterium in space, but careful measures show that the ratio of deuterium to hydrogen is probably in the range of 10^{-4} to 10^{-5}. From this value, a crude estimate of the density of the early universe can be inferred, and from that knowledge it is possible to predict the present-day density that would result. The calculation suggests a present density of about 10^{-31} g/cm^3, again pointing to an open universe. However, this test measures only the density of ordinary particles—protons and neutrons. If the dark matter is made of some unknown type of particle, then the density of matter could be much higher.

In summary, the amount of luminous matter, and even the amount of protons and neutrons, is far too small to halt the expansion of the universe. If ordinary matter were the only type present, the universe would be open. One of the major controversies in astronomy at the present time is whether or not there is enough of some exotic type of dark matter to increase the density of the universe to the critical density.

Other Cosmologies

The foregoing discussion is based on particular assumptions—the cosmological principle, conservation of mass energy, and general relativity theory with zero cosmological constant.

An alternative model of cosmology, which received wide attention in the 1950s, is the *steady-state theory.* The steady-state theory is based on a generalization of the cosmological principle called the *perfect cosmological principle,* in which it is assumed that on the large scale the universe is not only the same everywhere, but for all time. In the steady-state theory, as the universe expands and matter would otherwise thin out, new matter is continuously being created to keep the mean density the same at all times.

The steady-state theory predicts the rate at which matter is created to ensure that there will always be the same admixture of young and old stars and galaxies. The creation would occur so gradually (presumably as individual atoms coming into being here and there) that we would not notice it. The steady-state universe is infinite and eternal and has much philosophical appeal. But we have already seen (Chapter 15) that certain objects, especially quasars, change as the universe ages, which violates the perfect cosmological principle. Moreover, we shall see subsequently that there is direct evidence that the universe has evolved from a hot dense state, strongly supporting the idea of a big bang.

Although the steady-state theory is no longer fashionable, the important point to be learned from this brief description is that theories about even such exotic ideas as how the universe evolves can be tested by observation. If a theory does not stand up to the test—that is, if the predictions made by a theory are wrong—then the theory must be rejected.

16.3
THE BEGINNING

Abbé Georges Lemaître (1894–1966), a Belgian priest and cosmologist, was probably the first to propose a specific model for the big bang. He envisioned all of the matter of the universe starting in one great bulk he called the *primeval atom.* The primeval atom broke into a tremendous number of pieces, each of them further fragmenting, and so on, until what was left were the present atoms of the universe, created in a vast nuclear fission. In a popular account of his theory, Lemaître wrote, "The evolution of the world could be compared to a display of fireworks just ended—-some few red wisps, ashes and smoke. Standing on a well-cooled cinder we see the slow fading of the suns and we try to recall the vanished brilliance of the origin of the worlds." We know today much more about nuclear physics, and the primeval fission model is not correct. Yet Lemaître's vision was in some respects quite prophetic.

In the 1940s, American physicist George Gamow suggested a universe with the opposite kind of beginning—-nuclear fusion. He worked out the details with Ralph Alpher, and they published their results in 1948. (They added the name of physicist Hans Bethe to their paper, so that the coauthors would be Alpher, Bethe, and Gamow, a pun on the first three letters of the Greek alphabet: alpha, beta, and gamma.) Gamow's universe started with fundamental particles that built up the heavy elements by fusion in the big bang.

His ideas were close to our modern view except that the conditions in the primordial universe were not right for atoms to fuse to carbon and beyond, and only hydrogen and helium were formed in appreciable abundances. The heavier elements formed later in stars (see Chapter 12).

Figure 16.5 Standard model of the early universe showing how the temperature varies with time. The vertical line labeled A designates approximately the time at which neutrinos stop interacting with matter. B denotes the time when positrons and electrons annihilate. Helium synthesis occurs at time C, and the universe becomes transparent to radiation at time D.

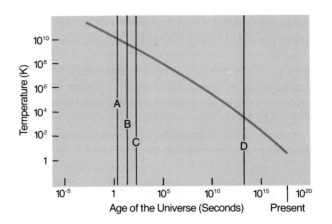

Standard Model of the Big Bang

The modern theory for the evolution of the early universe is called the *standard model* of the big bang. Three simple facts hold the key to tracing the changes that occurred during the first few minutes after the universe began: 1) the universe cools as it expands; 2) at high temperatures, pure energy can be converted to mass; and 3) the higher the temperature, the more massive the subatomic particles that are formed. Let's examine each of these facts in more detail.

The universe cools as it expands (Figure 16.5). In the first fraction of a second, the universe was unimaginably hot. By the time 1/100 of a second had elapsed, the temperature had dropped to 100 billion (10^{11}) K. After about 3 minutes, the temperature had reached about 1 billion (10^9) K, still some 70 times hotter than the interior of the Sun. After 700,000 years, the temperature was down to a mere 3000 K, and the universe has continued to cool since that time.

All of these temperatures but the last are derived from theoretical calculations, since (obviously) no one was there to measure them directly. As we shall see, however, we have actually detected the feeble glow of radiation emitted at a time when the universe was about 700,000 years old. Indeed, the fact that we have done so is one of the strongest arguments in favor of the validity of the big bang model.

At very early times, the universe was so hot that collisions of photons—that is, collisions of discrete units of pure electromagnetic energy—could produce material particles.

Suppose that two photons do collide and disappear, with all of their energy going into the production of two or more particles of matter. What kind of particles will be produced? Even at rest, a particle has an energy E associated with it. The amount of that energy can be calculated from Einstein's equation

$$E = mc^2,$$

where m is the mass of the particle and c is the speed of light. If two photons are to collide and produce two particles of mass m, then the photons must have a total energy equal to twice mc^2.

What was the universe like during the first few minutes after it began to expand?

But what determines the energy of a photon? Physicists have found that the higher the temperature, the higher the energy of a typical photon, and the more massive the particles that can be produced by the collision of two such photons. To take a specific example, at a temperature of 6 billion (6×10^9) K, the collision of two typical photons can create a positron and an electron. The much more massive proton can be created only in an environment that has a temperature in excess of 10^{14} K.

Keeping these three facts in mind, we will now trace the evolution of the universe from the time that it was about 0.01 second old and had a temperature of about 10^{11} K. Why not begin at the very beginning? The reason is that there are no theories that allow us to penetrate to a time when the universe was less than about 10^{-43} second old. When the universe was that young, the density was so high that the theory of general relativity is no longer applicable. We still have no theory that can deal with such extreme conditions.

When the universe was older than 10^{-43} second but still less than about 0.01 second old, and when the temperature was hotter than 10^{12} K, the universe was filled with strongly interacting subatomic particles. The theory of these particles is difficult to deal with, and it is only very recently that theoretical physicists have begun to speculate about what the universe might have been like at this time. We will look at some of those speculations later in this chapter.

The universe became a somewhat more familiar-sounding place by about 0.01 second after the beginning. At that time, it consisted of a "soup" of matter and radiation, and the matter included the protons and neutrons that constitute our material world. Each particle collided rapidly with other particles. The temperature was not high enough for the colliding photons to produce neutrons and protons, but it was high enough for the production of electrons and positrons. The neutrons and protons that were present were leftovers from an even younger and hotter universe. There may also have been a sea of exotic particles that would later play a role as "dark matter." It was a seething cauldron of a universe, with photons colliding and interchanging energy. Often the photons met with such a violent impact that they were destroyed, leaving in their wake an electron-positron pair. Neutrons were converted to protons and protons were converted to neutrons through collisions of particles. It was much too hot for protons and neutrons to combine to form heavier atomic nuclei.

By the time the universe was a little more than 1 second old, its temperature had dropped to a mere 10 billion K. Its density had dropped to the point where neutrinos no longer interacted with matter, but simply traveled freely through space. In fact, these neutrinos should now be all around us. Since they have been traveling through space unimpeded and unchanged since the universe was 1 second old, measurement of their properties would offer one of the best tests of the big bang model. Unfortunately, the very characteristic that makes them so useful, the fact that they interact so weakly with matter that they have survived unaltered since the first second of time, also renders them undetectable, at least with present techniques. Perhaps someday someone will devise a way to capture these messengers from the past.

The universe continued to expand and cool. When it was nearly 14 seconds old, the temperature had reached 3 billion K. Typical photons had too little energy to produce electron-positron pairs through collision, and so the electrons and positrons themselves began to collide and annihilate each other. Fortunately for us there is a slight excess of matter over antimatter, and we owe our existence to that excess. If the amount of matter and antimatter were equal, all particles would be annihilated—and we would not be here today!

Protons and neutrons began to form stable atomic nuclei only when the universe was about 3 minutes, 46 seconds old and when the temperature was down to 900 million K. The first step in building atomic nuclei is the collision of a neutron and proton to form deuterium (heavy hydrogen), and essentially all of the neutrons are used up by this reaction. At higher temperatures, the deuterium is immediately blasted apart by interactions with a photon before it can interact with a third particle to form a stable nucleus.

At the temperatures and densities reached between 3 and 4 minutes after the beginning, deuterium could survive long enough so that collisions could convert nearly all of it to helium, which has an atomic mass of 4 (2 protons and 2 neutrons). There is no stable particle of mass 5, however, and we believe that heavier elements were predominantly produced later deep in the interiors of stars.

According to models of the big bang, there are approximately 2 neutrons for every 14 protons at the time that helium nucleosynthesis occurs. If 2 neutrons and 2 protons are required to form a helium nucleus, then 12 protons, which are the nuclei of hydrogen atoms, remain. One prediction of the big bang model is, therefore, that there should be 1 helium atom for every 12 hydrogen atoms. In units of mass, this means that about 25 percent of the matter in the universe should be helium and about 75 percent should be hydrogen. Within the uncertainty of the observations, this ratio is precisely what is observed. Of course, a small enhancement of helium must have resulted from nucleosynthesis in stars, but we estimate that ten times as much helium was manufactured in the first 200 seconds of the universe as in all the generations of stars during the succeeding 10 to 15 billion years.

For the next few hundred thousand years, the universe was much like a stellar interior—hot and opaque, with radiation being scattered from one particle to another. By 700,000 years after the big bang, the temperature had dropped to about 3000 K and the density of atomic nuclei had dropped to about $1000/cm^3$. Under these conditions, the electrons and the nuclei combined to form stable atoms of hydrogen and helium, just as they do in the atmosphere of a star with a temperature of 3000 K. The universe became transparent, and matter and radiation no longer interacted. Each evolved in its separate way. One billion years after the big bang, stars and galaxies had probably begun to form. Deep in the interiors of stars matter was reheated, nuclear reactions were ignited, and the gradual synthesis of the heavier elements began.

We emphasize that the fireball must not be thought of as a localized explosion—like an exploding superstar. There were no boundaries and there was no site of the explosion. It was everywhere. In a sense, the fire-ball still exists. It has expanded greatly, but the original matter and radiation are still present and accounted for. Our bodies are composed of material that came from the fireball. We were and are still in the midst of it; it is all around us.

Discovery of the Cosmic Background Radiation

What happened to the radiation released when the universe became transparent at the tender age of about 700,000 years? That question was first considered in the late 1940s by Alpher and Robert Herman, both associates of Gamow. They realized that just before the universe became transparent it must have been radiating like a blackbody at a temperature of 3000 K. If we could have seen that radiation just after neutral atoms formed, it would have resembled radiation from a reddish star.

To observe the glow of the early universe, we must look out to such a great distance—10 to 15 billion LY—that we see the universe as it was when it was only about 700,000 years old. Those remote parts of the universe, because of its expansion, are receding from us at a speed within two parts in a million of that of light. When a blackbody approaches us, the Doppler shift shortens the wavelengths of its light and causes it to mimic a blackbody of higher temperature. When a blackbody recedes, it mimics a cooler blackbody. Alpher and Herman predicted that the glow from the fireball should now be at radio wavelengths and should resemble the radiation from a blackbody at a temperature of only 5 K—just a few degrees above absolute zero. But there was no way at the time they published their conclusion to observe such radiation from space, so the prediction was forgotten.

In the mid-1960s, in Holmdel, New Jersey, Arno Penzias and Robert Wilson of the Bell Laboratories were using a delicate microwave horn antenna to make careful measures of the absolute intensity of radio radiation coming from certain places in the Milky Way Galaxy. They were plagued with some unexpected background noise in the system that they could not get rid of. They checked everything and eliminated the Galaxy as a source, as well as the Sun, the sky, the ground, and even the equipment.

At one point they realized that a couple of pigeons had made their home in the antenna and nested up near the throat of the horn where it was warmer. Penzias and Wilson could chase the birds away while they observed, but they found that the birds left, as Penzias puts it, a layer of white sticky dielectric substance coating the inside of the horn. That substance would radiate, producing radio interference. They disassembled the horn and cleaned it, and the unwanted noise did diminish somewhat, but it was not eliminated completely.

Finally, Penzias and Wilson decided that they had to be detecting radiation from space. After discussions with other scientists, including most importantly Robert Dicke at Princeton, Penzias and Wilson realized that they had observed the predicted glow from the primeval fireball. They received the Nobel Prize for their work in 1978. And perhaps

What is the observational evidence that in the beginning the universe was very hot?

Figure 16.6 Observations with the Cosmic Background Explorer (COBE), which is shown here as it leaves the launch pad, have provided the most accurate measurements of the cosmic background radiation (CBR). *(NASA)*

almost equally fitting, just before his death in 1966, Lemaître learned about the discovery of his "vanished brilliance."

Properties of the Cosmic Background Radiation

The faint glow of radio radiation emitted when the universe became transparent is now called the **cosmic background radiation (CBR)**. The most accurate measurements of the CBR have been made with a satellite orbiting the Earth. This satellite, the Cosmic Background Explorer (COBE), was launched by NASA in 1989 (Figure 16.6). The first data received showed that the CBR closely matches that expected from a black body with a temperature of 2.735 K (Figure 16.7). This is exactly the result that one would expect if the CBR is indeed redshifted radiation emitted by a hot gas shortly after the universe began.

The first important conclusion from measurements of the CBR, therefore, is that the universe has evolved from a hot, uniform state. This observation provides direct support for the idea that we live in an evolving universe. It rules out the steady-state universe, which would not be filled with radiation from such a cosmic fireball.

A second result is that the CBR appears to be slightly hotter in one direction than in the exact opposite direction in the sky. This difference is the result of our own motion through space. When a blackbody is approached, its radiation is all Doppler-shifted to shorter wavelengths and

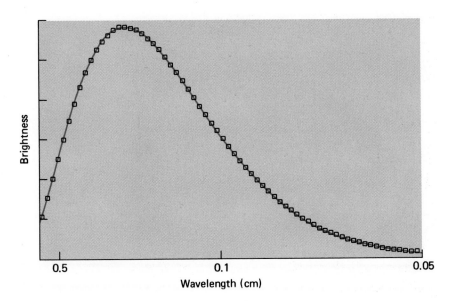

Figure 16.7 Measurements by COBE of the variation of intensity with wavelength show that the CBR follows the pattern expected for a blackbody with a temperature of 2.735 K.

resembles that from a slightly hotter blackbody. If a blackbody is moving away, its radiation appears like that from a slightly cooler object. The small temperature difference of the CBR viewed in two opposite directions on the sky indicates that the Sun, the Milky Way Galaxy, and all of the members of the Local Group are moving in the general direction of the constellation Hydra.

The motion of the Local Group toward Hydra is in addition to the motion of these galaxies that results from the overall expansion of the universe as a whole. As we saw in Chapter 15, this extra motion is probably caused by the gravitational attraction of an unusually dense concentration of matter, including dark matter and luminous matter, which pulls the Local Group toward Hydra.

A third conclusion from the COBE observations is that the early universe was a quiet place. The intensity of the CBR follows a blackbody curve with no excess radiation at any particular wavelength. If there had been violent events when the universe was very young, those events would have produced energy. That energy, added to the CBR, would have distorted the smooth curve shown in Figure 16.7. Astronomers had, for example, speculated that there might have been a generation of massive stars that formed and completed their evolution through the supernova phase prior to the formation of galaxies. The fact that the CBR follows a blackbody curve so precisely indicates that no more than about 1 percent of the hydrogen present in the universe could have been converted to helium inside stars prior to the time that galaxies formed.

A fourth result of CBR measurements was announced in the spring of 1992 and is potentially the most important discovery in cosmology since the detection of the CBR. It has been known for many years that the CBR is extremely *isotropic* on the small scale. Measurements made prior to the launch of COBE show that if we look in directions that differ

Figure 16.8 Microwave map of the whole sky made from one year's data obtained by COBE. Processing has removed all microwave radiation coming from sources other than the cosmic background radiation (CBR). The red (warmer) and blue (cooler) regions represent the temperature differences 0.001 percent above or below the average sky temperature of 2.735 K. These 15-billion-year-old temperature fluctuations also represent regions of differing density.

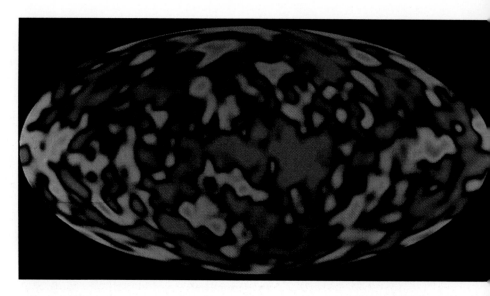

by less than a degree, any fluctuations in the intensity of the CBR are less than a few parts in 10^4.

But, according to theory, the temperature could not have been *perfectly* uniform when the CBR was emitted. All structures in the universe—galaxies, clusters of galaxies, and superclusters—evolve from tiny density fluctuations. Regions of higher-than-average density gravitationally attracted additional matter and grew into the galaxies and clusters that we see today. Regions of higher than average density in the early universe would appear to have lower-than-average temperature. Therefore, if the seeds of present-day galaxies existed at the time the CBR was emitted, we should see some slight changes in the temperature of the CBR as we look in different directions in the sky.

Scientists working with the data from the COBE satellite have now reported that their measurements are accurate enough to detect very subtle temperature differences present in the CBR (Figure 16.8). The temperature differences are very small—typically only 16×10^{-6} K, but they are real. The COBE results have recently been confirmed by measurements from a high-altitude balloon.

The regions of lower-than-average temperature come in a variety of sizes but even the smallest of the hot spots detected by COBE is far too large to be the precursor of an individual galaxy or even a supercluster of galaxies. New experiments either launched on balloons or located at the South Pole are now looking for even smaller hot spots that might be the seeds of galaxies or galaxy clusters. If these experiments are successful, then we can truly say that we have seen the birth of galaxies.

A Look Back to the Big Bang

Since the cosmic background radiation comes from the time when the fireball first became transparent, it is at the farthest point in space and time to which we can now observe. If we could see that radiation visu-

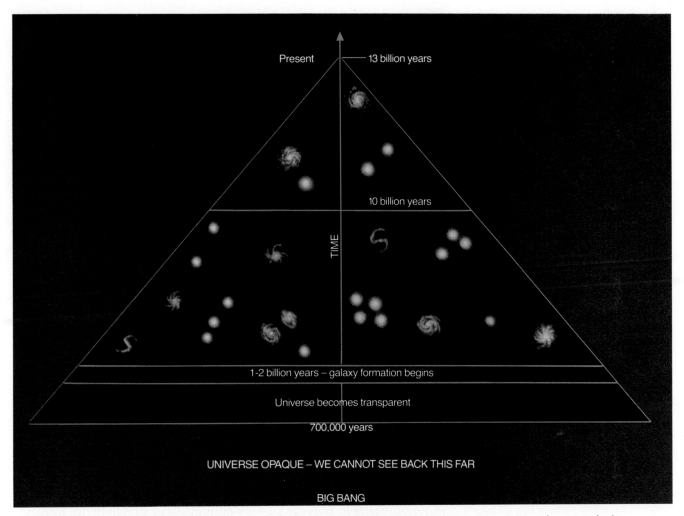

Figure 16.9   As we look to larger and larger distances and farther back into time, we see more and more galaxies. Ultimately, we see back to the point at which the universe became transparent, which occurred about 700,000 years after the expansion began. Because the universe was opaque before this time, we cannot directly observe earlier eras.

ally, it would be as if it were coming from an opaque wall. No radiation from a more distant source could ever reach us—for that source would have to lie farther back in time, where it would be behind that opaque wall.

As more time passes, we will see farther and farther away, and more galaxies will come into view (if we have a large enough telescope) for, as the length of time from the big bang becomes greater, we will be looking farther into the past to see the fireball, and hence farther away in space (Figure 16.9). Earlier, we could have seen (had we been here) only relatively nearer objects, and there were fewer galaxies between us and the threshold provided by the fireball itself. Thus, not only does the universe expand with time, but the part of it that is accessible to observation becomes greater as well.

16.4
THE NEW INFLATIONARY UNIVERSE

"The big bang is dead."

In recent years, many articles have appeared in newspapers with headlines like the sentence above. These headlines are very misleading. Over time, as our understanding of the universe grows, we incorporate new physics into our cosmological theories in order to explain the observations in more detail. Classical cosmology included the effects of gravity and attempted to describe the expansion of the universe and how the rate of expansion might change with time. The big bang model included more physics in order to explain the synthesis of elements in the first few minutes of the universe. Now, however, newer cosmological theories are being developed with the goal of trying to explain how structure—galaxies and clusters of galaxies—formed. The newer models are also trying to explore what happened during the first 0.01 second of the universe in an effort to explain some observations that the standard big bang model cannot explain. These newer theories are additions to the big bang model, not a substitute for it.

Problems with the Standard Big Bang Model

The big bang model successfully explains the relationship between velocity and distance that is observed for galaxies; it accounts for the cosmic background radiation; and it explains why about 25 percent of the mass of the universe is in the form of helium. There are, however, several important characteristics of the universe that the simple big bang model cannot explain, and we shall consider two of these in detail.

The first problem for which the big bang offers no explanation is the *uniformity* of the universe. From any direction we explore, the cosmic background radiation is the same to an accuracy that is better than 1 part in 10,000. There is, however, a maximum distance that light can have traveled since the time the universe began. This distance is called the *horizon distance*, because any two objects separated by this distance cannot ever have been in contact. No information and no physical process can propagate faster than the speed of light. One region of space separated by more than the horizon distance from another lies completely out of sight. Neither region could ever have received any information about the density or the temperature of the other.

If we measure the CBR in two opposite directions in the sky, we are observing regions that were separated by more than 90 times the horizon distance at the time the CBR was emitted. We can see both, but they can never have seen each other. Why then are their temperatures so precisely the same? According to the standard big bang model, energy has never been able to flow from one region to the other, and there is no reason why they should have identical temperatures. The only explanation is simply that the universe started out being absolutely uniform in temperature. Scientists are always very uncomfortable, however, when they

What properties of the universe cannot be explained by the standard models of the big bang?

TABLE 16.3 The Forces of Nature

Force or Interaction	Examples of Interacting Particles	Relative Strength Now	Range	Important Applications
Gravitation	All	10^{-38}	Whole universe	Motions of planets, stars, galaxies
Electromagnetic	All charged	10^{-2}	Whole universe	Atoms, molecules, electricity
Weak	Electrons	10^{-5}	10^{-17} m	Radioactive decay
Strong	Protons, neutrons	1	10^{-15} m	Nuclear forces

must appeal to a special set of initial conditions to account for what they see.

The second problem with the standard big bang model is that it does not explain why the density of matter in the universe is so close to the critical density. As we have seen, observations are unable to tell us whether the expansion of the universe will continue forever or will ultimately slow and perhaps even come to a halt and reverse itself. The interesting fact, however, is that the universe is so nearly balanced between these two possibilities that we cannot yet determine which is correct. There could have been, after all, so little matter that it would be obvious that the universe is open and that the expansion will continue forever. Alternatively, there could have been so much matter that it would be obvious that the universe is closed. Instead, the amount of matter present is within a factor of five to ten of the value that corresponds to precise balance between these two situations.

To understand the new theories that try to solve these problems, we must make a digression and talk about the forces acting on the tiniest particles in the universe. Then we will return to discussing the grand picture of how the universe might have evolved.

Grand Unified Theories

In the terminology of physics, an *interaction* is any process that affects the elementary particles—protons, neutrons, etc.—that make up the material universe. Such processes include the creation and annihilation of particles, radioactive decay, and absorption or scattering of energy. There are four types of interactions or forces that describe all known physical processes. These four are gravity, electromagnetism, the *weak interactions*, and the *strong interactions* (Table 16.3).

Although the force of gravity is the one that most obviously affects our everyday lives and appears strong to us, the force of gravity between two elementary particles, say two protons, is by far the weakest of the forces. Electromagnetism, which includes both magnetic and electric forces, holds atoms together and produces the electromagnetic radiation

Figure 16.10 The strength of the four forces depends on the temperature of the universe. This diagram shows that at very early times, when the temperature of the universe was very high, all four forces resembled one another and were indistinguishable. As the universe cooled, the forces took on separate and distinctive characteristics.

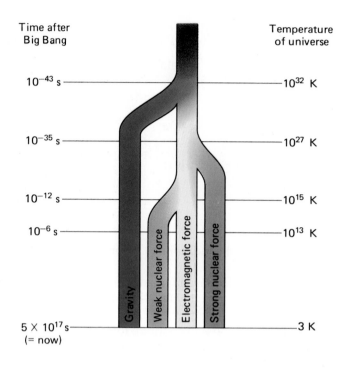

What four forces describe all known physical processes?

that we use to study the universe. The weak interactions are in fact much stronger than gravity but act only over very small distances—distances comparable to only one-hundredth the size of an atomic nucleus. The weak interaction is involved in radioactive decay and in reactions that result in the production of neutrinos. The strong interaction is also effective only over nuclear dimensions and is the force that holds protons and neutrons together in an atomic nucleus.

In exploring the fundamental nature of these four forces, physicists have developed so-called *grand unified theories (GUTs)*. In these theories, the strong, weak, and electromagnetic forces are not three independent forces but rather are different manifestations or aspects of what is in fact a single force. The theories predict that at high enough temperatures, there would in fact be only one force. At lower temperatures, however, a mechanism causes this single force to look like three different forces (Figure 16.10). The mechanism is given the name *spontaneous symmetry breaking*. Unfortunately, the temperatures at which the three forces are predicted to become one force are so high that they cannot be reached in any terrestrial laboratory. Only the early universe at times earlier than 10^{-35} sec was hot enough so that there was one force instead of three.

Some forms of the grand unified theories predict, in fact, that a remarkable event occurred about the time that the universe was 10^{-35} sec old and while spontaneous symmetry breaking was taking place. The equations of general relativity, combined with the special state of matter at that time, predict that gravity could briefly have been a repulsive force. We know that now in our own time gravity is an attractive force that slows the expansion of the universe, but for a brief instant near

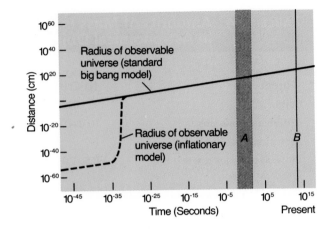

Figure 16.11   Radius of the observable universe as a function of time for the standard big bang model (*solid line*) and for the inflationary model (*dashed line*). The two models are the same for all times after 10^{-30} second. Electrons, positrons, and the lightest atomic nuclei are formed during the time interval labeled A. The universe becomes transparent to radiation at the time designated B.

10^{-35} sec after the expansion began, gravity could actually have accelerated the expansion.

A model universe in which this rapid early expansion occurs is called an **inflationary universe.** The inflationary universe is identical to the big bang universe for all time after the first 10^{-30} sec. Prior to that time, associated with the spontaneous symmetry breaking, there was a brief period of extraordinarily rapid expansion or inflation during which the scale of the universe increased by a factor of as much as 10^{75}; this increase is 10^{50} times more than that predicted by standard big bang models (Figure 16.11). As the universe expanded, its temperature dropped below the critical value at which all three forces behave in a symmetric fashion. In the cooler, asymmetric universe, the nuclear forces dominate the electromagnetic force, and continue to do so in our world today.

Prior to the inflation, all of the universe that we can now see was causally connected. That is, the horizon distance was large enough to include all of the universe that we can now observe. There was adequate time for the observable universe to homogenize itself and reach a uniform temperature.

In addition to accounting for the fact that we have determined the cosmic background radiation to be the same in all directions, GUTs make some predictions about the nature of the universe. For example, for reasonable choices for the parameters in the theory, it is possible to produce an excess of matter over antimatter, and we now live in a universe that contains far more matter than antimatter. (If the amounts were equal, the matter and antimatter would have annihilated each other and we wouldn't be here!)

Another prediction of the inflationary universe model is that the mean density of matter in the universe should precisely equal the critical density. (The only way to avoid this result is to assume that some special conditions occurred during the early expansion of the universe. Of course, the desire to avoid an appeal to special initial conditions was one of the original motivations for devising the inflationary theory.)

What predictions about the universe are made by the new inflationary model?

TABLE 16.4 Amounts of Mass Present in Astronomical Objects of Various Sizes	
Object	**Ratio of Measured Density to Critical Density**
Stars	< 0.01
Individual galaxies	0.01–0.03
Rich clusters of galaxies	0.2
Flow of galaxies toward the Great Attractor	0.5–1.0

Therefore, one of the most important tests for the inflationary models is to determine whether or not we live in a critical density universe.

Table 16.4 summarizes the amount of matter that is estimated to be present in various types of astronomical objects. Visible stars contribute less than 1 percent of the mass required to reach critical density. Even if we add to the stars and interstellar matter in galaxies the invisible dark matter that is detected through its gravitational influence on luminous objects, we only reach about 20 percent of the critical density.

The only observational evidence that the universe contains a critical density of matter comes from studies of flow of the Local Group and other galaxies toward the Great Attractor (Section 15.5), and these new results remain highly controversial. Over the next decade astronomers will attempt to determine accurately just what the density of matter really is.

If we do live in a critical density universe, most of the matter must be nonluminous dark matter. What is this dark matter made of? As we have already seen, observations of the abundance of deuterium show that it cannot consist of protons and neutrons. The mass could be in the form of neutrinos, but only if neutrinos have mass. So far experiments designed to show that neutrinos have a tiny mass have been inconclusive.

Theory has produced a number of other exotic possibilities. One of the most interesting is the WIMP (weakly interacting massive particle). The name comes about because such a particle, if it exists, would not respond to either electromagnetic or nuclear forces. A WIMP moving through matter would cause a disturbance only if it hit an atomic nucleus or an electron directly. The best chance of detecting a WIMP would be to look for the tiny vibration or perhaps the feeble radiation that would result from such a direct hit.

The hunt for dark matter is on—in huge accelerators where exotic new particles might be produced, in university laboratories around the world, and deep in underground mines where scientists are trying to trap elusive dark matter particles just as they once succeeded in capturing neutrinos. Scientists are taking a big gamble in devoting a substantial portion of their careers to searching for particles that may not even exist. The gamble is high but so is the reward. Detection of dark matter and the determination of its characteristics would revolutionize cosmological thought in the same way as did the discoveries of the expansion of the universe and of the cosmic background radiation.

Conclusion

Throughout this book we have traced the fascinating and often puzzling properties of the luminous matter in the universe. And now we find that visible matter may not even be the most important constituent of the universe. It may be that dark particles of a kind completely unknown in everyday experience have dominated the evolution of the universe. The grand unified theories that have been devised to account for the universe are beautiful and elegant, and very often in science beauty and simplicity have been a guide to truth. But the ultimate test must be whether the theories make verifiable predictions and whether those predictions are validated in laboratory experiments. The new cosmological theories are driving a host of developments in observational and experimental astrophysics. Over the next 20 years we should know whether current models contain truth as well as beauty. Whatever the ultimate judgment, it is quite clear that we will be able to probe much closer to the beginning of our universe than we would have thought possible only 20 years ago.

▌ SUMMARY

16.1 The universe is expanding. From the rate of expansion of the universe, we can estimate that all of the matter in the universe was concentrated in an infinitesimally small volume 10 to 15 billion years ago. The beginning of the expansion is what astronomers mean when they talk about the beginning of the universe. Einstein's equations for general relativity predict that the universe must be either expanding or contracting. At the time Einstein published his theory, scientists thought the universe was static, and so Einstein introduced into his equations a **cosmological constant** that would exactly balance gravity over large distances.

16.2 Modern cosmologies assume the **cosmological principle,** which states that the universe is homogeneous and isotropic over large volumes of space. The most widely accepted model of the universe is the **big bang** model. The factor that controls the evolution of the universe is the density of matter (and energy). If the density is high, then the rate of expansion will slow and possibly even reverse direction so that the galaxies all come together again (a **closed universe**). Observations suggest that the density is actually so low that the expansion will continue forever (an **open universe**). The age of the universe is probably between 10 and 15 billion years.

16.3 The early universe was so hot that the collision of photons could produce material particles. As the universe expanded and cooled, protons and neutrons formed first, and electrons and positrons formed subsequently. Fusion reactions then produced helium nuclei. Finally, the universe became cool enough to form hydrogen atoms. At this time the universe became transparent to radiation.

Scientists have detected the **cosmic background radiation (CBR)** from the hot early universe.

16.4 The big bang model does not explain why the cosmic background radiation has the same temperature in all directions. Neither does it explain why the density of the universe is so close to the critical density. The new **inflationary universe,** which makes use of **grand unified theories (GUTs)** of the electromagnetic, **strong,** and **weak** forces, is one theory that has been developed to try to explain these observations. One prediction of this new theory is that the density of the universe should be exactly equal to the critical density. This prediction can be true only if 99 percent of the matter in the universe is invisible.

▶ REVIEW QUESTIONS

1. Astronomers of today can actually observe galaxies as they were billions of years ago. Explain.

2. Describe three possible futures for the universe. What property of the universe determines which of these possibilities is correct?

3. Which formed first in the early universe—protons and neutrons or electrons and positrons? Why?

4. Which formed first—hydrogen nuclei or hydrogen atoms? Why?

5. Describe at least two characteristics of the universe that are explained by the big bang model.

6. Describe two properties of the universe that are not explained by the big bang cosmology.

▶ THOUGHT QUESTIONS

7. What is the most useful probe of the early evolution of the universe—a giant elliptical galaxy or an irregular galaxy like the Large Magellanic Cloud? Why?

8. What are the advantages and disadvantages of using quasars to probe the early history of the universe?

9. Consider the plot of radial velocities against the distances of remote clusters of galaxies. Are the measured distances and radial velocities that are plotted the present values of these quantities for the clusters? In each case, why? If not, try to describe how the diagram might differ if we did plot the present-day values of cluster distances and velocities.

10. Suppose that the universe will expand forever. Describe what will become of the radiation from the primeval fireball. What will the future evolution of galaxies be like?

11. Summarize the evidence for the existence of dark matter in the universe.

12. In this text we have discussed many motions of the Earth as it travels through space with the Sun. Describe as many of these motions as you can.

13. There are a variety of methods used to estimate the ages of various objects in the universe. Describe some of these methods and indicate how well they agree with one another and with the age of the universe itself as estimated by its expansion.

14. In the 19th century both geology and biology were based on the idea that evolution is a slow process. In the 20th century, we have come to the conclusion that violent events have played a significant role in shaping the evolution of the universe and everything in it, including the evolution of life on Earth. Discuss some of these violent events, starting with the big bang and including some events that have affected the Earth directly.

15. Since the time of Copernicus, each revolution in astronomy has moved humans farther from the center of the universe. Now it appears that we may not even be made of the most common form of matter. Trace the changes in scientific thought about the central nature of the Earth, the Sun, and the Milky Way Galaxy on a cosmic scale.

16. Construct a time line for the universe and indicate when various significant events occurred, from the beginning of the expansion to the formation of the Sun to the appearance of humans on Earth.

▶ PROBLEMS

17. Show that if $H = 25$ km/s per 1 million LY, then the maximum age of the universe is approximately 13 billion years. What would the maximum age be if H were 30 km/s per 1 million LY? Or 15 km/s per 1 million LY as some studies of supernovae suggest?

18. It is possible to derive the age of the universe, given the value of the Hubble constant and the distance to a galaxy. Consider a galaxy at a distance of 400 million LY, receding from us at a velocity v. If the Hubble constant is 25 km/s per 1 million LY, what is its velocity? How long ago was that galaxy right next door to our own Galaxy if it has always been receding at its present rate? Express your answer in years. Since the universe began when all galaxies were very close together, this number is an estimate for the age of the universe.

19. When Hubble and Humason first reported their results for the relation between distance and velocity for galaxies, they thought that the Hubble constant was equal to about 200 km/s per 1 million LY. What would be the age of the universe for this value of H? How does this age compare with the age of the oldest stars?

▶ SUGGESTIONS FOR ADDITIONAL READING

Books

Overbye, D. *Lonely Hearts of the Cosmos*. Harper Collins, 1991. A fascinating "inside tour" of modern cosmology by a skilled journalist.

Bartusiak, M. *Through a Universe Darkly*. Harper Collins, 1993. *Thursday's Universe*. 1986, Times Books, 1986. Well-written introductions to a number of topics in cosmology, by a science journalist.

Boslaugh, J. *Masters of Time: Cosmology at the End of Innocence*. Addison Wesley, 1992. Another journalist's view of developments in cosmology.

Ferris, T. *Coming of Age in the Milky Way*. Morrow, 1988. *The Red Limit*, 2nd ed. Morrow, 1983. Excellent histories of cosmology.

Gribbin, J. *In Search of the Big Bang*. Bantam, 1986. Fine layperson's introduction by a scientist who is also a noted science writer.

Lightman, A. & Brawer, R. *Origins: The Lives and Worlds of Modern Cosmologists*. Harvard U. Press, 1990. Interviews with active researchers in this field.

Harrison, E. *Cosmology*. Cambridge U. Press, 1981. An excellent college level textbook for beginning science majors.

Lederman, L. & Schramm, D. *From Quarks to the Cosmos*. W. H. Freeman/ Scientific American Library, 1989. A physicist and an astronomer examine the way particle physics can shed light on conditions in the early universe.

Riordan, M. & Schramm, D. *Shadows of Creation*. W. H. Freeman, 1991. An astronomer and a science writer discuss cosmology and dark matter.

Articles

On Cosmology in General

Davies, P. "Everyone's Guide to Cosmology" in *Sky & Telescope*, Mar. 1991, p. 250.

Jayawardhana, R. "The Age Paradox" in *Astronomy*, June 1993, p. 39. The conflict between various measurements of the age of the universe.

Rothman, T. "This Is the Way the World Ends" in *Discover*, July 1987, p. 82. On the future of the universe in the different models.

Mallove, E. "The Self-Reproducing Universe" in *Sky & Telescope*, Sept. 1988, p. 253. On the inflationary universe idea.

Bartusiak, M. "The Cosmic Burp: The Genesis of the Inflationary Universe Hypothesis" in *Mercury*, Mar./Apr. 1987, p. 34.

Shu, F. "The Expanding Universe and the Large-scale Geometry of Space-time" in *Mercury*, Nov./Dec. 1983, p. 162.

Freedman, W., *et al.* "The Expansion Rate and Size of the Universe" in *Scientific American*, Nov. 1992, p. 54.

Greenstein, G. "Through the Looking Glass" in *Astronomy*, Oct. 1989, p. 20. Inflation, GUT's, seed universes, etc.

Davies, P. "The First One Second of the Universe" in *Mercury*, May/June 1992, p. 82.

Odenwald, S. "To the Big Bang and Beyond" in *Astronomy*, May 1987, p. 90.

Halliwell, J. "Quantum Cosmology and the Creation of the Universe" in *Scientific American*, Dec. 1991.

On the Cosmic Background Radiation and the COBE Satellite

Ferris, T. "The Radio Sky and the Echo of Creation" in *Mercury*, Jan./Feb. 1984, p. 2. On the discovery of the cosmic background radiation.

Kippenhahn, R. "Light from the Depths of Time" in *Sky & Telescope*, Feb. 1987, p. 140. On the cosmic background.

Fienberg, R. "COBE Confronts the Big Bang" in *Sky & Telescope,* July 1992, p. 34.

Talcott, R. "COBE's Big Bang" in *Astronomy,* Aug. 1992, p. 42; Kanipe, J. "Beyond the Big Bang" in *Astronomy,* Apr. 1992, p. 30. Reports on results from the satellite.

Silk, J. "Probing the Primeval Fireball" in *Sky & Telescope,* June 1990, p. 600.

Brush, S. "How Cosmology Became a Science" in *Scientific American,* Aug. 1992. A historian examines the context of Penzias and Wilson's discovery.

▶ APPENDIX 1:   GLOSSARY

absolute zero.   A temperature of –273°C (or 0 K), where all molecular motion stops.

absorption spectrum.   Dark lines superimposed on a continuous spectrum.

accretion.   Gradual accumulation of mass, as by a planet forming by the building up of colliding particles in the solar nebula.

accretion disk.   A disk of hot material spiraling into a compact massive object (neutron star or black hole), usually detected by its emission of x rays.

active galactic nucleus.   A galaxy is said to have an active nucleus if unusually violent events are taking place in its center, emitting in the process very large quantities of electromagnetic radiation. Seyfert galaxies and quasars are examples of galaxies with active nuclei.

active galaxy.   A galaxy with a small bright nucleus that emits unusually large amounts of energy.

angular momentum.   A measure of the momentum associated with rotation about an axis.

aperture.   The diameter of an opening, or of the primary lens or mirror of a telescope.

aphelion.   Point in its orbit where a planet is farthest from the Sun.

apogee.   Point in its orbit where an Earth satellite is farthest from the Earth.

apparent brightness.   The brightness of a star as it appears to us, as opposed to its intrinsic brightness or luminosity.

asteroid.   An object orbiting the Sun that is smaller than a major planet but that shows no evidence of an atmosphere or of other types of activity associated with comets. Also called a minor planet.

asteroid belt.   The region of the solar system between the orbits of Mars and Jupiter in which most asteroids are located. The main belt, where the orbits are generally the most stable, extends from 2.2 to 3.3 AU from the Sun.

astronomical unit (AU).   Originally meant to be the semimajor axis of the orbit of the Earth; now defined in terms of the speed of light. The semimajor axis of the orbit of the Earth is 1.000 000 230 AU. One AU = 1.5×10^{11} m.

astrophysics.   The part of astronomy that deals principally with the physics of stars, stellar systems, and interstellar material.

aurora.   Light radiated by atoms and ions in the ionosphere, mostly in the magnetic polar regions; the polar lights.

barred spiral galaxy.   A spiral galaxy in which the nuclear bulge is elongated into a bar.

basalt.   Igneous rock, composed primarily of silicon, oxygen, iron, aluminum, and magnesium, produced by the cooling of lava. Basalts make up most of Earth's oceanic crust and are also found on other planets that have experienced extensive volcanic activity.

big bang theory.   A theory of cosmology in which the expansion of the universe is presumed to have begun with a primeval explosion.

binary star.   A double star; two stars revolving about each other.

blackbody.   A hypothetical perfect radiator, which absorbs and re-emits all radiation incident upon it.

black hole.   A hypothetical body whose velocity of escape is equal to or greater than the speed of light; thus no radiation can escape from it.

brown dwarf.   An object intermediate in size between a planet and a star. The approximate mass range is from about twice the mass of Jupiter up to the lower mass limit for self-sustaining nuclear reactions, which is 0.08 solar masses. Also called an *infrared dwarf.*

carbonaceous meteorite.   A primitive meteorite made primarily of silicates but often including chemically bound water, free carbon, and complex organic compounds. Also called carbonaceous chondrites.

carbonate.   Class of compounds containing carbon and oxygen, often found in sedimentary rock.

CBR.   See *cosmic background radiation.*

celestial equator.   A great circle on the celestial sphere 90° from the celestial poles; the circle of intersection of the celestial sphere with the plane of the Earth's equator.

celestial poles.   Points about which the celestial sphere appears to rotate; intersections of the celestial sphere with the Earth's polar axis.

celestial sphere.   Apparent sphere of the sky; a sphere of large radius centered on the observer. Directions of objects in the sky can be denoted by the position of those objects on the celestial sphere.

center of mass.   The average position of the various mass elements of a body or system, weighted according to their distances from that center of mass; that point in an isolated system that moves with constant velocity, according to Newton's first law of motion.

cepheid variable.   A yellow supergiant pulsating star. These stars vary periodically in brightness, and the relationship between their periods and luminosities is useful in deriving distances to them.

Chandrasekhar limit. The upper limit to the mass of a white dwarf (equals 1.4 times the mass of the Sun).

chromosphere. That part of the solar atmosphere that lies immediately above the photospheric layers.

closed universe. A universe with a density so high that its expansion will eventually halt, to be followed by gravitational collapse.

Clouds of Magellan. See *Magellanic Clouds.*

comet. A small body of icy and dusty matter that revolves about the Sun. When a comet comes near the Sun, some of its material vaporizes, forming a large *head* of tenuous gas, and often a *tail.*

constellation. A configuration of stars named for a particular object, person, or animal; or the area of the sky assigned to a particular configuration.

continental drift. A gradual drift of the continents over the surface of the Earth due to *plate tectonics.*

continuous spectrum. A spectrum of light comprised of radiation of a continuous range of wavelengths or colors rather than only certain discrete wavelengths.

convection. The transfer of energy by moving currents of a fluid containing that energy.

core (of a planet). The central part of a planet, consisting of higher-density material, often metallic.

corona. Outer atmosphere of the Sun, where the density is very low but temperatures are in the millions of degrees.

coronal hole. A region in the Sun's outer atmosphere where visible coronal streamers are absent.

cosmic background radiation (CBR). The microwave radiation coming from all directions that is believed to be the redshifted glow of the big bang.

cosmological constant. A term that arises in the development of general relativity, which represents a repulsive force in the universe. The cosmological constant is often assumed to be zero.

cosmological principle. The assumption that, on the large scale, the universe at any given time is the same everywhere.

cosmology. The study of the organization and evolution of the universe.

crater. A circular depression (from the Greek word for cup), generally of impact origin. The rarer volcanic craters are usually identified as such; crater by itself is used in this text to refer to an impact crater.

critical density. A density of matter and energy just sufficient to eventually halt the expansion of the universe.

crust (of Earth). The outer layer of the Earth.

dark matter. Nonluminous mass, whose presence can be inferred only because of its gravitational influence on luminous matter. Dark matter may constitute as much as 99 percent of all the mass in the universe. The composition of the dark matter is not known.

degenerate gas. A gas in which the allowable states for the electrons have been filled; it behaves according to different laws from those that apply to "perfect" gases.

density. The ratio of the mass of an object to its volume.

differentiation (geological). Gravitational separation or segregation of different kinds of material in different layers in the interior of a planet.

disk (of galaxy). The material in the plane of rotation of a spiral galaxy, including the spiral arms.

Doppler effect. Apparent change in wavelength of the radiation from a source due to its relative motion in the line of sight.

double star. Two stars revolving about each other.

Earth-approaching asteroid. An asteroid with an orbit that crosses the Earth's orbit or that will at some time cross the Earth's orbit as it evolves under the influence of the planet's gravity.

eccentricity (of ellipse). Ratio of the distance between the foci to the major axis.

eclipse. The cutting off of all or part of the light of one body by another; in planetary science, the passing of one body into the shadow of another.

eclipsing binary star. A binary star in which the plane of revolution of the two stars is nearly edge-on to our line of sight, so that the light of one star is periodically diminished by the other passing in front of it.

ecliptic. The apparent annual path of the Sun on the celestial sphere.

ejecta blanket. Rough, hilly region surrounding an impact crater made up of ejecta that has fallen back to the surface, usually extending one to three crater radii from the rim.

electromagnetic force. The force of attraction or repulsion exerted between two electric charges.

electromagnetic radiation. Radiation consisting of waves propagated through the building up and breaking down of electric and magnetic fields; these include radio, infrared, light, ultraviolet, x rays, and gamma rays.

electromagnetic spectrum. The whole array or family of electromagnetic waves.

electron. A negatively charged subatomic particle that normally moves about the nucleus of an atom.

element. A substance that cannot be decomposed, by chemical means, into simpler substances.

ellipse. A conic section: the curve of intersection of a circular cone and a plane cutting completely through the cone.

elliptical galaxy. A galaxy whose apparent brightness contours are ellipses and which contains no conspicuous interstellar material.

energy level (in an atom or ion). A particular level, or amount, of energy possessed by an atom or ion above the energy it possesses in its least energetic state.

equinox. One of the intersections of the ecliptic and celestial equator.

escape velocity. Upward speed sufficient to overcome the force of gravitation and to escape into space.

event horizon. The surface through which a collapsing star is hypothesized to pass when its velocity of escape is equal to the speed of light, that is, when the star becomes a black hole.

excited state (of an atom). Condition in which one or more electrons is in an energy level above the lowest level (the ground state).

extinction. Attenuation of light from a celestial body produced by the Earth's atmosphere or by interstellar absorption.

fall (of meteorites). Meteorites seen in the sky and recovered on the ground.

fault. In geology, a crack or break in the crust of a planet along which slippage or movement can take place, accompanied by seismic activity.

find (of meteorites). A meteorite that has been recovered but was not seen to fall.

flare. A sudden and temporary outburst of light from an extended region of the solar surface, accompanied by emission of x rays.

focus. Point where the rays of light converged by a mirror or lens meet.

focus of an ellipse. Mathematical point associated with a conic section. In an elliptical orbit, the Sun lies at one focus.

frequency. Number of vibrations per unit time; number of waves that cross a given point per unit time (in radiation).

fusion. The building up of heavier atomic nuclei from lighter ones.

galaxy. A large assemblage of stars; a typical galaxy contains millions to hundreds of billions of stars, and interstellar material.

gamma rays. Photons (of electromagnetic radiation) of energy higher than those of x rays; the most energetic form of electromagnetic radiation.

geocentric. Earth-centered.

giant planet. Any one of the planets Jupiter, Saturn, Uranus, and Neptune.

giant (star). A star of large luminosity and radius.

giant molecular cloud. See *molecular cloud*.

globular cluster. One of about 120 large star clusters that form a system of clusters centered on the center of the Galaxy.

grand unified theories (GUTs). Theories in physics that incorporate both general relativity and quantum mechanics, thus unifying the mathematical description of the microscopic and macroscopic universe.

granite. The type of igneous silicate rocks that make up most of the continental crust of the Earth.

gravitation. The universal force of attraction between all material objects.

greenhouse effect. The blanketing of infrared radiation near the surface of a planet by, for example, carbon dioxide in its atmosphere.

ground state. The lowest energy level of an electron.

GUTs. See *grand unified theories (GUTs)*.

half-life. The time required for half of the radioactive atoms in a sample to disintegrate.

halo (of galaxy). The outermost extent of our Galaxy or another, containing a sparse distribution of stars and globular clusters in a more or less spherical distribution.

heliocentric. Centered on the Sun.

helium flash. The nearly explosive ignition of helium in the triple-alpha process in the dense core of a red giant star.

Hertzsprung-Russell (HR) diagram. Plot of the luminosities of stars against their temperatures or spectral types.

highlands (lunar). The older, heavily cratered crust of the Moon, covering 83 percent of its surface.

Hubble constant. Constant of proportionality between the velocities of remote galaxies and their distances. The Hubble constant is thought to be approximately 25 km/s per million light years.

Hubble law. The law of the redshifts. The radial velocities of remote galaxies are proportional to their distances from us.

hydrostatic equilibrium. A balance between the weights of various layers, as in a star or the Earth's atmosphere, and the pressures that support them.

igneous rock. Any rock produced by cooling from a molten state.

inflationary universe. A theory of cosmology in which the universe is assumed to have undergone a phase of very rapid expansion during the first 10^{-30} second. After this period of rapid expansion, the big bang and inflationary models are identical.

infrared dwarf. See *brown dwarf*.

infrared radiation. Electromagnetic radiation of wavelength longer than the longest (red) wavelengths that can be perceived by the eye, but shorter than radio wavelengths.

inverse square law. The rule, which applies to both radiation and gravitation, that the effect decreases in proportion to the distance squared.

ion. An atom that has become electrically charged by the addition or loss of one or more electrons.

ionization. The process by which an atom gains or loses electrons.

ionosphere. The upper region of the Earth's atmosphere in which many of the atoms are ionized.

irons (meteorites). One of the three main types of meteorites, typically made of about 90 percent iron and 9 percent nickel, with traces of other elements.

irregular galaxy.   A galaxy without rotational symmetry; neither a spiral nor an elliptical galaxy.

isotope.   Any of two or more forms of the same element, whose atoms all have the same number of protons but different numbers of neutrons.

jovian planet.   One of the giant planets: Jupiter, Saturn, Uranus, or Neptune.

light curve.   A graph that displays the time variation in brightness of a variable or eclipsing binary star.

light year (LY).   The distance light travels in a vacuum in one year; $1 \text{ LY} = 9.46 \times 10^{15}$ m.

Local Group.   The cluster of galaxies to which our Galaxy belongs.

luminosity.   The rate of radiation of electromagnetic energy into space by a star or other object.

luminosity function.   The relative numbers or proportions of stars with different luminosities.

MACHO.   An acronym that stands for Massive Compact Halo Object. It has been hypothesized that much of the dark matter in the haloes of galaxies consists of compact objects, such as brown dwarfs and black holes. Such compact objects have been dubbed MACHOs.

Magellanic Clouds.   Two neighboring galaxies visible to the naked eye from southern latitudes.

magnetosphere.   The region around a planet in which its intrinsic magnetic field dominates over the interplanetary field carried by the solar wind; hence, the region within which charged particles can be trapped by the planetary magnetic field.

magnitude.   A measure of the apparent brightness of a star or other luminous object.

main sequence.   A sequence of stars on the HR diagram, containing the majority of stars, that runs diagonally from the upper left to the lower right.

mantle (of Earth).   The greatest part of the Earth's interior, lying between the crust and the core.

mare.   Latin for "sea"; name applied to the dark, relatively smooth features that cover 17 percent of the Moon.

mass extinction.   The sudden disappearance in the fossil record of a large number of species of life, to be replaced by new species in subsequent layers. Mass extinctions are indications of catastrophic changes in the environment, such as might be produced by a large impact on the Earth.

mass-luminosity relationship.   An empirical relation between the masses and luminosities of many (principally main-sequence) stars.

Maunder minimum.   The interval from 1645 to 1715 when solar activity was very low.

metamorphic rock.   Any rock produced by the physical and chemical alteration (without melting) of another rock that has been subjected to high temperature and pressure.

meteor.   The luminous phenomenon observed when a particle enters the Earth's atmosphere and burns up; popularly called a "shooting star."

meteor shower.   Many meteors appearing to radiate from a common point in the sky caused by the collision of the Earth with a swarm of meteoric particles.

meteorite.   A portion of interplanetary debris that survives passage through the atmosphere and strikes the ground.

microwave.   Short-wave radio wavelengths.

Milky Way.   The band of light encircling the sky, which is due to the many stars and diffuse nebulae lying near the plane of the Galaxy.

molecular cloud.   A relatively dense part of the interstellar medium in which complex molecules form, typically 100 LY in dimension; also called a giant molecular cloud or a cold molecular cloud.

molecule.   A combination of two or more atoms bound together; the smallest particle of a chemical compound or substance that exhibits the chemical properties of that substance.

nebula.   Cloud of interstellar gas or dust.

neutrino.   A fundamental particle that has little or no rest mass and no charge but that does have spin and energy. Many astrophysically important processes release neutrinos, from fusion reactions in stars to supernova explosions.

neutron.   A subatomic particle with no charge and with mass approximately equal to that of the proton.

neutron star.   A star of extremely high density composed almost entirely of neutrons. See *pulsar*.

nuclear bulge (of Galaxy).   The central, elliptical-shaped region in the center of our Galaxy, several thousand light years in size.

nucleosynthesis.   The building up of heavy elements from lighter ones by nuclear fusion.

nucleus (of atom).   The heavy part of an atom, composed mostly of protons and neutrons, and about which the electrons revolve.

nucleus (of comet).   The solid chunk of ice and dust in the head of a comet, typically a few kilometers in diameter.

nucleus (of galaxy).   Central concentration of matter at the center of a galaxy, typically less than 1 LY in diameter, and possibly containing a supermassive black hole.

Oort comet cloud.   The spherical region around the Sun from which most "new" comets come; new comets typically have aphelia at about 50,000 AU, or extending about a third of the way to the nearest other stars.

open cluster.   A comparatively loose or "open" cluster of stars, containing from a few dozen to a few thousand members, located in the spiral arms or disk of the Galaxy.

open universe.   A universe with a density low enough that it will expand forever.

optical window.   The part of the electromagnetic spectrum at visible and near-infrared wavelengths (from about 300 to 1300 nm) that is transmitted by the Earth's atmosphere.

oxidizing.   In chemistry, referring to conditions in which oxygen dominates over hydrogen, so that most other elements form compounds with oxygen. In very oxidizing conditions, such as are found in the atmosphere of the Earth, free oxygen gas (O_2) or even atomic oxygen (O) are present.

ozone.   A rare molecular form of oxygen that contains three oxygen atoms (O_3).

parallax.   An apparent displacement of an object due to a motion of the observer.

perfect gas laws.   Certain laws that describe the behavior of an ideal gas.

perigee.   The place in the orbit of an Earth satellite where it is closest to the center of the Earth.

perihelion.   The place in the orbit of an object revolving about the Sun where it is closest to the center of the Sun.

period-luminosity relationship.   An empirical relation between the periods and luminosities of cepheid variable stars.

photon.   A discrete unit of electromagnetic energy.

photosphere.   The region of the solar (or a stellar) atmosphere from which continuous radiation escapes into space.

photosynthesis.   The formation of carbohydrates in the chlorophyll-containing tissues of plants exposed to sunlight. In the process, oxygen is released to the atmosphere.

plage.   A bright region of the solar surface observed in the monochromatic light of some spectral line.

planetary nebula.   A shell of gas ejected from, and enlarging about, a certain kind of extremely hot star.

planetesimals.   The hypothetical objects, from tens to hundreds of kilometers in diameter, that formed in the solar nebula as an intermediate step between tiny grains and the larger planetary objects we see today. The comets and some asteroids may be left-over planetesimals

plate tectonics.   The motion of segments or plates of the outer layer of the Earth over the underlying mantle.

population (stellar).   The classification of stars based primarily on their composition. Population II stars have fewer heavy elements than population I stars.

positron.   Subatomic particle, like an electron but with a positive electric charge.

primitive meteorite.   A meteorite that has not been greatly altered chemically since its condensation

from the solar nebula; called in meteoritics a chondrite (either ordinary chondrite or carbonaceous chondrite).

primitive rock.   Any rock that has not experienced great heat or pressure and therefore remains representative of the original condensates from the solar nebula—never found on any object large enough to have undergone melting and differentiation.

proton.   A heavy subatomic particle that carries a positive charge; one of the two principal constituents of the atomic nucleus.

protostar.   A collapsing condensation in the interstellar medium that will evolve into a star; any very young star that has not yet reached the stage where it generates energy from thermonuclear reactions.

pulsar.   A radio source of small angular size that is observed to vary in regular periods that range from fractions of a second to minutes; a neutron star.

pulsating variable.   A variable star that pulsates in size and luminosity.

quasar.   A stellar-appearing object of very high redshift, presumed to be extragalactic and highly luminous; an active galactic nucleus.

RR Lyrae variable.   One of a class of giant pulsating stars with periods less than one day.

radar.   Radio (transmitted and received radiowave signals) used to measure the distance to, and motion of, a target object.

radial velocity.   The component of relative velocity that lies in the line of sight.

radiation.   A mode of energy transport whereby energy is transmitted through a vacuum; also the transmitted energy itself; see *electromagnetic radiation.*

radio galaxy.   An active galaxy which is unusually luminous at radio wavelengths.

radioactivity (radioactive decay).   The process by which certain kinds of atomic nuclei naturally decompose with the spontaneous emission of sub-atomic particles and gamma rays.

redshift.   A shift to longer wavelengths of the light from remote galaxies; presumed to be produced by a Doppler shift.

reducing.   In chemistry, referring to conditions in which hydrogen dominates over oxygen, so that most other elements form compounds with hydrogen. In very reducing conditions free hydrogen (H_2) is present and free oxygen (O_2) cannot exist.

reflecting telescope.   A telescope in which the principal optical component (objective) is a concave mirror.

refracting telescope.   A telescope in which the principal optical component (objective) is a lens or system of lenses.

resolution.   The degree to which fine details in an image are separated or resolved.

resonance.   An orbital condition in which one object is subject to periodic gravitational perturbations by another, most commonly arising when two objects orbiting a third have periods of revolution that are simple multiples or fractions of each other.

retrograde motion.   An apparent westward motion of a planet on the celestial sphere or with respect to the stars.

revolution.   The motion of one body around another.

rift zone.   In geology, a place where the crust is being torn apart by internal forces, generally associated with the injection of new material from the mantle and with the slow separation of tectonic plates.

rotation.   Turning of a body about an axis running through it.

runaway greenhouse effect.   A process whereby the heating of a planet leads to an increase in its atmospheric greenhouse effect and thus to further heating, thereby quickly altering the composition of its atmosphere and the temperature of its surface.

sedimentary rock.   Any rock formed by the deposition and cementing of fine grains of material. On Earth, sedimentary rocks are usually the result of erosion and weathering, followed by deposition in lakes or oceans.

seismic waves.   Vibrations traveling through the Earth's interior that result from earthquakes.

seismology (planetary).   The study of earthquakes and the conditions that produce them and of the internal structure of the Earth as deduced from analyses of seismic waves.

seismology (solar).   The study of small changes in the radial velocity of the Sun as a whole or of small regions on the surface of the Sun. Analyses of these velocity changes can be used to infer the internal structure of the Sun.

semimajor axis.   Half the major axis of an ellipse.

SETI.   The search for extraterrestrial intelligence, usually applied to searches for radio signals from other civilizations.

Seyfert galaxy.   A galaxy belonging to the class of those with active galactic nuclei; one whose nucleus shows bright emission lines.

silicate.   Any of the common classes of rocks composed of minerals rich in the elements silicon and oxygen (and often containing iron, magnesium, and other elements as well).

SNC meteorite.   One of a class of basaltic meteorites now believed by many planetary scientists to be impact-ejected fragments from Mars.

solar activity.   Phenomena of the solar atmosphere: sunspots, plages, flares, and related features.

solar cycle.   The pattern of variations in the numbers of sunspots and other types of solar activity, which repeats approximately every 11 years.

solar nebula.   The cloud of gas and dust from which the solar system formed.

solar wind.   A radial flow of plasma leaving the Sun, with a typical speed of 400 km/s.

solstice.   The point on the celestial sphere where the Sun reaches its greatest distance north or south of the celestial equator.

spectral class (or type).   A classification of a star according to the characteristics of its spectrum.

spectral lines.   Features in spectra, especially of gases, in which radiation at discrete wavelengths is either depleted (dark lines) or enhanced (bright lines).

spectroscopic binary star.   A binary star in which the components are not resolved optically, but whose binary nature is indicated by periodic variations in radial velocity, indicating orbital motion.

spectroscopy.   The study of spectra.

spectrum.   The array of colors or wavelengths obtained when light from a source is dispersed, as in passing it through a prism or grating.

spiral galaxy.   A flattened, rotating galaxy with pinwheel-like arms of interstellar material and young stars winding out from its nucleus.

standard candle.   An astronomical object of known luminosity; such an object can be used to determine distances.

Stefan-Boltzmann Law.   A formula from which the rate at which a blackbody radiates energy can be computed; the total rate of energy emission from a unit area of a blackbody is proportional to the fourth power of its absolute temperature.

stone (meteorite).   A meteorite composed mostly of stony material.

stony-iron (meteorite).   A type of meteorite that is a blend of nickel-iron and silicate materials.

stratosphere.   The layer of the Earth's atmosphere above the troposphere (where most weather takes place) and below the ionosphere.

strong force.   The force that binds together the parts of the atomic nucleus.

subduction zone.   In terrestrial geology, a region where one crustal plate is forced under another, generally associated with earthquakes, volcanic activity, and the formation of deep ocean trenches.

sunspot.   A temporary cool region in the solar photosphere that appears dark by contrast against the surrounding hotter photosphere.

sunspot cycle.   The semiregular 11-year period with which the frequency of sunspots fluctuates.

supercluster.   A large region of space (150 to 300 million LY across) where matter is concentrated into galaxies, groups of galaxies, and clusters of galaxies; a cluster of clusters of galaxies.

supergiant star.   A star of very high luminosity.

supernova.   An explosion that marks the final stage of evolution of a star. Type I supernovas are thought to occur when a white dwarf accretes enough matter to exceed the Chandrasekhar limit, collapses, and explodes. A type II supernova is thought to mark the final collapse of a massive star.

synchrotron radiation.   The radiation emitted by charged particles being accelerated in magnetic fields and moving at speeds near that of light.

tail (of comet).   Elongated tenuous atmosphere of gas and dust streaming away from a comet's nucleus, generally pointed away from the Sun.

tectonic.   Associated with forces acting in the solid crust of a planet.

terrestrial planets.   The inner planets: Mercury, Venus, Earth, and Mars. The Moon is also included among the terrestrial bodies.

thermal equilibrium.   A balance between the input and outflow of heat in a system.

thermal radiation.   The radiation emitted by any body or gas that is not at absolute zero.

thermonuclear reaction.   A nuclear reaction or transformation that results from encounters between nuclear particles that are given high velocities (by heating them).

triangulation.   Determining the distance to a remote object by measuring its direction as viewed from two locations. See *parallax.*

troposphere.   Lowest level of the Earth's atmosphere, where most weather takes place.

ultraviolet radiation.   Electromagnetic radiation of wavelengths shorter than the shortest (violet) wavelengths to which the eye is sensitive; radiation of wavelengths in the approximate range of 10 to 400 nm.

visual binary star.   A binary star in which the two components are telescopically resolved.

volatile materials.   Materials that are gaseous at fairly low temperatures. This is a *relative* term, usually applied to the gases in planetary atmospheres and to common ices (H_2O, CO_2, etc.), but also sometimes used for elements such as cadmium, zinc, lead, and rubidium that form gases at temperatures up to 1000 K. (These are called *volatile elements*, as opposed to refractory elements.)

wavelength.   The spacing of the crests or troughs in a wave train.

weak force.   The nuclear force involved in radioactive decay.

white dwarf.   A star that has exhausted most or all of its nuclear fuel and has collapsed to a very small size; believed to be near its final stage of evolution.

Wien's law.   Formula that relates the temperature of a blackbody to the wavelength at which it emits the greatest intensity of radiation.

x rays.   Photons of wavelengths intermediate between those of ultraviolet radiation and gamma rays.

Zeeman effect.   A splitting or broadening of spectral lines due to magnetic fields.

zenith.   The point on the celestial sphere opposite to the direction of gravity; or the direction opposite to that indicated by a plumb bob.

▶ APPENDIX 2: POWER OF TEN NOTATION

In astronomy and other sciences, it is often necessary to deal with very large or very small numbers. For example, the Earth is 150,000,000,000 m from the Sun, and the mass of the hydrogen atom is 0.00000000000000000000000000167 kg. Instead of writing and carrying so many zeros, the numbers are usually written as figures between 1 and 10 multiplied by the appropriate power of 10. For example, 150,000,000,000 is 1.5×10^{11}, and the mass of the hydrogen atom given above is written simply as 1.67×10^{-27} kg. Additional examples are given below.

one hundredth	=	$0.01 = 10^{-2}$
one tenth	=	$0.1 = 10^{-1}$
one	=	$1 = 10^{0}$
ten	=	$10 = 10^{1}$

one hundred	=	$100 = 10^{2}$
one thousand	=	$1000 = 10^{3}$
one million	=	$1,000,000 = 10^{6}$
one billion	=	$1,000,000,000 = 10^{9}$

The powers-of-ten notation is not only compact and convenient, it also simplifies arithmetic. To multiply two numbers expressed as powers of ten, you need only add the exponents. And to divide, you subtract the exponents. Following are several examples:

$$100 \times 100,000 = 10^{2} \times 10^{5} = 10^{2+5} = 10^{7}$$
$$0.01 \times 1,000,000 = 10^{-2} \times 10^{6} = 10^{6-2} = 10^{4}$$
$$1,000,000 \div 1000 = 10^{6} \div 10^{3} = 10^{6-3} = 10^{3}$$
$$100 \div 1,000,000 = 10^{2} \div 10^{6} = 10^{2-6} = 10^{-4}$$

▶ APPENDIX 3:   AMERICAN AND METRIC UNITS

In the American system of measure (originally developed in England), the fundamental units of length, mass, and time are the yard, pound, and second, respectively. There are also, of course, larger and smaller units, which include the ton (2000 lb), the mile (1760 yd), the rod ($16\frac{1}{2}$ ft), the inch ($\frac{1}{36}$ yd), the ounce ($\frac{1}{16}$ lb), and so on. Such units are inconvenient for conversion and arithmetic computation.

In science, therefore, it is more usual to use the metric system, which has been adopted in virtually all countries except the United States. The fundamental units of the metric system are

length: 1 meter (m)
mass: 1 kilogram (kg)
time: 1 second (s)

A meter was originally intended to be 1 ten-millionth of the distance from the equator to the North Pole along the surface of the Earth. It is about 1.1 yd. A kilogram is about 2.2 lb. The second is the same in metric and American units. The most commonly used quantities of length and mass of the metric system are the following:

Length

1 km	= 1 kilometer	= 1000 meters	= 0.6214 mile	
1 m	= 1 meter	= 1.094 yards	= 39.37 inches	
1 cm	= 1 centimeter	= 0.01 meter	= 0.3937 inch	
1 mm	= 1 millimeter	= 0.001 meter	= 0.01 cm	= 0.03937 inch
1 μm	= 1 micrometer	= 0.000 001 meter	= 0.0001 cm	
1 nm	= 1 nanometer	= 10^{-9} meter	= 10^{-7} cm	

also: 1 mile = 1.6093 km
1 inch = 2.5400 cm

Mass

1 metric ton	= 10^6 grams	= 1000 kg	= 2.2046×10^3 lb
1 kg		= 1000 grams	= 2.2046 lb
1 g	= 1 gram	= 0.0022046 lb	= 0.0353 oz
1 mg	= 1 milligram	= 0.001 g	

also: 1 lb = 0.4536 kg
1 oz = 28.3495 g

▶ APPENDIX 4: TEMPERATURE SCALES

Three temperature scales are in general use:

1. Fahrenheit (F); water freezes at 32°F and boils at 212°F.
2. Celsius or centigrade* (C); water freezes at 0°C and boils at 100°C.
3. Kelvin or absolute (K); water freezes at 273 K and boils at 373 K.

All molecular motion ceases at −459°F = −273°C = 0 K. Thus, Kelvin temperature is measured from this lowest possible temperature, called *absolute zero*. It is the temperature scale most often used in astronomy. Kelvins are degrees that have the same value as centigrade or Celsius degrees, since the difference between the freezing and boiling points of water is 100 degrees in each.

On the Fahrenheit scale, water boils at 212 degrees and freezes at 32 degrees; the difference is 180 degrees. Thus, to convert Celsius degrees or Kelvins to Fahrenheit, it is necessary to multiply by 180/100 = 9/5. To

*Celsius is now the name used for centigrade temperature; it has a more modern standardization but differs from the old centigrade scale by less than 0.1°.

convert from Fahrenheit to Celsius degrees or Kelvins, it is necessary to multiply by 100/180 = 5/9.

Example 1: What is 68°F in Celsius and in Kelvins?

$$68°F − 32°F = 36°F \text{ above freezing.}$$

$$\frac{5}{9} \times 36° = 20°;$$

thus,

$$68°F = 20°C = 293 \text{ K.}$$

Example 2: What is 37°C in Fahrenheit and in Kelvin?

$$37°C = 273° + 37° = 310 \text{ K;}$$

$$\frac{9}{5} \times 37° = 66.6 \text{ Fahrenheit degrees;}$$

thus,

$$37°C \text{ is } 66.6°F \text{ above freezing}$$

or

$$37°C = 32° + 66.6° = 98.6°F.$$

▶ APPENDIX 5: SOME USEFUL CONSTANTS

PHYSICAL CONSTANTS

speed of light	c	$= 2.9979 \times 10^8$ m/s
constant of gravitation	G	$= 6.672 \times 10^{-11}$ N m^2/kg^2
Planck's constant	h	$= 6.626 \times 10^{-34}$ joules
mass of hydrogen atom	m_H	$= 1.673 \times 10^{-27}$ kg
mass of electron	m_e	$= 9.109 \times 10^{-31}$ kg
charge on electron	e	$= 4.803 \times 10^{-10}$ eu
Stefan-Boltzmann constant	σ	$= 5.670 \times 10^{-8}$ joule/m^2·deg^4
constant in Wien's law	$\lambda_{max}T$	$= 2.898 \times 10^{-3}$ m·deg
energy equivalent of 1 ton TNT	E	$= 4.2 \times 10^9$ joules

ASTRONOMICAL CONSTANTS

astronomical unit	AU	$= 1.496 \times 10^{11}$ m
light year	LY	$= 9.461 \times 10^{15}$ m
sidereal year	yr	$= 3.158 \times 10^7$ s
mass of Earth	M_E	$= 5.977 \times 10^{24}$ kg
equatorial radius of Earth	R_E	$= 6.378 \times 10^6$ m
mass of Sun	M_S	$= 1.989 \times 10^{30}$ kg
equatorial radius of Sun	R_S	$= 6.960 \times 10^8$ m
luminosity of Sun	L_S	$= 3.83 \times 10^{26}$ watts
solar constant (at Earth)	S	$= 1.37 \times 10^3$ watts/m^2

APPENDIX 6: SOME NUCLEAR REACTIONS OF IMPORTANCE IN ASTRONOMY

Given here are the series of thermonuclear reactions that are most important in stellar interiors. The subscript to the left of a nuclear symbol is the atomic number; the superscript to the left is the atomic mass number. The symbols for the positive electron (positron) and electron are e^+ and e^-, respectively, for the neutrino is ν, and for a photon (generally of gamma-ray energy) is γ.

1. The Proton-Proton Chains

(Important below 15×10^6 K)

There are three ways the proton-proton chain can be completed. The first (a_1, b_1, c_1) is the most important, but depending on the physical conditions in the stellar interior, some energy is released by one or both of the following alternatives: a_1, b_1, c_2, d_2, e_2, and a_1, b_1, c_2, d_3, e_3, f_3.

(a_1) ${}_1^1\text{H} + {}_1^1\text{H} \rightarrow {}_1^2\text{H} + e^+ + \nu$
(b_1) ${}_1^2\text{H} + {}_1^1\text{H} \rightarrow {}_2^3\text{He} + \gamma$
(c_1) ${}_2^3\text{He} + {}_2^3\text{He} \rightarrow {}_2^4\text{He} + 2{}_1^1\text{H}$
or (c_2) ${}_2^3\text{He} + {}_2^4\text{He} \rightarrow {}_4^7\text{Be} + \gamma$

(d_2) ${}_4^7\text{Be} + e^- \rightarrow {}_3^7\text{Li} + \nu$
(e_2) ${}_3^7\text{Li} + {}_1^1\text{H} \rightarrow 2{}_2^4\text{He}$
or (d_3) ${}_4^7\text{Be} + {}_1^1\text{H} \rightarrow {}_5^8\text{B} + \gamma$
(e_3) ${}_5^8\text{B} \rightarrow {}_4^8\text{Be} + e^+ + \nu$
(f_3) ${}_4^8\text{Be} \rightarrow 2{}_2^4\text{He}$

2. The Carbon-Nitrogen Cycle

(Important above 15×10^6 K)
(a) ${}_6^{12}\text{C} + {}_1^1\text{H} \rightarrow {}_7^{13}\text{N} + \gamma$
(b) ${}_7^{13}\text{N} \rightarrow {}_6^{13}\text{C} + e^+ + \nu$
(c) ${}_6^{13}\text{C} + {}_1^1\text{H} \rightarrow {}_7^{14}\text{N} + \gamma$
(d) ${}_7^{14}\text{N} + {}_1^1\text{H} \rightarrow {}_8^{15}\text{O} + \gamma$
(e) ${}_8^{15}\text{O} \rightarrow {}_7^{15}\text{N} + e^+ + \nu$
(f) ${}_7^{15}\text{N} + {}_1^1\text{H} \rightarrow {}_6^{12}\text{C} + {}_2^4\text{He}$

3. The Triple-Alpha Process

(Important above 10^8 K)
(a) ${}_2^4\text{He} + {}_2^4\text{He} \rightarrow {}_4^8\text{Be} + \gamma$
(b) ${}_2^4\text{He} + {}_4^8\text{Be} \rightarrow {}_6^{12}\text{C} + \gamma$

► APPENDIX 7: ORBITAL DATA FOR THE PLANETS

Planet	Semimajor Axis		Sidereal Period		Mean Orbital Speed (km/s)	Orbital Eccentricity	Inclination of Orbit to Ecliptic (°)
	AU	10^6 km	Tropical Years	Days			
Mercury	0.3871	57.9	0.24085	87.97	47.9	0.206	7.004
Venus	0.7233	108.2	0.61521	224.70	35.0	0.007	3.394
Earth	1.0000	149.6	1.000039	365.26	29.8	0.017	0.0
Mars	1.5237	227.9	1.88089	686.98	24.1	0.093	1.850
(Ceres)	2.7671	414	4.603		17.9	0.077	10.6
Jupiter	5.2028	778	11.86		13.1	0.048	1.308
Saturn	9.538	1427	29.46		9.6	0.056	2.488
Uranus	19.191	2871	84.07		6.8	0.046	0.774
Neptune	30.061	4497	164.82		5.4	0.010	1.774
Pluto	39.529	5913	248.6		4.7	0.248	17.15

Adapted from *The Astronomical Almanac* (U.S. Naval Observatory), 1981.

APPENDIX 8: PHYSICAL DATA FOR THE PLANETS

Planet	Diameter (km)	Diameter (Earth = 1)	Mass (Earth = 1)	Mean Density (g/cm³)	Rotation Period (days)	Inclination of Equator to Orbit (°)	Surface Gravity (Earth = 1)	Velocity of Escape (km/s)
Mercury	4878	0.38	0.055	5.43	58.6	0.0	0.38	4.3
Venus	12,104	0.95	0.82	5.24	−243.0	177.4	0.91	10.4
Earth	12,756	1.00	1.00	5.52	0.997	23.4	1.00	11.2
Mars	6794	0.53	0.107	3.9	1.026	25.2	0.38	5.0
Jupiter	142,796	11.2	317.8	1.3	0.41	3.1	2.53	60
Saturn	120,000	9.41	94.3	0.7	0.43	26.7	1.07	36
Uranus	52,400	4.11	14.6	1.3	−0.65	97.9	0.92	21
Neptune	50,450	3.81	17.2	1.5	0.72	29	1.18	24
Pluto	2310	0.17	0.0025	2.1	6.387	118	0.09	1

▶ APPENDIX 9:   SATELLITES OF THE PLANETS

Planet	Satellite Name	Discovery	Semimajor Axis (km × 1000)	Period (days)	Diameter (km)	Mass (10²⁰ kg)	Density (g/cm³)
Earth	Moon	—	384	27.32	3476	735	3.3
Mars	Phobos	Hall (1877)	9.4	0.32	23	1×10^{-4}	2.0
	Deimos	Hall (1877)	23.5	1.26	13	2×10^{-5}	1.7
Jupiter	Metis	Voyager (1979)	128	0.29	20	—	—
	Adrastea	Voyager (1979)	129	0.30	40	—	—
	Amalthea	Barnard (1892)	181	0.50	200	—	—
	Thebe	Voyager (1979)	222	0.67	90	—	—
	Io	Galileo (1610)	422	1.77	3630	894	3.6
	Europa	Galileo (1610)	671	3.55	3138	480	3.0
	Ganymede	Galileo (1610)	1070	7.16	5262	1482	1.9
	Callisto	Galileo (1610)	1883	16.69	4800	1077	1.9
	Leda	Kowal (1974)	11,090	239	15	—	—
	Himalia	Perrine (1904)	11,480	251	180	—	—
	Lysithea	Nicholson (1938)	11,720	259	40	—	—
	Elara	Perrine (1905)	11,740	260	80	—	—
	Ananke	Nicholson (1951)	21,200	631 (R)	30	—	—
	Carme	Nicholson (1938)	22,600	692 (R)	40	—	—
	Pasiphae	Melotte (1908)	23,500	735 (R)	40	—	—
	Sinope	Nicholson (1914)	23,700	758 (R)	40	—	—
Saturn	Unnamed	Voyager (1985)	118.2	0.48	15?	3×10^{-5}	—
	Pan	Voyager (1985)	133.6	0.58	20	3×10^{-5}	—
	Atlas	Voyager (1980)	137.7	0.60	40	—	—
	Prometheus	Voyager (1980)	139.4	0.61	80	—	—
	Pandora	Voyager (1980)	141.7	0.63	100	—	—
	Janus	Dolefus (1966)	151.4	0.69	190	—	—
	Epimetheus	Fountain, Larson (1980)	151.4	0.69	120	—	—
	Mimas	Herschel (1789)	186	0.94	394	0.4	1.2
	Enceladus	Herschel (1789)	238	1.37	502	0.8	1.2
	Tethys	Cassini (1684)	295	1.89	1048	7.5	1.3
	Telesto	Reitsema et al. (1980)	295	1.89	25	—	—
	Calypso	Pascu et al. (1980)	295	1.89	25	—	—
	Dione	Cassini (1684)	377	2.74	1120	11	1.4
	Helene	Lecacheux, Laques (1980)	377	2.74	30	—	—
	Rhea	Cassini (1672)	527	4.52	1530	25	1.3
	Titan	Huygens (1655)	1222	15.95	5150	1346	1.9
	Hyperion	Bond, Lassell (1848)	1481	21.3	270	—	—

APPENDIX 9 (CONTINUED)

Planet	Satellite Name	Discovery	Semimajor Axis (km × 1000)	Period (days)	Diameter (km)	Mass (10²⁰ kg)	Density (g/cm³)
	Iapetus	Cassini (1671)	3561	79.3	1435	19	1.2
	Phoebe	Pickering (1898)	12,950	550 (R)	220	—	—
Uranus	Cordelia	Voyager (1986)	49.8	0.34	40?	—	—
	Ophelia	Voyager (1986)	53.8	0.38	50?	—	—
	Bianca	Voyager (1986)	59.2	0.44	50?	—	—
	Cressida	Voyager (1986)	61.8	0.46	60?	—	—
	Desdemona	Voyager (1986)	62.7	0.48	60?	—	—
	Juliet	Voyager (1986)	64.4	0.50	80?	—	—
	Portia	Voyager (1986)	66.1	0.51	80?	—	—
	Rosalind	Voyager (1986)	69.9	0.56	60?	—	—
	Belinda	Voyager (1986)	75.3	0.63	60?	—	—
	Puck	Voyager (1985)	86.0	0.76	170	—	—
	Miranda	Kuiper (1948)	130	1.41	485	0.8	1.3
	Ariel	Lassell (1851)	191	2.52	1160	13	1.6
	Umbriel	Lassell (1851)	266	4.14	1190	13	1.4
	Titania	Herschel (1787)	436	8.71	1610	35	1.6
	Oberon	Herschel (1787)	583	13.5	1550	29	1.5
Neptune	Naiad	Voyager (1989)	48	0.30	50	—	
	Thalassa	Voyager (1989)	50	0.31	90	—	
	Despina	Voyager (1989)	53	0.33	150	—	
	Galatea	Voyager (1989)	62	0.40	150	—	
	Larissa	Voyager (1989)	74	0.55	200	—	
	Proteus	Voyager (1989)	118	1.12	400	—	
	Triton	Lassell (1846)	355	5.88 (R)	2720	220	2.1
	Nereid	Kuiper (1949)	5511	360	340	—	—
Pluto	Charon	Christy (1978)	19.7	6.39	1200	—	—

Star	Distance (LY)	Radial Velocity (km/s)	Spectra of Components			Visual Magnitudes of Components			Visual Luminosities of Components (L_s)		
			A	B	C	A	B	C	A	B	C
Sun			G2V			−26.8			1.0		
Proxima Centauri*	4.3	−16	M5V			+11.05			5.8×10^{-5}		
α Centauri	4.4	−22	G2V	K0V		−0.01	+1.33		1.4	0.44	
Barnard's Star	5.9	−108	M5V			+9.54			4.4×10^{-4}		
Wolf 359	7.7	+13	M8V			+13.53			1.7×10^{-5}		
Lalande 21185	8.2	−84	M2V			+7.50			5.2×10^{-3}		
Luyten 726-8	8.5	+30	M5.5V	M5.5V		+12.45	+12.95		6.3×10^{-5}	4.0×10^{-5}	
Sirius	8.6	−8	A1V	wd		−1.46	+8.68		2.3	1.9×10^{-3}	
Ross 154	9.5	−4	M4.5V			+10.6			4.0×10^{-4}		
Ross 248	10.2	−81	M6V			+12.29			$+1 \times 10^{-4}$		
ε Eridani	10.7	+16	K2V			+3.73			0.30		
Ross 128	10.8	−13	M5V			+11.10			3.3×10^{-4}		
Luyten 789-6	10.8	−60	M6V			−12.18			1.20×10^{-4}		
61 Cygni	11.1	−64	K5V	K7V		+5.22	+6.03		0.076	0.036	
ε Indi	11.2	−40	K5V			+4.68			0.13		
τ Ceti	11.3	−16	G8V			+3.50			0.44		
Procyon	11.4	−3	F5IV-V	wd		+0.37	+10.7		7.6	5.2×10^{-4}	
BD + 59°1915	11.5	+5	M4V	M5V		+8.90	+9.69		2.8×10^{-3}	1.4×10^{-3}	
BD + 43°44	11.6	+17	M1V	M6V		+8.07	+11.04		6.3×10^{-3}	4.0×10^{-4}	
CD − 36°15693	11.7	+10	M2V			+7.36			0.012		
G51-15	11.9		MV			+14.8			1.3×10^{-5}		
Luyten 725-32	12.4		M5V			+11.5			3.0×10^{-4}		
BD + 5°1668	12.4	+26	M5V			+9.82			1.3×10^{-3}		
CD − 39°14192	12.6	+21	M0V			+6.67			0.025		
Kapteyn's Star	12.7	+245	M0V			+8.81			4.0×10^{-3}		
Kruger 60	12.8	−26	M3V	M4.5V		+9.85	+11.3		1.4×10^{-3}	4.0×10^{-4}	
Ross 614	13.4	+24	M7V	?		+11.07	+14.8		4.8×10^{-4}	1.6×10^{-5}	
BD − 12°4523	13.7	−13	M5V			+10.12			1.2×10^{-3}		
Wolf 424	13.9	−5	M5.5V	M6V		+13.16	+13.4		8.3×10^{-5}	6.9×10^{-5}	
v. Maanen's Star	14.1	+54	wd			+12.37			1.6×10^{-4}		
CD − 37°15492	14.5	+23	M3V			+8.63			5.8×10^{-3}		
Luyten 1159-16	14.7		M8V			+12.27			2.3×10^{-4}		
BD + 50°1725	15.0	−26	K7V			+6.59			0.040		
CD − 46°11540	15.1		M4V			+9.36			3.3×10^{-3}		
CD − 49°13515	15.2	+8	M3V			+8.67			6.3×10^{-3}		
CD − 44°11909	15.3		M5V			+11.2			6.3×10^{-4}		
BD + 68°946	15.3	−22	M3.5V			+9.15			4.0×10^{-3}		
G158 − 27	15.4		MV			+13.7			6.3×10^{-5}		
G208-44/45	15.5		MV	MV		+13.4	+14.0		8.3×10^{-5}	4.8×10^{-5}	
Ross 780	15.6	+9	M5V			+10.7			1.6×10^{-3}		
40 Eridani	15.7	−43	K0V	wd	M4.5V	+4.43	+9.53	+11.17	0.33	3.0×10^{-3}	6.9×10^{-4}
Luyten 145-141	15.8		wd			+11.44			5.2×10^{-4}		
BD + 20°2465	16.1	+11	M4.5V			+9.43			3.3×10^{-3}		
70 Ophiuchi	16.1	−7	K1V	K5V		+4.2	+6.0		0.44	0.83	
BD + 43°4305	16.3	−2	M4.5V			+10.2			1.7×10^{-3}		

*Proxima Centauri is sometimes considered an outlying member of the α Centauri system.
Adapted from data supplied by the U.S. Naval Observatory.

Appendix 11: The Twenty Brightest Stars

Star	Distance* (LY)	Spectra of Components			Visual Magnitudes of Components			Visual Luminosities of Components (L_s)		
		A	B	C	A	B	C	A	B	C
Sirius	8.8	A1V	wd		−1.46	+8.7		+23	$+1.9 \times 10^{-3}$	
Canopus	98	F01b-II			−0.72			−1450		
α Centauri	4.4	G2V	K0V		−0.01	+1.3		+1.45	+0.44	
Arcturus	37	K2IIIp			−0.06			−110		
Vega	26	A0V			+0.04			+52		
Capella	46	GIII	M1V	M5V	+0.05	+10.2	+13.7	−160	+0.013	$+5.2 \times 10^{-4}$
Rigel	800	B8 Ia	B9		+0.14	+6.6		4.4×10^4	−120	
Procyon	11	F5IV-V	wd		+0.37	+10.7		+7.6	$+5.2 \times 10^{-4}$	
Betelgeuse	500	M2Iab			+0.41v			1.3×10^4		
Achernar	65	B5V			+0.51			−209		
β Centauri	300	B1III	?		+0.63	+4		−3600	−174	
Altair	17	A7IV-V			+0.77			+11		
α Crucis	400	B1IV	B3		+1.39	+1.9		−3300	−2100	
Aldebaran	52	K5III	M2V		+0.86	+13		−100	$+1.3 \times 10^{-3}$	
Spica	260	B1V			+0.91v			−2300		
Antares	400	MIIb	B4eV		+0.92v	+5.1		−5250	−110	
Pollux	39	K0III			+1.16			+40		
Fomalhaut	23	A3V	K4V		+1.19	+6.5		+13	+0.1	
Deneb	1400	A2Ia			+1.26			4.8×10^4		
β Crucis	500	B0.5IV			+1.28v			−5800		

*Distances of the more remote stars have been estimated from their spectral types and apparent magnitudes and are only approximate.
Note: Several of the components listed are themselves spectroscopic binaries. A "v" after a magnitude denotes that the star is variable, in which case the magnitude at median light is given. A "p" after a spectral type indicates that the spectrum is peculiar. An "e" after a spectral type indicates that emission lines are present. When the luminosity classification is rather uncertain, a range is given.

APPENDIX 12:   VARIABLE STARS

Type of Variable	Kind of Star	Peak Luminosity (L_S)	Period (days)	Description	Example
Cepheids (type I)	F and G supergiants	10^3 to 10^4	3 to 50	Regular pulsation as a stage in the late evolution of moderately massive stars of solar-type composition. Period-luminosity relation exists.	δ Cep
Cepheids type (II)	F and G supergiants	about 10^3	5 to 30	Regular pulsation as a stage in the late evolution of moderately massive stars depleted in metals. Period-luminosity relation exists.	W Vir
RR Lyrae	A and F giants	about 50	0.5 to 1	Very regular, small-amplitude pulsations as a stage in the late evolution of stars depleted in metals.	RR Lyr
Long-period	M red giants	10^2 to 10^4	80 to 600	Large-amplitude, semiperiodic variations in evolved, luminous red giants that are losing mass. Much of luminosity is in infrared.	o Ceti
Novae	O to A binaries	10^4 to 10^5	—	Eruptive event with ejection of shell due to explosive hydrogen fusion in the atmosphere of one of two binaries exchanging mass. Star brightens in a few days by as much as 10,000 times.	GK Per
Supernovae (type II)	Massive red supergiants	10^8 to 10^{10}	—	Catastrophic ejection of most of the mass from a collapsing stellar core of a massive, evolved star. Star brightens in a few days by 10^6 or more, can outshine an entire galaxy at maximum.	SN1987A

APPENDIX 13: THE LOCAL GROUP OF GALAXIES

Galaxy	Type	Distance (1000 LY)	Diameter (1000 LY)	Luminosity (L_s)	Radial Velocity (km/s)	Mass (Solar Masses)
Our Galaxy	Sb	—	100	(2.1×10^{10})	—	2×10^{11}
Large Magellanic Cloud	Irr I	175	30	1×10^9	+276	2.5×10^{10}
Small Magellanic Cloud	Irr I	210	25	3.3×10^8	+168	
Ursa Minor system	E4 (dwarf)	220	3	(3.3×10^5)		
Sculptor system	E3 (dwarf)	270	7	4.4×10^6		$(2 \text{ to } 4 \times 10^6)$
Draco system	E2 (dwarf)	330	4.5	(8.3×10^5)		
Carina system	E3 (dwarf)	(550)	4.8	(8.3×10^5)		
Fornax system	E3 (dwarf)	800	15	2.3×10^7	+39	$(1.2 \text{ to } 2 \times 10^7)$
Leo II system	E0 (dwarf)	750	5.2	8.3×10^5		(1.1×10^6)
Leo I system	E4 (dwarf)	900	5	1.2×10^6		
NGC 6822	Irr I	1500	9	6.9×10^7	−32	
NGC 147	E6	1900	10	5.2×10^7		
NGC 185	E2	1900	8	6.9×10^7	−305	
NGC 205	E5	2700	16	3.3×10^8	−239	
NGC 221 (M32)	E3	2700	8	3.3×10^8	−214	
IC 1613	Irr I	2700	16	6.3×10^7	−238	
Andromeda galaxy (NGC 224; M31)	Sb	2700	130	2.5×10^{10}	−266	3×10^{11}
And I	E0 (dwarf)	(2700)	1.6	(2.1×10^6)		
And II	E0 (dwarf)	(2700)	2.3	(2.1×10^6)		
And III	E3 (dwarf)	(2700)	0.9	(2.1×10^6)		
NGC 598 (M33)	Sc	3000	60	3.0×10^9	−189	8×10^9

▶ APPENDIX 14:   THE CHEMICAL ELEMENTS

Element	Symbol	Atomic Number	Atomic Weight* (Chemical Scale)	Number of Atoms per 10^{12} Hydrogen Atoms
Hydrogen	H	1	1.0080	1×10^{12}
Helium	He	2	4.003	8×10^{10}
Lithium	Li	3	6.940	2×10^{3}
Beryllium	Be	4	9.013	3×10^{1}
Boron	B	5	10.82	9×10^{2}
Carbon	C	6	12.011	4.5×10^{8}
Nitrogen	N	7	14.008	9.2×10^{7}
Oxygen	O	8	16.0000	7.4×10^{8}
Fluorine	F	9	19.00	3.1×10^{4}
Neon	Ne	10	20.183	1.3×10^{8}
Sodium	Na	11	22.991	2.1×10^{6}
Magnesium	Mg	12	24.32	4.0×10^{7}
Aluminum	Al	13	26.98	3.1×10^{6}
Silicon	Si	14	28.09	3.7×10^{7}
Phosphorus	P	15	30.975	3.8×10^{5}
Sulfur	S	16	32.066	1.9×10^{7}
Chlorine	Cl	17	35.457	1.9×10^{5}
Argon	Ar(A)	18	39.944	3.8×10^{6}
Potassium	K	19	39.100	1.4×10^{5}
Calcium	Ca	20	40.08	2.2×10^{6}
Scandium	Sc	21	44.96	1.3×10^{3}
Titanium	Ti	22	47.90	8.9×10^{4}
Vanadium	V	23	50.95	1.0×10^{4}
Chromium	Cr	24	52.01	5.1×10^{5}
Manganese	Mn	25	54.94	3.5×10^{5}
Iron	Fe	26	55.85	3.2×10^{7}
Cobalt	Co	27	58.94	8.3×10^{4}
Nickel	Ni	28	58.71	1.9×10^{6}
Copper	Cu	29	63.54	1.9×10^{4}
Zinc	Zn	30	65.38	4.7×10^{4}
Gallium	Ga	31	69.72	1.4×10^{3}
Germanium	Ge	32	72.60	4.4×10^{3}
Arsenic	As	33	74.91	2.5×10^{2}
Selenium	Se	34	78.96	2.3×10^{3}
Bromine	Br	35	79.916	4.4×10^{2}
Krypton	Kr	36	83.80	1.7×10^{3}
Rubidium	Rb	37	85.48	2.6×10^{2}

*Where mean atomic weights have not been well determined, the atomic mass numbers of the most stable isotopes are given in parentheses.

APPENDIX 14 (CONTINUED)

Element	Symbol	Atomic Number	Atomic Weight* (Chemical Scale)	Number of Atoms per 10^{12} Hydrogen Atoms
Strontium	Sr	38	87.63	8.8×10^2
Yttrium	Y	39	88.92	2.5×10^2
Zirconium	Zr	40	91.22	4.0×10^2
Niobium (Columbium)	Nb(Cb)	41	92.91	2.6×10^1
Molybdenum	Mo	42	95.95	9.3×10^1
Technetium	Tc(Ma)	43	(99)	—
Ruthenium	Ru	44	101.1	68
Rhodium	Rh	45	102.91	13
Palladium	Pd	46	106.4	51
Silver	Ag	47	107.880	20
Cadmium	Cd	48	112.41	63
Indium	In	49	114.82	75
Tin	Sn	50	118.70	1.4×10^2
Antimony	Sb	51	121.76	13
Tellurium	Te	52	127.61	1.8×10^2
Iodine	I(J)	53	126.91	33
Xenon	Xe(X)	54	131.30	1.6×10^2
Cesium	Cs	55	132.91	14
Barium	Ba	56	137.36	1.6×10^2
Lenthanum	La	57	138.92	17
Cerium	Ce	58	140.13	43
Praseodymium	Pr	59	140.92	6
Neodymium	Nd	60	144.27	31
Promethium	Pm	61	(147)	—
Samarium	Sm(Sa)	62	150.35	10
Europium	Eu	63	152.0	4
Gadolinium	Gd	64	157.26	13
Terbium	Tb	65	158.93	2
Dysprosium	Dy(Ds)	66	162.51	15
Holmium	Ho	67	164.94	3
Erbium	Er	68	167.27	9
Thulium	Tm(Tu)	69	168.94	2
Ytterbium	Yb	70	173.04	8
Lutecium	Lu(Cp)	71	174.99	2
Hafnium	Hf	72	178.50	6
Tantalum	Ta	73	180.95	1
Tungsten	W	74	183.86	5

*Where mean atomic weights have not been well determined, the atomic mass numbers of the most stable isotopes are given in parentheses.

APPENDIX 14 (CONTINUED)

Element	Symbol	Atomic Number	Atomic Weight* (Chemical Scale)	Number of Atoms per 10^{12} Hydrogen Atoms
Rhenium	Re	75	186.22	2
Osmium	Os	76	190.2	27
Iridium	Ir	77	192.2	24
Platinum	Pt	78	195.09	56
Gold	Au	79	197.0	6
Mercury	Hg	80	200.61	19
Thallium	Tl	81	204.39	8
Lead	Pb	82	207.21	1.2×10^2
Bismuth	Bi	83	209.00	5
Polonium	Po	84	(209)	—
Astatine	At	85	(210)	—
Radon	Rn	86	(222)	—
Francium	Fr(Fa)	87	(223)	—
Radium	Ra	88	226.05	—
Actinium	Ac	89	(227)	—
Thorium	Th	90	232.12	1
Protactinium	Pa	91	(231)	—
Uranium	U(Ur)	92	238.07	1
Neptunium	Np	93	(237)	—
Plutonium	Pu	94	(244)	—
Americium	Am	95	(243)	—
Curium	Cm	96	(248)	—
Berkelium	Bk	97	(247)	—
Californium	Cf	98	(251)	—
Einsteinium	E	99	(254)	—
Fermium	Fm	100	(253)	—
Mendeleevium	Mv	101	(256)	—
Nobelium	No	102	(253)	—

*Where mean atomic weights have not been well determined, the atomic mass numbers of the most stable isotopes are given in parentheses.

APPENDIX 15:   THE CONSTELLATIONS

Constellation (Latin name)	Genitive Case Ending	English Name or Description	Abbreviation
Andromeda	Andromedae	Princess of Ethiopia	And
Antlia	Antliae	Air pump	Ant
Apus	Apodis	Bird of paradise	Aps
Aquarius	Aquarii	Water bearer	Aqr
Aquila	Aquilae	Eagle	Aql
Ara	Arae	Altar	Ara
Aries	Arietis	Ram	Ari
Auriga	Aurigae	Charioteer	Aur
Boötes	Boötis	Herdsman	Boo
Caelum	Caeli	Graving tool	Cae
Camelopardus	Camelopardis	Giraffe	Cam
Cancer	Cancri	Crab	Cnc
Canes Venatici	Canum Venaticorum	Hunting dogs	CVn
Canis Major	Canis Majoris	Big dog	CMa
Canis Minor	Canis Minoris	Little dog	CMi
Capricornus	Capricorni	Sea goat	Cap
Carina*	Carinae	Keel of Argonauts' ship	Car
Cassiopeia	Cassiopeiae	Queen of Ethiopia	Cas
Centaurus	Centauri	Centaur	Cen
Cepheus	Cephei	King of Ethiopia	Cep
Cetus	Ceti	Sea monster (whale)	Cet
Chamaeleon	Chamaeleontis	Chameleon	Cha
Circinus	Circini	Compasses	Cir
Columba	Columbae	Dove	Col
Coma Berenices	Comae Berenices	Berenice's hair	Com
Corona Australis	Coronae Australis	Southern crown	CrA
Corona Borealis	Coronae Borealis	Northern crown	CrB
Corvus	Corvi	Crow	Crv
Crater	Crateris	Cup	Crt
Crux	Crucis	Cross (southern)	Cru
Cygnus	Cygni	Swan	Cyg
Delphinus	Delphini	Porpoise	Del

*The four constellations Carina, Puppis, Pyxis, and Vela originally form the single constellation, Argo Navis.

APPENDIX 15 (CONTINUED)

Constellation (Latin name)	Genitive Case Ending	English Name or Description	Abbreviation
Dorado	Doradus	Swordfish	Dor
Draco	Draconis	Dragon	Dra
Equuleus	Equulei	Little horse	Equ
Eridanus	Eridani	River	Eri
Fornax	Fornacis	Furnace	For
Gemini	Geminorum	Twins	Gem
Grus	Gruis	Crane	Gru
Hercules	Herculis	Hercules, son of Zeus	Her
Horologium	Horologii	Clock	Hor
Hydra	Hydrae	Sea serpent	Hya
Hydrus	Hydri	Water snake	Hyi
Indus	Indi	Indian	Ind
Lacerta	Lacertae	Lizard	Lac
Leo	Leonis	Lion	Leo
Leo Minor	Leonis Minoris	Little lion	LMi
Lepus	Leporis	Hare	Lep
Libra	Librae	Balance	Lib
Lupus	Lupi	Wolf	Lup
Lynx	Lyncis	Lynx	Lyn
Lyra	Lyrae	Lyre or harp	Lyr
Mensa	Mensae	Table Mountain	Men
Microscopium	Microscopii	Microscope	Mic
Monoceros	Monocerotis	Unicorn	Mon
Musca	Muscae	Fly	Mus
Norma	Normae	Carpenter's level	Nor
Octans	Octantis	Octant	Oct
Ophiuchus	Ophiuchi	Holder of serpent	Oph
Orion	Orionis	Orion, the hunter	Ori
Pavo	Pavonis	Peacock	Pav
Pegasus	Pegasi	Pegasus, the winged horse	Peg
Perseus	Persei	Perseus, hero who saved Andromeda	Per

*The four constellations Carina, Puppis, Pyxis, and Vela originally form the single constellation, Argo Navis.

APPENDIX 15 (CONTINUED)

Constellation (Latin name)	Genitive Case Ending	English Name or Description	Abbreviation
Phoenix	Phoenicis	Phoenix	Phe
Pictor	Pictoris	Easel	Pic
Pisces	Piscium	Fishes	Psc
Piscis Austrinus	Piscis Austrini	Southern fish	PsA
Puppis*	Puppis	Stern of the Argonauts' ship	Pup
Pyxis* (=Malus)	Pyxidus	Compass on the Argonauts' ship	Pyx
Reticulum	Reticuli	Net	Ret
Sagitta	Sagittae	Arrow	Sge
Sagittarius	Sagittarii	Archer	Sgr
Scorpius	Scorpii	Scorpion	Sco
Sculptor	Sculptoris	Sculptor's tools	Scl
Scutum	Scuti	Shield	Sct
Serpens	Serpentis	Serpent	Ser
Sextans	Sextantis	Sextant	Sex
Taurus	Tauri	Bull	Tau
Telescopium	Telescopii	Telescope	Tel
Triangulum	Trianguli	Triangle	Tri
Triangulum Australe	Trianguli Australis	Southern triangle	TrA
Tucana	Tucanae	Toucan	Tuc
Ursa Major	Ursae Majoris	Big bear	UMa
Ursa Minor	Ursae Minoris	Little bear	VMi
Vela*	Velorum	Sail of the Argonauts' ship	Vel
Virgo	Virginis	Virgin	Vir
Volans	Volantis	Flying fish	Vol
Vulpecula	Vulpeculae	Fox	Vul

*The four constellations Carina, Puppis, Pyxis, and Vela originally form the single constellation, Argo Navis.

▶ INDEX

NORTHERN HORIZON

EASTERN HORIZON

WESTERN HORIZON

SOUTHERN HORIZON

THE NIGHT SKY IN JANUARY

Latitude of chart is 34° N, but it is practical throughout the continental United States.

To use: Hold chart vertically and turn it so the direction you are facing shows at the bottom.

Chart time (Local Standard):

10 p.m. First of month
9 p.m. Middle of month
8 p.m. Last of month

Star Chart from *GRIFFITH OBSERVER*, Griffith Observatory, Los Angeles

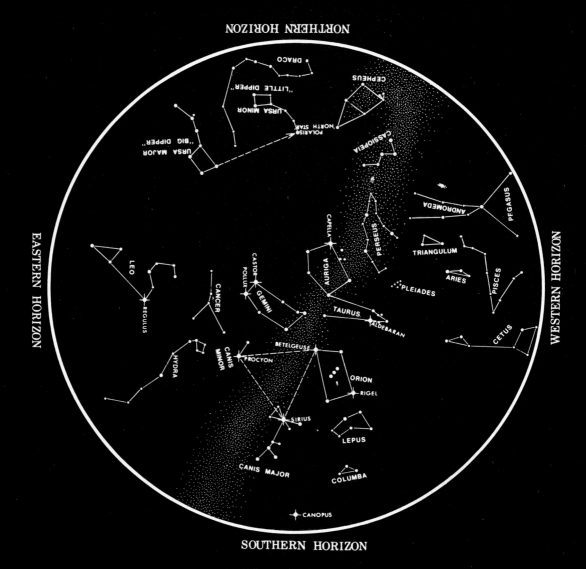

THE NIGHT SKY IN FEBRUARY

Latitude of chart is 34°N, but it is
practical throughout the continental
United States.

To use: Hold chart vertically and turn
it so the direction you are facing
shows at the bottom.

Chart time (Local Standard):

10 p.m. First of month

9 p.m. Middle of month

8 p.m. Last of month

Star Chart from *GRIFFITH OBSERVER*, Griffith Observatory, Los Angeles

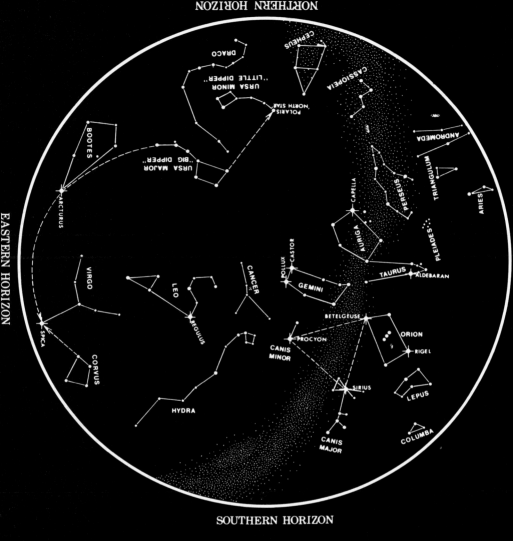

THE NIGHT SKY IN MARCH

Latitude of chart is 34°N, but it is practical throughout the continental United States.

To use: Hold chart vertically and turn it so the direction you are facing shows at the bottom.

Chart time (Local Standard):

10 p.m. First of month
9 p.m. Middle of month
8 p.m. Last of month

Star Chart from *GRIFFITH OBSERVER*, Griffith Observatory, Los Angeles

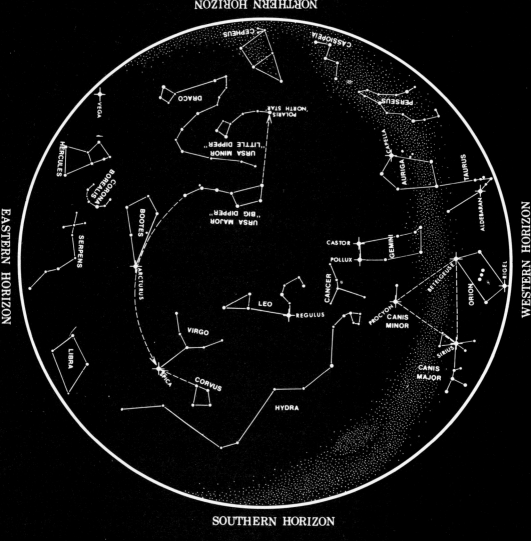

THE NIGHT SKY IN APRIL

Latitude of chart is 34°N, but it is
practical throughout the continental
United States.

To use: Hold chart vertically and turn
it so the direction you are facing
shows at the bottom.

Chart time (Local Standard):

10 p.m. First of month

9 p.m. Middle of month

8 p.m. Last of month

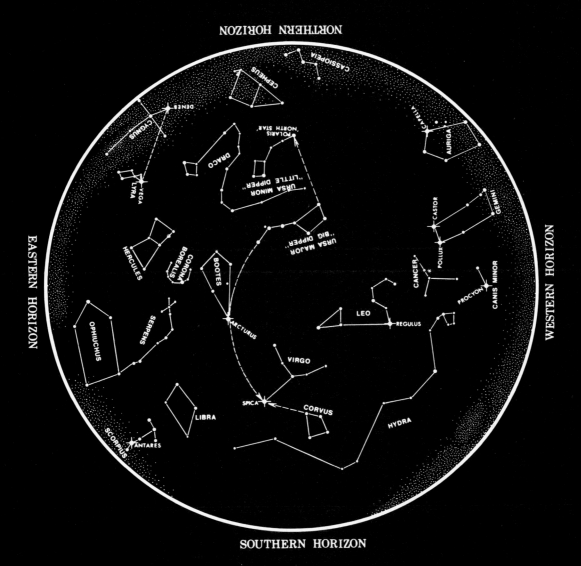

NORTHERN HORIZON

EASTERN HORIZON

WESTERN HORIZON

SOUTHERN HORIZON

THE NIGHT SKY IN MAY

Latitude of chart is 34°N, but it is practical throughout the continental United States.

To use: Hold chart vertically and turn it so the direction you are facing shows at the bottom.

Chart time (Local Standard):

10 p.m. First of month

9 p.m. Middle of month

8 p.m. Last of month

Star Chart from *GRIFFITH OBSERVER*, Griffith Observatory, Los Angeles

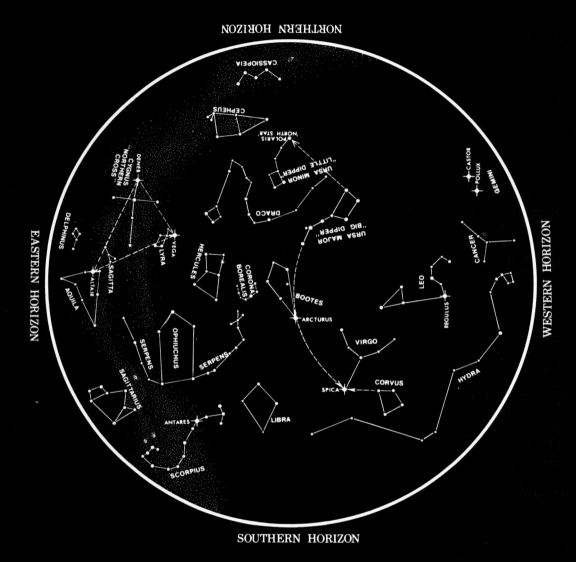

NORTHERN HORIZON

EASTERN HORIZON

WESTERN HORIZON

SOUTHERN HORIZON

THE NIGHT SKY IN JUNE

Latitude of chart is 34°N, but it is practical throughout the continental United States.

To use: Hold chart vertically and turn it so the direction you are facing shows at the bottom.

Chart time (Local Standard):
10 p.m. First of month
9 p.m. Middle of month
8 p.m. Last of month

Star Chart from *GRIFFITH OBSERVER*, Griffith Observatory, Los Angeles

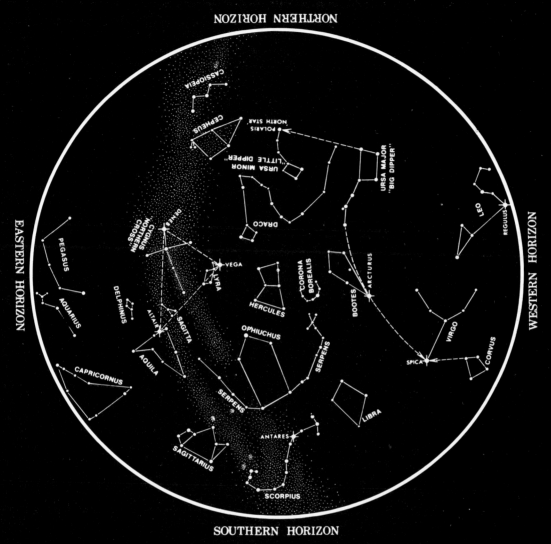

NORTHERN HORIZON

EASTERN HORIZON

WESTERN HORIZON

SOUTHERN HORIZON

THE NIGHT SKY IN JULY

Latitude of chart is 34° N, but it is practical throughout the continental United States.

To use: Hold chart vertically and turn it so the direction you are facing shows at the bottom.

Chart time (Local Standard):

10 p.m. First of month

9 p.m. Middle of month

8 p.m. Last of month

Star Chart from *GRIFFITH OBSERVER*, Griffith Observatory, Los Angeles

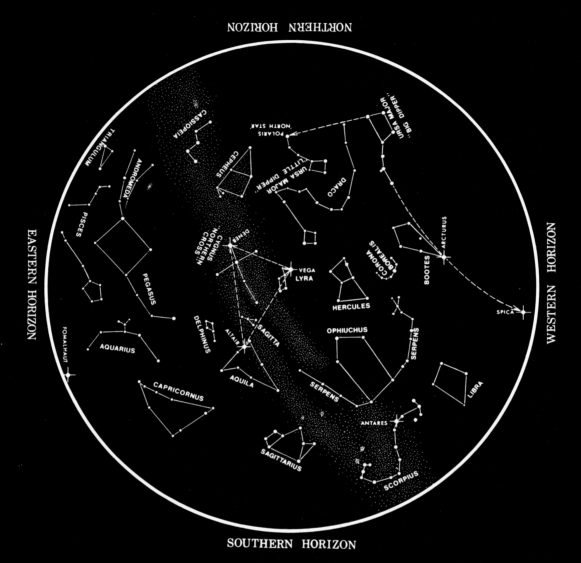

NORTHERN HORIZON

EASTERN HORIZON

WESTERN HORIZON

SOUTHERN HORIZON

THE NIGHT SKY IN AUGUST

Latitude of chart is 34°N, but it is practical throughout the continental United States.

To use: Hold chart vertically and turn it so the direction you are facing shows at the bottom.

Chart time (Local Standard):

10 p.m. First of month
9 p.m. Middle of month
8 p.m. Last of month

Star Chart from *GRIFFITH OBSERVER*, Griffith Observatory, Los Angeles

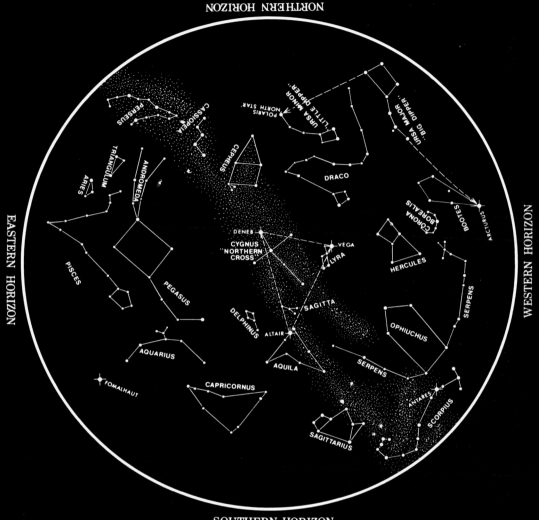

NORTHERN HORIZON

EASTERN HORIZON

WESTERN HORIZON

SOUTHERN HORIZON

THE NIGHT SKY IN SEPTEMBER

Latitude of chart is 34°N, but it is practical throughout the continental United States.

To use: Hold chart vertically and turn it so the direction you are facing shows at the bottom.

Chart time (Local Standard):

10 p.m. First of month

9 p.m. Middle of month

8 p.m. Last of month

Star Chart from *GRIFFITH OBSERVER*, Griffith Observatory, Los Angeles

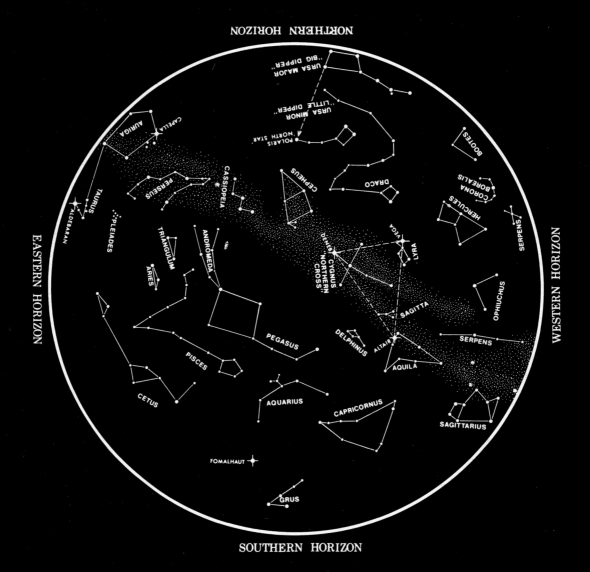

THE NIGHT SKY IN OCTOBER

Latitude of chart is 34° N, but it is
practical throughout the continental
United States.

To use: Hold chart vertically and turn
it so the direction you are facing
shows at the bottom.

Chart time (Local Standard):

10 p.m. First of month

9 p.m. Middle of month

8 p.m. Last of month

Star Chart from *GRIFFITH OBSERVER*, Griffith Observatory, Los Angeles

NORTHERN HORIZON

EASTERN HORIZON

WESTERN HORIZON

SOUTHERN HORIZON

THE NIGHT SKY IN NOVEMBER

Latitude of chart is 34°N, but it is practical throughout the continental United States.

To use: Hold chart vertically and turn it so the direction you are facing shows at the bottom.

Chart time (Local Standard):

10 p.m. First of month

9 p.m. Middle of month

8 p.m. Last of month

Star Chart from *GRIFFITH OBSERVER*, Griffith Observatory, Los Angeles

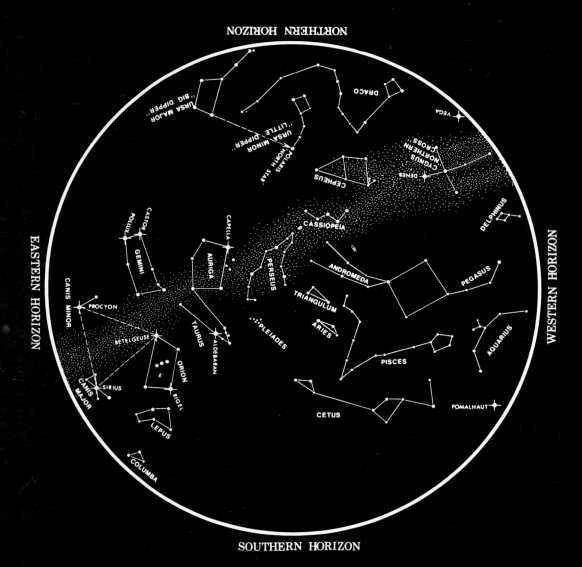

THE NIGHT SKY IN DECEMBER

Latitude of chart is 34°N, but it is practical throughout the continental United States.

To use: Hold chart vertically and turn it so the direction you are facing shows at the bottom.

Chart time (Local Standard):
10 p.m. First of month
9 p.m. Middle of month
8 p.m. Last of month

Star Chart from *GRIFFITH OBSERVER*, Griffith Observatory, Los Angeles